CLASSICAL DYNAMICS
OF PARTICLES AND SYSTEMS
SECOND EDITION

CLASSICAL DYNAMICS

OF PARTICLES AND SYSTEMS

SECOND EDITION

JERRY B. MARION

UNIVERSITY OF MARYLAND
COLLEGE PARK

ACADEMIC PRESS NEW YORK AND LONDON

ACADEMIC PRESS, INC.
111 Fifth Avenue, New York, New York 10003

United Kingdom Edition published by
ACADEMIC PRESS, INC. (LONDON) LTD.
Berkeley Square House, London WIX 6BA

LIBRARY OF CONGRESS CATALOG CARD NUMBER: 78-107545

PRINTED IN THE UNITED STATES OF AMERICA

Contents

Chapter 8. Central-Force Motion

Chapter 9. Kinematics of Two-Particle Collisions

Chapter 10. The Special Theory of Relativity

Chapter 11. Motion in a Noninertial Reference Frame

Chapter 12. Dynamics of Rigid Bodies

Chapter 13. Coupled Oscillations

Chapter 14. Vibrating Strings

Chapter 15. The Wave Equation in One Dimension

Appendix A. Taylor's Theorem 524

Preface to the Second Edition

The preparation of this second edition has afforded the opportunity to make several additions and to take advantage of suggestions to improve a number of parts of the text. A few errors have been located and corrected and several superfluous paragraphs have been deleted here and there. Certain material has been rearranged to provide a more orderly flow of the development. Because of the great importance of oscillatory phenomena in almost all phases of modern engineering and physics, some additional material is now included (for example, Laplace transform methods and more extensive applications to electrical oscillations). The discussion of vector methods has been condensed and now includes only those techniques of immediate usefulness for the purposes of this book. Newtonian mechanics (including potential theory) is covered in a single chapter. Relativity theory is placed later in the book and includes all of the applications to mechanics that were formerly scattered through several chapters. Some new problems have been added, particularly in the chapters on oscillations, and certain problems deemed too difficult have been deleted.

Preface to the First Edition

This book presents a modern and reasonably complete account of the classical mechanics of particles, systems of particles, and rigid bodies for physics students at the advanced undergraduate level. It is designed as a text for a one-year, three-hour course in mechanics. With careful planning and appropriate omissions,* however, the essential topics can be covered in a one-semester, four-hour course. Such topics as nonlinear oscillations, Liouville's theorem, the three-body problem, and relativistic collisions, as well as some others, might well be omitted if the course is not of sufficient length to permit their inclusion. It is hoped, however, that this material will not in general be dropped by the wayside, since these are mainly the "fun" topics in classical physics. Alternatively, the book may be used in the mechanics portion of courses in mathematical physics or in theoretical physics. Students taking these courses will have completed an introductory course in physics, and mathematics through integral calculus.

The purpose of this book is threefold: (a) To present a modern treatment of classical mechanical systems in such a way that the transition to the quantum theory of physics can be made with the least possible difficulty. To this end, modern notation and terminology are used throughout. Also, special note is made of concepts which are important to modern physics. (b) To acquaint the student with new mathematical techniques wherever possible, and to give him sufficient practice in solving problems so that he may become reasonably proficient in their use. (c) To impart to the student, at the crucial period in his career between "introductory" and "advanced" physics, some degree of sophistication in handling both the formalism of the theory and the operational technique of problem solving.

New mathematical methods are developed in this volume as the occasion demands. However, it is expected that while taking a course using this book students will also be studying advanced mathematics in a separate course. Mathematical rigor must be learned and appreciated by students of physics, but where the continuity of the physics would be disturbed by insisting on complete generality and mathematical rigor, the *physics* has been given precedence.

Vector methods are developed in the first chapter and are used throughout the book. It is assumed that the student has been exposed to the "directed line segment" approach to vectors. Therefore, a more fundamental viewpoint

* Sections which can be omitted without destroying the continuity are marked thusly: ■

is adopted here and vector analysis is developed by considering the properties of coordinate transformations. This has the advantage of building a firm basis for the later transition to tensor methods (Chapters 12 and 13). Appendices on the use of complex numbers and the solution of ordinary differential equations, as well as other mathematical techniques, are provided for the benefit of those students whose background is somewhat deficient, and for review by more advanced students.

Throughout the development of material in this volume, frequent examples are included. The reader will find that in these examples, as well as in the derivations, a generous amount of detail has been given, and that " it-may-be-shown-that's " have been kept to a minimum. However, some of the more lengthy and tedious algebraic manipulations have been omitted when necessary, to insure that the student does not lose sight of the development underway. Problems also form an integral portion of the text and a student should work a substantial fraction of those given here if he expects to have an adequate command of the material.

One aspect of this book that the author found particularly enjoyable was the preparation of the historical footnotes. The history of physics has been almost eliminated from present-day curricula, and as a result, the student is frequently unaware of the background of a particular topic even to the point of being unfamiliar with the names of the giants of mathematics and physics who struggled with the developments of the subject. These footnotes are therefore included to whet the appetite and to encourage the student to inquire into the history of his field. In general, references to the original literature have been omitted except for relatively modern papers (in English!) to which the student may be expected to have access.

Lists of " Suggested References " are to be found at the end of each chapter. These have been categorized according to subject matter and level of difficulty. The lists are frequently extensive; it is, of course, not expected that a student will be able to consult each reference, but a sufficient number is given so that there will be a reasonable probability that we will be able to locate at least some source of collateral material. Relatively recent texts which will probably be more accessible and appealing to the student make up the bulk of the lists.

The author wishes to express his gratitude to the University of Maryland Computer Science Center which extended to him the use of the IBM 7090/1401 computer for the purpose of calculating many of the curves which appear in this book.

CLASSICAL DYNAMICS

OF PARTICLES AND SYSTEMS

SECOND EDITION

CHAPTER 1

Matrices, Vectors, and Vector Calculus

1.1 Introduction

The discussion of physical phenomena can be carried out in a most concise and elegant fashion through the use of vector methods.* In the application of physical "laws" to particular situations, it is clear that the results must be independent of whether we choose a rectangular coordinate system or a bipolar cylindrical coordinate system as well as independent of the exact choice of origin for the coordinates. The use of vectors gives us this independence from the special features of the coordinate system which we happen to use. We can be assured, therefore, that a given physical law will still be correctly represented regardless of the fact that we have chosen a particular coordinate system as being most convenient for the description of a particular problem. In addition, the use of vector notation provides an extremely compact method of expressing even the most complicated results.

* To Josiah Willard Gibbs (1839–1903) goes much of the credit for the development of vector analysis, largely carried out in the period 1880–1882. Much of the present-day vector notation was originated by Oliver Heaviside (1850–1925), an English electrical engineer, and dates from about 1893.

In elementary treatments of vectors, the discussion may start with the statement that "a vector is a quantity that can be represented as a directed line segment." To be sure, this type of development will yield correct results, and it is even beneficial in order to impart a certain feeling for the physical nature of a vector. We shall assume that the reader is familiar with this type of development, but we shall forego the approach here since we wish to emphasize the relationship that a vector bears to a coordinate transformation. For this purpose it will be convenient to introduce matrices and matrix notation to describe not only the transformation but the vector as well. We shall also introduce and use here a type of notation which is readily adapted to the use of tensors, although we shall not encounter these objects until the normal course of events requires their use (Chapter 12).

A complete exposition of vector methods is not attempted here; rather, we shall consider only those topics that are necessary for a study of mechanical systems. Thus, in this chapter the fundamentals of matrix and vector algebra and vector calculus are treated. For a more complete discussion of vector analysis from the standpoint of application to physical situations, see Marion, *Principles of Vector Analysis* (Ma65a). Many questions of detail and several of the more involved proofs have been omitted from this chapter. The reader is referred to Ma65a for this additional material.

1.2 The Concept of a Scalar

Consider the array of particles shown in Fig. 1-1a. Each particle of the array is labeled according to its mass, say, in grams. The coordinate axes are shown so that it is possible to specify a particular particle by a pair of numbers (x, y). The mass M of the particle at (x, y) can be expressed as

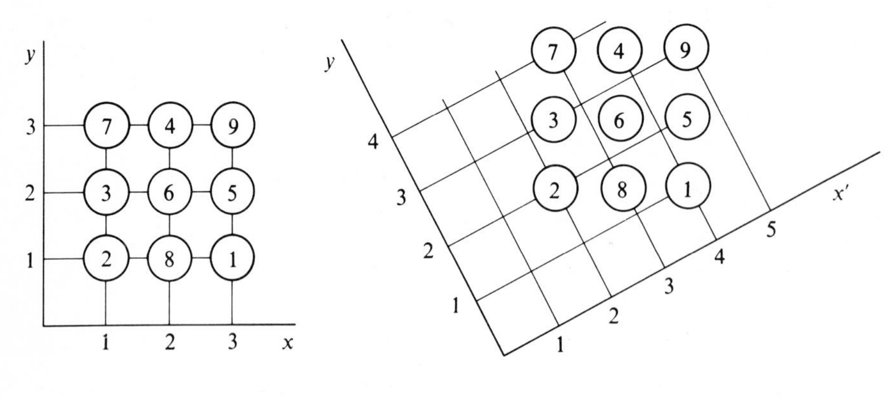

FIG. 1-1a

FIG. 1-1b

$M(x, y)$; thus, the mass of the particle at $x = 2$, $y = 3$ can be written as $M(x = 2, y = 3) = 4$. Now, consider the axes rotated and displaced in the manner shown in Fig. 1-1b. It is evident that the 4-gm mass is now located at $x' = 4$, $y' = 3.5$; i.e., the mass is specified by $M(x' = 4, y' = 3.5) = 4$. And in general,

$$M(x, y) = M(x', y') \tag{1.1}$$

since the mass of any particle is not affected by a change in the coordinate axes. Quantities which have the property that they are *invariant under coordinate transformations*, i.e., that they obey an equation of the type (1.1), are termed *scalars*.

Although it is possible to give the mass of a particle (or the temperature, or the speed, etc.) relative to any coordinate system by the same number, it is clear that there are some physical properties associated with the particle (such as the direction of motion of the particle or the direction of a force that may act on the particle) which cannot be specified in such a simple manner. The description of these more complicated quantities requires the use of *vectors*. Just as a scalar is defined as a quantity that remains invariant under a coordinate transformation, a vector may also be defined in terms of transformation properties. We shall begin by considering the way in which the coordinates of a point change when the coordinate system undergoes a rotation about its origin.

1.3 Coordinate Transformations

Consider a point P which has coordinates* (x_1, x_2, x_3) with respect to a certain coordinate system. Next, consider a different coordinate system which can be generated from the original system by a simple rotation; let the coordinates of the point P with respect to the new coordinate system be (x_1', x_2', x_3'). The situation is illustrated for a two-dimensional case in Fig. 1-2.

The new coordinate x_1' is the sum of the projection of x_1 onto the x_1'-axis (the line $\overline{Oa}$) plus the projection of x_2 onto the x_1'-axis (the line $\overline{ab} + \overline{bc}$). That is,

$$\begin{aligned} x_1' &= x_1 \cos\theta + x_2 \sin\theta \\ &= x_1 \cos\theta + x_2 \cos\left(\frac{\pi}{2} - \theta\right) \end{aligned} \tag{1.2a}$$

* The labeling of the axes will be given by x_1, x_2, x_3 instead of x, y, z in order to simplify the notation when summations are performed. For the moment the discussion is limited to Cartesian (or rectangular) coordinate systems.

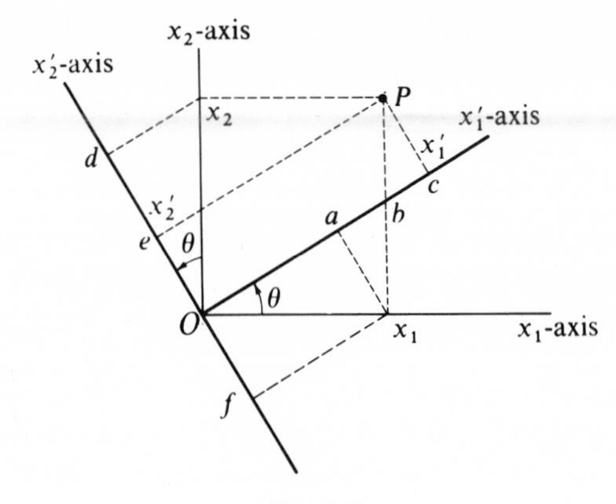

FIG. **1-2**

The coordinate x_2' is the sum of similar projections: $x_2' = \overline{Od} - \overline{de}$, but the line $\overline{de}$ is just as equal to the line $\overline{Of}$. Therefore,

$$x_2' = -x_1 \sin\theta + x_2 \cos\theta$$
$$= x_1 \cos\left(\frac{\pi}{2} + \theta\right) + x_2 \cos\theta \tag{1.2b}$$

Let us introduce the following notation: we will write the angle between the x_1'-axis and the x_1-axis as (x_1', x_1), and, in general, the angle between the x_i'-axis and the x_j-axis will be denoted by (x_i', x_j). Furthermore, we will define the cosine of (x_i', x_j) to be λ_{ij}:

$$\lambda_{ij} \equiv \cos(x_i', x_j) \tag{1.3}$$

Therefore, we have for the case of Fig. 1-2,

$$\left.\begin{aligned}
\lambda_{11} &= \cos(x_1', x_1) = \cos\theta \\
\lambda_{12} &= \cos(x_1', x_2) = \cos\left(\frac{\pi}{2} - \theta\right) = \sin\theta \\
\lambda_{21} &= \cos(x_2', x_1) = \cos\left(\frac{\pi}{2} + \theta\right) = -\sin\theta \\
\lambda_{22} &= \cos(x_2', x_2) = \cos\theta
\end{aligned}\right\} \tag{1.4}$$

The equations of transformation now become

$$x_1' = x_1 \cos(x_1', x_1) + x_2 \cos(x_1', x_2)$$
$$= \lambda_{11} x_1 + \lambda_{12} x_2 \tag{1.5a}$$
$$x_2' = x_1 \cos(x_2', x_1) + x_2 \cos(x_2', x_2)$$
$$= \lambda_{21} x_1 + \lambda_{22} x_2 \tag{1.5b}$$

Thus, in general, for three dimensions we have

$$\left.\begin{aligned} x_1' &= \lambda_{11}x_1 + \lambda_{12}x_2 + \lambda_{13}x_3 \\ x_2' &= \lambda_{21}x_1 + \lambda_{22}x_2 + \lambda_{23}x_3 \\ x_3' &= \lambda_{31}x_1 + \lambda_{32}x_2 + \lambda_{33}x_3 \end{aligned}\right\} \tag{1.6}$$

Or, in summation notation,

$$\boxed{x_i' = \sum_{j=1}^{3} \lambda_{ij}x_j, \qquad i = 1, 2, 3} \tag{1.7}$$

The inverse transformation is

$$\begin{aligned} x_1 &= x_1' \cos(x_1', x_1) + x_2' \cos(x_2', x_1) + x_3' \cos(x_3', x_1) \\ &= \lambda_{11}x_1' + \lambda_{21}x_2' + \lambda_{31}x_3' \end{aligned}$$

or, in general,

$$\boxed{x_i = \sum_{j=1}^{3} \lambda_{ji}x_j', \qquad i = 1, 2, 3} \tag{1.8}$$

The quantity λ_{ij} is just the *direction cosine* of the x_i'-axis relative to the x_j-axis. It is convenient to arrange the λ_{ij} into a square array called a *matrix*. The symbol $\boldsymbol{\lambda}$ will be used to denote the totality of the individual elements λ_{ij} when arranged in the following manner:

$$\boldsymbol{\lambda} = \begin{pmatrix} \lambda_{11} & \lambda_{12} & \lambda_{13} \\ \lambda_{21} & \lambda_{22} & \lambda_{23} \\ \lambda_{31} & \lambda_{32} & \lambda_{33} \end{pmatrix} \tag{1.9}$$

Once the direction cosines which relate the two sets of coordinate axes are found, Eqs. (1.7) and (1.8) give the general rules for specifying the coordinates of a point in either system.

When $\boldsymbol{\lambda}$ is defined in the manner above, and specifies the transformation properties of the coordinates of a point, it is called a *transformation matrix* or a *rotation matrix*.

1.4 Properties of Rotation Matrices

To begin the discussion it is necessary to recall two trigonometric results. Consider, as in Fig. 1-3a, a line segment which extends in a certain direction in space. We choose an origin for our coordinate system which lies at some point on the line. The line then makes certain definite angles with each of the coordinate axes; we let the angles made with the x_1-, x_2-, x_3-axes be α, β, γ. The quantities of interest are the cosines of these angles: $\cos\alpha$, $\cos\beta$, $\cos\gamma$. These quantities are called the *direction cosines* of the line. The first result which we shall need is the identity

$$\cos^2\alpha + \cos^2\beta + \cos^2\gamma = 1 \tag{1.10}$$

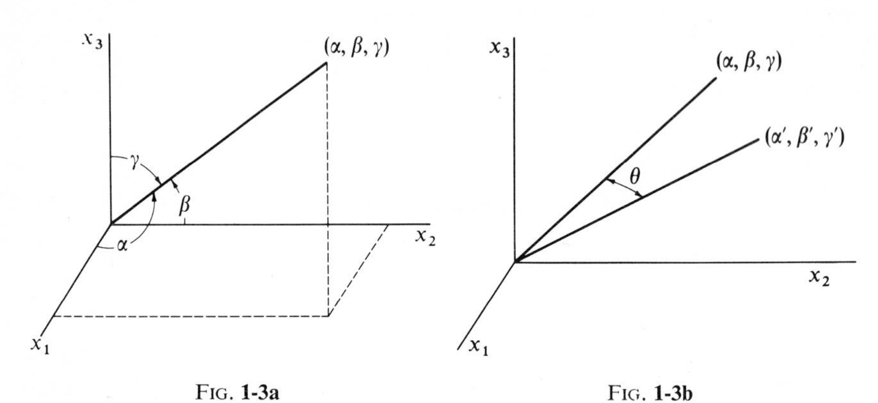

FIG. **1-3a** FIG. **1-3b**

Secondly, if we have two lines with direction cosines $\cos\alpha$, $\cos\beta$, $\cos\gamma$ and $\cos\alpha'$, $\cos\beta'$, $\cos\gamma'$, then the cosine of the angle θ between these lines (see Fig. 1-3b) is given by

$$\cos\theta = \cos\alpha\cos\alpha' + \cos\beta\cos\beta' + \cos\gamma\cos\gamma' \tag{1.11}$$

Let us now take a set of axes x_1, x_2, x_3 and perform an arbitrary rotation about some axis through the origin, In the new position we label the axes x_1', x_2', x_3'. The coordinate rotation may be specified by giving the cosines of all of the angles between the various axes, i.e., by the λ_{ij}.

Not all of the nine quantities λ_{ij} are independent; there are, in fact, six relations that exist among the λ_{ij}, so that only three are independent. These six relations are found by using the trigonometric results stated above [Eqs. (1.10) and (1.11)].

First, the x_1'-axis may be considered alone to be a line in the (x_1, x_2, x_3) coordinate system; then, the direction cosines of this line are $(\lambda_{11}, \lambda_{12}, \lambda_{13})$. Similarly, the direction cosines of the x_2'-axis in the (x_1, x_2, x_3) system are given by $(\lambda_{21}, \lambda_{22}, \lambda_{23})$. Since the angle between the x_1'-axis and the x_2'-axis is $\pi/2$, we have, from Eq. (1.11),

$$\lambda_{11}\lambda_{21} + \lambda_{12}\lambda_{22} + \lambda_{13}\lambda_{23} = \cos\theta = \cos(\pi/2) = 0$$

or,*

$$\sum_j \lambda_{1j}\lambda_{2j} = 0$$

*All summations here are understood to run from 1 to 3.

And, in general,

$$\sum_j \lambda_{ij}\lambda_{kj} = 0, \qquad i \neq k \tag{1.12a}$$

Equation (1.12a) gives three (one for each value of i or k) of the six relations among the λ_{ij}.

Since the sum of the squares of the direction cosines of a line equals unity [Eq. (1.10)], we have for the x'_1-axis in the (x_1, x_2, x_3) system,

$$\lambda_{11}^2 + \lambda_{12}^2 + \lambda_{13}^2 = 1$$

or,

$$\sum_j \lambda_{1j}^2 = \sum_j \lambda_{1j}\lambda_{1j} = 1$$

And, in general,

$$\sum_j \lambda_{ij}\lambda_{kj} = 1, \qquad i = k \tag{1.12b}$$

which are the remaining three relations among the λ_{ij}.

The results given by Eqs. (1.12a) and (1.12b) may be combined as

$$\boxed{\sum_j \lambda_{ij}\lambda_{kj} = \delta_{ik}} \tag{1.13}$$

where δ_{ik} is the *Kronecker delta symbol**:

$$\delta_{ik} = \begin{cases} 0, & \text{if} \quad i \neq k \\ 1, & \text{if} \quad i = k \end{cases} \tag{1.14}$$

The validity of Eq. (1.13) depends upon the fact that the coordinate axes in each of the systems are mutually perpendicular. Such systems are said to be *orthogonal* and Eq. (1.13) is the *orthogonality condition.* The transformation matrix $\boldsymbol{\lambda}$ which specifies the rotation of any orthogonal coordinate system must then obey the relation (1.13).

If we were to consider the x_i-axes as lines in the x'_i coordinate system and perform a calculation analogous to that above, we would find the relation

$$\boxed{\sum_i \lambda_{ij}\lambda_{ik} = \delta_{jk}} \tag{1.15}$$

The two orthogonality relations which we have derived [Eqs. (1.13) and (1.15)] appear to be different. [Note that in Eq. (1.13) the summation is over the *second* indices of the λ_{ij}, whereas in Eq. (1.15) the summation is over the

* Introduced by Leopold Kronecker (1823–1891).

first indices.] Thus, it seems that we have an overdetermined system: twelve equations* in nine unknowns. Such is not the case, however, since Eqs. (1.13) and (1.15) are not actually "different." In fact, the validity of either of these equations implies the validity of the other. This is quite clear on physical grounds (since the transformations between the two coordinate systems in either direction are equivalent), and we omit a formal proof.† We regard either Eq. (1.13) or (1.15) as providing the orthogonality relations for our systems of coordinates.

In the preceding discussion regarding the transformation of coordinates and the properties of rotation matrices, we adopted the point of view that we would consider the point P to be fixed and allow the coordinate axes to be rotated. This interpretation is not unique; we could equally well have maintained the axes fixed and allowed the point to rotate (always keeping constant the distance to the origin). In either event, the transformation matrix will be the same. For example, consider the two cases illustrated in Figs. 1-4a and 1-4b. In Fig. 1-4a, the axes x_1 and x_2 serve as reference axes and the x_1'- and

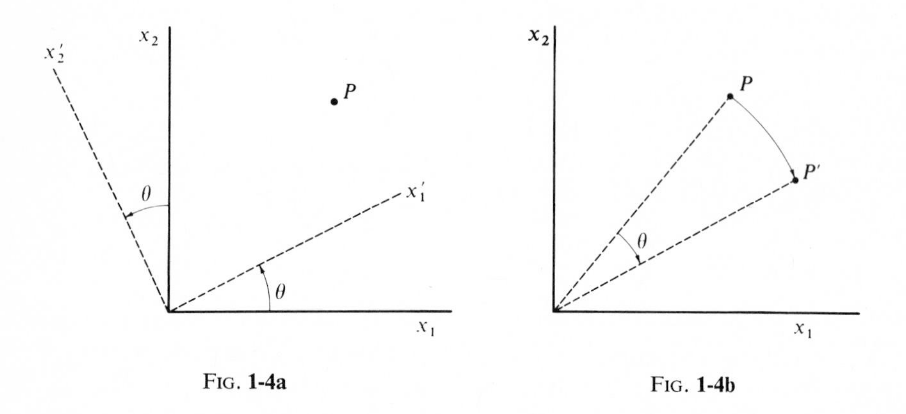

FIG. **1-4a** FIG. **1-4b**

x_2'-axes have been obtained by a rotation through an angle θ. Therefore, the coordinates of the point P with respect to the rotated axes may be found from [see Eqs. (1.2a) and (1.2b)]

$$\left.\begin{aligned} x_1' &= x_1 \cos\theta + x_2 \sin\theta \\ x_2' &= -x_1 \sin\theta + x_2 \cos\theta \end{aligned}\right\} \tag{1.16}$$

On the other hand, if the axes are fixed and the point P is allowed to rotate (as in Fig. 1-4b) through an angle θ about the origin (but in the opposite sense

* Recall that each of the orthogonality relations represents six equations.
† It is given in Marion (Ma65a, Section 1.16).

from that of the rotated axes), then the coordinates of P' will be exactly those given by Eqs. (1.16). Therefore, we may elect to say either that the transformation acts on the *point* giving a new state of the point which is expressed with respect to a fixed coordinate system (Fig. 1-4b), or that the transformation acts on the *frame of reference* (the coordinate system), as in Fig. 1-4a. The interpretations are mathematically entirely equivalent.

1.5 Matrix Operations*

The matrix λ given in Eq. (1.9) has equal numbers of rows and columns and is therefore called a *square matrix*. It is not necessary that a matrix be square. In fact, the coordinates of a point may be written as a *column* matrix:

$$\mathsf{x} = \begin{pmatrix} x_1 \\ x_2 \\ x_3 \end{pmatrix} \tag{1.17a}$$

or as a *row* matrix:

$$\mathsf{x} = (x_1 \quad x_2 \quad x_3) \tag{1.17b}$$

We must now establish rules whereby it is possible to multiply two matrices. These rules must be consistent with Eqs. (1.7) and (1.8) when we choose to express the x_i and the x_i' in matrix form. Let us take a *column matrix* for the coordinates. Then we have the following equivalent expressions:

$$x_i' = \sum_j \lambda_{ij} x_j \tag{1.18a}$$

$$\mathsf{x}' = \lambda\mathsf{x} \tag{1.18b}$$

$$\begin{pmatrix} x_1' \\ x_2' \\ x_3' \end{pmatrix} = \begin{pmatrix} \lambda_{11} & \lambda_{12} & \lambda_{13} \\ \lambda_{21} & \lambda_{22} & \lambda_{23} \\ \lambda_{31} & \lambda_{32} & \lambda_{33} \end{pmatrix} \begin{pmatrix} x_1 \\ x_2 \\ x_3 \end{pmatrix} \tag{1.18c}$$

$$\left.\begin{aligned} x_1' &= \lambda_{11}x_1 + \lambda_{12}x_2 + \lambda_{13}x_3 \\ x_2' &= \lambda_{21}x_1 + \lambda_{22}x_2 + \lambda_{23}x_3 \\ x_3' &= \lambda_{31}x_1 + \lambda_{32}x_2 + \lambda_{33}x_3 \end{aligned}\right\} \tag{1.18d}$$

* The theory of matrices was first extensively developed by A. Cayley in 1855, but many of these ideas were due to Sir William Rowan Hamilton who had discussed "linear vector operators" in 1852. The term "matrix" was first used by J. J. Sylvester, 1850.

Equations (1.18) completely specify the operation of matrix multiplication for the case of a matrix of 3 rows and 3 columns operating on a matrix of 3 rows and 1 column. [In order to be consistent with standard matrix convention we choose x and x' to be column matrices; multiplication of the type shown in Eq. (1.18c) is not defined if x and x' are row matrices.*] We must now extend our definition of multiplication to include matrices with arbitrary numbers of rows and columns.

The multiplication of a matrix A and a matrix B is defined only if the number of *columns* of A is equal to the number of *rows* of B. (The number of *rows* of A and the number of *columns* of B is each arbitrary; however, we shall have occasion to consider only cases for which A is a square matrix.) Therefore, in analogy with Eq. (1.18a), the product AB is given by

$$\boxed{\begin{aligned} &\mathsf{C} = \mathsf{AB} \\ &C_{ij} = [\mathsf{AB}]_{ij} = \sum_k A_{ik} B_{kj} \end{aligned}} \tag{1.19}$$

It is evident from Eq. (1.19) that matrix multiplication is not commutative. Thus, if A and B are both square matrices, then the sums

$$\sum_k A_{ik} B_{kj} \qquad \text{and} \qquad \sum_k B_{ik} A_{kj}$$

are both defined, but, in general, they will not be equal.

▶ Example 1.5 The Noncommutativity of Matrices

If A and B are the matrices

$$\mathsf{A} = \begin{pmatrix} 2 & 1 \\ -1 & 3 \end{pmatrix}; \qquad \mathsf{B} = \begin{pmatrix} -1 & 2 \\ 4 & -2 \end{pmatrix} \tag{1}$$

then

$$\mathsf{AB} = \begin{pmatrix} 2 & 2 \\ 13 & -8 \end{pmatrix} \tag{2}$$

but

$$\mathsf{BA} = \begin{pmatrix} -4 & 5 \\ 10 & -2 \end{pmatrix} \tag{3}$$

*Although whenever we operate on x with the λ matrix, the coordinate matrix x must be expressed as a column matrix, we may also write x as a row matrix, (x_1, x_2, x_3), for other applications.

1.6 Further Definitions

A *transposed matrix* is a matrix derived from an original matrix by the interchange of rows and columns. The *transpose* of a matrix A is denoted by A^t. According to the definition, we have

$$\boxed{\lambda_{ij}^t = \lambda_{ji}} \tag{1.20}$$

Evidently,

$$(\lambda^t)^t = \lambda \tag{1.21}$$

Equation (1.8) may therefore be written as any of the following equivalent expressions:

$$x_i = \sum_j \lambda_{ji} x_j' \tag{1.22a}$$

$$x_i = \sum_j \lambda_{ij}^t x_j' \tag{1.22b}$$

$$\mathsf{x} = \lambda^t \mathsf{x}' \tag{1.22c}$$

$$\begin{pmatrix} x_1 \\ x_2 \\ x_3 \end{pmatrix} = \begin{pmatrix} \lambda_{11} & \lambda_{21} & \lambda_{31} \\ \lambda_{12} & \lambda_{22} & \lambda_{32} \\ \lambda_{13} & \lambda_{23} & \lambda_{33} \end{pmatrix} \begin{pmatrix} x_1' \\ x_2' \\ x_3' \end{pmatrix} \tag{1.22d}$$

The *identity matrix* is defined to be that matrix which when multiplied by another matrix, leaves the latter unaffected. Thus,

$$\mathsf{1A} = \mathsf{A}; \qquad \mathsf{B1} = \mathsf{B} \tag{1.23}$$

That is,

$$\mathsf{1A} = \begin{pmatrix} 1 & 0 \\ 0 & 1 \end{pmatrix} \begin{pmatrix} A_1 \\ A_2 \end{pmatrix} = \begin{pmatrix} A_1 \\ A_2 \end{pmatrix} = \mathsf{A}$$

Let us consider the orthogonal rotation matrix λ for the case of two dimensions,

$$\lambda = \begin{pmatrix} \lambda_{11} & \lambda_{12} \\ \lambda_{21} & \lambda_{22} \end{pmatrix}$$

Then,

$$\begin{aligned} \lambda\lambda^t &= \begin{pmatrix} \lambda_{11} & \lambda_{12} \\ \lambda_{21} & \lambda_{22} \end{pmatrix} \begin{pmatrix} \lambda_{11} & \lambda_{21} \\ \lambda_{12} & \lambda_{22} \end{pmatrix} \\ &= \begin{pmatrix} \lambda_{11}^2 + \lambda_{12}^2 & \lambda_{11}\lambda_{21} + \lambda_{12}\lambda_{22} \\ \lambda_{21}\lambda_{11} + \lambda_{22}\lambda_{12} & \lambda_{21}^2 + \lambda_{22}^2 \end{pmatrix} \end{aligned}$$

Using the orthogonality relation [Eq. (1.13)], we find

$$\lambda_{11}^2 + \lambda_{12}^2 = \lambda_{21}^2 + \lambda_{22}^2 = 1$$
$$\lambda_{21}\lambda_{11} + \lambda_{22}\lambda_{12} = \lambda_{11}\lambda_{21} + \lambda_{12}\lambda_{22} = 0$$

so that for the special case of the orthogonal rotation matrix $\mathsf{\lambda}$ we have*

$$\mathsf{\lambda\lambda}^t = \begin{pmatrix} 1 & 0 \\ 0 & 1 \end{pmatrix} = \mathsf{1} \tag{1.24}$$

The *inverse* of a matrix is defined as that matrix which when multiplied by the original matrix produces the identity matrix; the inverse of the matrix $\mathsf{\lambda}$ is denoted by $\mathsf{\lambda}^{-1}$:

$$\mathsf{\lambda\lambda}^{-1} = \mathsf{1} \tag{1.25}$$

By comparing Eqs. (1.24) and (1.25), we find

$$\boxed{\mathsf{\lambda}^t = \mathsf{\lambda}^{-1}} \qquad \text{for orthogonal matrices} \tag{1.26}$$

Therefore, the transpose and the inverse of the rotation matrix $\mathsf{\lambda}$ are identical. In fact, the transpose of *any* orthogonal matrix is equal to its inverse.

To summarize some of the rules of matrix algebra:

(a) Matrix multiplication is not commutative in general:

$$\mathsf{AB} \neq \mathsf{BA} \tag{1.27a}$$

The special case of the multiplication of a matrix and its inverse is commutative:

$$\mathsf{AA}^{-1} = \mathsf{A}^{-1}\mathsf{A} = \mathsf{1} \tag{1.27b}$$

The identity matrix always commutes:

$$\mathsf{1A} = \mathsf{A1} = \mathsf{A} \tag{1.27c}$$

(b) Matrix multiplication is associative:

$$[\mathsf{AB}]\mathsf{C} = \mathsf{A}[\mathsf{BC}] \tag{1.28}$$

(c) Matrix addition is performed by adding corresponding elements of the two matrices. The components of C from the addition $\mathsf{C} = \mathsf{A} + \mathsf{B}$ are

$$C_{ij} = A_{ij} + B_{ij} \tag{1.29}$$

Addition is, of course, defined only if A and B have the same dimensions.

* This result is not valid for matrices in general; it is true only for *orthogonal* matrices.

1.7 Geometrical Significance of Transformation Matrices

Consider the case in which the coordinate axes are rotated counterclockwise* through an angle of 90° about the x_3-axis, as in Fig. 1-5. In such a rotation, $x_1' = x_2$, $x_2' = -x_1$, $x_3' = x_3$.

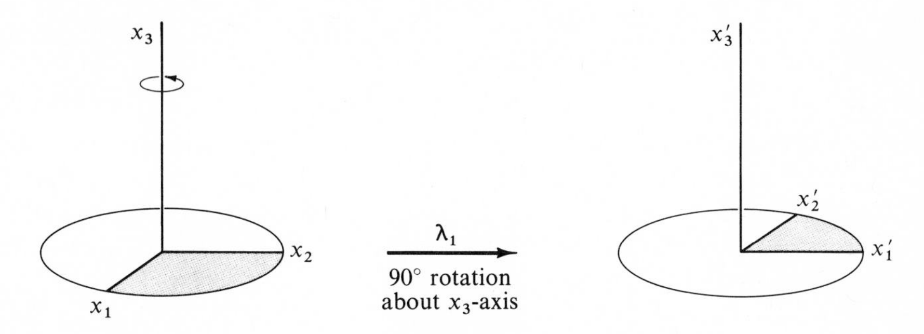

FIG. 1-5

The only nonvanishing cosines are

$$\cos(x_1', x_2) = 1 = \lambda_{12}$$
$$\cos(x_2', x_1) = -1 = \lambda_{21}$$
$$\cos(x_3', x_3) = 1 = \lambda_{33}$$

So that the λ matrix for this case is

$$\lambda_1 = \begin{pmatrix} 0 & 1 & 0 \\ -1 & 0 & 0 \\ 0 & 0 & 1 \end{pmatrix} \tag{1.30}$$

Next, consider the counterclockwise rotation through 90° about the x_1-axis, as in Fig. 1-6. We have $x_1', = x_1$, $x_2' = x_3$, $x_3' = -x_2$, and the transformation matrix is

$$\lambda_2 = \begin{pmatrix} 1 & 0 & 0 \\ 0 & 0 & 1 \\ 0 & -1 & 0 \end{pmatrix} \tag{1.31}$$

* The sense of the rotation is determined by looking along the positive portion of the axis of rotation at the plane being rotated. This definition is then consistent with the "right-hand rule" in which the positive direction is the direction of advance of a right-hand screw when turned in the same sense.

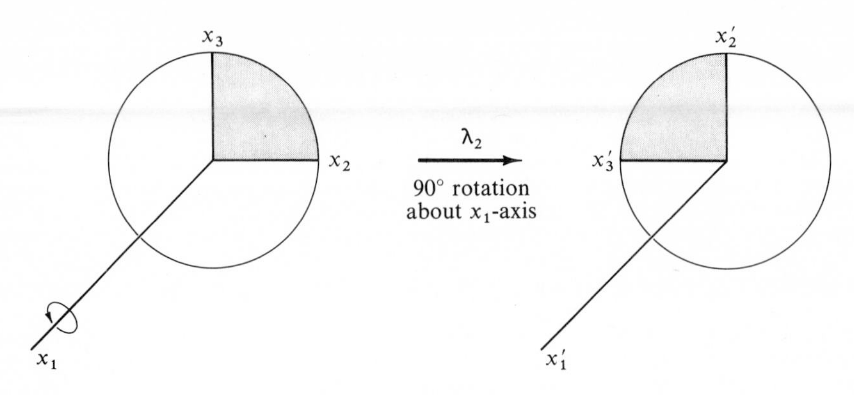

FIG. 1-6

In order to find the transformation matrix for the combined transformation for rotation about the x_3-axis, followed by rotation about the x_1-axis, we have (see Fig. 1-7)

$$\mathbf{x}' = \boldsymbol{\lambda}_1 \mathbf{x} \tag{1.32a}$$

and

$$\mathbf{x}'' = \boldsymbol{\lambda}_2 \mathbf{x}' \tag{1.32b}$$

or,

$$\mathbf{x}'' = \boldsymbol{\lambda}_2 \boldsymbol{\lambda}_1 \mathbf{x} \tag{1.33a}$$

$$\begin{pmatrix} x_1'' \\ x_2'' \\ x_3'' \end{pmatrix} = \begin{pmatrix} 1 & 0 & 0 \\ 0 & 0 & 1 \\ 0 & -1 & 0 \end{pmatrix} \begin{pmatrix} 0 & 1 & 0 \\ -1 & 0 & 0 \\ 0 & 0 & 1 \end{pmatrix} \begin{pmatrix} x_1 \\ x_2 \\ x_3 \end{pmatrix} = \begin{pmatrix} 0 & 1 & 0 \\ 0 & 0 & 1 \\ 1 & 0 & 0 \end{pmatrix} \begin{pmatrix} x_1 \\ x_2 \\ x_3 \end{pmatrix} = \begin{pmatrix} x_2 \\ x_3 \\ x_1 \end{pmatrix} \tag{1.33b}$$

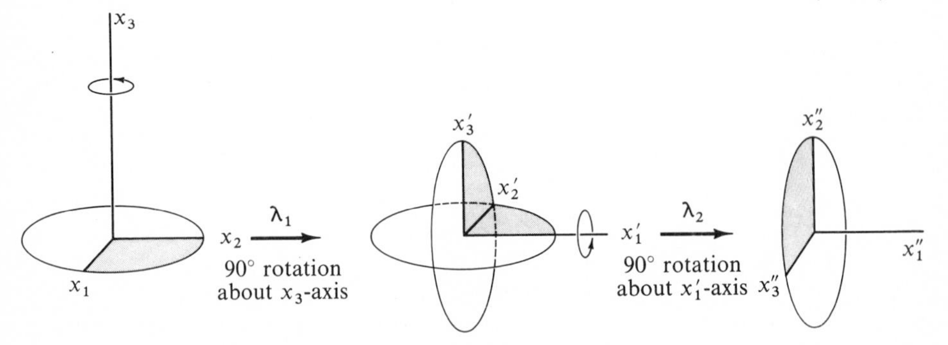

FIG. 1-7

Therefore, the two rotations described above may be represented by a single transformation matrix:

$$\lambda_3 = \lambda_2 \lambda_1 = \begin{pmatrix} 0 & 1 & 0 \\ 0 & 0 & 1 \\ 1 & 0 & 0 \end{pmatrix} \tag{1.34}$$

and the final orientation is specified by $x_1'' = x_2$, $x_2'' = x_3$, $x_3'' = x_1$. Note that the order in which the transformation matrices operate on $\mathbf{x}$ is important since the multiplication is not commutative. In the other order,

$$\begin{aligned} \lambda_4 &= \lambda_1 \lambda_2 \\ &= \begin{pmatrix} 0 & 1 & 0 \\ -1 & 0 & 0 \\ 0 & 0 & 1 \end{pmatrix} \begin{pmatrix} 1 & 0 & 0 \\ 0 & 0 & 1 \\ 0 & -1 & 0 \end{pmatrix} \\ &= \begin{pmatrix} 0 & 0 & 1 \\ -1 & 0 & 0 \\ 0 & -1 & 0 \end{pmatrix} \neq \lambda_3 \end{aligned} \tag{1.35}$$

and an entirely different orientation results. Figure 1-8 illustrates the different final orientations of a parallelepiped that undergoes rotations which correspond to two rotation matrices λ_A, λ_B when successive rotations are made in different order. The upper portion of the figure represents the matrix product $\lambda_B \lambda_A$, while the lower portion represents the product $\lambda_A \lambda_B$.

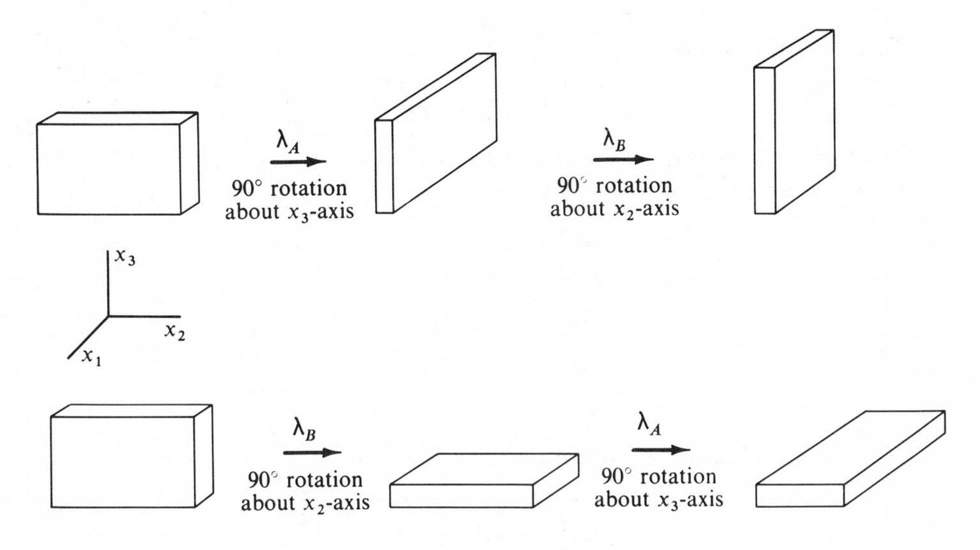

FIG. **1-8**

Next, consider the coordinate rotation pictured in Fig. 1-9 (which is the same as that in Fig. 1-2). The elements of the transformation matrix in two dimensions are given by the following cosines:

$$\cos(x_1', x_1) = \cos\theta = \lambda_{11}$$

$$\cos(x_1', x_2) = \cos\left(\frac{\pi}{2} - \theta\right) = \sin\theta = \lambda_{12}$$

$$\cos(x_2', x_1) = \cos\left(\frac{\pi}{2} + \theta\right) = -\sin\theta = \lambda_{21}$$

$$\cos(x_2', x_2) = \cos\theta = \lambda_{22}$$

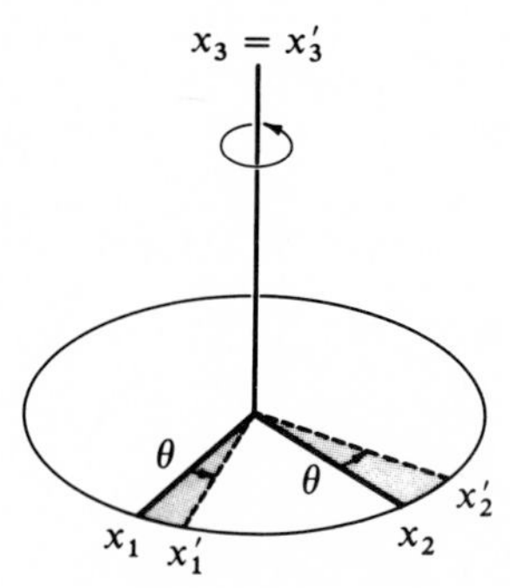

FIG. 1-9

Therefore, the matrix is

$$\lambda_5 = \begin{pmatrix} \cos\theta & \sin\theta \\ -\sin\theta & \cos\theta \end{pmatrix} \tag{1.36}$$

If this rotation were a three-dimensional rotation, in which $x_3' = x_3$, then we would have the following additional cosines:

$$\cos(x_1', x_3) = 0 = \lambda_{13}$$
$$\cos(x_2', x_3) = 0 = \lambda_{23}$$
$$\cos(x_3', x_3) = 1 = \lambda_{33}$$
$$\cos(x_3', x_1) = 0 = \lambda_{31}$$
$$\cos(x_3', x_2) = 0 = \lambda_{32}$$

and the three-dimensional transformation matrix is

$$\lambda_{5'} = \begin{pmatrix} \cos\theta & \sin\theta & 0 \\ -\sin\theta & \cos\theta & 0 \\ 0 & 0 & 1 \end{pmatrix} \tag{1.36a}$$

As a final example, consider the transformation which results in the reflection through the origin of all of the axes, as in Fig. 1-10. Such a transformation is called an *inversion*. Then, $x_1' = -x_1$, $x_2' = -x_2$, $x_3' = -x_3$, and

$$\lambda_6 = \begin{pmatrix} -1 & 0 & 0 \\ 0 & -1 & 0 \\ 0 & 0 & -1 \end{pmatrix} \tag{1.37}$$

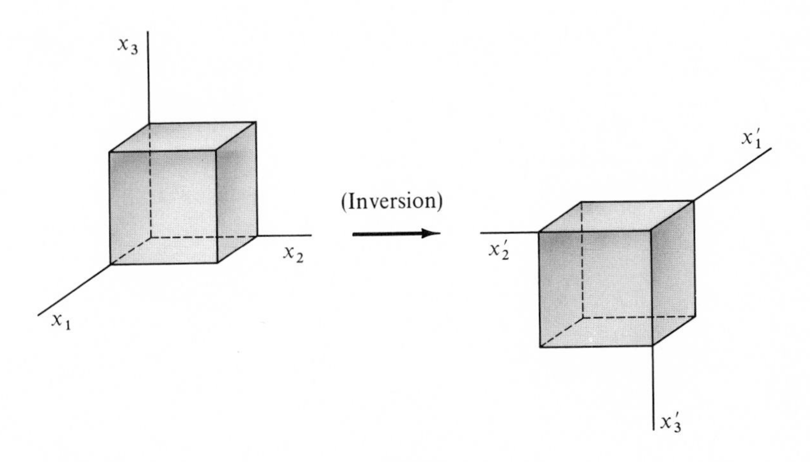

FIG. **1-10**

In the preceding examples, the transformation matrix λ_3 was defined to be the result of two successive rotations, each of which was an orthogonal transformation: $\lambda_3 = \lambda_2 \lambda_1$. It is possible to prove in the following manner that the successive application of orthogonal transformations always results in an orthogonal transformation. We write

$$x_i' = \sum_j \lambda_{ij} x_j ; \qquad x_k'' = \sum_i \mu_{ki} x_i'$$

Combining these expressions, we obtain

$$\begin{aligned} x_k'' &= \sum_j \left(\sum_i \mu_{ki} \lambda_{ij} \right) x_j \\ &= \sum_j [\mu\lambda]_{kj} x_j \end{aligned}$$

Thus, the transformation from x_i to x_i'' is accomplished by operating on x_i with the $(\mu\lambda)$ matrix. The combined transformation will then be shown to be orthogonal if $(\mu\lambda)^t = (\mu\lambda)^{-1}$. Now, the transpose of a product matrix is the product of the transposed matrices taken in reverse order (see Problem 1-3); i.e., $(\mathsf{AB})^t = \mathsf{B}^t\mathsf{A}^t$. Therefore,

$$(\mu\lambda)^t = \lambda^t \mu^t \tag{1.38}$$

But, since λ and μ are orthogonal, $\lambda^t = \lambda^{-1}$ and $\mu^t = \mu^{-1}$. And then multiplying the above equation by $\mu\lambda$ from the right, we obtain

$$\begin{aligned}(\mu\lambda)^t\mu\lambda &= \lambda^t\mu^t\mu\lambda \\ &= \lambda^t 1 \lambda \\ &= \lambda^t\lambda \\ &= 1 \\ &= (\mu\lambda)^{-1}\mu\lambda\end{aligned}$$

Hence,

$$(\mu\lambda)^t = (\mu\lambda)^{-1} \tag{1.39}$$

and the $\mu\lambda$ matrix is orthogonal.

The determinants of all of the rotation matrices in the examples above can be calculated according to the standard rule for the evaluation of determinants of second or third order:

$$|\lambda| = \begin{vmatrix} \lambda_{11} & \lambda_{12} \\ \lambda_{21} & \lambda_{22} \end{vmatrix} = \lambda_{11}\lambda_{22} - \lambda_{12}\lambda_{21} \tag{1.40}$$

$$\begin{aligned}|\lambda| &= \begin{vmatrix} \lambda_{11} & \lambda_{12} & \lambda_{13} \\ \lambda_{21} & \lambda_{22} & \lambda_{23} \\ \lambda_{31} & \lambda_{32} & \lambda_{33} \end{vmatrix} \\ &= \lambda_{11}\begin{vmatrix} \lambda_{22} & \lambda_{23} \\ \lambda_{32} & \lambda_{33} \end{vmatrix} - \lambda_{12}\begin{vmatrix} \lambda_{21} & \lambda_{23} \\ \lambda_{31} & \lambda_{33} \end{vmatrix} + \lambda_{13}\begin{vmatrix} \lambda_{21} & \lambda_{22} \\ \lambda_{31} & \lambda_{32} \end{vmatrix}\end{aligned} \tag{1.41}$$

where the third-order determinant has been expanded in minors of the first row. Therefore, we find

$$|\lambda_1| = |\lambda_2| = \cdots = |\lambda_5| = 1$$

but

$$|\lambda_6| = -1$$

Thus, all of those transformations which result from *rotations starting from the original set of axes* have determinants equal to $+1$. An *inversion*, however, cannot be generated by any series of rotations, and the determinant of an inversion matrix is equal to -1. Orthogonal transformations, the determinant of whose matrices is $+1$, are called *proper rotations*; those with determinant equal to -1 are called *improper rotations*. That *all* orthogonal matrices must have a determinant equal to either $+1$ or -1 is proved in Marion (Ma65a, Section 1.7). Henceforth, we shall confine our attention to the effect of proper rotations; we shall not be concerned with the special properties of vectors that are manifest in improper rotations. (These latter effects are discussed in detail in *Principles of Vector Analysis*.)

1.8 Definitions of a Scalar and a Vector in Terms of Transformation Properties

Consider a coordinate transformation of the type

$$x_i' = \sum_j \lambda_{ij} x_j \tag{1.42}$$

with

$$\sum_j \lambda_{ij} \lambda_{kj} = \delta_{ik} \tag{1.43}$$

If, under such a transformation, a quantity φ is unaffected, then φ is called a *scalar* (or *scalar invariant.*)

If a set of quantities (A_1, A_2, A_3) is transformed from the x_i system to the x_i' system by means of a transformation matrix $\boldsymbol{\lambda}$ with the result

$$\boxed{A_i' = \sum_j \lambda_{ij} A_j} \tag{1.44}$$

then the quantities A_i transform as the coordinates of a point [i.e., according to Eq. (1.42)], and the quantity $\mathbf{A} = (A_1, A_2, A_3)$ is termed a *vector.*

1.9 Elementary Scalar and Vector Operations

In the following, $\mathbf{A}$ and $\mathbf{B}$ are vectors (with components A_i and B_i) and φ, ψ, and ξ are scalars.

Addition

$$A_i + B_i = B_i + A_i \qquad \text{Commutative law} \tag{1.45}$$

$$A_i + (B_i + C_i) = (A_i + B_i) + C_i \qquad \text{Associative law} \tag{1.46}$$

$$\varphi + \psi = \psi + \varphi \qquad \text{Commutative law} \tag{1.47}$$

$$\varphi + (\psi + \xi) = (\varphi + \psi) + \xi \qquad \text{Associative law} \tag{1.48}$$

Multiplication by a scalar ξ

$$\xi\mathbf{A} = \mathbf{B} \qquad \text{is a vector} \tag{1.49}$$

$$\xi\varphi = \psi \qquad \text{is a scalar} \tag{1.50}$$

Equation (1.49) can be proved as follows:

$$\begin{aligned} B_i' &= \sum_j \lambda_{ij} B_j = \sum_j \lambda_{ij} \xi A_j \\ &= \xi \sum_j \lambda_{ij} A_j = \xi A_i' \end{aligned} \tag{1.51}$$

and $\xi \mathbf{A}$ transforms as a vector. Similarly, $\xi\varphi$ transforms as a scalar.

1.10 The Scalar Product of Two Vectors

The multiplication of two vectors **A** and **B** to form the *scalar product* is defined to be

$$\boxed{\mathbf{A} \cdot \mathbf{B} = \sum_i A_i B_i} \tag{1.52}$$

where the dot between **A** and **B** is used to denote *scalar* multiplication; this operation is sometimes called the *dot product*.*

The vector **A** has components A_1, A_2, A_3, and the magnitude (or length) of **A** is given by

$$|\mathbf{A}| = +\sqrt{A_1^2 + A_2^2 + A_3^2} \equiv A \tag{1.53}$$

where the magnitude is indicated by $|\mathbf{A}|$ or, if there is no possibility of confusion, simply by A. Dividing both sides of Eq. (1.52) by AB, we have

$$\frac{\mathbf{A} \cdot \mathbf{B}}{AB} = \sum_i \frac{A_i}{A} \frac{B_i}{B} \tag{1.54}$$

Now, from Fig. 1-11, we see that A_1/A is the cosine of the angle α between the vector **A** and the x_1-axis. In general, A_i/A and B_i/B are the direction cosines Λ_i^A and Λ_i^B of the vectors **A** and **B**:

$$\frac{\mathbf{A} \cdot \mathbf{B}}{AB} = \sum_i \Lambda_i^A \Lambda_i^B \tag{1.55}$$

The sum $\sum_i \Lambda_i^A \Lambda_i^B$ is just the cosine of the angle between **A** and **B** [see Eq. (1.11)]:

$$\cos(\mathbf{A}, \mathbf{B}) = \sum_i \Lambda_i^A \Lambda_i^B$$

or,

$$\boxed{\mathbf{A} \cdot \mathbf{B} = AB \cos(\mathbf{A}, \mathbf{B})} \tag{1.56}$$

* Older notation includes (**AB**), (**A . B**), and (**A, B**).

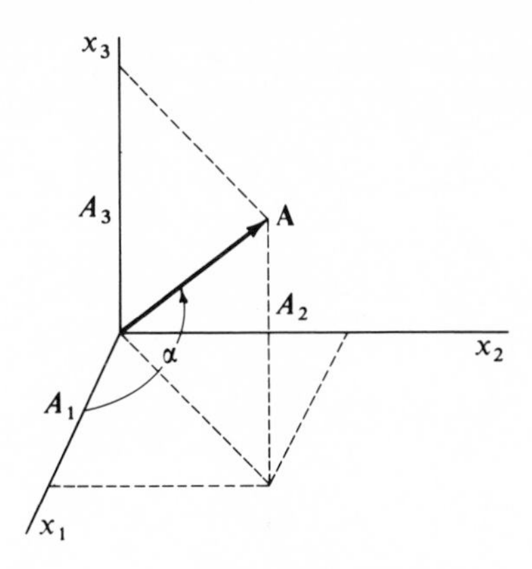

FIG. **1-11**

That the product $\mathbf{A} \cdot \mathbf{B}$ is indeed a scalar may be shown as follows. **A** and **B** transform as vectors:

$$A_i' = \sum_j \lambda_{ij} A_j ; \quad B_i' = \sum_k \lambda_{ik} B_k \tag{1.57}$$

Therefore, the product $\mathbf{A}' \cdot \mathbf{B}'$ becomes

$$\begin{aligned} \mathbf{A}' \cdot \mathbf{B}' &= \sum_i A_i' B_i' \\ &= \sum_i \left(\sum_j \lambda_{ij} A_j \right) \left(\sum_k \lambda_{ik} B_k \right) \end{aligned}$$

Rearranging the summations, we can write

$$\mathbf{A}' \cdot \mathbf{B}' = \sum_{j,k} \left(\sum_i \lambda_{ij} \lambda_{ik} \right) A_j B_k$$

But, according to the orthogonality condition, the term in parentheses is just δ_{ik}. Thus,

$$\begin{aligned} \mathbf{A}' \cdot \mathbf{B}' &= \sum_j \left(\sum_k \delta_{jk} A_j B_k \right) \\ &= \sum_j A_j B_j \\ &= \mathbf{A} \cdot \mathbf{B} \end{aligned} \tag{1.58}$$

Since the value of the product is unaltered by the coordinate transformation, the product must be a scalar.

Notice that the distance from the origin to the point (x_1, x_2, x_3) defined by the vector **A** is given by

$$|\mathbf{A}| = \sqrt{\mathbf{A} \cdot \mathbf{A}} = \sqrt{x_1^2 + x_2^2 + x_3^2} = \sqrt{\sum_i x_i^2}$$

Similarly, the distance from the point (x_1, x_3, x_3) to another point $(\bar{x}_1, \bar{x}_2, \bar{x}_3)$ defined by the vector **B** is

$$\sqrt{\sum_i (x_i - \bar{x}_i)^2} = \sqrt{(\mathbf{A} - \mathbf{B}) \cdot (\mathbf{A} - \mathbf{B})} = |\mathbf{A} - \mathbf{B}|$$

That is, we can define the vector connecting any point with any other point as the difference of the vectors which define the individual points, as in Fig. 1-12. The distance between the points is then the magnitude of the difference

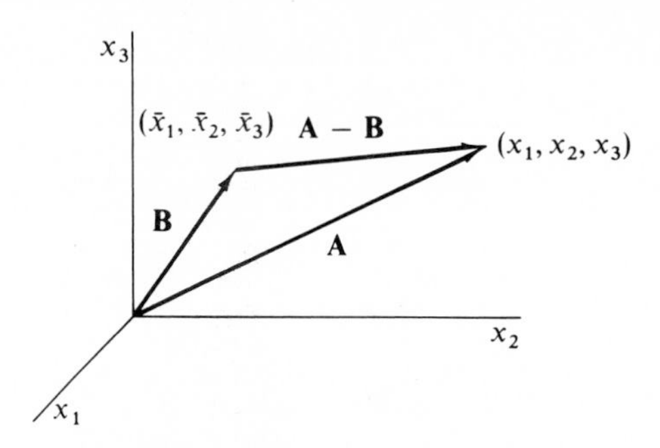

FIG. 1-12

vector, and since this magnitude is the square root of a scalar product it is invariant to a coordinate transformation. This is an important fact and can be summarized by the statement that *orthogonal transformations are distance-preserving transformations.* It is also clear that the angle between two vectors will be preserved under an orthogonal transformation. These two results are obviously essential if we are to successfully apply transformation theory to physical situations.

The scalar product obeys the commutative and distributive laws:

$$\mathbf{A} \cdot \mathbf{B} = \sum_i A_i B_i = \sum_i B_i A_i = \mathbf{B} \cdot \mathbf{A} \tag{1.59}$$

$$\begin{aligned} \mathbf{A} \cdot (\mathbf{B} + \mathbf{C}) &= \sum_i A_i(B + C)_i = \sum_i A_i(B_i + C_i) \\ &= \sum_i (A_i B_i + A_i C_i) = (\mathbf{A} \cdot \mathbf{B}) + (\mathbf{A} \cdot \mathbf{C}) \end{aligned} \tag{1.60}$$

1.11 The Vector Product of Two Vectors

We next consider another method for the combination of two vectors—the so-called *vector product* (or *cross product*). First, we assert that this operation does, in fact, produce a vector.* The vector product of **A** and **B**

* The product considered here actually produces an *axial* vector, but the term *vector product* will be used in order to be consistent with popular usage. [See Marion (Ma65a, Section 1.15).]

is denoted by a bold cross $\times$,*

$$\mathbf{C} = \mathbf{A} \times \mathbf{B} \tag{1.61}$$

where $\mathbf{C}$ is the vector that we assert results from this operation. The components of $\mathbf{C}$ are defined by the relation

$$\boxed{C_i = \sum_{j,k} \varepsilon_{ijk} A_j B_k} \tag{1.62}$$

where the symbol ε_{ijk} is the *permutation symbol* (or *Levi-Civita density*) and has the following properties:

$$\varepsilon_{ijk} = \left.\begin{array}{ll} 0, & \text{if any index is equal to any other index} \\ +1, & \text{if } i, j, k \text{ form an } \textit{even} \text{ permutation of } 1, 2, 3 \\ -1, & \text{if } i, j, k \text{ form an } \textit{odd} \text{ permutation of } 1, 2, 3 \end{array}\right\} \tag{1.63}$$

Thus,

$$\varepsilon_{122} = \varepsilon_{313} = \varepsilon_{211} = 0, \text{ etc.}$$
$$\varepsilon_{123} = \varepsilon_{231} = \varepsilon_{312} = +1$$
$$\varepsilon_{132} = \varepsilon_{213} = \varepsilon_{321} = -1$$

Using the above notation the components of $\mathbf{C}$ can be explicitly evaluated. For $i = 1$, the only nonvanishing ε_{ijk} are ε_{123} and ε_{132}, i.e., for $j, k = 2, 3$ in either order. Therefore,

$$C_1 = \sum_{j,k} \varepsilon_{1jk} A_j B_k = \varepsilon_{123} A_2 B_3 + \varepsilon_{132} A_3 B_2$$
$$= A_2 B_3 - A_3 B_2 \tag{1.64a}$$

Similarly,

$$C_2 = A_3 B_1 - A_1 B_3 \tag{1.64b}$$

$$C_3 = A_1 B_2 - A_2 B_1 \tag{1.64c}$$

Consider now the expansion of the quantity $[AB \sin(\mathbf{A}, \mathbf{B})]^2 = (AB \sin\theta)^2$

$$\begin{aligned} A^2 B^2 \sin^2\theta &= A^2 B^2 - A^2 B^2 \cos^2\theta \\ &= \left(\sum_i A_i^2\right)\left(\sum_i B_i^2\right) - \left(\sum_i A_i B_i\right)^2 \\ &= (A_2 B_3 - A_3 B_2)^2 + (A_3 B_1 - A_1 B_3)^2 + (A_1 B_2 - A_2 B_1)^2 \end{aligned} \tag{1.65}$$

where the last equality requires some algebra. Identifying the components of $\mathbf{C}$ in the last expression, we can write

$$(AB \sin\theta)^2 = C_1^2 + C_2^2 + C_3^2 = |\mathbf{C}^2| = C^2 \tag{1.66}$$

* Older notation includes [**AB**], [**A** . **B**], and [**A** $\wedge$ **B**].

Taking the positive square root of both sides of this equation,

$$C = AB \sin \theta \tag{1.67}$$

This equation states that if $\mathbf{C} = \mathbf{A} \times \mathbf{B}$, then the magnitude of $\mathbf{C}$ is equal to the product of the magnitudes of $\mathbf{A}$ and $\mathbf{B}$ multiplied by the sine of the angle between them. Geometrically, $AB \sin \theta$ is the area of the parallelogram defined by the vectors $\mathbf{A}$ and $\mathbf{B}$ and the angle between them, as in Fig. 1-13.

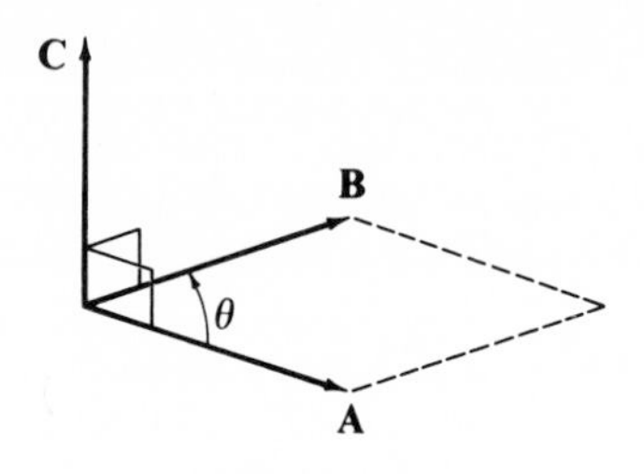

FIG. 1-13

Now, $\mathbf{A} \times \mathbf{B}$ (i.e., $\mathbf{C}$) is perpendicular to the plane defined by $\mathbf{A}$ and $\mathbf{B}$ because $\mathbf{A} \cdot (\mathbf{A} \times \mathbf{B}) = 0$ and $\mathbf{B} \cdot (\mathbf{A} \times \mathbf{B}) = 0$. Since a plane area can be represented by a vector normal to the plane and of magnitude equal to the area, evidently $\mathbf{C}$ is such a vector. The positive direction of $\mathbf{C}$ is chosen to be the direction of advance of a right-hand screw when rotated from $\mathbf{A}$ to $\mathbf{B}$.

The definition of the vector product is now complete; components, magnitude, and geometrical interpretation have been given. We may therefore reasonably expect that $\mathbf{C}$ is indeed a vector. The ultimate test, however, is to examine the transformation properties of $\mathbf{C}$. This is carried out in Marion (Ma65a, Section 1.14), where it is shown that $\mathbf{C}$ does in fact transform as a vector under a proper rotation.

We should note the following properties of the vector product which result from the definitions:

$$\text{(a)} \quad \mathbf{A} \times \mathbf{B} = -\mathbf{B} \times \mathbf{A} \tag{1.68a}$$

but, in general,

$$\text{(b)} \quad \mathbf{A} \times (\mathbf{B} \times \mathbf{C}) \neq (\mathbf{A} \times \mathbf{B}) \times \mathbf{C} \tag{1.68b}$$

Another important result is (see Problem 1-16)

$$\mathbf{A} \times (\mathbf{B} \times \mathbf{C}) = (\mathbf{A} \cdot \mathbf{C})\mathbf{B} - (\mathbf{A} \cdot \mathbf{B})\mathbf{C} \tag{1.69}$$

▶ Example 1.11 Derivation of a Vector Identity

The customary method of proving complicated vector relations is to know the result beforehand and to expand both sides of the equation by components, eventually obtaining an identity. By using the ε_{ijk} notation, however, it is possible to derive even the most complicated results in a straightforward manner. In order to evaluate the product $(\mathbf{A} \times \mathbf{B}) \cdot (\mathbf{C} \times \mathbf{D})$, for example, we first write the ith component of each vector product using Eq. (1.62):

$$(\mathbf{A} \times \mathbf{B})_i = \sum_{j,k} \varepsilon_{ijk} A_j B_k \tag{1}$$

$$(\mathbf{C} \times \mathbf{D})_i = \sum_{l,m} \varepsilon_{ilm} C_l D_m \tag{2}$$

The scalar product is then computed according to Eq. (1.52):

$$(\mathbf{A} \times \mathbf{B}) \cdot (\mathbf{C} \times \mathbf{D}) = \sum_i \left(\sum_{j,k} \varepsilon_{ijk} A_j B_k \right) \left(\sum_{l,m} \varepsilon_{ilm} C_l D_m \right) \tag{3}$$

Rearranging the summations, we have

$$(\mathbf{A} \times \mathbf{B}) \cdot (\mathbf{C} \times \mathbf{D}) = \sum_{l,m} \left(\sum_i \varepsilon_{jki} \varepsilon_{lmi} \right) A_j B_k C_l D_m \tag{4}$$

where the indices of the ε's have been permuted (twice each so that no sign change occurs) to place in the third position the index over which the sum is carried out. We can now make use of an important property of the ε_{ijk} (see Problem 1-16):

$$\boxed{\sum_k \varepsilon_{ijk} \varepsilon_{lmk} = \delta_{il} \delta_{jm} - \delta_{im} \delta_{jl}} \tag{5}$$

Therefore, Eq. (4) becomes

$$(\mathbf{A} \times \mathbf{B}) \cdot (\mathbf{C} \times \mathbf{D}) = \sum_{\substack{j,k \\ l,m}} (\delta_{jl} \delta_{km} - \delta_{jm} \delta_{kl}) A_j B_k C_l D_m \tag{6}$$

Carrying out the summations over j and k, the Kronecker deltas reduce the expression to

$$(\mathbf{A} \times \mathbf{B}) \cdot (\mathbf{C} \times \mathbf{D}) = \sum_{l,m} (A_l B_m C_l D_m - A_m B_l C_l D_m) \tag{7}$$

This can be rearranged to obtain

$$(\mathbf{A} \times \mathbf{B}) \cdot (\mathbf{C} \times \mathbf{D}) = \left(\sum_l A_l C_l \right) \left(\sum_m B_m D_m \right) - \left(\sum_l B_l C_l \right) \left(\sum_m A_m D_m \right) \tag{8}$$

Since each term in parentheses on the right-hand side is just a scalar product, we have finally,

$$(\mathbf{A} \times \mathbf{B}) \cdot (\mathbf{C} \times \mathbf{D}) = (\mathbf{A} \cdot \mathbf{C})(\mathbf{B} \cdot \mathbf{D}) - (\mathbf{B} \cdot \mathbf{C})(\mathbf{A} \cdot \mathbf{D}) \tag{9}$$

1.12 Unit Vectors

It is frequently desirable to be able to describe a vector in terms of the components along the three coordinate axes together with a convenient specification of these axes. For this purpose, we introduce the *unit vectors* along the rectangular axes*; $\mathbf{e}_1$, $\mathbf{e}_2$, $\mathbf{e}_3$. Therefore, the following ways of expressing the vector **A** are equivalent:

$$\mathbf{A} = (A_1, A_2, A_3) \quad \text{or} \quad \mathbf{A} = \mathbf{e}_1 A_1 + \mathbf{e}_2 A_2 + \mathbf{e}_3 A_3 = \sum_i \mathbf{e}_i A_i \tag{1.70}$$

The components of the vector **A** are obtained by projection onto the axes:

$$A_i = \mathbf{e}_i \cdot \mathbf{A} \tag{1.71}$$

We have seen [Eq. (1.56)] that the scalar product of two vectors has a magnitude equal to the product of the individual magnitudes multiplied by the cosine of the angle between the vectors:

$$\mathbf{A} \cdot \mathbf{B} = AB \cos(\mathbf{A}, \mathbf{B}) \tag{1.72}$$

Therefore, the scalar product of any two unit vectors (because they are orthogonal) becomes

$$\boxed{\mathbf{e}_i \cdot \mathbf{e}_j = \delta_{ij}} \tag{1.73}$$

Moreover, the orthogonality of the $\mathbf{e}_i$ requires the vector product to be

$$\mathbf{e}_i \times \mathbf{e}_j = \mathbf{e}_k, \qquad i, j, k \text{ in cyclic order} \tag{1.74}$$

The permutation symbol can be used to express this result as

$$\boxed{\mathbf{e}_i \times \mathbf{e}_j = \mathbf{e}_k \varepsilon_{ijk}} \tag{1.74a}$$

The vector product $\mathbf{C} = \mathbf{A} \times \mathbf{B}$, for example, can now be expressed as

$$\boxed{\mathbf{C} = \sum_{i,j,k} \varepsilon_{ijk} \mathbf{e}_i A_j B_k} \tag{1.75}$$

* Many variants of the symbols for the unit vectors are in use. Perhaps the most common set is **i**, **j**, **k** or $\hat{\mathbf{i}}$, $\hat{\mathbf{j}}$, $\hat{\mathbf{k}}$; others include $\mathbf{i}_1$, $\mathbf{i}_2$, $\mathbf{i}_3$ and $\boldsymbol{I}_x$, $\boldsymbol{I}_y$, $\boldsymbol{I}_k$.

By direct expansion and comparison with Eq. (1.75), we can verify a determinental expression for the vector product:

$$\mathbf{C} = \mathbf{A} \times \mathbf{B} = \begin{vmatrix} \mathbf{e}_1 & \mathbf{e}_2 & \mathbf{e}_3 \\ A_1 & A_2 & A_3 \\ B_1 & B_2 & B_3 \end{vmatrix} \tag{1.76}$$

The following identities are stated without proof:

$$\mathbf{A} \cdot (\mathbf{B} \times \mathbf{C}) = \mathbf{B} \cdot (\mathbf{C} \times \mathbf{A}) = \mathbf{C} \cdot (\mathbf{A} \times \mathbf{B}) \equiv \mathbf{ABC} \tag{1.77}$$

$$\mathbf{A} \times (\mathbf{B} \times \mathbf{C}) = (\mathbf{A} \cdot \mathbf{C})\mathbf{B} - (\mathbf{A} \cdot \mathbf{B})\mathbf{C} \tag{1.78}$$

$$\left.\begin{aligned} (\mathbf{A} \times \mathbf{B}) \cdot (\mathbf{C} \times \mathbf{D}) &= \mathbf{A} \cdot [\mathbf{B} \times (\mathbf{C} \times \mathbf{D})] \\ &= \mathbf{A} \cdot [(\mathbf{B} \cdot \mathbf{D})\mathbf{C} - (\mathbf{B} \cdot \mathbf{C})\mathbf{D}] \\ &= (\mathbf{A} \cdot \mathbf{C})(\mathbf{B} \cdot \mathbf{D}) - (\mathbf{A} \cdot \mathbf{D})(\mathbf{B} \cdot \mathbf{C}) \end{aligned}\right\} \tag{1.79}$$

$$\left.\begin{aligned} (\mathbf{A} \times \mathbf{B}) \times (\mathbf{C} \times \mathbf{D}) &= [(\mathbf{A} \times \mathbf{B}) \cdot \mathbf{D}]\mathbf{C} - [(\mathbf{A} \times \mathbf{B}) \cdot \mathbf{C}]\mathbf{D} \\ &= (\mathbf{ABD})\mathbf{C} - (\mathbf{ABC})\mathbf{D} \end{aligned}\right\} \tag{1.80}$$

1.13 Differentiation of a Vector with Respect to a Scalar

If a scalar function $\varphi = \varphi(s)$ is differentiated with respect to the scalar variable s, then since neither part of the derivative can change under a coordinate transformation, the derivative itself cannot change and must therefore be a scalar. That is, in the x_i and x_i' coordinate systems, $\varphi = \varphi'$ and $s = s'$, so that $d\varphi = d\varphi'$ and $ds = ds'$. Hence,

$$d\varphi/ds = d\varphi'/ds' = (d\varphi/ds)' \tag{1.81}$$

In a similar manner, we can formally define the differentiation of a vector $\mathbf{A}$ with respect to a scalar s. The components of $\mathbf{A}$ transform according to

$$A_i' = \sum_j \lambda_{ij} A_j \tag{1.82}$$

Therefore, upon differentiation, we obtain (since the λ_{ij} are independent of s')

$$\frac{dA_i'}{ds'} = \frac{d}{ds'} \sum_j \lambda_{ij} A_j = \sum_j \lambda_{ij} \frac{dA_j}{ds'}$$

Since s and s' are identical, we have

$$\frac{dA_i'}{ds'} = \left(\frac{dA_i}{ds}\right)' = \sum_j \lambda_{ij} \left(\frac{dA_j}{ds}\right) \tag{1.83}$$

Thus, the quantities dA_j/ds transform as do the components of a vector and, hence, *are* the components of a vector which we can write as $d\mathbf{A}/ds$.

We can give a geometrical interpretation to the vector $d\mathbf{A}/ds$ as follows. First, in order for $d\mathbf{A}/ds$ to exist, it is necessary that $\mathbf{A}$ be a continuous function of the variable $s : \mathbf{A} = \mathbf{A}(s)$. Suppose this function to be represented by the continuous curve Γ in Fig. 1-14; at the point P the variable has the

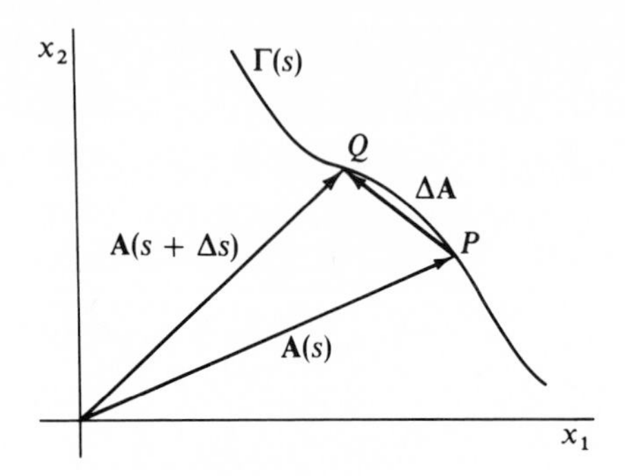

FIG. 1-14

value s, while at Q it has the value $s + \Delta s$. The derivative of $\mathbf{A}$ with respect to s is then given in standard fashion by

$$\begin{aligned}\frac{d\mathbf{A}}{ds} &= \lim_{\Delta s \to 0} \frac{\Delta \mathbf{A}}{\Delta s} \\ &= \lim_{\Delta s \to 0} \frac{\mathbf{A}(s + \Delta s) - \mathbf{A}(s)}{\Delta s}\end{aligned} \tag{1.84a}$$

The derivatives of vector sums and products obey the rules of ordinary vector calculus; for example,

$$\frac{d}{ds}(\mathbf{A} + \mathbf{B}) = \frac{d\mathbf{A}}{ds} + \frac{d\mathbf{B}}{ds} \tag{1.84b}$$

$$\frac{d}{ds}(\mathbf{A} \cdot \mathbf{B}) = \mathbf{A} \cdot \frac{d\mathbf{B}}{ds} + \frac{d\mathbf{A}}{ds} \cdot \mathbf{B} \tag{1.84c}$$

$$\frac{d}{ds}(\mathbf{A} \times \mathbf{B}) = \mathbf{A} \times \frac{d\mathbf{B}}{ds} + \frac{d\mathbf{A}}{ds} \times \mathbf{B} \tag{1.84d}$$

$$\frac{d}{ds}(\varphi \mathbf{A}) = \varphi \frac{d\mathbf{A}}{ds} + \frac{d\varphi}{ds} \mathbf{A} \tag{1.84e}$$

and similarly for total differentials and for partial derivatives.

1.14 Examples of Derivatives—Velocity and Acceleration

Of particular importance in the development of the dynamics of point particles (and of systems of particles) is the representation of the motion of these particles by means of vectors. In order to do this, we require vectors to represent the position, the velocity, and the acceleration of a given particle. It is customary to specify the *position* of a particle with respect to a certain reference frame by a vector $\mathbf{r}$, which in general will be a function of time: $\mathbf{r} = \mathbf{r}(t)$. The *velocity* vector $\mathbf{v}$ and the *acceleration* vector $\mathbf{a}$ are defined according to

$$\mathbf{v} \equiv \frac{d\mathbf{r}}{dt} = \dot{\mathbf{r}} \tag{1.85}$$

$$\mathbf{a} \equiv \frac{d\mathbf{v}}{dt} = \frac{d^2\mathbf{r}}{dt^2} = \ddot{\mathbf{r}} \tag{1.86}$$

where a single dot above a symbol denotes the first time derivative and two dots denote the second time derivative. In rectangular coordinates the expression for $\mathbf{r}$, $\mathbf{v}$, and $\mathbf{a}$ are

$$\mathbf{r} = x_1\mathbf{e}_1 + x_2\mathbf{e}_2 + x_3\mathbf{e}_3 = \sum_i x_i\mathbf{e}_i \tag{1.87}$$

$$\mathbf{v} = \dot{\mathbf{r}} = \sum_i \dot{x}_i\mathbf{e}_i = \sum_i \frac{dx_i}{dt}\mathbf{e}_i \tag{1.88}$$

$$\mathbf{a} = \dot{\mathbf{v}} = \ddot{\mathbf{r}} = \sum_i \ddot{x}_i\mathbf{e}_i = \sum_i \frac{d^2x_i}{dt^2}\mathbf{e}_i \tag{1.89}$$

The calculation of these quantities in rectangular coordinates is straightforward since the unit vectors $\mathbf{e}_i$ are constant in time. In nonrectangular coordinate systems, however, the unit vectors at the position of the particle as it moves in space are not necessarily constant in time, and the components of the time derivatives of $\mathbf{r}$ are no longer simple relations as above. We shall not have occasion to discuss general curvilinear coordinate systems here, but *plane polar* coordinates, *spherical* coordinates, and *cylindrical* coordinates are of sufficient importance to warrant a discussion here of velocity and acceleration in these coordinate systems.*

* Refer to the figures in Appendix F for the geometry of these coordinate systems. For a general discussion of curvilinear coordinates, see Marion (Ma65a, Chapter 4).

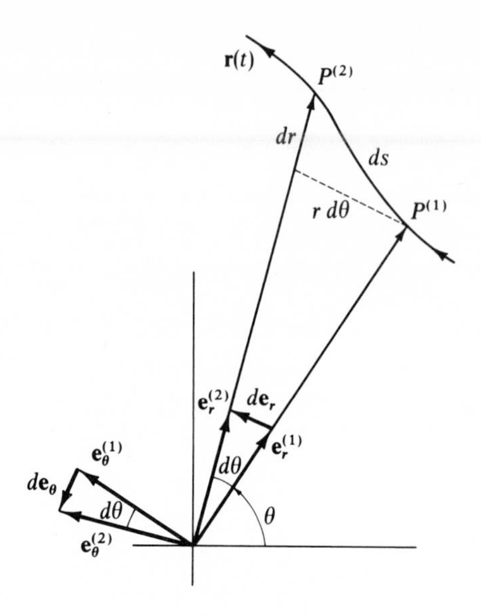

FIG. **1-15**

In order to express **v** and **a** in plane polar coordinates, consider the situation in Fig. 1-15. Here, a point moves along the curve $\mathbf{r}(t)$ and in the time interval $t_2 - t_1 = dt$ moves from $P^{(1)}$ to $P^{(2)}$. The unit vectors, $\mathbf{e}_r$ and $\mathbf{e}_\theta$, which are orthogonal, change from $\mathbf{e}_r^{(1)}$ to $\mathbf{e}_r^{(2)}$ and from $\mathbf{e}_\theta^{(1)}$ to $\mathbf{e}_\theta^{(2)}$. The change in $\mathbf{e}_r$ is

$$\mathbf{e}_r^{(2)} - \mathbf{e}_r^{(1)} = d\mathbf{e}_r \tag{1.90}$$

which is a vector normal to $\mathbf{e}_r$ (and, therefore, in the direction of $\mathbf{e}_\theta$). Similarly, the change in $\mathbf{e}_\theta$ is

$$\mathbf{e}_\theta^{(2)} - \mathbf{e}_\theta^{(1)} = d\mathbf{e}_\theta \tag{1.91}$$

which is a vector normal to $\mathbf{e}_\theta$. We can then write

$$d\mathbf{e}_r = d\theta\ \mathbf{e}_\theta \tag{1.92}$$

and

$$d\mathbf{e}_\theta = -d\theta\ \mathbf{e}_r \tag{1.93}$$

where the minus sign enters the second relation since $d\mathbf{e}_\theta$ is directed *opposite* to $\mathbf{e}_r$ (see Fig. 1-15). Dividing each side of Eqs. (1.92) and (1.93) by dt, we have

$$\boxed{\dot{\mathbf{e}}_r = \dot{\theta}\mathbf{e}_\theta} \tag{1.94}$$

$$\boxed{\dot{\mathbf{e}}_\theta = -\dot{\theta}\mathbf{e}_r} \tag{1.95}$$

If we express $\mathbf{v}$ as

$$\mathbf{v} = \frac{d\mathbf{r}}{dt} = \frac{d}{dt}(r\mathbf{e}_r) = \dot{r}\mathbf{e}_r + r\dot{\mathbf{e}}_r \tag{1.96}$$

we have immediately, using Eq. (1.94),

$$\boxed{\mathbf{v} = \dot{\mathbf{r}} = \dot{r}\mathbf{e}_r + r\dot{\theta}\mathbf{e}_\theta} \tag{1.97}$$

so that the velocity is resolved into a *radial* component $\dot{r}$ and an *angular* (or *transverse*) component $r\dot{\theta}$.

A second differentiation yields the acceleration:

$$\begin{aligned}
\mathbf{a} &= \frac{d}{dt}(\dot{r}\mathbf{e}_r + r\dot{\theta}\mathbf{e}_\theta) \\
&= \ddot{r}\mathbf{e}_r = \dot{r}\dot{\mathbf{e}}_r + \dot{r}\dot{\theta}\mathbf{e}_\theta + r\ddot{\theta}\mathbf{e}_\theta + r\dot{\theta}\dot{\mathbf{e}}_\theta \\
&= (\ddot{r} - r\dot{\theta}^2)\mathbf{e}_r + (r\ddot{\theta} + 2\dot{r}\dot{\theta})\mathbf{e}_\theta
\end{aligned} \tag{1.98}$$

so that the acceleration is resolved into a radial component $(\ddot{r} - r\dot{\theta}^2)$ and an angular (or transverse) component $(r\ddot{\theta} + 2\dot{r}\dot{\theta})$.

The expressions for ds^2, v^2, and $\mathbf{v}$ in the three most important coordinate systems are (see also Appendix F):

Rectangular coordinates

$$\left.\begin{aligned}
ds^2 &= dx_1^2 + dx_2^2 + dx_3^2 \\
v^2 &= \dot{x}_1^2 + \dot{x}_2^2 + \dot{x}_3^2 \\
\mathbf{v} &= \dot{x}_1\mathbf{e}_1 + \dot{x}_2\mathbf{e}_2 + \dot{x}_3\mathbf{e}_3
\end{aligned}\right\} \tag{1.99}$$

Spherical coordinates

$$\left.\begin{aligned}
ds^2 &= dr^2 + r^2\,d\theta^2 + r^2\sin^2\theta\,d\varphi^2 \\
v^2 &= \dot{r}^2 + r^2\dot{\theta}^2 + r^2\sin^2\theta\dot{\varphi}^2 \\
\mathbf{v} &= \dot{r}\mathbf{e}_r + r\dot{\theta}\mathbf{e}_\theta + r\sin\theta\dot{\varphi}\mathbf{e}_\varphi
\end{aligned}\right\} \tag{1.100}$$

[The expressions for plane polar coordinates result from Eqs. (1.100) by setting $d\varphi = 0$.]

Cylindrical coordinates

$$\left.\begin{aligned}
ds^2 &= dr^2 + r^2\,d\theta^2 + dz^2 \\
v^2 &= \dot{r}^2 + r^2\dot{\theta}^2 + \dot{z}^2 \\
\mathbf{v} &= \dot{r}\mathbf{e}_r + r\dot{\theta}\mathbf{e}_\theta + \dot{z}\mathbf{e}_z
\end{aligned}\right\} \tag{1.101}$$

1.15 Angular Velocity

A point or a particle which is moving arbitrarily in space may always be considered, *at a given instant*, to be moving in a plane, circular path about a certain axis. That is, the path which a particle describes during an infinitesimal time interval δt may be represented as an infinitesimal arc of a circle. The line which passes through the center of the circle and is perpendicular to the instantaneous direction of motion is called the *instantaneous axis of rotation*. As the particle moves in the circular path, the rate of change of the angular position is called the *angular velocity*:

$$\omega = \frac{d\theta}{dt} = \dot{\theta} \tag{1.102}$$

Consider a particle that moves instantaneously in a circle of radius R about an axis perpendicular to the plane of motion, as in Fig. 1-16. Let the radius

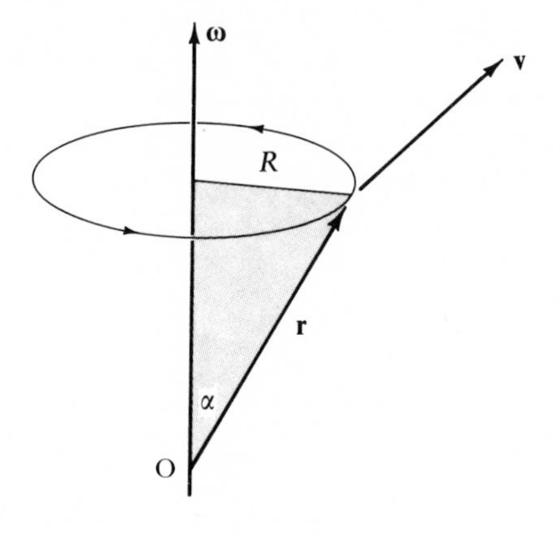

FIG. 1-16

vector **r** of the particle be drawn from an origin located at an arbitrary point O on the axis of rotation. The time rate of change of the radius vector is the linear velocity vector of the particle, $\dot{\mathbf{r}} = \mathbf{v}$, and for motion in a circle of radius R, the *magnitude* of the linear velocity is given by

$$v = R\frac{d\theta}{dt} = R\omega \tag{1.103}$$

The *direction* of the linear velocity **v** is, of course, perpendicular to **r** and in the plane of the circle.

Now, it would be very convenient if we could devise a vector representation of the angular velocity (say, $\boldsymbol{\omega}$) so that all of the quantities of interest in the motion of the particle could be described on a common basis. We can define a *direction* for the angular velocity in the following manner. If the particle moves instantaneously in a plane, then the normal to that plane defines a precise direction in space, or, rather, *two* directions. We may elect to choose as *positive* that direction which corresponds to the direction of advance of a right-hand screw when turned in the same sense as the rotation of the particle (see Fig. 1-16). We can also write the magnitude of the linear velocity by noting that $R = r \sin \alpha$. Thus,

$$v = r\omega \sin \alpha \tag{1.104}$$

Having defined a direction and a magnitude for the angular velocity, we note that if we write

$$\boxed{\mathbf{v} = \boldsymbol{\omega} \times \mathbf{r}} \tag{1.105}$$

then both of these definitions are satisfied and we have the desired vector representation of the angular velocity.

We should note at this point an important distinction between finite and infinitesimal rotations. Whereas an *infinitesimal rotation* can be represented by a vector (actually, an *axial* vector), a *finite rotation* cannot be so represented. The impossibility of describing a finite rotation by a vector is a result of the fact that such rotations do not commute (see the example of Fig. 1-8) and therefore in general different results will be obtained depending on the order in which the rotations are made. To illustrate this statement, consider the successive application of two finite rotations, described by the rotation matrices $\boldsymbol{\lambda}_1$ and $\boldsymbol{\lambda}_2$. Let us associate the vectors $\mathbf{A}$ and $\mathbf{B}$ in a one-to-one manner with these rotations. The vector sum is $\mathbf{C} = \mathbf{A} + \mathbf{B}$, which is equivalent to the matrix $\boldsymbol{\lambda}_3 = \boldsymbol{\lambda}_2 \boldsymbol{\lambda}_1$. But since vector addition is commutative, we also have $\mathbf{C} = \mathbf{B} + \mathbf{A}$, with $\boldsymbol{\lambda}_4 = \boldsymbol{\lambda}_1 \boldsymbol{\lambda}_2$. We know, however, that matrix operations are not commutative, so that in general, $\boldsymbol{\lambda}_3 \neq \boldsymbol{\lambda}_4$. Hence, the vector $\mathbf{C}$ is not unique, and therefore we cannot associate a vector with a finite rotation.

Now, as we shall show, *infinitesimal* rotations do not suffer from this defect of noncommutation. We are therefore led to expect that an infinitesimal rotation can be represented by a vector. Although this expectation is in fact true, the ultimate test of the vector nature of a quantity is contained in its transformation properties. We shall give only a qualitative argument here; the complete proof is carried out in Marion (Ma65a, Section 2.5).

Refer to Fig. 1-17; if the radius vector of a point changes from $\mathbf{r}$ to $\mathbf{r} + \delta\mathbf{r}$, then the geometrical situation is correctly represented if we write

$$\delta\mathbf{r} = \delta\boldsymbol{\theta} \times \mathbf{r} \tag{1.106}$$

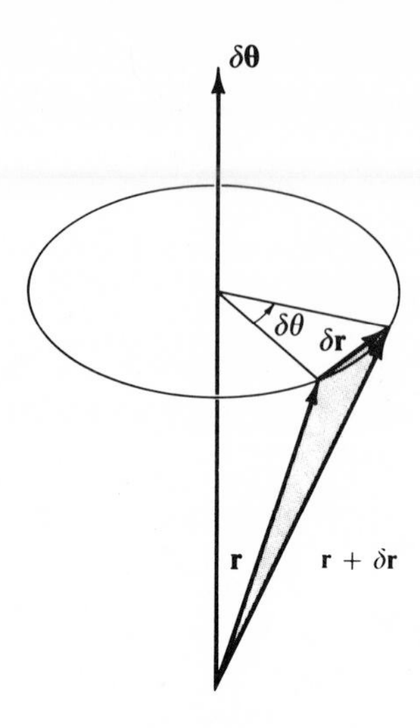

FIG. 1-17

where $\delta\boldsymbol{\theta}$ is a quantity whose magnitude is equal to the infinitesimal rotation angle and which has a direction along the instantaneous axis of rotation. The mere fact that Eq. (1.110) correctly describes the situation illustrated in Fig. 1-17 is not in itself sufficient to establish that $\delta\boldsymbol{\theta}$ is a vector. (We reiterate that the true test must be based on the transformation properties of $\delta\boldsymbol{\theta}$.) But if it can be shown that two infinitesimal rotation "vectors," $\delta\boldsymbol{\theta}_1$ and $\delta\boldsymbol{\theta}_2$, actually *commute*, then the sole objection to representing a finite rotation by a vector will have been removed.

Let us consider that a rotation $\delta\boldsymbol{\theta}_1$ takes $\mathbf{r}$ into $\mathbf{r} + \delta\mathbf{r}_1$, where $\delta\mathbf{r}_1 = \delta\boldsymbol{\theta}_1 \times \mathbf{r}$. If this is followed by a second rotation $\delta\boldsymbol{\theta}_2$ around a different axis, the initial radius vector for this rotation is $\mathbf{r} + \delta\mathbf{r}_1$. Thus,

$$\delta\mathbf{r}_2 = \delta\boldsymbol{\theta}_2 \times (\mathbf{r} + \delta\mathbf{r}_1),$$

and the final radius vector for $\delta\boldsymbol{\theta}_1$ followed by $\delta\boldsymbol{\theta}_2$ is

$$\mathbf{r} + \delta\mathbf{r}_{12} = \mathbf{r} + [\delta\boldsymbol{\theta}_1 \times \mathbf{r} + \delta\boldsymbol{\theta}_2 \times (\mathbf{r} + \delta\mathbf{r}_1)]$$

That is, neglecting second-order infinitesimals,

$$\delta\mathbf{r}_{12} = \delta\boldsymbol{\theta}_1 \times \mathbf{r} + \delta\boldsymbol{\theta}_2 \times \mathbf{r} \tag{1.107}$$

Similarly, if $\delta\boldsymbol{\theta}_2$ is followed by $\delta\boldsymbol{\theta}_1$, we have

$$\mathbf{r} + \delta\mathbf{r}_{21} = \mathbf{r} + [\delta\boldsymbol{\theta}_2 \times \mathbf{r} + \delta\boldsymbol{\theta}_1 \times (\mathbf{r} + \delta\mathbf{r}_2)]$$

or,

$$\delta\mathbf{r}_{21} = \delta\boldsymbol{\theta}_2 \times \mathbf{r} + \delta\boldsymbol{\theta}_1 \times \mathbf{r} \tag{1.108}$$

Clearly, $\delta \mathbf{r}_{12}$ and $\delta \mathbf{r}_{21}$ are equal, so that the rotation "vectors" $\delta \boldsymbol{\theta}_1$ and $\delta \boldsymbol{\theta}_2$ do commute. It therefore seems reasonable that $\delta \boldsymbol{\theta}$ in Eq. (1.106) is indeed a vector.

It is the fact that $\delta \boldsymbol{\theta}$ is a vector that allows angular velocity to be represented by a vector, since angular velocity is the ratio of an infinitesimal rotation angle to an infinitesimal time:

$$\boldsymbol{\omega} = \frac{\delta \boldsymbol{\theta}}{\delta t}$$

Therefore, dividing Eq.(1.106) by δt, we have

$$\frac{\delta \mathbf{r}}{\delta t} = \frac{\delta \boldsymbol{\theta}}{\delta t} \times \mathbf{r}$$

or, in passing to the limit, $\delta t \to 0$,

$$\mathbf{v} = \boldsymbol{\omega} \times \mathbf{r}$$

as before.

1.16 The Gradient Operator

We now turn to the discussion of the most important member of a class called *vector differential operators*—the *gradient operator.*

Consider a scalar φ which is an explicit function of the coordinates x_j, and, moreover, is a continuous, single-valued function of these coordinates throughout a certain region of space. Then, under a coordinate transformation that carries the x_i into the x_i', $\varphi'(x_1', x_2', x_3') = \varphi(x_1, x_2, x_3)$, and by the chain rule of differentiation, we can write

$$\frac{\partial \varphi'}{\partial x_1'} = \sum_j \frac{\partial \varphi}{\partial x_j} \frac{\partial x_j}{\partial x_1'} \tag{1.109}$$

Similarly for $\partial \varphi / \partial x_2'$ and $\partial \varphi' / \partial x_3'$, so that in general we have

$$\frac{\partial \varphi'}{\partial x_i'} = \sum_j \frac{\partial \varphi}{\partial x_j} \frac{\partial x_j}{\partial x_i'} \tag{1.110}$$

Now, the inverse coordinate transformation is

$$x_j = \sum_k \lambda_{kj} x_k' \tag{1.111}$$

Differentiating,

$$\frac{\partial x_j}{\partial x'_i} = \frac{\partial}{\partial x'_i}\left(\sum_k \lambda_{kj} x'_k\right) = \sum_k \lambda_{kj}\left(\frac{\partial x'_k}{\partial x'_i}\right) \tag{1.112}$$

But the term in the last parentheses is just δ_{ik}, so that

$$\frac{\partial x_j}{\partial x'_i} = \sum_k \lambda_{kj}\,\delta_{ik} = \lambda_{ij} \tag{1.113}$$

Substituting Eq. (1.113) into Eq. (1.110), we obtain

$$\frac{\partial \varphi'}{\partial x'_i} = \sum_j \lambda_{ij}\frac{\partial \varphi}{\partial x_j} \tag{1.114}$$

Since it follows the correct transformation equation, the function $\partial\varphi/\partial x_j$ is the jth component of a vector which is termed the *gradient* of the function φ. Note that even though φ is a *scalar*, the *gradient* of φ is a *vector*. The gradient of φ is written either as **grad** φ or as $\nabla\varphi$ ("del" φ).

Since the function φ is an arbitrary scalar function, it is convenient to define the differential operator described above in terms of the *gradient operator*:

$$(\mathbf{grad})_i = \nabla_i = \frac{\partial}{\partial x_i} \tag{1.115}$$

We can express the complete vector operator as

$$\boxed{\mathbf{grad} = \boldsymbol{\nabla} = \sum_i \mathbf{e}_i \frac{\partial}{\partial x_i}} \tag{1.116}$$

The gradient operator can (a) operate directly on a scalar function, as in $\boldsymbol{\nabla}\varphi$; (b) be used in a scalar product with a vector function as in $\boldsymbol{\nabla}\cdot\mathbf{A}$ (the *divergence* of **A**); or (c) be used in a vector product with a vector function as in $\boldsymbol{\nabla}\times\mathbf{A}$ (the *curl* of **A**).

In order to give a geometrical interpretation of the gradient of a scalar function, consider, as in Fig. 1-18, the topographical map of a conical hill of circular cross section. The circles represent lines of constant altitude. Let φ denote the altitude at any point, $\varphi = \varphi(x_1, x_2, x_3)$. Then,

$$d\varphi = \sum_i \frac{\partial\varphi}{\partial x_i}\,dx_i = \sum_i (\mathbf{grad}\,\varphi)_i\,dx_i \tag{1.117}$$

The components of the displacement vector $d\mathbf{s}$ are the incremental displacements in the direction of the three orthogonal axes:

$$d\mathbf{s} = (dx_1, dx_2, dx_3) \tag{1.118}$$

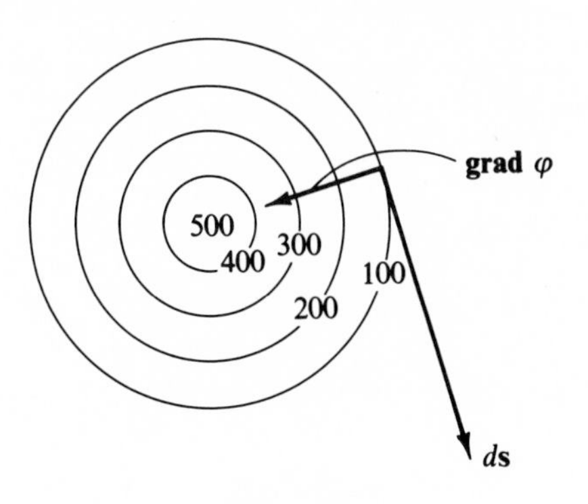

FIG. 1-18

Therefore,

$$d\varphi = (\mathbf{grad}\ \varphi) \cdot d\mathbf{s} \tag{1.119}$$

Now, let $d\mathbf{s}$ be directed tangentially along one of the isolatitude lines, (i.e., along a line for which $\varphi = \text{const.}$), as indicated in Fig. 1-18. Since $\varphi = \text{const.}$ for this case, $d\varphi = 0$. But, since neither **grad** φ nor $d\mathbf{s}$ is in general zero, it follows that they must therefore be perpendicular to each other. Thus, **grad** φ is normal to the line (or in three dimensions, to the surface) for which $\varphi = \text{const.}$

The maximum value of $d\varphi$ results when **grad** φ and $d\mathbf{s}$ are in the same direction; then,

$$(d\varphi)_{\max} = |\mathbf{grad}\ \varphi|\ ds, \qquad \text{for} \quad \mathbf{grad}\ \varphi \parallel d\mathbf{s}$$

or,

$$|\mathbf{grad}\ \varphi| = \left(\frac{d\varphi}{ds}\right)_{\max} \tag{1.120}$$

Therefore, **grad** φ is in the direction of the greatest change in φ.

These results are summarized as follows:

(a) The vector **grad** φ is, at any point, normal to the lines or surfaces for which $\varphi = \text{const.}$

(b) The vector **grad** φ has the direction of the maximum change in φ.

(c) Since any direction in space can be specified in terms of the unit vector $\mathbf{n}$ in that direction, the rate of change of φ in the direction of $\mathbf{n}$ (the *directional derivative* of φ) can be found from $\mathbf{n} \cdot \mathbf{grad}\ \varphi \equiv \partial\varphi/\partial n$.

The successive operation of the gradient operator produces

$$\boldsymbol{\nabla} \cdot \boldsymbol{\nabla} = \sum_i \frac{\partial}{\partial x_i} \frac{\partial}{\partial x_i} = \sum_i \frac{\partial^2}{\partial x_i^2} \tag{1.121}$$

This important product operator is called the *Laplacian** and is also written

$$\boxed{\nabla^2 = \sum_i \frac{\partial^2}{\partial x_i^2}} \tag{1.122}$$

When the Laplacian operates on a scalar we have, for example,

$$\nabla^2 \psi = \sum_i \frac{\partial^2 \psi}{\partial x_i^2} \tag{1.123}$$

The other important vector differential operators, the *divergence* and the *curl*, do not often arise† in the discussion of topics in dynamics to which this book is devoted and so we will not pursue further the development of vector operators here. [See Appendix F and Marion (Ma65a, Chapters 2 and 3).]

1.17 Integration of Vectors

The vector which results from the volume integration of a vector function $\mathbf{A} = \mathbf{A}(x_i)$ throughout a volume V is given by‡

$$\int_V \mathbf{A}\, dv = \left(\int_V A_1\, dv, \quad \int_V A_2\, dv, \quad \int_V A_3\, dv \right) \tag{1.124}$$

Thus, the integration of the vector $\mathbf{A}$ throughout V is accomplished simply by performing three separate, ordinary integrations.

The integral over a surface S of the projection of a vector function $\mathbf{A} = \mathbf{A}(x_i)$ onto that surface is defined to be‡

$$\int_S \mathbf{A} \cdot d\mathbf{a}$$

where $d\mathbf{a}$ is an element of area of the surface, as in Fig. 1-19. We write $d\mathbf{a}$ as a vector quantity since we may attribute to it not only a magnitude da but also a direction corresponding to the normal to the surface at the point in question. If the unit normal vector is $\mathbf{n}$, then

$$d\mathbf{a} = \mathbf{n}\, da \tag{1.125}$$

*After Pierre Simon Laplace (1749–1827); the notation ∇^2 is due to Sir William Rowan Hamilton.

† We will refer to a theorem involving the *curl* at only one point (Section 2.5).

‡ The symbol $\int_V$ actually represents a *triple* integral over a certain volume V. Similarly, the symbol $\int_S$ stands for a *double* integral over a certain surface S.

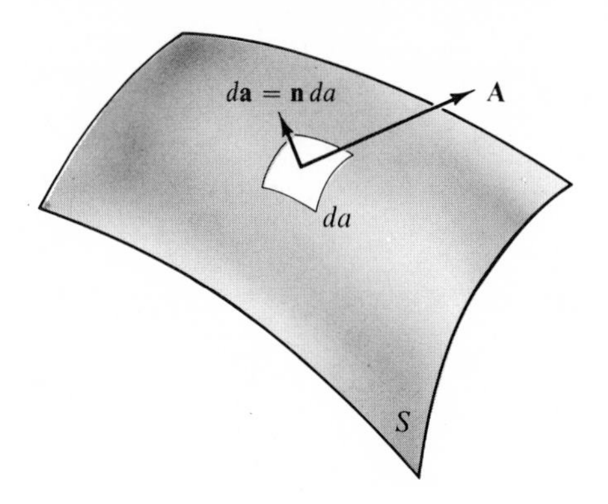

FIG. 1-19

Thus, the components of $d\mathbf{a}$ are the projections of the element of area on the three mutually perpendicular planes defined by the rectangular axes:

$$da_1 = dx_2\, dx_3, \qquad \text{etc.} \tag{1.126}$$

Therefore, we have

$$\int_S \mathbf{A} \cdot d\mathbf{a} = \int_S \mathbf{A} \cdot \mathbf{n}\, da \tag{1.127}$$

or,

$$\int_S \mathbf{A} \cdot d\mathbf{a} = \int_S \sum_i A_i\, da_i \tag{1.128}$$

Equation (1.127) states that the integral of $\mathbf{A}$ over the surface S is the integral of the normal component of $\mathbf{A}$ over this surface.

The normal to a surface may be taken to lie in either of two possible directions ("up" or "down"); thus, the sign of $\mathbf{n}$ is ambiguous. If the surface is *closed*, we adopt the convention that the *outward* normal is positive.

The *line integral* of a vector function $\mathbf{A} = \mathbf{A}(x_i)$ along a given path extending from the point B to the point C is given by the integral of the component of $\mathbf{A}$ along the path:

$$\int_{BC} \mathbf{A} \cdot d\mathbf{s} = \int_{BC} \sum_i A_i\, dx_i \tag{1.129}$$

The quantity $d\mathbf{s}$ is an element of length along the given path, as in Fig. 1-20. The direction of $d\mathbf{s}$ is taken to be positive along the direction in which the path is traversed. In Fig. 1-20, at point P the angle between $d\mathbf{s}$ and $\mathbf{A}$ is less than $\pi/2$, so that $\mathbf{A} \cdot d\mathbf{s}$ is positive at this point. At point Q the angle is greater than $\pi/2$ and the contribution to the integral at this point is negative.

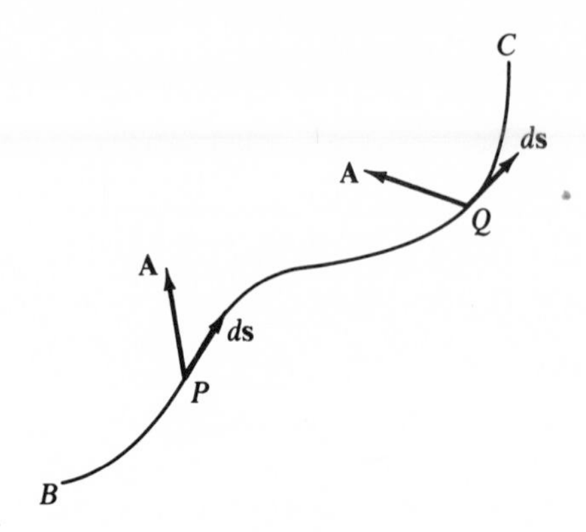

FIG. 1-20

Suggested References

Texts on vector methods abound. A standard, often-quoted work is that of Phillips (Ph33). A brief and readable account of vector algebra will be found in Constant (Co54, Chapter 1). A detailed treatment of vector algebra is given by Barnett and Fujii (Ba63). Spiegel's discussion (Sp59, Chapters 1 and 3) is extensive and contains many worked problems and examples. The recent book by Lindgren (Li64) is particularly good, as is that by Davis (Da61). Arfken (Ar66) has a good discussion.

A good introduction to matrix methods is given by Yefimov (Ye64), and matrix methods in vector analysis are emphasized by Eisenman (Ei63).

An extensive, although pedantic, discussion of vectors and matrices will be found in Jeffreys and Jeffreys (Je46, Chapters 2 and 4); the ε_{ijk} notation is used.

Formal approaches to matrices, vectors, and linear vector spaces are given by Dettman (De62, Chapter 1) and by Corben and Stehle (Co60, Chapter 1 and Appendix II). A brief discussion from a modern veiwpoint is also to be found in Goldstein (Go50, Chapter 4).

One of the better expositions of vector algebra from the mathematician's viewpoint is presented by Kaplan (Ka52, Chapter 1). This text is highly recommended, not only for material on vector methods, but as a general mathematics reference for students of physics.

An authoritative account of general vector methods which is heavy going in spots, is given by Morse and Feshbach (Mo53, Chapter 1).

Vector calculus is discussed in most of the books referred to above. See particularly Constant (Co54, Chapter 2), Kaplan (Ka52, Chapters 3 and 5), Lindgren (Li64), and Spiegel (Sp59, Chapters 3, 4, and 5).

Books on electromagnetism are frequently excellent sources for material on vector calculus; good, brief discussions are presented, for example, by Corson and Lorrain (Co62, Chapter 1) and by Owen (Ow63, Appendix A). Owen's notation is quite similar to that used here.

Problems

1-1. Find the transformation matrix which rotates a rectangular coordinate system through an angle of 120° about an axis making equal angles with the original three coordinate axes.

1-2. Prove Eqs. (1.10) and (1.11) from trigonometric considerations.

1-3. Show
(a) $(\mathsf{AB})^t = \mathsf{B}^t\mathsf{A}^t$.
(b) $(\mathsf{AB})^{-1} = \mathsf{B}^{-1}\mathsf{A}^{-1}$.

1-4. Show by direct expansion that $|\lambda|^2 = 1$. For simplicity, take λ to be a two-dimensional transformation matrix.

1-5. Show that Eq. (1.15) can be obtained by using the requirement that the transformation leave unchanged the length of a line segment.

1-6. Consider a unit cube with one corner at the origin and three adjacent sides lying along the three axes of a rectangular coordinate system. Find the vectors which describe the diagonals of the cube. What is the angle between any pair of diagonals?

1-7. Let $\mathbf{A}$ be a vector from the origin to a point P fixed in space. Let $\mathbf{r}$ be a vector from the origin to a variable point $Q(x_1, x_2, x_3)$. Show that

$$\mathbf{A} \cdot \mathbf{r} = A^2$$

is the equation of a plane perpendicular to $\mathbf{A}$ and passing through the point P.

1-8. Show that the *triple scalar product* $(\mathbf{A} \times \mathbf{B}) \cdot \mathbf{C}$ can be written as

$$(\mathbf{A} \times \mathbf{B}) \cdot \mathbf{C} = \begin{vmatrix} A_1 & A_2 & A_3 \\ B_1 & B_2 & B_3 \\ C_1 & C_2 & C_3 \end{vmatrix}$$

Show also that the product is unaffected by an interchange of the scalar and vector product operations or by a change in the order of $\mathbf{A}$, $\mathbf{B}$, $\mathbf{C}$, as long as they are in cyclic order; that is,

$$(\mathbf{A} \times \mathbf{B}) \cdot \mathbf{C} = \mathbf{A} \cdot (\mathbf{B} \times \mathbf{C}) = \mathbf{B} \cdot (\mathbf{C} \times \mathbf{A}) = (\mathbf{C} \times \mathbf{A}) \cdot \mathbf{B}, \qquad \text{etc.}$$

We may therefore use the notation $\mathbf{ABC}$ to denote the triple scalar product. Finally, give a geometrical interhretation of $\mathbf{ABC}$ by computing the volume of the parallelepiped defined by the three vectors $\mathbf{A}$, $\mathbf{B}$, $\mathbf{C}$.

1-9. Let **a**, **b**, **c** be three constant vectors drawn from the origin to the points A, B, C. What is the distance from the origin to the plane defined by the points A, B, C? What is the area of the triangle ABC?

1-10. If **X** is an unknown vector which satisfies the following relations involving the known vectors **A** and **B** and the scalar φ,

$$\mathbf{A} \times \mathbf{X} = \mathbf{B}; \qquad \mathbf{A} \cdot \mathbf{X} = \varphi$$

express **X** in terms of **A**, **B**, φ, and the magnitude of **A**.

1-11. Obtain the cosine law of plane trigonometry by interpreting the product $(\mathbf{A} + \mathbf{B}) \cdot (\mathbf{A} + \mathbf{B})$ and the expansion of the product.

1-12. Obtain the sine law of plane trigonometry by interpreting the product $\mathbf{A} \times \mathbf{B}$ and the alternate representation $(\mathbf{A} - \mathbf{B}) \times \mathbf{B}$.

1-13. Derive the following expressions by using vector algebra:

(a) $\cos(A - B) = \cos A \cos B + \sin A \sin B$

(b) $\sin(A - B) = \sin A \cos B - \cos A \sin B$

1-14. Show that

(a) $\sum_{i,j} \varepsilon_{ijk} \delta_{ij} = 0$ (b) $\sum_{j,k} \varepsilon_{ijk} \varepsilon_{ljk} = 2\delta_{il}$ (c) $\sum_{i,j,k} \varepsilon_{ijk} \varepsilon_{ijk} = 6$

1-15. Show that (see also Problem 1-8)

$$\mathbf{ABC} = \sum_{i,j,k} \varepsilon_{ijk} A_i B_j C_k$$

1-16. Evaluate the sum $\sum_k \varepsilon_{ijk} \varepsilon_{lmk}$ (which contains 81 terms) by considering the result for all possible combinations of i, j, l, m, viz.,

(a) $i = j$ (b) $i = l$

(c) $i = m$ (d) $j = l$

(e) $j = m$ (f) $l = m$

(g) $i \neq l$ or m (h) $j \neq l$ or m

Show that

$$\sum_k \varepsilon_{ijk} \varepsilon_{lmk} = \delta_{il} \delta_{jm} - \delta_{im} \delta_{jl}$$

and then use this result to prove

$$\mathbf{A} \times (\mathbf{B} \times \mathbf{C}) = (\mathbf{A} \cdot \mathbf{C})\mathbf{B} - (\mathbf{A} \cdot \mathbf{B})\mathbf{C}$$

1-17. Use the ε_{ijk} notation and derive the identity

$$(\mathbf{A} \times \mathbf{B}) \times (\mathbf{C} \times \mathbf{D}) = \mathbf{C}(\mathbf{ABD}) - \mathbf{D}(\mathbf{ABC})$$

1-18. Let $\mathbf{A}$ be an arbitrary vector and let $\mathbf{e}$ be a unit vector in some fixed direction. Show that

$$\mathbf{A} = \mathbf{e}(\mathbf{A} \cdot \mathbf{e}) + \mathbf{e} \times (\mathbf{A} \times \mathbf{e})$$

What is the geometrical significance of each of the two terms of the expansion?

1-19. Find the components of the acceleration vector $\mathbf{a}$ in spherical and in cylindrical coordinates.

1-20. A particle moves with $v = \text{const.}$ along the curve $r = k(1 + \cos\theta)$ (a *cardioid*). Find $\ddot{\mathbf{r}} \cdot \mathbf{e}_r = \mathbf{a} \cdot \mathbf{e}_r$, $|\mathbf{a}|$, and $\dot{\theta}$.

1-21. If $\mathbf{r}$ and $\dot{\mathbf{r}} = \mathbf{v}$ are both explicit functions of time, show that

$$\frac{d}{dt}[\mathbf{r} \times (\mathbf{v} \times \mathbf{r})] = r^2\mathbf{a} + (\mathbf{r} \cdot \mathbf{v})\mathbf{v} - (v^2 + \mathbf{r} \cdot \mathbf{a})\mathbf{r}$$

1-22. Show that

$$\mathbf{grad}(\ln|\mathbf{r}|) = \frac{\mathbf{r}}{r^2}$$

1-23. Find the angle between the surfaces defined by $r^2 = 9$ and $x + y + z^2 = 1$ at the point $(2, -2, 1)$.

1-24. Show that $\mathbf{grad}(\varphi\psi) = \varphi\ \mathbf{grad}\ \psi + \psi\ \mathbf{grad}\ \varphi$.

1-25. Show that

(a) $\mathbf{grad}\ r^n = nr^{(n-2)}\ \mathbf{r}$

(b) $\mathbf{grad}\ f(r) = \dfrac{\mathbf{r}}{r}\dfrac{df}{dr}$

(c) $\nabla^2(\ln r) = \dfrac{1}{r^2}$

1-26. Show that

$$\int (2\mathbf{r} \cdot \dot{\mathbf{r}} + 2\dot{\mathbf{r}} \cdot \ddot{\mathbf{r}})\, dt = r^2 + \dot{r}^2 + \text{const.}$$

where $\mathbf{r}$ is the vector from the origin to the point (x_1, x_2, x_3). The quantities r and $\dot{r}$ are the magnitudes of the vectors $\mathbf{r}$ and $\dot{\mathbf{r}}$, respectively.

1-27. Show that

$$\int \left(\frac{\dot{\mathbf{r}}}{r} - \frac{\mathbf{r}\dot{r}}{r^2} \right) dt = \frac{\mathbf{r}}{r} + \mathbf{C}$$

where $\mathbf{C}$ is a constant vector.

1-28. Evaluate the integral

$$\int \mathbf{A} \times \ddot{\mathbf{A}}\, dt$$

1-29. Show that the volume common to the intersecting cylinders defined by $x^2 + y^2 = a^2$ and $x^2 + z^2 = a^2$ is $V = 16a^3/3$.

CHAPTER 2

Newtonian Mechanics

2.1 Introduction

The science of mechanics seeks to provide a precise and consistent description of the dynamics of particles and systems of particles. That is, we attempt to discover a set of physical laws which provide us with a method for mathematically describing the motions of bodies and aggregates of bodies. In order to do this, we need to introduce certain fundamental concepts. It is implicit in Newtonian theory that the concept of *distance* is intuitively understandable from a geometrical viewpoint. Furthermore, *time* is considered to be an absolute quantity, capable of precise definition by an arbitrary observer. In relativity theory, however, we must modify these Newtonian ideas (see Chapter 10). The combination of the concepts of distance and time allows us to define the *velocity* and *acceleration* of a particle. The third fundamental concept, *mass*, requires some elaboration which we shall give in connection with the discussion of Newton's laws.

The physical laws which we introduce must be based on experimental fact. A *physical law* may be characterized by the statement that it "might have been otherwise." Thus, there is no *a priori* reason to expect that the gravitational attraction between two bodies must vary exactly as the inverse square of the distance between them. But experiment indicates that this is so. Once a set of experimental data has been correlated and a postulate has been formulated regarding the phenomena to which the data refer, then various implications can be worked out. If these implications are all verified by experiment, there is reason to believe that the postulate is generally true. The postulate then assumes the status of a *physical law*. If some experiments are found to be in disagreement with the predictions of the law, then the theory must be modified in order to be consistent with all known facts.

Newton has provided us with the fundamental laws of mechanics. We shall state these laws in modern terms and discuss their meaning and then proceed to derive the implications of the laws in various situations. It must be noted, however, that the logical structure of the science of mechanics is not a straightforward issue. The line of reasoning that is followed here in interpreting Newton's laws is not the only one possible.* We shall not pursue in any detail the philosophy of mechanics, but will give only sufficient elaboration of Newton's laws to allow us to continue with the discussion of classical dynamics. The latter portion of this chapter is concerned with gravitational attraction and potentials. We will devote our attention to the motion of *particles* for the next several chapters; in Chapter 12 the topic of *rigid-body* motion will be considered and in Chapters 13, 14, and 15 we will consider the dynamics of *systems* of particles. This chapter concludes with a brief discussion of the fundamental and practical limitations of Newton's laws.

2.2 Newton's Laws

We begin by simply stating in conventional form Newton's laws of mechanics†:

I. A body remains at rest or in uniform motion unless acted upon by a force.
II. A body acted upon by a force moves in such a manner that the time rate of change of momentum equals the force.
III. If two bodies exert forces on each other, these forces are equal in magnitude and opposite in direction.

* Ernst Mach (1838–1916) expressed his views in his famous book, *The Science of Mechanics*, first published in 1883. A translation of a later edition is available (Ma60). Interesting discussions are also given, for example, by Lindsay and Margenau (Li36) and by Feather (Fe59).

† Enunciated by Sir Isaac Newton (1642–1727) in his *Principia*, 1687. Galileo had previously generalized the results of his mechanics experiments with statements equivalent to the First and Second Laws, although he was unable to complete the description of dynamics because he did not appreciate the significance of the Third Law and therefore lacked a precise meaning of *force*.

These laws are so familiar that we sometimes tend to lose sight of their true significance (or lack of it) as physical laws. The First Law, for example, is meaningless without the concept of "force." In fact, standing alone, the First Law conveys a precise meaning only for *zero force*; that is, a body which remains at rest or in uniform (i.e., unaccelerated, rectilinear) motion is subject to no force whatsoever. A body which moves in this manner is termed a *free body* (or *free particle*). The question of the frame of reference with respect to which the "uniform motion" is to be measured will be discussed in the following section.

In pointing out the lack of content in the First Law, Sir Arthur Eddington* has made the somewhat facetious statement that all the law actually says is that "every particle continues in its state of rest or uniform motion in a straight line except insofar as it doesn't." While such a comment is hardly fair to Newton, who meant something very definite by his statement, it does emphasize the point that the First Law by itself provides us with only a qualitative notion regarding "force."

An explicit statement concerning "force" is provided by the Second Law in which force is related to the time rate of change of *momentum*. Momentum was appropriately defined by Newton (although he used the term "quantity of motion") to be the product of mass and velocity, so that

$$\mathbf{p} \equiv m\mathbf{v} \tag{2.1}$$

Therefore, the Second Law can be expressed as

$$\mathbf{F} = \frac{d\mathbf{p}}{dt} = \frac{d}{dt}(m\mathbf{v}) \tag{2.2}$$

The definition of force becomes complete and precise only when "mass" is defined. Thus, the First and Second Laws are not really "laws" in the usual sense of the term as used in physics; rather, they may be considered as *definitions*. The Third Law, on the other hand, is indeed a *law*. It is a statement concerning the real physical world and contains all of the physics in Newton's laws of motion.†

We must hasten to add, however, that the Third Law is not a *general* law of Nature. The law applies only in the event that the force exerted by one (point) object on another (point) object is directed along the line connecting the objects. Such forces are called *central forces*; the Third Law applies whether a central force is attractive or repulsive. Gravitational and electrostatic forces are central forces, so Newton's Laws can be used in problems

* Sir Arthur Eddington (Ed30, p. 124).

† The reasoning presented here, viz., that the First and Second Laws are actually definitions and that the Third Law contains the physics, is not the only possible interpretation. Lindsay and Margenau (Li36), for example, present the first two Laws as physical *laws* and then derive the Third Law as a consequence.

involving these types of forces. Sometimes, elastic forces (which are actually macroscopic manifestations of microscopic electrostatic forces) are central in character. For example, two point objects connected by a straight spring or elastic string are subject to forces that obey the Third Law. Any force that depends on the velocities of the interacting bodies is non-central in character, and the Third Law does not apply in such a situation. Velocity-dependent forces are characteristic of interactions that propagate with finite velocity. Thus, the force between *moving* electric charges does not obey the Third Law since the force propagates with the velocity of light. Even the gravitational force between *moving* bodies is velocity-dependent, but the effect is small and difficult to detect; the only observable effect is the precession of the perihelia of the inner planets (see Section 8.10). In this book we shall be concerned exclusively with gravitational and elastic forces; the accuracy of the Third Law is quite sufficient for all such discussions.

In order to demonstrate the significance of the Third Law, let us paraphrase it in the following way, which incorporates the appropriate definition of mass:

III′. If two bodies constitute an ideal, isolated system, then the accelerations of these bodies are always in opposite directions and the ratio of the magnitudes of the accelerations is constant. This constant ratio is the inverse ratio of the masses of the bodies.

With this statement we are now in a position to give a practical definition of mass and therefore to give precise meaning to the equations that summarize Newtonian dynamics. If we consider two isolated bodies, 1 and 2, then the Third Law states that

$$\mathbf{F}_1 = -\mathbf{F}_2 \tag{2.3}$$

Using the definition of force as given by the Second Law, we have

$$\frac{d\mathbf{p}_1}{dt} = -\frac{d\mathbf{p}_2}{dt} \tag{2.4a}$$

or,

$$m_1\left(\frac{d\mathbf{v}_1}{dt}\right) = m_2\left(-\frac{d\mathbf{v}_2}{dt}\right) \tag{2.4b}$$

and, since acceleration is the time derivative of velocity,

$$m_1(\mathbf{a}_1) = m_2(-\mathbf{a}_2) \tag{2.4c}$$

Hence,

$$\frac{m_2}{m_1} = -\frac{a_1}{a_2} \tag{2.5}$$

where the negative sign indicates only that the two acceleration vectors are oppositely directed.

We can always select, say, m_1 as the *unit mass* and then by comparing the ratio of accelerations when m_1 is allowed to interact with any other body, the mass of the other body can be determined. Of course, in order to measure the accelerations, one must have appropriate clocks and measuring rods; in addition, it is necessary to choose a suitable coordinate system or reference frame. The question of a "suitable reference frame" is discussed in the next section.

One of the more common methods of determining the mass of an object is by *weighing*—for example, by comparing its weight to that of a standard by means of a beam balance. This procedure makes use of the fact that in a gravitational field the *weight* of a body is just the gravitational force acting on the body. That is, Newton's equation $\mathbf{F} = m\mathbf{a}$ becomes $\mathbf{W} = m\mathbf{g}$, where $\mathbf{g}$ is the acceleration due to gravity. The validity of using this procedure rests on a fundamental assumption, viz., that the mass m which appears in Newton's equation and which is defined according to the statement III′ above, is equal to the mass m which appears in the gravitational force equation. These two masses are called the *inertial mass* and *gravitational mass*, respectively. The definitions may be stated as follows:

Inertial mass: That mass which determines the acceleration of a body under the action of a given force.

Gravitational mass: That mass which determines the gravitational forces between a body and other bodies.

Galileo was the first to test the equivalence of inertial and gravitational mass in his (perhaps apocryphal) experiment with falling weights at the Tower of Pisa. Newton also considered the problem and measured the periods of pendula of equal lengths but with bobs of different materials. Neither found any difference, but the methods were quite crude.* In 1890 Eötvös† devised an ingenious method to test the equivalence of inertial and gravitational masses. Using two objects made of different materials, he compared the effect of the Earth's gravitational force (i.e., the *weight*) with the effect of the inertial force due to the Earth's rotation. The experiment involved the use of a *null* method employing a sensitive torsion balance and was therefore capable of high accuracy. More recent experiments (notably those of Dicke‡), using essentially the same method, have improved the accuracy, and it has now been established that inertial and gravitational mass are identical to within a few

* In Newton's experiment, he could have detected a difference of only 1 part in 10^3.

† Roland von Eötvös (1848–1919), a Hungarian baron; his research in gravitational problems led to the development of a gravimeter which was of use in geological studies.

‡ R. H. Dicke, *Science* **124**, 621 (1959).

parts in 10^{11}. This result is of considerable importance in the general theory of relativity.* The assertion of the *exact* equality of inertial and gravitational mass is termed the *principle of equivalence*.

Newton's Third Law is stated in terms of two bodies which constitute an isolated system. It is, of course, impossible to achieve such an ideal condition; every body in the Universe interacts with every other body, although the force of interaction may be far too weak to be of any practical importance if great distances are involved. Newton avoided the question of how one is to disentangle the desired effects from all of the extraneous effects. But this practical difficulty only serves to emphasize the enormity of Newton's assertion made in the Third Law. It is a tribute to the depth of his perception and physical insight that the conclusion based on limited observations has successfully borne the test of experiment for almost 300 years. It has only been within this century that measurements have been made with sufficient detail to reveal certain discrepancies with the predictions of Newtonian theory. The pursuit of these details has led to the development of relativity theory and quantum mechanics.†

2.3 Frames of Reference

Newton realized that in order for the laws of motion to have meaning, a reference frame must be chosen with respect to which the motion of bodies can be measured. A reference frame is called an *inertial frame* if Newton's laws are indeed valid in that frame. That is, if a body subject to no external force is found to move in a straight line with constant velocity (or to remain at rest), then the coordinate system used to establish this fact is an inertial reference frame. This is a clear-cut operational definition and one that also follows from the general theory of relativity.

Now, if Newton's laws are valid in one reference frame, then they are also valid in any reference frame that is in uniform motion (i.e., is not accelerated) with respect to the first system.‡ This is a result of the fact that the equation $\mathbf{F} = m\ddot{\mathbf{r}}$ involves the second time derivative of $\mathbf{r}$ so that a change of coordinates involving a constant velocity will not influence the equation; this result is called *Galilean invariance*, or the *principle of Newtonian relativity*.

Relativity theory has shown us that the concepts of *absolute rest* and an *absolute* inertial reference frame are meaningless. Therefore, even though we

* See, for example, the discussion by Bergmann (Be46) or Weber (We61). An analysis of the Eötvös experiment is also given in Weber's book (Chapter 1).

† See also Section 2.12.

‡ In Chapter 11 we shall discuss the modification of Newton's equations that must be made if it is desired to describe the motion of a body with respect to a *noninertial* frame of reference, i.e., a frame which is accelerated with respect to an inertial frame.

conventionally adopt a reference frame which is fixed with respect to the "fixed" stars—and, indeed, in such a frame the Newtonian equations are valid to a high degree of accuracy—such a frame is, in fact, not an absolute inertial frame. We may, however, consider the "fixed" stars to define a reference frame that approximates an "absolute" inertial frame to an extent quite sufficient for our present purposes.

Although the "fixed"-star reference frame is a conveniently definable system and one that is suitable for many purposes, it must be emphasized that the fundamental definition of an inertial frame makes no mention of stars, "fixed" or otherwise. If a body subject to no force moves with constant velocity in a certain coordinate system, that system is, by definition, an inertial frame.

Since the precise description of the motion of a real physical object in the real physical world is an almost impossible task, we always resort to idealizations and approximations of varying degree. That is, we ordinarily neglect the lesser forces on a body if these forces do not significantly affect the body's motion. The simplest possible situation would be one in which the experiment is carried out in a complete "void" so that the motion of a particle is not perturbed by any extraneous forces. However, since we must always postulate some fiducial marks on which to base a reference system (e.g., the distant "fixed" stars), we already introduce thereby an approximation to the ideal case, viz., we assume that the influence of distant matter is nil. Such an approximation will not, of course, seriously affect the description of the motion of the particle, but because of the fundamental necessity of always providing a hook on which to hang the reference frame, it is conceptually impossible to consider measurements of the motion of a *completely free* particle. Nevertheless, the free-particle concept is a useful one for many purposes.

If we wish to describe the motion of, say, a free particle, and if we choose for this purpose some coordinate system in an inertial frame, then we require that the (vector) equation of motion of the particle be independent of the *position* of the origin of the coordinate system and independent of its *orientation* in space. That is, we require that in an inertial frame *space* must have the properties of *homogeneity* and *isotropy*. We further require that *time* be homogeneous; that is, a free particle which moves with a certain constant velocity in the coordinate system during a certain time interval must not, during a later time interval, be found to be moving with a different velocity.

We can illustrate the importance of these properties by the following example. Consider, as in Fig. 2-1, a free particle which moves along a certain path AC. Let us choose for the description of the motion of the particle a rectangular coordinate system whose origin moves in a circle, as shown; for simplicity, we let the orientation of the axes be fixed in space. The particle moves with a velocity $\mathbf{v}_p$ relative to an inertial reference frame. If the

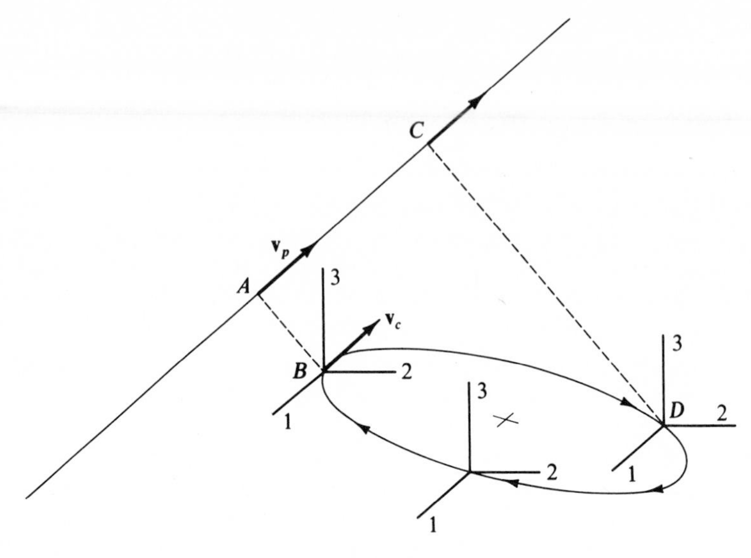

FIG. 2-1

coordinate system moves with a linear velocity $\mathbf{v}_c$ when at the point B, and if $\mathbf{v}_c = \mathbf{v}_p$, then to an observer in the moving coordinate system, the particle (at A) will appear to be *at rest*. At some later time, however, when the particle is at C and the coordinate system is at D, the particle will appear to be accelerating away from the observer. We must, therefore, conclude that the rotating coordinate system does not qualify as an inertial reference frame.

The observations described above are not sufficient, however, to decide whether time is homogeneous. To reach such a conclusion, it is necessary that repeated measurements be made in identical situations at various times; identical results would indicate the homogeneity of time.

Newton's equations do not describe the motion of bodies in noninertial systems. We can, however, devise a method whereby it is possible to describe the motion of a particle by means of a rotating coordinate system, but, as we shall see in Chapter 11, the resulting equations contain several terms which do not appear in the simple Newtonian equation, $\mathbf{F} = m\mathbf{a}$. For the moment, then, we shall restrict our attention to inertial frames of reference for the description of the dynamics of particles.

2.4 The Equation of Motion for a Particle

Newton's equation, $\mathbf{F} = d\mathbf{p}/dt$, can be expressed alternatively as

$$\mathbf{F} = \frac{d}{dt}(m\mathbf{v}) = m\frac{d\mathbf{v}}{dt} = m\ddot{\mathbf{r}} \tag{2.6}$$

if we assume that the mass m does not vary with the time. This is a second-order differential equation for $\mathbf{r} = \mathbf{r}(t)$ which can be integrated if the function $\mathbf{F}$ is known. The specification of the initial values of $\mathbf{r}$ and $\dot{\mathbf{r}} = \mathbf{v}$ then allows the evaluation of the two arbitrary constants of integration. Many examples of problems of this type can be found in elementary texts.

It should be emphasized that the force $\mathbf{F}$ in Eq. (2.6) is not necessarily constant and, indeed, it may consist of several distinct parts. For example, if a particle falls in a constant gravitational field, the gravitational force will be $\mathbf{F}_g = m\mathbf{g}$, where $\mathbf{g}$ is the acceleration of gravity. If, in addition, there is a retarding force $\mathbf{F}$, which is some function of the instantaneous velocity, then the total force is

$$\begin{aligned} \mathbf{F} &= \mathbf{F}_g + \mathbf{F}_r \\ &= m\mathbf{g} + \mathbf{F}_r(v) \end{aligned} \tag{2.7}$$

It is frequently sufficient to consider that $\mathbf{F}_r(v)$ is simply proportional to some power of the velocity. In general, *real* retarding forces are more complicated, but the power-law approximation is useful in many instances in which the velocity does not vary greatly. (Even more to the point is the fact that if $F_r \propto v^n$, then the equation of motion can usually be integrated directly, whereas, if the true velocity dependence were used, numerical integration would probably be necessary!) With this type of approximation, we can then write

$$\mathbf{F} = m\mathbf{g} - mkv^n \frac{\mathbf{v}}{v} \tag{2.7a}$$

where k is a positive constant that specifies the strength of the retarding force and where $\mathbf{v}/v$ is a unit vector in the direction of $\mathbf{v}$. Experimentally, it is found that for a relatively small object moving in air, $n \cong 1$ for velocities less than about 2400 cm/sec ($\sim$80 ft/sec); for higher velocities but below the velocity of sound ($\sim$33,000 cm/sec or 1100 ft/sec) the retarding force is approximately proportional to the square of the velocity.*

The effect of air resistance is of obvious importance in military ballistics, as, for example, in the calculation of the trajectory of an artillery shell. Therefore, extensive tabulations have been made of the velocity as a function of flight time for projectiles of various sorts. A set of tables that is popular in this country is that of Ingalls.† One part of these tables gives the performance for a projectile 1 in. in diameter and weighing 1 lb, and having a particular

* The motion of a particle in a medium in which there is a resisting force proportional to the velocity or to the square of the velocity (or to a linear combination of the two) was examined by Newton in his *Principia* (1687). The extension to any power of the velocity was made by Johann Bernoulli in 1711. The term *Stokes' law of resistance* is sometimes applied to a resisting force that is proportional to the velocity; *Newton's law of resistance* is a retarding force proportional to the square of the velocity.

† Col. James M. Ingalls, U.S. Army Artillery Circular M, 1918.

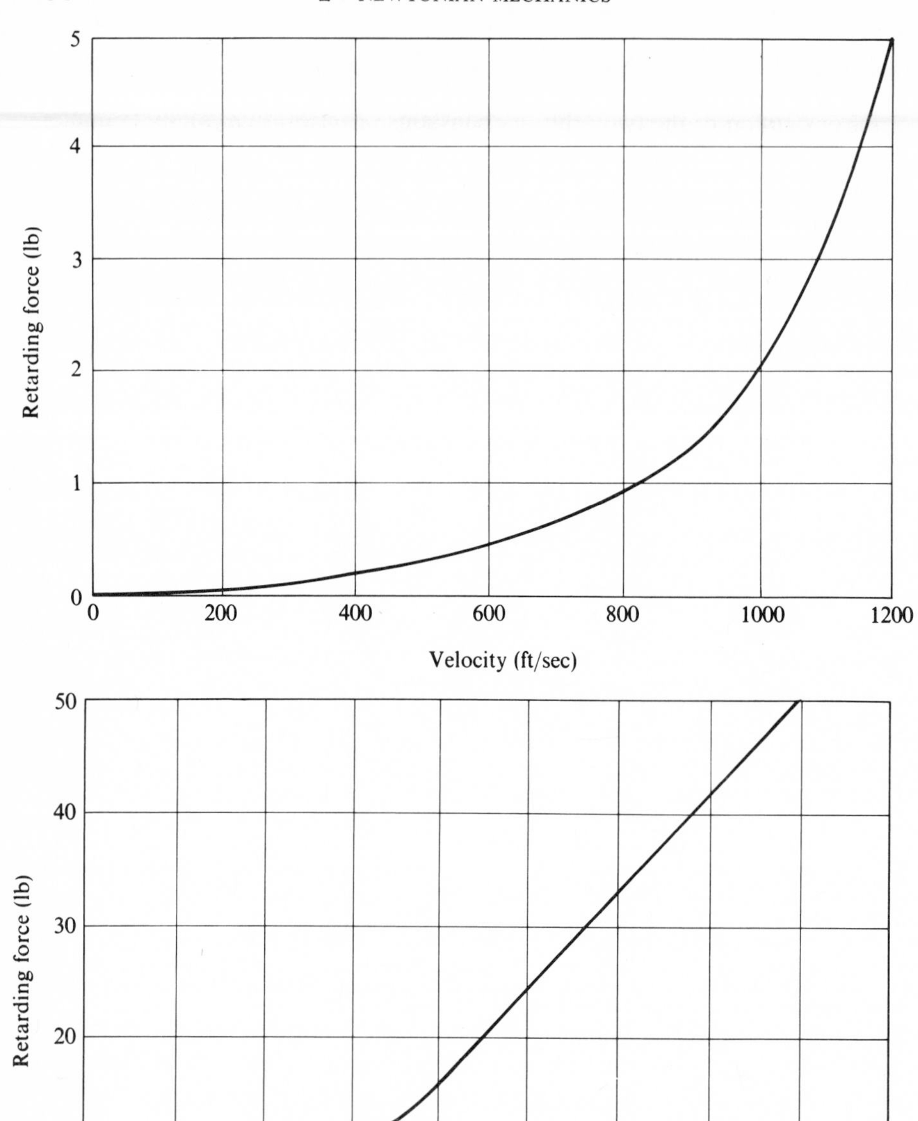
Retarding force (lb)
0
1
2
3
4
5
0
200
400
600
800
1000
1200
Velocity (ft/sec)
Retarding force (lb)
0
10
20
30
40
50
0
1000
2000
3000
4000
Velocity (ft/sec)

Fig. 2-2

shape of pointed nose.* Figure 2-2 shows the retarding force (lb) as a function of velocity (ft/sec) for this particular shell. The upper curve shows the velocity region from zero to about 1200 ft/sec. It is evident that an equation of at least second degree is necessary to describe this function. Indeed, segments of the curve (of 400 ft/sec or so) can be quite well represented by parabolas. The lower curve shows a more extensive velocity region and indicates that for velocities greater than about 2000 ft/sec, the retarding force varies approximately linearly with velocity. Note the change in shape of the curve in the region just above the velocity of sound.

Several examples of the motion of a particle subjected to various forces are given below.

▶ Example 2.4(a) Horizontal Motion of a Particle in a Resisting Medium

As the simplest example of the resisted motion of a particle, let us consider the horizontal motion in a medium in which the retarding force is proportional to the velocity. Such a situation would apply, for example, to a sliding particle subject to friction. The Newtonian equation, $\mathbf{F} = m\mathbf{a}$, provides us with the equation of motion:

$$ma = m\frac{dv}{dt} = -kmv \tag{1}$$

where kmv is the magnitude of the resisting force ($k = \text{const.}$). Then,

$$\int \frac{dv}{v} = -k\int dt$$

$$\ln v = -kt + C_1 \tag{2}$$

The integration constant in Eq. (2) can be evaluated if we prescribe the initial condition $v(t=0) \equiv v_0$. Then $C_1 = \ln v_0$, and

$$v = v_0 e^{-kt} \tag{3}$$

We can integrate this equation to obtain the displacement x as a function of time:

$$v = \frac{dx}{dt} = v_0 e^{-kt}$$

$$x = v_0 \int e^{-kt}\, dt = -\frac{v_0}{k} e^{-kt} + C_2 \tag{4}$$

The initial condition $x(t=0) \equiv 0$ implies $C_2 = v_0/k$. Therefore,

$$x = \frac{v_0}{k}(1 - e^{-kt}) \tag{5}$$

* Actually, a 2-caliber radius ogive. Depending on the exact shape of the nose, the retarding force at any velocity may be a factor of 2 greater (for a more blunt nose) or a factor of 2 less (for a more pointed nose) than that for a 2-caliber ogive.

This result shows that x asymptotically approaches the value v_0/k as $t \to \infty$.

We can also obtain the velocity as a function of displacement by writing

$$\frac{dv}{dx} = \frac{dv}{dt}\frac{dt}{dx} = \frac{dv}{dt} \cdot \frac{1}{v} \tag{6}$$

so that

$$v\frac{dv}{dx} = \frac{dv}{dt} = -kv \tag{7}$$

or,

$$\frac{dv}{dx} = -k \tag{8}$$

from which we find, by using the same initial conditions,

$$v = v_0 - kx \tag{9}$$

Therefore, the velocity decreases linearly with displacement.

▶ Example 2.4(b) Vertical Motion of a Particle in a Resisting Medium

We next treat the case of a particle undergoing vertical motion in a medium in which there again exists a retarding force proportional to the velocity. Let us consider that the particle is projected downward with an initial velocity v_0 from a height h in a constant gravitational field. The equation of motion is

$$F = m\frac{dv}{dt} = -mg - kmv \tag{1}$$

where $-kmv$ represents an *upward* force since we take z and $v = \dot{z}$ to be positive upwards and the motion is downwards; i.e., $v < 0$ so that $-kmv > 0$. From Eq. (1) we have

$$\frac{dv}{kv + g} = -dt \tag{2}$$

Integrating Eq. (2) and setting $v(t = 0) \equiv v_0$, we have

$$v = \frac{dz}{dt} = -\frac{g}{k} + \frac{kv_0 + g}{k} e^{-kt} \tag{3}$$

Integrating once more and evaluating the constant by setting $z(t = 0) \equiv h$, we find

$$z = h - \frac{gt}{k} + \frac{kv_0 + g}{k^2}(1 - e^{kt}) \tag{4}$$

Equation (3) shows that as the time becomes very long, the velocity approaches the limiting value $-g/k$; this is called the *terminal velocity*, v_t. We note that Eq. (1) yields the same result since the force will vanish, and hence there will be no further acceleration, when $v = -g/k$. If the initial velocity exceeds the terminal velocity in magnitude, then the body immediately begins to slow down and v approaches the terminal velocity from the opposite direction. Figure 2-3 illustrates these results.

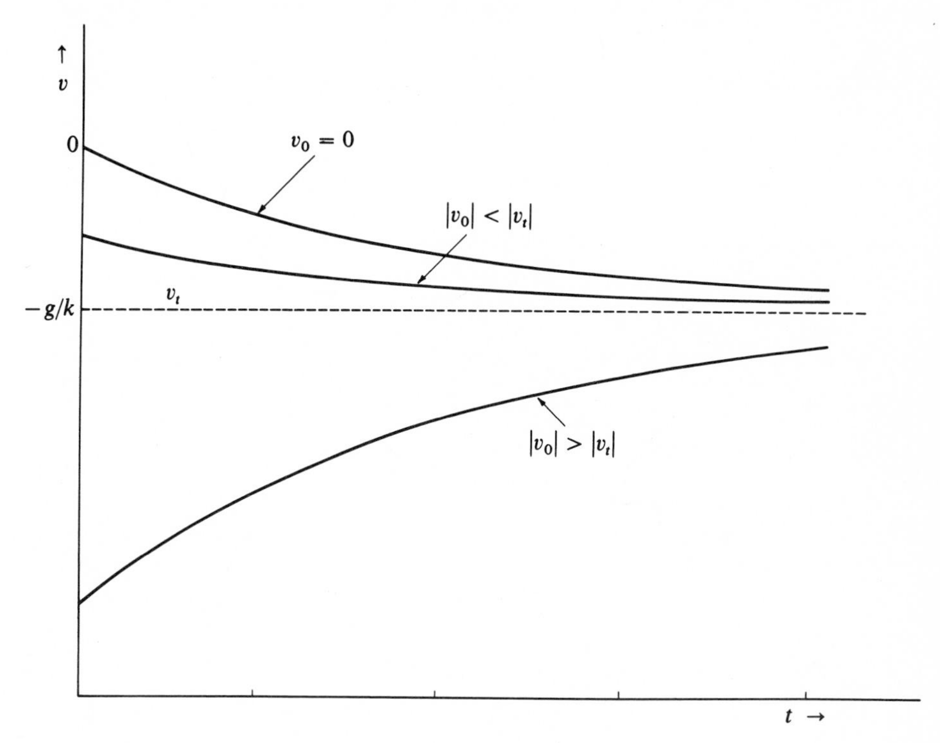

FIG. 2-3

▶ Example 2.4(c) Motion of a Projectile in a Resisting Medium—Approximate Solution by Series Expansion

Let us next consider the motion of a projectile in the atmosphere; the air resistance will give rise to a retarding force which we will assume is proportional to the instantaneous velocity. Since we can separate the motion into horizontal and vertical components, this example combines the results of the preceding two examples. We wish to calculate the decrease in range that is brought about by the existence of the air resistance. The situation is shown in Fig. 2-4 for several values of the retarding force constant k (in units of some standard, k_0).

If we take R and R' to be the range without* and with air resistance, respectively,

* The parabolic solution for the path of an unresisted projectile was obtained by Galileo in his *Discorsi*, 1638.

then the change of range due to air resistance is $R - R' \equiv \Delta R$. If α is the angle of inclination of the initial motion, the initial conditions can then be stated as

$$\left.\begin{aligned} x(t=0) &= 0 = y(t=0) \\ \dot{x}(t=0) &= v_0 \cos\alpha \equiv U \\ \dot{y}(t=0) &= v_0 \sin\alpha \equiv V \end{aligned}\right\} \tag{1}$$

The equations of motion are

$$m\ddot{x} = -km\dot{x} \tag{2}$$

$$m\ddot{y} = -km\dot{y} - mg \tag{3}$$

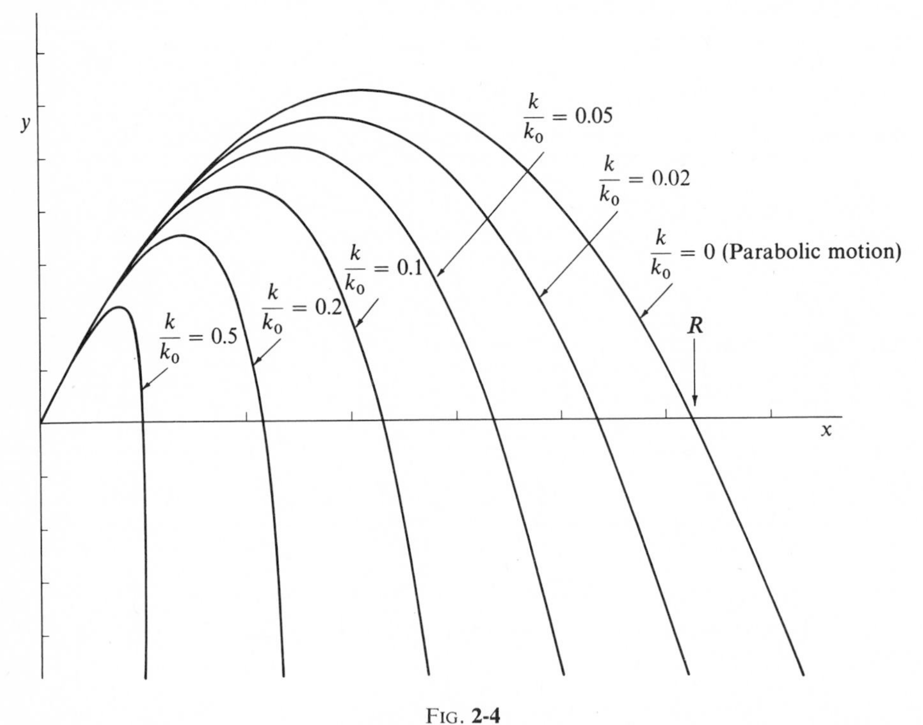

FIG. **2-4**

Equation (2) is exactly that which we used in Example 2.4(a). Therefore, the solution is

$$x = \frac{U}{k}(1 - e^{-kt}) \tag{4}$$

Similarly, Eq. (3) is the same as the equation of the motion in Example 2.4(b). We can use the solution found in that example by letting $h = 0$. (The fact that we considered the particle to be projected *downward* in Example 2.4(b) is of no consequence; the sign of the initial velocity will automatically take this into account.) Therefore,

$$y = -\frac{gt}{k} + \frac{kV + g}{k^2}(1 - e^{-kt}) \tag{5}$$

The range R' can be found by first calculating the time T required for the entire trajectory and then substituting this value into Eq. (4) for x. The time can be found by noting that $y = 0$ at the end of the trajectory; i.e., $y(t = T) = 0$. From Eq. (5) we find

$$T = \frac{kV + g}{gk}(1 - e^{-kT}) \tag{6}$$

This is a transcendental equation and therefore we cannot obtain an analytic expression for T; we must resort to an approximation procedure. We first expand the exponential [see Eq. (D.34), Appendix D]:

$$T = \frac{kV + g}{gk}(kT - \tfrac{1}{2}k^2T^2 + \tfrac{1}{6}k^3T^3 - \cdots) \tag{7}$$

Upon dividing through by T, this equation can be rearranged to yield

$$T = \frac{2V/g}{(1 + kV/g)} + \frac{1}{3}kT^2 \tag{8}$$

Equation (8) shows (as does the elementary treatment in which air resistance is neglected) that in the limit $k \to 0$, the time of flight of the projectile is

$$T_0 = \frac{2V}{g}, \qquad k \to 0 \tag{9}$$

Therefore, if we consider the case in which k is small (but nonvanishing), then the flight time will be *approximately* equal to T_0. If we then use this approximate value to calculate the second-order term in Eq. (8), we have

$$T \cong \frac{2V}{g}\left(1 - \frac{kV}{g}\right) + \frac{1}{3}k\left(\frac{2V}{g}\right)^2 \tag{10}$$

so that

$$T \cong \frac{2V}{g}\left(1 - \frac{kV}{3g}\right) \tag{11}$$

which is the desired approximate expression for the flight time.

Next, we write the equation for x [Eq. (4)] in expanded form:

$$x = \frac{U}{k}(kt - \tfrac{1}{2}k^2t^2 + \tfrac{1}{6}k^3t^3 - \cdots) \tag{12}$$

Since $x(t = T) \equiv R'$, we have approximately for the range,

$$R' \cong U(T - \tfrac{1}{2}kT^2) \tag{13}$$

We can now evaluate this expression by using the value of T from Eq. (11). If we retain only terms linear in k, we find

$$R' \cong \frac{2UV}{g}\left(1 - \frac{4kV}{3g}\right) \tag{14}$$

Now, the quantity $2UV/g$ can be written as [using Eqs. (1)]

$$\frac{2UV}{g} = \frac{2v_0^2}{g} \sin \alpha \cos \alpha = \frac{v_0^2}{g} \sin 2\alpha \tag{15}$$

which will be recognized as the range R of the projectile when air resistance is neglected. Therefore,

$$R' \cong R\left(1 - \frac{4kV}{3g}\right) \tag{16}$$

so that the change of range, correct to first order in k, is

$$\Delta R = R - R' \cong \frac{4kVR}{3g} \tag{17}$$

or,

$$\Delta R \cong \frac{4kv_0^3}{3g^2} \sin \alpha \sin 2\alpha \tag{18}$$

The change of range ΔR is shown as a function of k/k_0 in Fig. 2.5 which is based on the results of the case illustrated in Fig. 2.4. (The dots on the curve refer to the trajectories that are shown in Fig. 2-4.) It is evident that the linear approximation for ΔR tends to become quite inaccurate for $k/k_0 \gtrsim 0.02$.

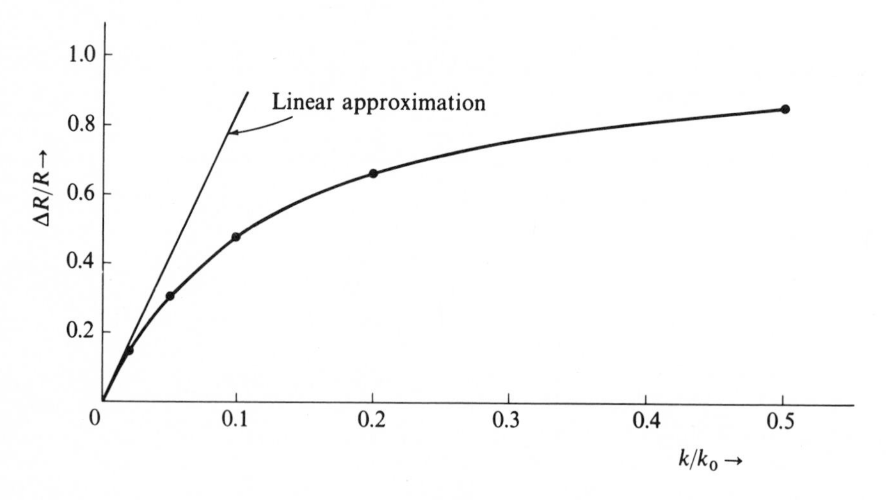

FIG. 2-5

▶ **Example 2.4(d) Motion with Variable Mass—The "Rocket Equation"**

Let the velocity of the rocket with respect to a certain fixed coordinate system be $\mathbf{v}(t)$, and let the mass of the rocket (and unconsumed fuel) be $m(t)$. We shall assume that the exhaust gases which are ejected by the rocket in order to propel it forward are all ejected at a constant velocity $\mathbf{V}$ relative to the rocket. The velocity of the exhaust gases relative to the fixed coordinate system is therefore $\mathbf{v}(t) + \mathbf{V}$. The rate of change of momentum of the rocket is

$$\frac{d}{dt}\mathbf{p}_r = \frac{d}{dt}(m\mathbf{v}) = m\dot{\mathbf{v}} + \dot{m}\mathbf{v} \tag{1}$$

Also, the rate of change of momentum of the exhaust gases is

$$\frac{d}{dt}\mathbf{p}_e = (\mathbf{v} + \mathbf{V})|\dot{m}| = -(\mathbf{v} + \mathbf{V})\dot{m} \tag{2}$$

where $\dot{m} < 0$ since the mass is decreasing. The rate of change of the total momentum is just the resultant *external* force $\mathbf{F}$ acting on the system:

$$\begin{aligned}\mathbf{F} &= \frac{d}{dt}(\mathbf{p}_r + \mathbf{p}_e)\\ &= m\dot{\mathbf{v}} + \dot{m}\mathbf{v} - (\mathbf{v} + \mathbf{V})\dot{m}\\ &= m\dot{\mathbf{v}} - \mathbf{V}\dot{m}\end{aligned} \tag{3}$$

or, equivalently,

$$(\mathbf{F} + \mathbf{V}\dot{m}) = m\dot{\mathbf{v}} \tag{4}$$

The additional "force" term, $\mathbf{V}\dot{m}$, which occurs because the mass of the rocket is not constant in time, is called the *thrust*.

For a free-space situation, in which there is no external force, the equation of motion can be integrated to give

$$\mathbf{v}(t) = \mathbf{V} \ln m + \text{const.} \tag{5}$$

If the initial conditions are $\mathbf{v}(0) = \mathbf{v}_0$ and $m(0) = m_0$, then we have

$$\mathbf{v} - \mathbf{v}_0 = -\mathbf{V} \ln \frac{m_0}{m} \tag{6}$$

Now, if the rocket originally consisted of a frame of mass M_0 and fuel mass M, then $m_0 = M_0 + M$. Therefore, at the time t_b at which the fuel is completely consumed ("burnout"), the mass is $m(t_b) = M_0$. This is the point at which maximum velocity is reached. Thus,

$$\mathbf{v}_{\max} = \mathbf{v}_0 - \mathbf{V} \ln \left(1 + \frac{M}{M_0}\right) \tag{7}$$

(Recall that **v** and **V** are oppositely directed.) Therefore, if the rocket is to reach velocities that exceed V, it is necessary that $M \gg M_0$; that is, the fuel-to-payload ratio must be quite large.

2.5 Conservation Theorems

We now turn to a detailed discussion of the Newtonian mechanics of a single particle and derive the important theorems regarding conserved quantities. It must be emphasized that we are here not *proving* the conservation of the various quantities. We are merely deriving the consequences of Newton's laws of dynamics. These implications must be put to the test of experiment and their verification then supplies a measure of confirmation of the original dynamical laws. The fact that these conservation theorems have indeed been found to be valid in many instances furnishes an important part of the proof for the correctness of Newton's laws, at least in classical physics.*

The first of the conservation theorems concerns the linear momentum of a particle. If the particle is *free*, i.e., if the particle experiences no force, then Eq. (2.2) becomes simply $\dot{\mathbf{p}} = 0$. Therefore, **p** is a vector constant in time, and *the linear momentum of a free particle is conserved.*

Note that this result is derived from a vector equation, $\dot{\mathbf{p}} = 0$, and therefore applies for each component of the linear momentum. In order to state the result in other terms, we let **s** be some constant vector such that $\mathbf{F} \cdot \mathbf{s} = 0$, independent of time. Then,

$$\dot{\mathbf{p}} \cdot \mathbf{s} = \mathbf{F} \cdot \mathbf{s} = 0$$

or, integrating with respect to time,

$$\mathbf{p} \cdot \mathbf{s} = \text{const.} \tag{2.8}$$

which states that the *component of linear momentum in a direction in which the force vanishes is constant in time.*

The *angular momentum* **L** of a particle with respect to an origin from which **r** is measured is defined to be

$$\boxed{\mathbf{L} \equiv \mathbf{r} \times \mathbf{p}} \tag{2.9}$$

The *torque* or *moment of force* **N** with respect to the same origin is defined to be

$$\boxed{\mathbf{N} \equiv \mathbf{r} \times \mathbf{F}} \tag{2.10}$$

* For further comment on this point, see the discussion at the end of Section 2.6.

or,

$$\mathbf{N} = \mathbf{r} \times m\dot{\mathbf{v}} = \mathbf{r} \times \dot{\mathbf{p}}$$

Now,

$$\dot{\mathbf{L}} = \frac{d}{dt}(\mathbf{r} \times \mathbf{p}) = (\dot{\mathbf{r}} \times \mathbf{p}) + (\mathbf{r} \times \dot{\mathbf{p}})$$

but

$$\dot{\mathbf{r}} \times \mathbf{p} = \dot{\mathbf{r}} \times m\mathbf{v} = m(\dot{\mathbf{r}} \times \dot{\mathbf{r}}) \equiv 0$$

so that

$$\boxed{\dot{\mathbf{L}} = \mathbf{r} \times \dot{\mathbf{p}} = \mathbf{N}} \tag{2.11}$$

If there are no torques acting on a particle (i.e., if $\mathbf{N} = 0$), then $\dot{\mathbf{L}} = 0$ and $\mathbf{L}$ is a vector constant in time. This is the second important conservation theorem: *the angular momentum of a particle subject to no torque is conserved.*

If work is done on a particle by a force $\mathbf{F}$ in transforming the particle from condition 1 to condition 2, then this work is defined to be

$$\boxed{W_{12} \equiv \int_1^2 \mathbf{F} \cdot d\mathbf{r}} \tag{2.12}$$

Now,

$$\begin{aligned}\mathbf{F} \cdot d\mathbf{r} &= m\frac{d\mathbf{v}}{dt} \cdot \frac{d\mathbf{r}}{dt}\,dt = m\frac{d\mathbf{v}}{dt} \cdot \mathbf{v}\,dt \\ &= \frac{m}{2}\frac{d}{dt}(\mathbf{v} \cdot \mathbf{v})\,dt = \frac{m}{2}\frac{d}{dt}(v^2)\,dt \\ &= d(\tfrac{1}{2}mv^2)\end{aligned} \tag{2.13}$$

Therefore, the integrand in Eq. (2.12) is an exact differential, and

$$W_{12} = (\tfrac{1}{2}mv^2)\Big|_1^2 = \tfrac{1}{2}m(v_2^2 - v_1^2) = T_2 - T_1 \tag{2.14}$$

where $T \equiv \frac{1}{2}mv^2$ is the *kinetic energy* of the particle. If $T_1 > T_2$ then $W_{12} < 0$ and the particle has done work with a resulting decrease in kinetic energy. It is important to realize that the force $\mathbf{F}$ appearing in Eq. (2.12) is the *total* (i.e., net resultant) force on the particle.

Let us now examine the integral appearing in Eq. (2.12) from a different standpoint. In many physical problems the force **F** has the property that the work required to move a particle from one position to another without any change in kinetic energy is dependent only upon the original and final positions and not upon the exact path taken by the particle. This property is exhibited, for example, by a constant gravitational force field. Thus, if a particle of mass m is raised through a height h (by *any* path), then an amount of work mgh has been done on the particle, and the particle has the capacity to do an equal amount of work in returning to its original position. This *capacity to do work* is called the *potential energy* of the particle.*

We may define the potential energy of a particle in terms of the work required to transport the particle from a position 1 to a position 2 (with no net change in kinetic energy):

$$\int_1^2 \mathbf{F} \cdot d\mathbf{r} = U_1 - U_2 \tag{2.15}$$

That is, the work done in moving the particle is simply the difference in the potential energy U at the two points. Equation (2.15) can be reproduced if we write **F** as the gradient of the scalar function U†:

$$\boxed{\mathbf{F} = -\mathbf{grad}\, U} \tag{2.16}$$

Then,

$$\begin{aligned} \int_1^2 \mathbf{F} \cdot d\mathbf{r} &= -\int_1^2 (\mathbf{grad}\, U) \cdot d\mathbf{r} \\ &= -\int_1^2 \sum_i \frac{\partial U}{\partial x_i}\, dx_i \\ &= -\int_1^2 dU = U_1 - U_2 \end{aligned} \tag{2.17}$$

In most systems of interest, the potential energy is a function of position and, possibly, the time: $U = U(\mathbf{r})$ or $U = U(\mathbf{r}, t)$. We shall not consider cases in which the potential energy is a function of the velocity.‡

* Potential energy and the general concept of a potential are discussed in detail beginning in Section 2.7.

† The necessary and sufficient condition that permits a vector function to be represented by the gradient of a scalar function is that the *curl* of the vector function vanish identically. [See Marion (Ma65a, Chapter 3).]

‡ Velocity-dependent potentials are necessary in certain situations, e.g., in electromagnetism (the so-called *Liénard–Wiechert potentials*).

It is important to realize that the potential energy is defined only to within an additive constant. That is, the force defined by $-\mathbf{grad}\, U$ is no different from that defined by $-\mathbf{grad}(U + \text{const.})$. Therefore, potential energy has no absolute meaning; only *differences* of potential energy are physically meaningful [as in Eq. (2.15)].

If we choose a certain inertial frame of reference for the description of a mechanical process, the laws of motion are the same as in any other reference frame which is in uniform motion relative to the original frame. The velocity of a particle is in general different depending on which inertial reference frame is chosen as the basis for the description of the motion. Therefore, we find that it is impossible to ascribe an *absolute kinetic energy* to a particle in much the same way that it is impossible to assign any absolute meaning to potential energy. Both of these limitations are the result of the fact that the selection of an *origin* of the coordinate system used for the description of physical processes is always arbitrary. James Clerk Maxwell has summarized the situation as follows*:

> We must, therefore, regard the energy of a material system as a quantity of which we may ascertain the increase or diminution as the system passes from one definite condition to another. The absolute value of the energy in the standard condition is unknown to us, and it would be of no value to us if we did know it, as all phenomena depend on the variations of energy and not on its absolute value.

Now, the *total energy* of a particle is defined to be the sum of the kinetic and potential energies:

$$\boxed{E \equiv T + U} \tag{2.18}$$

The total time derivative of E is

$$\frac{dE}{dt} = \frac{dT}{dt} + \frac{dU}{dt} \tag{2.19}$$

In order to evaluate the time derivatives appearing on the right-hand side of this equation, we first note that Eq. (2.13) can be written as

$$\mathbf{F} \cdot d\mathbf{r} = d(\tfrac{1}{2}mv^2) = dT \tag{2.20}$$

Dividing through by dt,

$$\frac{dT}{dt} = \mathbf{F} \cdot \frac{d\mathbf{r}}{dt} = \mathbf{F} \cdot \dot{\mathbf{r}} \tag{2.21}$$

* *Matter and Motion*, Cambridge Univ. Press, 1877; reprinted by Dover, New York, p. 91.

Also, we have

$$\begin{aligned}\frac{dU}{dt} &= \sum_i \frac{\partial U}{\partial x_i}\frac{dx_i}{dt} + \frac{\partial U}{\partial t} \\ &= \sum_i \frac{\partial U}{\partial x_i}\dot{x}_i + \frac{\partial U}{\partial t} \\ &= (\textbf{grad}\ U)\cdot\dot{\mathbf{r}} + \frac{\partial U}{\partial t}\end{aligned} \tag{2.22}$$

Substituting Eqs. (2.21) and (2.22) into (2.20), we find

$$\begin{aligned}\frac{dE}{dt} &= \mathbf{F}\cdot\dot{\mathbf{r}} + (\textbf{grad}\ U)\cdot\dot{\mathbf{r}} + \frac{\partial U}{\partial t} \\ &= (\mathbf{F} + \textbf{grad}\ U)\cdot\dot{\mathbf{r}} + \frac{\partial U}{\partial t} \\ &= \frac{\partial U}{\partial t}\end{aligned} \tag{2.23}$$

since the term $\mathbf{F} + \textbf{grad}\ U$ vanishes in view of the definition of the potential energy [Eq. (2.16)].

If U is not an explicit function of the time (i.e., if $\partial U/\partial t = 0$; recall that we do not consider velocity-dependent potentials), then the force field represented by $\mathbf{F}$ is said to be *conservative*. Under these conditions, we have the third important conservation theorem: *the total energy E of a particle in a conservative force field is a constant in time.**

2.6 Conservation Theorems for a System of Particles

We now extend our discussion from a single particle to a system of n particles. These particles may form a loose aggregate, such as a group of pellets, or a rigid body in which the constituent particles are restrained from moving relative to one another. The mass of this system is denoted by M:

$$M = \sum_\alpha m_\alpha \tag{2.24}$$

where the summation over α (as in all summations below which are carried out over Greek indices) runs from $\alpha = 1$ to $\alpha = n$.

* The general law of conservation of energy was formulated in 1847 by Hermann von Helmholtz (1821–1894). His conclusion was based largely on the calorimetric experiments of James Prescott Joule (1818–1889) which were begun in 1840.

If the vector connecting the origin with the αth particle is $\mathbf{r}_\alpha$, then the vector which defines the position of the center of mass of the system is

$$\boxed{\mathbf{R} = \frac{1}{M}\sum_\alpha m_\alpha \mathbf{r}_\alpha} \tag{2.25}$$

Since it is often convenient to specify the position of a particle with respect to the center of mass (see Fig. 2-6), we define this vector $\bar{\mathbf{r}}_\alpha$ as

$$\bar{\mathbf{r}}_\alpha = \mathbf{r}_\alpha - \mathbf{R} \tag{2.26}$$

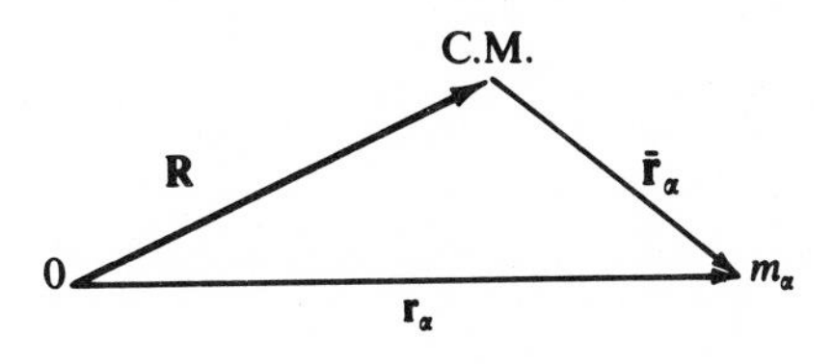

FIG. 2-6

If a certain group of particles is considered to constitute a *system*, then the resultant force which acts on a particle within the system (say, the αth particle) is in general composed of two parts. One part is the resultant of all forces whose origin lies outside of the system; this is called the *external force*, $\mathbf{F}_\alpha^{(e)}$. The other part is the resultant of the forces which arise from the interaction of all of the other $n - 1$ particles with the αth particle; this is called the *internal force*, $\mathbf{f}_\alpha$. Now, $\mathbf{f}_\alpha$ is given by the vector sum of all of the individual forces $\mathbf{f}_{\alpha\beta}$,

$$\sum_\beta \mathbf{f}_{\alpha\beta} = \mathbf{f}_\alpha \tag{2.27}$$

where $\mathbf{f}_{\alpha\beta}$ represents the force *on* the αth particle *due* to the βth particle. Therefore, the total force acting on the αth particle is

$$\mathbf{F}_\alpha = \mathbf{F}_\alpha^{(e)} + \mathbf{f}_\alpha \tag{2.28}$$

In addition, according to Newton's Third Law,* we have

$$\mathbf{f}_{\alpha\beta} = -\mathbf{f}_{\beta\alpha} \tag{2.29}$$

Newton's second law for the αth particle can be written as

$$\dot{\mathbf{p}}_\alpha = m_\alpha \ddot{\mathbf{r}}_\alpha = \mathbf{F}_\alpha^{(e)} + \mathbf{f}_\alpha \tag{2.30}$$

or,

$$\frac{d^2}{dt^2}(m_\alpha \mathbf{r}_\alpha) = \mathbf{F}_\alpha^{(e)} + \sum_\beta \mathbf{f}_{\alpha\beta} \tag{2.30a}$$

* Recall that the Third Law (and therefore all of the succeeding analysis) is not valid for moving charged particles; electromagnetic forces are *velocity dependent*.

Summing this expression over α, we have

$$\frac{d^2}{dt^2}\sum_{\alpha} m_\alpha \mathbf{r}_\alpha = \sum_{\alpha} \mathbf{F}_\alpha^{(e)} + \sum_{\substack{\alpha \\ \alpha \neq \beta}}\sum_{\beta} \mathbf{f}_{\alpha\beta} \tag{2.31}$$

where the terms $\alpha = \beta$ do not enter in the second sum on the right-hand side since $\mathbf{f}_{\alpha\alpha} \equiv 0$. The summation on the left-hand side just yields $M\mathbf{R}$ [see Eq. (2.25)] and the second time derivative is $M\ddot{\mathbf{R}}$. The first term on the right-hand side is the sum of all of the external forces and can be written as

$$\sum_{\alpha} \mathbf{F}_\alpha^{(e)} \equiv \mathbf{F} \tag{2.32}$$

The second term on the right-hand side in Eq. (2.31) can be expressed as*

$$\sum_{\substack{\alpha \\ \alpha \neq \beta}}\sum_{\beta} \mathbf{f}_{\alpha\beta} \equiv \sum_{\alpha,\, \beta \neq \alpha} \mathbf{f}_{\alpha\beta} = \sum_{\alpha < \beta} (\mathbf{f}_{\alpha\beta} + \mathbf{f}_{\beta\alpha})$$

which vanishes† according to Eq. (2.29), Thus, we have the important result

$$M\ddot{\mathbf{R}} = \mathbf{F} \tag{2.33}$$

which states that the *center of mass of a system moves as if it were a single particle, of mass equal to the total mass of the system, acted upon by the total external force, and independent of the nature of the internal forces* (as long as they follow $\mathbf{f}_{\alpha\beta} = -\mathbf{f}_{\beta\alpha}$).

* This equation can easily be verified by explicitly calculating both sides for a single case, e.g., $n = 3$. The last summation symbol means "sum over all α and β subject to the restriction $\alpha < \beta$."

† Note that we can prove the vanishing of

$$\sum_{\substack{\alpha \\ \alpha \neq \beta}}\sum_{\beta} \mathbf{f}_{\alpha\beta}$$

by appealing to the following argument. Since the summations are carried out over both α and β, these indices are dummies; in particular, we may interchange α and β without affecting the sum. Using the more compact notation, we have

$$\sum_{\alpha,\beta \neq \alpha} \mathbf{f}_{\alpha\beta} = \sum_{\beta,\alpha \neq \beta} \mathbf{f}_{\beta\alpha}$$

But, by hypothesis, $\mathbf{f}_{\alpha\beta} = -\mathbf{f}_{\beta\alpha}$, so that

$$\sum_{\alpha,\beta \neq \alpha} \mathbf{f}_{\alpha\beta} = -\sum_{\alpha,\beta \neq \alpha} \mathbf{f}_{\alpha\beta}$$

And if a quantity is equal to its negative, it must vanish identically.

The total linear momentum of the system is

$$\mathbf{P} = \sum_\alpha m_\alpha \dot{\mathbf{r}}_\alpha = \frac{d}{dt}\sum_\alpha m_\alpha \mathbf{r}_\alpha = \frac{d}{dt}(M\mathbf{R}) = M\dot{\mathbf{R}} \tag{2.34}$$

and

$$\dot{\mathbf{P}} = M\ddot{\mathbf{R}} = \mathbf{F} \tag{2.35}$$

Thus, the total linear momentum of the system is conserved if there is no external force. We may further note that Eq. (2.34) states that *the linear momentum of the system is the same as if a single particle of mass M were located at the position of the center of mass and moving in the manner in which the center of mass moves.*

The angular momentum of the αth particle about the origin is given by Eq. (2.9):

$$\mathbf{L}_\alpha = \mathbf{r}_\alpha \times \mathbf{p}_\alpha \tag{2.36}$$

Summing this expression over α, and using Eq. (2.26), we have

$$\begin{aligned}\mathbf{L} = \sum_\alpha \mathbf{L}_\alpha &= \sum_\alpha (\mathbf{r}_\alpha \times \mathbf{p}_\alpha) = \sum_\alpha (\mathbf{r}_\alpha \times m_\alpha \dot{\mathbf{r}}_\alpha)\\ &= \sum_\alpha (\bar{\mathbf{r}}_\alpha + \mathbf{R}) \times m_\alpha(\dot{\bar{\mathbf{r}}}_\alpha + \dot{\mathbf{R}})\\ &= \sum_\alpha m_\alpha[(\bar{\mathbf{r}}_\alpha \times \dot{\bar{\mathbf{r}}}_\alpha) + (\bar{\mathbf{r}}_\alpha \times \dot{\mathbf{R}}) + (\mathbf{R} \times \dot{\bar{\mathbf{r}}}_\alpha) + (\mathbf{R} \times \dot{\mathbf{R}})]\end{aligned} \tag{2.37}$$

The middle two terms can be written as

$$\left(\sum_\alpha m_\alpha \bar{\mathbf{r}}_\alpha\right) \times \dot{\mathbf{R}} + \mathbf{R} \times \frac{d}{dt}\left(\sum_\alpha m_\alpha \bar{\mathbf{r}}_\alpha\right)$$

which vanishes since

$$\begin{aligned}\sum_\alpha m_\alpha \bar{\mathbf{r}}_\alpha &= \sum_\alpha m_\alpha(\mathbf{r}_\alpha - \mathbf{R}) = \sum_\alpha m_\alpha \mathbf{r}_\alpha - \mathbf{R}\sum_\alpha m_\alpha\\ &= M\mathbf{R} - M\mathbf{R} \equiv 0\end{aligned}$$

That is, $\sum_\alpha m_\alpha \bar{\mathbf{r}}_\alpha$ specifies the position of the center of mass in the center-of-mass coordinate system, and is therefore a null vector. Thus,

$$\begin{aligned}\mathbf{L} &= M\mathbf{R} \times \dot{\mathbf{R}} + \sum_\alpha \bar{\mathbf{r}}_\alpha \times \bar{\mathbf{p}}_\alpha\\ &= \mathbf{R} \times \mathbf{P} + \sum_\alpha \bar{\mathbf{r}}_\alpha \times \bar{\mathbf{p}}_\alpha\end{aligned} \tag{2.38}$$

and the total angular momentum is the sum of the angular momentum of the center of mass about the origin and the angular momentum of the system about the position of the center of mass.

The time derivative of the angular momentum of the αth particle is, from Eq. (2.11),

$$\dot{\mathbf{L}}_\alpha = \mathbf{r}_\alpha \times \dot{\mathbf{p}}_\alpha \tag{2.39}$$

and, using Eq. (2.30a), we have

$$\dot{\mathbf{L}}_\alpha = \mathbf{r}_\alpha \times \left(\mathbf{F}_\alpha^{(e)} + \sum_\beta \mathbf{f}_{\alpha\beta}\right) \tag{2.40}$$

Summing this expression over α, we have

$$\dot{\mathbf{L}} = \sum_\alpha \dot{\mathbf{L}}_\alpha = \sum_\alpha (\mathbf{r}_\alpha \times \mathbf{F}_\alpha^{(e)}) + \sum_{\alpha,\,\beta \neq \alpha} (\mathbf{r}_\alpha \times \mathbf{f}_{\alpha\beta}) \tag{2.41}$$

It is easy to verify that the last term may be written as

$$\sum_{\alpha,\,\beta \neq \alpha} (\mathbf{r}_\alpha \times \mathbf{f}_{\alpha\beta}) = \sum_{\alpha < \beta} [(\mathbf{r}_\alpha \times \mathbf{f}_{\alpha\beta}) + (\mathbf{r}_\beta \times \mathbf{f}_{\beta\alpha})] \tag{2.42}$$

Now, the vector connecting the αth and βth particles is defined to be

$$\mathbf{r}_{\alpha\beta} \equiv \mathbf{r}_\alpha - \mathbf{r}_\beta \tag{2.43}$$

and then since $\mathbf{f}_{\alpha\beta} = -\mathbf{f}_{\beta\alpha}$, we have

$$\begin{aligned}\sum_{\alpha,\,\beta \neq \alpha} (\mathbf{r}_\alpha \times \mathbf{f}_{\alpha\beta}) &= \sum_{\alpha < \beta} (\mathbf{r}_\alpha - \mathbf{r}_\beta) \times \mathbf{f}_{\alpha\beta} \\ &= \sum_{\alpha < \beta} (\mathbf{r}_{\alpha\beta} \times \mathbf{f}_{\alpha\beta})\end{aligned} \tag{2.44}$$

But, since we have limited the discussion to the case of central forces, $\mathbf{f}_{\alpha\beta}$ is directed along the line joining m_α with m_β, i.e., along $\mathbf{r}_{\alpha\beta}$. Hence,

$$\mathbf{r}_{\alpha\beta} \times \mathbf{f}_{\alpha\beta} \equiv 0 \tag{2.45}$$

and

$$\dot{\mathbf{L}} = \sum_\alpha (\mathbf{r}_\alpha \times \mathbf{F}_\alpha^{(e)}) \tag{2.46}$$

The right-hand side of this expression is just the sum of all of the external torques:

$$\dot{\mathbf{L}} = \sum_\alpha \mathbf{N}_\alpha^{(e)} = \mathbf{N}^{(e)} \tag{2.47}$$

Thus, *if the external torques about a given axis vanish, then the total angular momentum of the system about that axis remains constant in time.*

Note also that the term

$$\sum_\beta \mathbf{r}_\alpha \times \mathbf{f}_{\alpha\beta}$$

is the torque on the αth particle due to all of the internal forces, i.e., it is the *internal torque*. Since the sum of this quantity over all of the particles α vanishes [see Eq. (2.44)],

$$\sum_{\alpha,\,\beta\neq\alpha} (\mathbf{r}_\alpha \times \mathbf{f}_{\alpha\beta}) = \sum_{\alpha<\beta} (\mathbf{r}_{\alpha\beta} \times \mathbf{f}_{\alpha\beta}) = 0$$

we can then state that the *total internal torque must vanish if the internal forces are central in character*, i.e., if $\mathbf{f}_{\alpha\beta} = -\mathbf{f}_{\beta\alpha}$, and the *angular momentum of an isolated system cannot be altered without the application of external forces.*

The final conservation theorem, that of energy, may be derived for a system of particles as follows. Consider the work done on the system in moving it from a configuration 1, in which all of the coordinates $\mathbf{r}_\alpha$ are specified, to a configuration 2, in which the coordinates $\mathbf{r}_\alpha$ have some different specification. (Note that the individual particles may just be rearranged in such a process, and that, for example, the position of the center of mass could remain stationary.) In analogy with Eq. (2.12), we write

$$W_{12} = \sum_\alpha \int_1^2 \mathbf{F}_\alpha \cdot d\mathbf{r}_\alpha \tag{2.48}$$

Using a procedure similar to that used to obtain Eq. (2.14), we have

$$W_{12} = \sum_\alpha \int_1^2 d(\tfrac{1}{2}m_\alpha v_\alpha^2) = T_2 - T_1 \tag{2.49}$$

where

$$T = \sum_\alpha T_\alpha = \sum_\alpha \tfrac{1}{2}m_\alpha v_\alpha^2 \tag{2.50}$$

Using the relation [cf. Eq. (2.26)]

$$\mathbf{r}_\alpha = \bar{\mathbf{r}}_\alpha + \mathbf{R} \tag{2.51}$$

we have

$$\begin{aligned}\dot{\mathbf{r}}_\alpha \cdot \dot{\mathbf{r}}_\alpha = v_\alpha^2 &= (\dot{\bar{\mathbf{r}}}_\alpha + \dot{\mathbf{R}}) \cdot (\dot{\bar{\mathbf{r}}}_\alpha + \dot{\mathbf{R}}) \\ &= (\dot{\bar{\mathbf{r}}}_\alpha \cdot \dot{\bar{\mathbf{r}}}_\alpha) + 2(\dot{\bar{\mathbf{r}}}_\alpha \cdot \dot{\mathbf{R}}) + (\dot{\mathbf{R}} \cdot \dot{\mathbf{R}}) \\ &= \bar{v}_\alpha^2 + 2(\dot{\bar{\mathbf{r}}}_\alpha \cdot \dot{\mathbf{R}}) + V^2\end{aligned} \tag{2.52}$$

where $\bar{\mathbf{v}} \equiv \dot{\bar{\mathbf{r}}}$ and where V is the velocity of the center of mass. Then,

$$\begin{aligned}T &= \sum_\alpha \tfrac{1}{2}m_\alpha v_\alpha^2 \\ &= \sum_\alpha \tfrac{1}{2}m_\alpha \bar{v}_\alpha^2 + \sum_\alpha \tfrac{1}{2}m_\alpha V^2 + \dot{\mathbf{R}} \cdot \frac{d}{dt}\sum_\alpha m_\alpha \bar{\mathbf{r}}_\alpha\end{aligned} \tag{2.53}$$

But, by the same argument used above, $\sum_\alpha m_\alpha \bar{\mathbf{r}}_\alpha = 0$, and the last term vanishes. Thus,

$$\boxed{T = \sum \tfrac{1}{2} m_\alpha \bar{v}_\alpha^2 + \tfrac{1}{2} M V^2} \tag{2.54}$$

which states that the *total kinetic energy of the system is equal to the sum of the kinetic energy of a particle of mass M moving with the velocity of the center of mass and the kinetic energy of motion of the individual particles relative to the center of mass.*

Next, the total force in Eq. (2.48) can be separated as in Eq. (2.30):

$$W_{12} = \sum_\alpha \int_1^2 \mathbf{F}_\alpha^{(e)} \cdot d\mathbf{r}_\alpha + \sum_{\alpha,\, \beta \neq \alpha} \int_1^2 \mathbf{f}_{\alpha\beta} \cdot d\mathbf{r}_\alpha \tag{2.55}$$

If the forces $\mathbf{F}_\alpha^{(e)}$ and $\mathbf{f}_{\alpha\beta}$ are conservative, then they are derivable from potential functions, and we can write

$$\left.\begin{aligned} \mathbf{F}_\alpha^{(e)} &= -\mathbf{grad}_\alpha\, U_\alpha \\ \mathbf{f}_{\alpha\beta} &= -\mathbf{grad}_\alpha\, \overline{U}_{\alpha\beta} \end{aligned}\right\} \tag{2.56}$$

where U_α and $\overline{U}_{\alpha\beta}$ are the potential functions, but which do not necessarily have the same form. The notation $\mathbf{grad}_\alpha$ means that the gradient operation is performed with respect to the coordinates of the αth particle.

The first term in Eq. (2.55) becomes

$$\begin{aligned} \sum_\alpha \int_1^2 \mathbf{F}_\alpha^{(e)} \cdot d\mathbf{r}_\alpha &= -\sum_\alpha \int_1^2 (\mathbf{grad}_\alpha\, U_\alpha) \cdot d\mathbf{r}_\alpha \\ &= -\sum_\alpha U_\alpha \Big|_1^2 \end{aligned} \tag{2.57}$$

The second term in Eq. (2.55) is*

$$\begin{aligned} \sum_{\alpha,\, \beta \neq \alpha} \int_1^2 \mathbf{f}_{\alpha\beta} \cdot d\mathbf{r}_\alpha &= \sum_{\alpha < \beta} \int_1^2 (\mathbf{f}_{\alpha\beta} \cdot d\mathbf{r}_\alpha + \mathbf{f}_{\beta\alpha} \cdot d\mathbf{r}_\beta) \\ &= \sum_{\alpha < \beta} \int_1^2 \mathbf{f}_{\alpha\beta} \cdot (d\mathbf{r}_\alpha - d\mathbf{r}_\beta) = \sum_{\alpha < \beta} \int_1^2 \mathbf{f}_{\alpha\beta} \cdot d\mathbf{r}_{\alpha\beta} \end{aligned} \tag{2.58}$$

where, following the definition in Eq. (2.43), $d\mathbf{r}_{\alpha\beta} = d\mathbf{r}_\alpha - d\mathbf{r}_\beta$.

Now, $\overline{U}_{\alpha\beta}$ is a function only of the distance between m_α and m_β, and therefore $\overline{U}_{\alpha\beta}$ depends on six quantities, viz., the three coordinates of m_α

* Note that, unlike the term $\sum_{\alpha,\, \beta \neq \alpha} \mathbf{t}_{\alpha\beta}$ which appears in Eq. (3.31), the term $\sum_{\alpha,\, \beta \neq \alpha} \int_1^2 \mathbf{f}_{\alpha\beta} \cdot d\mathbf{r}_\alpha$ is *not* antisymmetric in α and β and therefore does not, in general, vanish.

(the $x_{\alpha, i}$) and the three coordinates of m_β (the $x_{\beta, i}$). The total derivative of $\bar{U}_{\alpha\beta}$ is therefore the sum of six partial derivatives and is given by

$$d\bar{U}_{\alpha\beta} = \sum_i \left(\frac{\partial \bar{U}_{\alpha\beta}}{\partial x_{\alpha, i}} \, dx_{\alpha, i} + \frac{\partial \bar{U}_{\alpha\beta}}{\partial x_{\beta, i}} \, dx_{\beta, i} \right) \tag{2.59}$$

where the $x_{\beta, i}$ are held constant in the first term and the $x_{\alpha, i}$ are held constant in the second. Thus,

$$d\bar{U}_{\alpha\beta} = (\mathbf{grad}_\alpha \, \bar{U}_{\alpha\beta}) \cdot d\mathbf{r}_\alpha + (\mathbf{grad}_\beta \, \bar{U}_{\alpha\beta}) \cdot d\mathbf{r}_\beta \tag{2.60}$$

Now,

$$\mathbf{grad}_\alpha \, \bar{U}_{\alpha\beta} = -\mathbf{f}_{\alpha\beta} \tag{2.61a}$$

But $\bar{U}_{\alpha\beta} = \bar{U}_{\beta\alpha}$, so that

$$\mathbf{grad}_\beta \, \bar{U}_{\alpha\beta} = \mathbf{grad}_\beta \, \bar{U}_{\beta\alpha} = -\mathbf{f}_{\beta\alpha} = \mathbf{f}_{\alpha\beta} \tag{2.61b}$$

Therefore,

$$\begin{aligned} d\bar{U}_{\alpha\beta} &= -\, \mathbf{f}_{\alpha\beta} \cdot (d\mathbf{r}_\alpha - d\mathbf{r}_\beta) \\ &= -\mathbf{f}_{\alpha\beta} \cdot d\mathbf{r}_{\alpha\beta} \end{aligned} \tag{2.62}$$

Using this result in Eq. (2.58), we have

$$\sum_{\alpha, \beta \neq \alpha} \int_1^2 \mathbf{f}_{\alpha\beta} \cdot d\mathbf{r}_\alpha = -\sum_{\alpha < \beta} \int_1^2 d\bar{U}_{\alpha\beta} = -\sum_{\alpha < \beta} \bar{U}_{\alpha\beta} \Big|_1^2 \tag{2.63}$$

Combining Eqs. (2.57) and (2.63) to evaluate W_{12}, we find

$$W_{12} = -\sum_\alpha U_\alpha \Big|_1^2 - \sum_{\alpha < \beta} \bar{U}_{\alpha\beta} \Big|_1^2 \tag{2.64}$$

This equation was obtained under the assumption that both the external and internal forces are derivable from potentials. In such a case, the *total potential energy* for the system can be written as

$$U = \sum_\alpha U_\alpha + \sum_{\alpha < \beta} \bar{U}_{\alpha\beta} \tag{2.65}$$

Then,

$$W_{12} = -U \Big|_1^2 = U_1 - U_2 \tag{2.66}$$

Combining this result with Eq. (2.49), we have

$$T_2 - T_1 = U_1 - U_2$$

or,

$$T_1 + U_1 = T_2 + U_2$$

so that

$$\boxed{E_1 = E_2} \tag{2.67}$$

which expresses the conservation of energy for the system. This result is valid for a system in which all of the forces are derivable from potentials which do not depend explicitly on the time; we say that such a system is *conservative*.

In Eq. (2.65), the term

$$\sum_{\alpha<\beta} \bar{U}_{\alpha\beta}$$

is the *internal potential energy* of the system. If the system is a rigid body in which the constituent particles are restrained to maintain their relative positions, then, in any process in which the body is involved, the internal potential energy will remain constant. In such a case the internal potential energy can be ignored when computing the total potential energy of the system. This amounts simply to redefining the position of zero potential energy, but this position is arbitrarily chosen anyway. That is, as pointed out previously, it is only the *difference* in potential energy that is physically significant; the absolute value of the potential energy is an arbitrary quantity.

It must be reiterated that we have not *proved* the conservation laws of linear momentum, angular momentum, and energy. We have only derived various consequences of Newton's laws. That is, *if* these laws are valid in a certain situation, then momentum and energy will be conserved. But we have become so enamored with these conservation theorems that we have elevated them to the status of "laws" and we have come to *insist* that they be valid in any physical theory, even those that apply to situations in which Newtonian mechanics is not valid, as, for example, in the interaction of moving charges or in quantum-mechanical systems. That is, we do not actually have conservation "laws" in such situations, but rather conservation *postulates* which we force upon the theory. For example, if we have two, isolated moving electric charges, then the electromagnetic forces between them are not conservative. We therefore endow the electromagnetic field with a certain amount of energy in order that energy conservation will be valid. This procedure is satisfactory only if the consequences do not contradict any experimental fact, and this is indeed the case for moving charges. We therefore extend the usual conception of energy to include "electromagnetic energy" in order to satisfy our preconceived notion that energy must be conserved. While this may seem an arbitrary and drastic step to take, nothing, it is said, succeeds as does success, and these conservation "laws" have been the most successful set of principles in physics. The refusal to relinquish energy and momentum conservation led Pauli to postulate in 1930 the existence of the *neutrino* in order to account for the "missing" energy and momentum in radioactive β decay. This postulate allowed Fermi to construct a successful theory of β decay in 1934, but direct observation of the neutrino was not made until 1953 when Reines and Cowan performed their famous experiment. Therefore, by adhering to the conviction that energy and momentum must be conserved, a new elementary particle was discovered, one that is of great importance in modern theories of nuclear physics. This discovery is only one of the many advances in the understanding of the properties of matter that have resulted directly from the application of the conservation laws.

2.7 The Universal Law of Gravitation

Newton's universal law of gravitation* states that at a distance r from a particle of mass M a second particle of mass m experiences an attractive gravitational force†

$$\boxed{\mathbf{F} = -\gamma \frac{mM}{r^2} \mathbf{e}_r} \tag{2.68}$$

where $\mathbf{e}_r$ is the unit vector from M toward m. (With this definition of $\mathbf{e}_r$, the minus sign insures that the force is *attractive*.) In the form of Eq. (2.68), the law strictly applies only to *point particles*. If one or both of the particles is replaced by a body which has a certain extension, we must make an additional hypothesis before we can calculate the force. We must assume that the gravitational force field is a *linear field*. That is, we assume that it is possible to calculate the net gravitational force on a particle due to many other particles by simply taking the vector sum of all of the individual forces. For a body which consists of a continuous distribution of matter, the sum becomes an integral (see Fig. 2-7):

$$\mathbf{F} = -\gamma m \int_V \frac{\rho(\mathbf{r}')\mathbf{e}_r}{r^2} \, dv' \tag{2.69}$$

where $\rho(\mathbf{r}')$ is the mass density and dv' is the element of volume at the position defined by the vector $\mathbf{r}'$.

* Newton had formulated and numerically checked his gravitation law in 1666, but the result was not published until the appearance of his *Principia*, in 1687. (See footnote on p. 82). A laboratory verification of the law and a determination of the value of γ was made in 1798 by Henry Cavendish (1731–1810), using a method devised by the Rev. John Michell. (Cavendish's results were actually expressed in terms of the average density of the Earth rather than as a value of γ.)

† The numerical value of the gravitation constant γ is $6.668 \pm 0.005 \times 10^{-8}$ dyne–cm^2/gm^2.

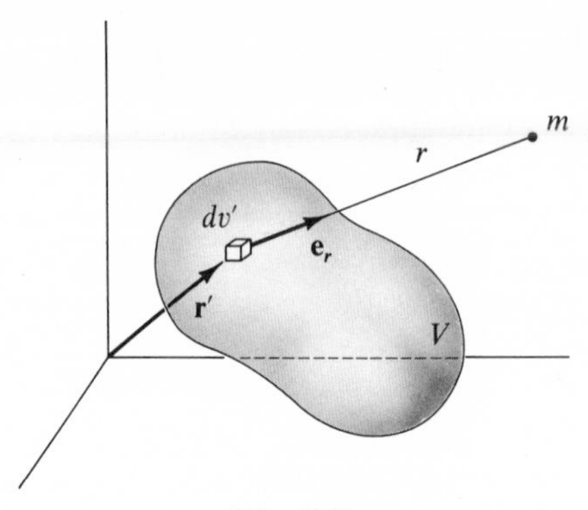

FIG. 2-7

If both the body of mass M and the body of mass m have extension, then a second integration over the volume of m will be necessary in order to compute the total gravitational force.

The *gravitational field vector* $\mathbf{g}$ is defined to be the vector that represents the force per unit mass exerted on a particle in the field of a body of mass M. Thus,

$$\mathbf{g} = \frac{\mathbf{F}}{m} = -\gamma \frac{M}{r^2} \mathbf{e}_r \tag{2.70}$$

or,

$$\boxed{\mathbf{g} = -\gamma \int_V \frac{\rho(\mathbf{r}')\mathbf{e}_r}{r^2} \, dv'} \tag{2.71}$$

The quantity $\mathbf{g}$ has the dimensions of *force per unit mass* or *acceleration*. In fact, near the surface of the Earth, the magnitude of $\mathbf{g}$ is just the quantity that we call the *gravitational acceleration constant*. A measurement with a simple pendulum (or some more sophisticated variation) is sufficient to show that $|\mathbf{g}|$ is approximately 980 cm/sec^2 at the surface of the Earth.

2.8 The Gravitational Potential

The gravitational field vector $\mathbf{g}$ varies as $1/r^2$ and therefore satisfies the requirement* that permits $\mathbf{g}$ to be represented as the gradient of a scalar function. Hence, we can write

$$\boxed{\mathbf{g} \equiv -\mathbf{grad}\, \Phi} \tag{2.72}$$

* That is, **curl** $\mathbf{g} \equiv 0$. [See Marion (Ma65a, Section 3.7).]

where Φ is called the *gravitational potential* and has dimensions of (*force per unit mass*) × (*distance*) or *energy per unit mass*.

Since **g** has only a radial variation, the potential Φ can have at most a variation with r. Therefore, using Eq. (2.70) for **g**, we have

$$\textbf{grad}\,\Phi = \frac{d\Phi}{dr}\,\mathbf{e}_r = \gamma\frac{M}{r^2}\,\mathbf{e}_r$$

Upon integrating, we obtain

$$\boxed{\Phi = -\gamma\frac{M}{r}} \tag{2.73}$$

The possible constant of integration has been suppressed since the potential is undetermined to within an additive constant. That is, only *differences* in potential are meaningful, not particular values. We usually remove the ambiguity in the value of the potential by arbitrarily requiring that $\Phi \to 0$ as $r \to \infty$; then, Eq. (2.73) correctly gives the potential for this condition.

The potential due to a continuous distribution of matter is

$$\Phi = -\gamma\int_V \frac{\rho(\mathbf{r}')}{r}\,dv' \tag{2.74}$$

Similarly, if the mass is distributed only over a thin shell (i.e., a *surface* distribution), then

$$\Phi = -\gamma\int_S \frac{\rho_s}{r}\,da' \tag{2.75}$$

where ρ_s is the surface density of mass (or *areal mass density*).

Finally, if there is a *line source* with linear mass density ρ_l, then

$$\Phi = -\gamma\int_\Gamma \frac{\rho_l}{r}\,ds' \tag{2.76}$$

The physical significance of the gravitational potential function becomes clear if we consider the work per unit mass dW which must be done on a body in a gravitational field in order to displace it a distance $d\mathbf{r}$. Now, work is equal to the scalar product of the force and the displacement. Thus, for the work done *on* the body per unit mass, we have

$$\begin{aligned} dW &= -\mathbf{g}\cdot d\mathbf{r} = (\textbf{grad}\,\Phi)\cdot d\mathbf{r} \\ &= \sum_i \frac{\partial\Phi}{\partial x_i}\,dx_i = d\Phi \end{aligned} \tag{2.77}$$

since Φ is a function only of the coordinates of the point at which it is measured: $\Phi = \Phi(x_1, x_2, x_3) = \Phi(x_i)$. Therefore, the amount of work per unit mass which must be done on a body to move it from one position to another in a gravitational field is equal to the difference in potential at the two points.

If the final position is further from the source mass M than the initial position, work has been done *on* the unit mass. Now, the positions of the two points are arbitrary, and we may take one of them to be at infinity. If we define the potential to be zero at infinity, then we may interpret Φ at any point to be the work per unit mass required to bring the body from infinity to that point. The *potential energy* is, of course, equal to the mass of the body multiplied by the potential Φ. If U is the potential energy, then

$$U = m\Phi \tag{2.78}$$

and the force on a body is given by the negative of the gradient of the potential energy of that body,

$$\boxed{\mathbf{F} = -\mathbf{grad}\ U} \tag{2.79}$$

which is just the expression we have previously used [Eq. (2.16)].

We note that both the potential and the potential energy *increase* when work is done *on* the body. (The potential, according to our definition, is always negative and only approaches its maximum value, viz., zero, as r tends to infinity.)

A certain potential energy exists whenever a body is placed in the gravitational field of a source mass. This potential energy resides, of course, in the *field*,* but it is customary, under these circumstances to speak of the potential energy "of the body." We shall continue this practice here. We may also consider the source mass itself to have an intrinsic potential energy. This potential energy is equal to the gravitational energy released when the body was formed, or, conversely, is equal to the energy that must be supplied (i.e., the work that must be done) to disperse the mass over the sphere at infinity. For example, when interstellar gas condenses to form a star, the gravitational energy that is released goes largely into the initial heating of the star. As the temperature increases, energy is radiated away as electromagnetic radiation. In all the problems which we shall treat, the structure of the bodies will be considered to remain unchanged during the process which we are studying. Thus, there will be no change in the intrinsic potential energy and it may be neglected for the purposes of whatever calculation we are making.

2.9 Lines of Force and Equipotential Surfaces

Let us consider a mass that gives rise to a gravitational field which can be described by a field vector $\mathbf{g}$. Let us draw a line outward from the surface

* See, however, the remarks at the end of Section 2.6 regarding the energy in a field.

of the mass in such a way that the direction of the line at every point is the same as the direction of **g** at that point. This line will extend from the surface of the mass to infinity. Such a line is called a *line of force.*

By drawing similar lines from every small increment of surface area of the mass we can indicate the direction of the force field at any arbitrary point in space. It is clear that the lines of force for a single point mass are all straight lines extending from the mass to infinity. Defined in this way, the lines of force are related only to the *direction* of the force field at any point. We may consider, however, that the *density* of such lines, i.e., the number of lines passing through a unit volume in space, is proportional to the *magnitude* of the force in that volume. Therefore, the lines-of-force picture is a convenient way to visualize both the magnitude and the direction (i.e., the *vector* property) of the field.

Now, the potential function is defined at every point in space (except at the position of a point mass). Therefore, the equation

$$\Phi = \Phi(x_1, x_2, x_2) = \text{const.} \tag{2.80}$$

defines a surface on which the potential is constant. Such a surface is called an *equipotential surface.* Since the field vector **g** is equal to the gradient of Φ, then **g** can have no component *along* an equipotential surface. It therefore follows that every line of force must be normal to every equipotential surface.

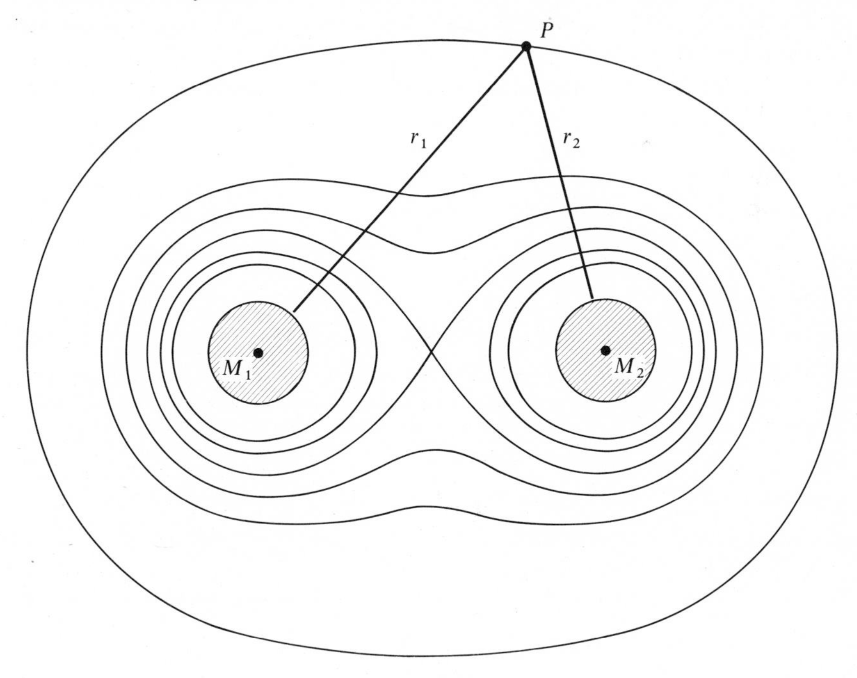

FIG. **2-8**

Thus, the field does no work on a body that moves along an equipotential surface. Since the potential function is single-valued, no two equipotential surfaces can intersect. The surfaces of equal potential that surround a single, isolated point mass (or any spherically symmetric mass) are all spheres. Consider two point masses M that are separated by a certain distance. If r_1 is the distance from one mass to some point in space and if r_2 is the distance from the other mass to the same point, then

$$\Phi = -\gamma M\left(\frac{1}{r_1} + \frac{1}{r_2}\right) = \text{const.}$$

defines the equipotential surfaces. Several of these surfaces are shown in Fig. 2-8 for this two-particle system. In three dimensions the surfaces are generated by rotating this diagram around the line connecting the two masses.

2.10 The Gravitational Potential of a Spherical Shell

One of the important problems of gravitational theory concerns the calculation of the gravitational force due to a homogeneous sphere. Clearly, this problem is a special case of the more general calculation for a homogeneous spherical shell. A solution to the problem of the shell can be obtained by directly computing the force (see Problem 2.23), but it is easier to proceed via the potential method.

We consider a shell with outer radius a and inner radius b (as in Fig. 2-9) and calculate the potential at the point P a distance R from the center of the

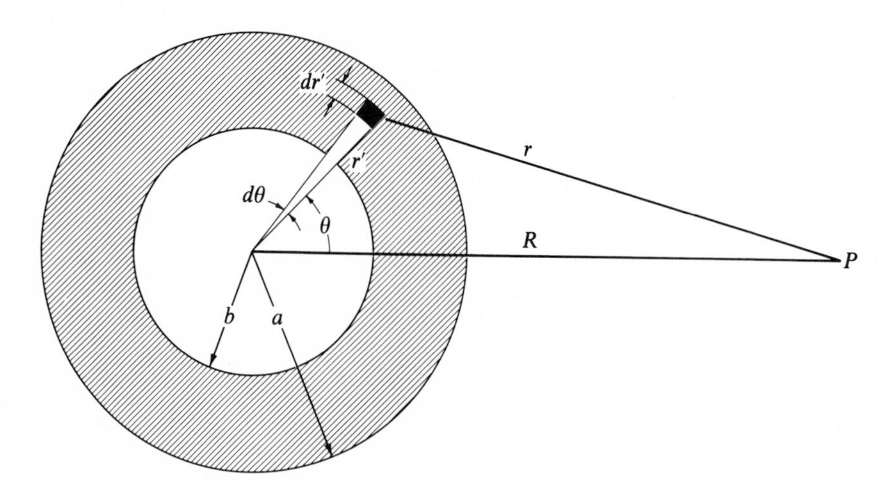

FIG. 2-9

shell. Since the problem clearly has symmetry about the line connecting the center of the sphere and the field point P, the azimuthal angle φ is not shown in Fig. 2-9, and we can immediately integrate over $d\varphi$ in the expression for the potential. Thus,

$$\Phi = -\gamma \int_V \frac{\rho(r')}{r}\, dv'$$

$$= -2\pi\gamma\rho \int_b^a r'^2\, dr' \int_0^\pi \frac{\sin\theta}{r}\, d\theta \tag{2.81}$$

According to the law of cosines,

$$r^2 = r'^2 + R^2 - 2r'R\cos\theta \tag{2.82}$$

Since R is a constant, then for a given r' we may differentiate this equation and obtain

$$2r\, dr = 2r'R \sin\theta\, d\theta$$

or,

$$\frac{\sin\theta}{r}\, d\theta = \frac{dr}{r'R} \tag{2.83}$$

Substituting this expression into Eq. (2.81), we have

$$\Phi = -\frac{2\pi\gamma\rho}{R} \int_b^a r'\, dr' \int_{r_{\min}}^{r_{\max}} dr \tag{2.84}$$

The limits on the integral over dr depend upon the location of the point P. If P is *outside* the shell, then

$$\Phi(R > a) = -\frac{2\pi\gamma\rho}{R} \int_b^a r'\, dr' \int_{R-r'}^{R+r'} dr$$

$$= -\frac{4\pi\gamma\rho}{R} \int_b^a r'^2\, dr'$$

$$= -\frac{4}{3}\frac{\pi\gamma\rho}{R}(a^3 - b^3) \tag{2.85}$$

But the mass M of the shell is

$$M = \tfrac{4}{3}\pi\rho(a^3 - b^3) \tag{2.86}$$

so that the potential is

$$\boxed{\Phi(R > a) = -\frac{\gamma M}{R}} \tag{2.87}$$

This important result states that the potential at any point outside a spherically symmetric distribution of matter (shell or solid) is independent of the size of the distribution. Therefore, for the purpose of calculating the potential (or the force), all of the mass can be considered to be concentrated at the center.*

If the field point lies inside the shell, then

$$
\begin{aligned}
\Phi(R < b) &= -\frac{2\pi\gamma\rho}{R}\int_b^a r'\,dr' \int_{r'-R}^{r'+R} dr \\
&= -4\pi\gamma\rho\int_b^a r'\,dr' \\
&= -2\pi\gamma\rho(a^2 - b^2)
\end{aligned}
\tag{2.88}
$$

The potential is therefore constant and independent of position within the shell.

Finally, if we wish to calculate the potential for points *within* the shell, it is only necessary to replace the lower limit of integration in the expression for $\Phi(R < b)$ by the variable R and to replace the upper limit of integration in the expression for $\Phi(R > a)$ by R, and add the results. We find

$$
\begin{aligned}
\Phi(b < R < a) &= -\frac{4\pi\gamma\rho}{3R}(R^3 - b^3) - 2\pi\gamma\rho(a^2 - R^2) \\
&= -4\pi\gamma\rho\left(\frac{a^2}{2} - \frac{b^3}{3R} - \frac{R^2}{6}\right)
\end{aligned}
\tag{2.89}
$$

We see that if $R \to a$, then Eq. (2.89) yields the same result as does Eq. (2.87) for the same limit. Similarly, Eqs. (2.89) and (2.88) produce the same result for the limit $R \to b$. Therefore, the potential is *continuous*. If the potential were not continuous at some point, the gradient of the potential, and hence the force, would be infinite at that point. Since infinite forces do not represent physical reality, we conclude that potential functions must always be continuous.

The magnitude of the field vector $\mathbf{g}$ may be computed from $g = -d\Phi/dR$ for each of the three regions. The results are

$$
\left.
\begin{aligned}
g(R < b) &= 0 \\
g(b < R < a) &= \frac{4\pi\gamma\rho}{3}\left(\frac{b^3}{R^2} - R\right) \\
g(R > a) &= -\gamma\frac{M}{R^2}
\end{aligned}
\right\}
\tag{2.90}
$$

We see that not only is the potential continuous, but so is the field vector

* This result is the key to Newton's gravitation theory. Although Newton had correctly formulated the theory in 1666, he could not justify his method of numerical calculation (in which he considered the Earth and the moon as point masses) until he had proven this theorem (1685). The theory, thus substantiated, was published in the *Principia*, 1687.

(and hence the force). The *derivative* of the field vector, however, is not continuous across the outer and inner surfaces of the shell.

All of these results for the potential and the field vector can be summarized as in Fig. 2-10.

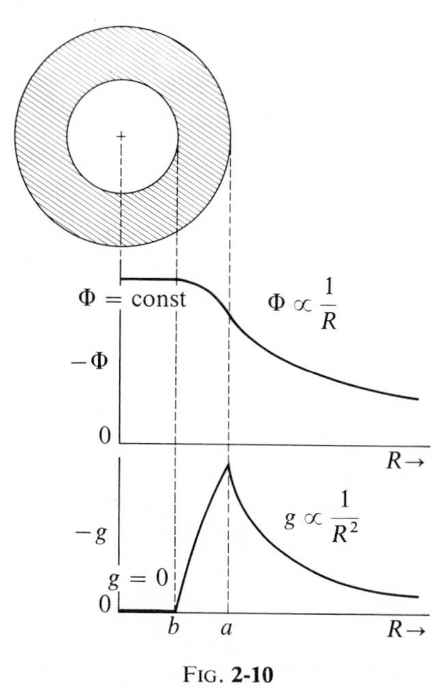

FIG. 2-10

2.11 When Is the Potential Concept Useful?

The use of potentials to describe the effects of "action-at-a-distance" forces is an extremely important and powerful technique. One should not, however, lose sight of the fact that the ultimate justification for the use of a potential is to provide a convenient means for the calculation of the force on a body, for it is the *force* and not the *potential* that is the physically meaningful quantity. Thus, in some problems it may be easier to calculate the force directly, rather than to compute a potential and then take the gradient. The advantage of using the potential method is that the potential is a *scalar* quantity* and one does not need to deal with the added complication of sorting out the components of a vector until the gradient operation is performed. In direct calculations of the force, the components must be carried

* We shall see in Chapter 7 another example of a scalar function from which vector results may be obtained. This is the *Lagrangian function* which, to emphasize the similarity, sometimes (mostly in older treatments) is called the *kinetic potential.*

through the entire computation. It is apparent, then, that some skill is necessary in choosing the particular approach to use. For example, if a problem has a particular symmetry which, from physical considerations, allows one to determine that the force has a certain direction, then the choice of that direction as one of the coordinate directions reduces the vector calculation to a simple scalar calculation. In such a case the direct calculation of the force may be sufficiently straightforward to obviate the necessity of proceeding via the potential method. Every problem for which a force is required must be examined in an attempt to discover the easiest method of computation.

2.12 Limitations of Newtonian Mechanics

In this chapter we have introduced such concepts as *position*, *time*, *momentum*, and *energy*. The implication is that these are all measurable quantities and that they can be specified with any desired accuracy, depending only upon the degree of sophistication of our measuring instruments. Indeed, this implication appears to be verified by our experience with all macroscopic objects. At any given instant of time, for example, we can measure with great precision the position of, say, a planet in its orbit about the Sun. A series of such measurements allows the determination (also with great precision) of the planet's velocity at any given position.

When we attempt to make precision measurements on microscopic objects, however, we find that there is a fundamental limitation to the accuracy of the results. For example, we can conceive of the measurement of the position of an electron by scattering a light photon from the electron. The wave character of the photon precludes an *exact* measurement, and the position of the electron will be determined only within some uncertainty Δx that is related to the *extent* (i.e., the wavelength) of the photon. By the very act of measurement, however, we have induced a change in the state of the electron since the scattering of the photon will impart momentum to the electron. This momentum will be uncertain by an amount Δp. The product $\Delta x\, \Delta p$ is a measure of the precision with which we can simultaneously determine the position and the momentum of the electron; $\Delta x \to 0$, $\Delta p \to 0$ implies a measurement with all imaginable precision. It was shown by Heisenberg in 1927 that this product must always be larger than a certain minimum value.* That is, we cannot simultaneously specify both the position *and* the momentum of the electron with infinite precision, for if $\Delta x \to 0$, then we must have $\Delta p \to \infty$ in order that Heisenberg's *uncertainty principle be* satisfied.

The minimum value of $\Delta x\, \Delta p$ is of the order of 10^{-27} erg-sec. This is

* This result also applies to the measurement of *energy* at a particular *time*, in which case the product of the uncertainties is $\Delta E\, \Delta t$ (which has the same dimensions as $\Delta x\, \Delta p$).

extremely small by macroscopic standards, so that for laboratory-scale objects there is no practical difficulty in performing simultaneous measurements of position and momentum. Therefore, Newton's laws can be applied as if position and momentum were precisely definable. Because of the uncertainty principle, however, Newtonian mechanics cannot be applied to microscopic systems. In order to overcome these fundamental difficulties in the Newtonian system, a new method of dealing with microscopic phenomena was developed, beginning in 1926. The work of Schrödinger, Heisenberg, Born, Dirac, and others subsequently placed this new discipline on a firm foundation. Thus, Newtonian mechanics is found to be perfectly adequate for the description of large-scale phenomena, but the new mechanics (*quantum mechanics*) is necessary for the analysis of processes in the atomic domain. As the size of the system is increased, quantum mechanics goes over into the limiting form of Newtonian mechanics.

In addition to the fundamental limitations of Newtonian mechanics as applied to microscopic objects, there is another inherent difficulty in the Newtonian scheme which rests upon the concept of *time*. In the Newtonian view, time is an *absolute* concept; i.e., it is supposed that it is always possible to determine unambiguously whether two events have occurred simultaneously or whether one has preceded the other. In order to be able always to decide on the time sequence of events, it is necessary that the two observers of the events be in instantaneous communication, either through some system of signals or by establishing two exactly synchronous clocks at the points of observation. But the setting of two clocks into exact synchronism requires the knowledge of the time of transit of a signal *in one direction* from one observer to the other. (This could be accomplished if we already had two synchronous clocks, but this is a circular argument.) When we actually measure signal velocities, however, we always obtain an *average* velocity for propagation in opposite directions, and to devise an experiment to measure the velocity in only *one* direction inevitably leads to the introduction of some new assumption which we cannot verify prior to the experiment.

Now, we know that instantaneous communication by signaling is impossible since interactions between material bodies propagate with finite velocity and an interaction of some sort must occur in order for a signal to be transmitted. The *maximum* velocity with which any signal can be propagated is that of light in free space, viz., $c \cong 3 \times 10^{10}$ cm/sec.*

The difficulties with establishing a time scale between separate points in space leads us to believe that time is, after all, not an absolute concept and that space and time are somehow intimately related. The solution to the dilemma was found in 1904–1905 by Lorentz, Poincaré, and Einstein and is

* More precisely, $c = 2.997925 \pm 0.000001 \times 10^{10}$ cm/sec.

embodied in the *theory of relativity*. Some of the notions of this theory are discussed in Chapter 10.

Newtonian mechanics is therefore subject to fundamental limitations when small distances or high velocities are encountered. There is also a *practical* limitation that occurs when the number of bodies which constitute the system is large. As we shall see in Chapter 8, we cannot obtain a general solution in closed form for the motion of a system of more than two interacting bodies even for the relatively simple case of gravitational interaction. In order to calculate the motion in a three-body system, we must resort to a numerical approximation procedure. Although such a method is in principle capable of any desired accuracy, the labor involved is considerable. The motion in even more complex systems (for example, the system composed of all the major objects in the Solar System) can likewise be computed, but the procedure rapidly becomes too unwieldy to be of much use for any larger system. To calculate the motion of the individual molecules in, say, a cubic centimeter of gas which contains $\approx 10^{19}$ molecules, is clearly out of the question. A successful method of calculating the *average* properties of such systems was developed in the latter part of the 19th century by Boltzmann, Maxwell, Gibbs, Liouville, and others. These procedures allowed the dynamics of systems to be calculated from probability theory and a *statistical mechanics* was evolved. Some comments regarding the formulation of statistical concepts in mechanics will be found in Section 7.15.

Suggested References

The foundations of Newtonian mechanics are presented, for example, by Lindsay and Margenau (Li36, Chapter 3) and by Feather (Fe59, Chapter 8). The standard work, of course, is that of Mach (Ma60). Newton's *Principia* can still be read with profit; translations are available (for example, by F. Cajori, Univ. of California Press, 1946).

There is a plethora of texts which treat Newtonian mechanics at all levels and from many viewpoints. Some of these in the intermediate level category are those by Constant (Co54, Chapters 5–8), Joos and Freeman (Jo50, Chapters 5 and 6), Konopinski (Ko69, Chapter 2), Page (Pa52, Chapter 1), and Symon (Sy60, Chapters 1 and 4). Many examples are to be found in Becker (Be54, Chapters 1, 6, and 8).

At the advanced level, the treatments by Corben and Stehle (Co60, Chapter 2) and by Goldstein (Go50, Chapter 1) are concise and elegant.

General potential theory is concerned with harmonic functions and solutions to Laplace's equation. The standard treatise is that of Kellogg (Ke29), a complete and generally readable account. Also at the advanced level are the brief treatments by Jeffreys and Jeffreys (Je46, Chapter 6) and by Menzel (Me53, Sections 11–19).

At the intermediate level, see the book by Ramsey (Ra40), although it is a bit old-fashioned. A brief, modern discussion is given by Symon (Sy60, Chapter 6).

An interesting historical view of gravitational problems is presented by Feather (Fe59, Chapter 10).

Problems

2-1. Suppose that the force acting on a particle is factorable into one of the following forms:

(a) $F(x_i, t) < f(x_i)g(t)$ (b) $F(\dot{x}_i)g = f(\dot{x}_i)g(t)$ (c) $F(x_i, \dot{x}_i) = f(x_i)g(\dot{x}_i)$

For which cases are the equations of motion integrable?

2-2. A particle of mass m moves under the influence of a force $\mathbf{F}$ on the surface of a sphere of radius R. Write the equation of motion.

2-3. If a projectile is fired from the origin of the coordinate system with an initial velocity v_0 and in a direction making an angle α with the horizontal, calculate the time required for the projectile to cross a line passing through the origin and making an angle $\beta < \alpha$ with the horizontal.

2-4. A projectile is fired with a velocity v_0 such that it passes through two points both a distance h above the horizontal. Show that if the gun is adjusted for maximum range, the separation of the points is

$$d = \frac{v_0}{g}\sqrt{v_0^2 - 4gh}$$

2-5. Consider a projectile fired vertically in a constant gravitational field. For the same initial velocities, compare the times required for the projectile to reach its maximum height

(a) for zero resisting force,

(b) for a resisting force proportional to the instantaneous velocity of the projectile.

2-6. A particle is projected vertically upward in a constant gravitational field with an initial velocity v_0. Show that if there is a retarding force proportional to the square of the instantaneous velocity, then the velocity of the particle when it returns to the initial position is

$$\frac{v_0 v_t}{\sqrt{v_0^2 + v_t^2}}$$

where v_t is the terminal velocity.

2-7. Consider a particle of mass m whose motion starts from rest in a constant gravitational field. If a resisting force proportional to the square of the velocity (i.e., kmv^2) is encountered, show that the distance s through which the particle falls in accelerating from v_0 to v_1 is given by

$$s(v_0 \rightarrow v_1) = \frac{1}{2k} \ln \left[\frac{g - kv_0^2}{g - kv_1^2} \right]$$

2-8. A projectile is fired with an initial velocity v_0. Find the expression for the maximum range when the distance is measured up a slope which makes an angle β with the horizontal. (The projectile is fired from a point on the slope.)

2-9. A particle moves in a medium under the influence of a force equal to $mk(v^3 + a^2 v)$, where k and a are constants. Show that for any value of the initial velocity the particle will never move a distance greater than $\pi/2ka$ and that the particle comes to rest only for $t \rightarrow \infty$.

2-10. A water droplet falling in the atmosphere is spherical in shape. As the droplet passes through a cloud it acquires mass at a rate proportional to its cross-sectional area. Consider a droplet of initial radius r_0 that enters a cloud with a velocity v_0. Assume no resistive force and show

(a) that the radius increases linearly with the time and

(b) that if r_0 is negligibly small then the velocity increases linearly with the time within the cloud.

2-11. A particle of mass m slides down an inclined plane under the influence of gravity. If the motion is resisted by a force $f = kmv^2$, show that the time required to move a distance d after starting from rest is

$$t = \frac{\cosh^{-1}(e^{kd})}{\sqrt{kg \sin \theta}}$$

where θ is the angle of inclination of the plane.

2-12. A particle is projected with an initial velocity v_0 up a slope which makes an angle α with the horizontal. Assume frictionless motion and find the time required for the particle to return to its starting position.

2-13. If a projectile moves in such a way that its distance from the point of projection is always increasing, find the maximum angle above the horizontal with which the particle could have been projected. (Assume no air resistance.)

2-14. A gun fires a projectile of the type to which the curves of Fig. 2-2 apply. The muzzle velocity is 400 ft/sec. Through what angle must the barrel be elevated if it is desired to hit a target on the same horizontal plane as the gun and 3000 ft away? Compare the results with those for the case of no retardation.

2-15. Show directly that the time rate of change of the angular momentum about the origin for a projectile fired from the origin is equal to the moment of force (or torque) about the origin.

2-16. The force of attraction between two particles is given by

$$\mathbf{f}_{12} = k\left[(\mathbf{r}_2 - \mathbf{r}_1) - \frac{r}{v_0}(\dot{\mathbf{r}}_2 - \dot{\mathbf{r}}_1)\right]$$

where k is a constant, v_0 is a constant velocity, and $r \equiv |\mathbf{r}_2 - \mathbf{r}_1|$. Calculate the internal torque for the system; why does this quantity not vanish? Is the system conservative?

2-17. The motion of a charged particle in an electromagnetic field can be obtained from the *Lorentz equation** for the force on a particle in such a field. If the electric field vector is **E** and the magnetic induction vector is **B**, then the force on a particle of mass m which carries a charge e and which has a velocity **v** is given by (in Gaussian units)

$$\mathbf{F} = e\mathbf{E} + \frac{e}{c}\mathbf{v} \times \mathbf{B}$$

where v is the velocity of light and where it is assumed that $v \ll c$.

(a) If there is no electric field and if the particle enters the magnetic field in a direction perpendicular to the lines of magnetic flux, show that the trajectory is a circle with radius

$$r = \frac{cmv}{eB} = \frac{v}{\omega_c}$$

where $\omega_c \equiv eB/mc$ is the *cyclotron frequency*.

(b) Choose the z-axis to lie in the direction of **B** and let the plane containing **E** and **B** be the y-z plane. Thus,

$$\mathbf{B} = B\mathbf{e}_z\,; \qquad \mathbf{E} = E_y\mathbf{e}_y + E_z\mathbf{e}_z$$

* See for example, Marion, *Classical Electromagnetic Radiation* (Ma65b, Section 1.6).

Show that the z-component of the motion is given by

$$z(t) = z_0 + \dot{z}_0 t + \frac{eE_z}{2m} t^2$$

where

$$z(0) \equiv z_0 \qquad \text{and} \qquad \dot{z}(0) \equiv \dot{z}_0$$

(c) Continue the calculation and obtain expressions for $x(t)$ and $y(t)$. Show that the *time averages* of these velocity components are

$$\langle \dot{x} \rangle = \frac{cE_y}{B}; \qquad \langle \dot{y} \rangle = 0$$

(Show that the motion is periodic and then average over one complete period.)

(d) Integrate the velocity equations found in (c) and show that (with an appropriate choice of the initial conditions)

$$x(t) = \frac{A}{\omega_c} \sin \omega_L t + \frac{cE_y}{B} t; \qquad y(t) = \frac{A}{\omega_c} (\cos \omega_L t - 1)$$

These are the parametric equations of a *trochoid*. Sketch the projection of the trajectory on the x-y plane for the cases (i) $A > |cE_y/B|$, (ii) $A < |cE_y/B|$, and (iii) $A = |cE_y/B|$. (The last case yields a *cycloid*.)

2-18. Sketch the equipotential surfaces and the lines of force for two point masses separated by a certain distance. Next, consider one of the masses to have a fictitious mass $-M$. Sketch the equipotential surfaces and lines of force for this case. To what kind of physical situation does this set of equipotentials and field lines apply? (Note that the lines of force have *direction*; indicate this with appropriate arrows.)

2-19. If the field vector is independent of the radial distance within a sphere, find the function which describes the density $\rho = \rho(r)$ of the sphere.

2-20. Under the assumption that air resistance is unimportant, calculate the minimum velocity which a particle must have at the surface of the Earth in order to escape from the Earth's gravitational field. Obtain a numerical value for the result. (This velocity is called the *escape velocity*.)

2-21. A particle is attracted toward a center of force according to the relation $F = -mk^2/x^3$. Show that the time required for the particle to reach the force center from a distance d is d^2/k.

2-22. A particle falls to the Earth starting from rest at a great height. Neglect air resistance and show that it requires approximately $\frac{9}{11}$ of the total time of fall to traverse the first half of the distance.

2-23. Compute directly the gravitational force on a unit mass at a point exterior to a homogeneous sphere of matter.

2-24. Show that the gravitational potential of a thin circular disk of radius a and areal density ρ is given by

$$\Phi(z) = -2\pi\gamma\rho_s[\sqrt{z^2 + a^2} - |z|]$$

for the case that the field point is on the axis of the disk and a distance z away.

2-25. Calculate the gravitational potential due to a thin rod of length l and mass M at a distance R from the center of the rod and in a direction perpendicular to the rod.

2-26. Calculate the gravitational field vector due to a homogeneous cylinder at exterior points on the axis of the cylinder. Perform the calculation

(a) by computing the force directly and
(b) by computing the potential first.

2-27. Calculate the potential due to a thin circular ring of radius a and mass M for points lying in the plane of the ring and exterior to it. The result can be expressed as an elliptic integral.* Assume that the distance from the center of the ring to the field point is large compared with the radius of the ring. Expand the expression for the potential and find the first correction term.

2-28. Find the potential at off-axis points due to a thin circular ring of radius a and mass M. Let R be the distance from the center of the ring to the field point, and let θ be the angle between the line connecting the center of the ring with the field point and the axis of the ring. Assume $R \gg a$ so that terms of order $(a/R)^3$ and higher may be neglected.

2-29. Consider a massive body of arbitrary shape and a spherical surface which is exterior to the body. Show that the average value of the potential due to the body taken over the spherical surface is equal to the value of the potential at the center of the sphere.

* See Appendix E.4 for a list of some elliptic integrals.

CHAPTER 3

Linear Oscillations

3.1 Introduction

We begin by considering the oscillatory motion of a particle that is constrained to move in one dimension. We assume that there exists a position of stable equilibrium for the particle and we designate this point as the origin. If the particle is displaced from the origin (in either direction), a certain force tends to restore the particle to its original position. This force is in general some complicated function of the displacement and perhaps of the particle's velocity or even of some higher time derivative of the position coordinate. We consider here only cases in which the restoring force F is a function only of the displacement: $F = F(x)$.

We shall assume that the function $F(x)$ which describes the restoring force possesses continuous derivatives of all orders so that the function can be expanded in a Taylor series:

$$F(x) = F_0 + x\left(\frac{dF}{dx}\right)_0 + \frac{1}{2}x^2\left(\frac{d^2F}{dx^2}\right)_0 + \cdots \tag{3.1}$$

where F_0 is the value of $F(x)$ at the origin ($x = 0$), and $(d^nF/dx^n)_0$ is the value of the nth derivative at the origin. Since the origin is defined to be the equilibrium point, F_0 must vanish, Then, if we confine our attention to displacements of the particle that are sufficiently small, we can neglect all terms involving x^2 and higher powers of x. We have, therefore, the approximate relation:

$$\boxed{F(x) = -kx} \tag{3.2}$$

where we have substituted $k \equiv -(dF/dx)_0$. Since the restoring force is always directed toward the equilibrium position (the origin), the derivative $(dF/dx)_0$ is negative and therefore k is a positive constant. Only the first power of the displacement occurs in $F(x)$, so that the restoring force in this approximation is a *linear* force.

Physical systems that can be described in terms of Eq. (3.2) are said to obey *Hooke's Law.** One of the classes of physical processes which can be treated by applying Hooke's Law is that involving elastic deformations. As long as the displacements are small and the elastic limits are not exceeded, a linear restoring force can be used for problems of stretched springs, elastic springs, bending beams, etc. But, it must be emphasized that such calculations are only approximate, since essentially every real restoring force in Nature is more complicated than the simple Hooke's Law force. We must always bear in mind the fact that linear forces are only useful approximations and their validity is limited to cases in which the amplitudes of the oscillations are small. (See, however, Problem 3-8.) The more realistic (and more complicated) case of nonlinear oscillations will be discussed in Chapter 5.

3.2 The Simple Harmonic Oscillator

The equation of motion for the simple harmonic oscillator may be obtained by substituting the Hooke's Law force into the Newtonian equation, $F = ma$. Thus,

$$-kx = m\ddot{x} \tag{3.3}$$

If we define

$$\omega_0^2 \equiv k/m \tag{3.4}$$

Eq. (3.3) becomes

$$\boxed{\ddot{x} + \omega_0^2 x = 0} \tag{3.5}$$

* Robert Hooke (1635–1703). The equivalent of this force law was originally announced by Hooke in 1676 in the form of a Latin cryptogram: *CEIIINOSSSTTUV, ut tensio sic vis.*

According to the results of Appendix C, the solution of this equation can be expressed in either of the forms

$$x(t) = A \sin(\omega_0 t - \delta) \tag{3.6a}$$

$$x(t) = A \cos(\omega_0 t - \phi) \tag{3.6b}$$

where the phases δ and ϕ differ, of course, by $\pi/2$. (An alteration of the phase angle corresponds to a change of the instant that we designate $t = 0$, the origin of the time scale.) Equations (3.6) exhibit the well-known sinusoidal behavior of the displacement of the simple harmonic oscillator.

The relationship between the total energy of the oscillator and the amplitude of its motion can be obtained as follows. Using Eq. (3.6a) for $x(t)$, we find for the kinetic energy,

$$\begin{aligned} T = \tfrac{1}{2}m\dot{x}^2 &= \tfrac{1}{2}m\omega_0^2 A^2 \cos^2(\omega_0 t - \delta) \\ &= \tfrac{1}{2}kA^2 \cos^2(\omega_0 t - \delta) \end{aligned} \tag{3.7}$$

The potential energy may be obtained by calculating the work required to displace the particle a distance x. The incremental amount of work dW that is necessary to move the particle by an amount dx against the restoring force F is

$$dW = -F\,dx = kx\,dx \tag{3.8}$$

Integrating from 0 to x and setting the work done on the particle equal to the potential energy, we have

$$U = \tfrac{1}{2}kx^2 \tag{3.9}$$

Then,

$$U = \tfrac{1}{2}kA^2 \sin^2(\omega_0 t - \delta) \tag{3.10}$$

Combining the expressions for T and U to find the total energy E, we have

$$\boxed{E = T + U = \tfrac{1}{2}kA^2} \tag{3.11}$$

so that the total energy is proportional to the *square of the amplitude*; this is a general result for linear systems. Notice also that E is independent of the time; that is, energy is conserved. (Energy conservation is, of course, guaranteed because we have been considering a system without frictional losses.)

The period τ_0 of the motion is defined to be the time interval between successive repetitions of the particle's position and direction of motion. This occurs when the argument of the sine in Eq. (3.6a) increases by 2π:

$$\omega_0 \tau_0 = 2\pi \tag{3.12}$$

or,

$$\tau_0 = 2\pi\sqrt{\frac{m}{k}} \tag{3.13}$$

From this expression, as well as from Eq. (3.6a), it is clear that ω_0 represents the *angular frequency* of the motion which is related to the frequency ν_0 by*

$$\omega_0 = 2\pi\nu_0 = \sqrt{\frac{k}{m}} \tag{3.14}$$

$$\nu_0 = \frac{1}{\tau_0} = \frac{1}{2\pi}\sqrt{\frac{k}{m}} \tag{3.15}$$

Note that the period of the simple harmonic oscillator is independent of the amplitude (or total energy); a system which exhibits this property is said to be *isochronous*.

3.3 Phase Diagrams

The state of motion of a one-dimensional oscillator, such as that discussed in the preceding sections, will be completely specified as a function of time if *two* quantities are given—the displacement $x(t)$ and the velocity $\dot{x}(t)$. (*Two* quantities are needed because the differential equation for the motion is of *second* order.) The quantities $x(t)$ and $\dot{x}(t)$ may be considered to be the coordinates of a point in a two-dimensional space, called *phase space*. (In two dimensions, the phase space is, of course, a phase *plane*, but for a general oscillator with n degrees of freedom the phase space is a $2n$-dimensional space.) As the time varies, the point $P(x, \dot{x})$ which describes the state of the oscillating particle will move along a certain phase path in the phase plane. For different initial conditions of the oscillator, the motion will be described by different phase paths. Any given path represents the complete time history of the oscillator for a certain set of initial conditions. The totality of all possible phase paths constitutes the *phase portrait* or the *phase diagram* of the oscillator.†

* Henceforth we shall adhere to the convention of denoting angular frequencies by ω (units: radians per unit time) and frequencies by ν (units: vibrations per unit time or Hertz, Hz). Usually, ω will be referred to as a "frequency" for brevity, although "angular frequency" is to be understood.

† These considerations are not restricted to oscillating particles or oscillating systems. The concept of phase space is applied extensively in various fields of physics, particularly in statistical mechanics.

According to the results of the preceding section, we have, for the simple harmonic oscillator,

$$x(t) = A \sin(\omega_0 t - \delta) \tag{3.16a}$$

$$\dot{x}(t) = A\omega_0 \cos(\omega_0 t - \delta) \tag{3.16b}$$

If we eliminate t from these equations, we find for the equation of the path

$$\frac{x^2}{A^2} + \frac{\dot{x}^2}{A^2\omega_0^2} = 1 \tag{3.17}$$

This equation represents a family of ellipses,* several of which are shown in Fig. 3-1. Now, we know that the total energy E of the oscillator is $\frac{1}{2}kA^2$ [Eq. (3.11)], and since $\omega_0^2 = k/m$, Eq. (3.17) can be written as

$$\frac{x^2}{2E/k} + \frac{\dot{x}^2}{2E/m} = 1 \tag{3.18}$$

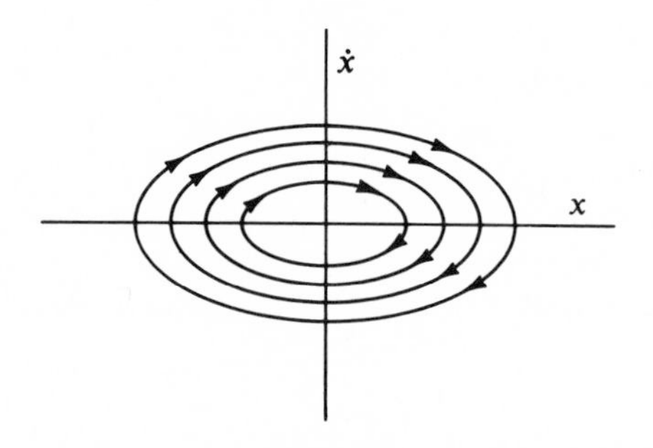

FIG. 3-1

That is, each phase path corresponds to a definite total energy of the oscillator. This result is expected because the system is conservative (i.e., $E = \text{const.}$).

No two phase paths of the oscillator can cross. If they could cross, this would imply that for a given set of initial conditions, $x(t_0)$, $\dot{x}(t_0)$, (i.e., the coordinates of the crossing point), the motion could proceed along different phase paths. But this is impossible since the solution of the differential equation is *unique*.

If the coordinate axes of the phase plane are chosen as in Fig. 3-1, the motion of the *representative point* $P(x, \dot{x})$ will always be in a clockwise direction, since for $x > 0$ the velocity $\dot{x}$ is always decreasing and for $x < 0$ the velocity is always increasing.

* The ordinate of the phase plane is sometimes chosen to be $\dot{x}/\omega_0$ instead of $\dot{x}$; the phase paths are then *circles*.

In order to obtain Eqs. (3.16) for $x(t)$ and $\dot{x}(t)$, it is necessary to integrate a second-order differential equation:

$$\frac{d^2x}{dt^2} + \omega_0^2 x = 0 \tag{3.19}$$

The equation for the phase path, however, can be obtained by a simpler procedure, since Eq. (3.19) can be replaced by the pair of equations

$$\frac{dx}{dt} = \dot{x}; \qquad \frac{d\dot{x}}{dt} = -\omega_0^2 x \tag{3.20}$$

If we divide the second of these equations by the first, we obtain

$$\frac{d\dot{x}}{dx} = -\omega_0^2 \frac{x}{\dot{x}} \tag{3.21}$$

This is a *first-order* differential equation for $\dot{x} = \dot{x}(x)$, the solution to which is just Eq. (3.17). For the case of the simple harmonic oscillator, there is, of course, no difficulty in obtaining the general solution for the motion by solving the second-order equation, but in more complicated situations it is sometimes considerably easier to find directly the equation of the phase path $\dot{x} = \dot{x}(x)$ without proceeding through the calculation of $x(t)$.

Although it is always possible, for any system, to write an equation for $d\dot{x}/dx$ of the form

$$\frac{d\dot{x}}{dx} = f(x, \dot{x}) \tag{3.22}$$

this equation may not be integrable in terms of elementary functions. In such a case, one may proceed to construct the phase diagram by the *method of isoclines*. The quantity $d\dot{x}/dx$ represents the slope of the phase path, $\dot{x} = \dot{x}(x)$. Therefore, if ξ is some constant, the equation

$$f(x, \dot{x}) = \xi \tag{3.23}$$

is the equation of a line that intersects every phase path at a point where the slope is equal to ξ. Thus, by choosing a range of values for ξ, solving Eq. (3.23), and plotting the corresponding lines (the isoclines), it is possible to construct the desired phase diagram. For the simple harmonic oscillator, Eq. (3.23) becomes

$$-\omega_0^2 \frac{x}{\dot{x}} = \xi \tag{3.24}$$

so that the isoclines are all straight lines passing through the origin. Several of these isoclines are shown in Fig. 3-2. Clearly, the curves which intersect the isoclines with the appropriate slopes are all ellipses.

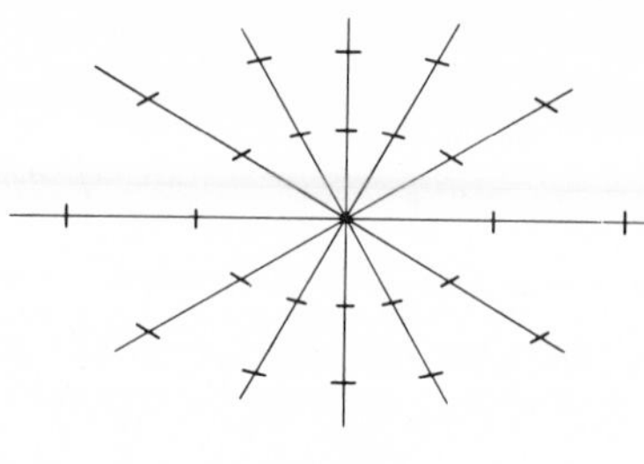

FIG. 3-2

3.4 Harmonic Oscillations in Two Dimensions

We next consider the motion of a particle which is allowed two degrees of freedom. We take the restoring force to be proportional to the distance of the particle from a force center located at the origin and to be directed toward the origin:

$$\mathbf{F} = -k\mathbf{r} \tag{3.25}$$

which can be resolved in polar coordinates into the components

$$\left.\begin{aligned} F_x &= -kr\cos\theta = -kx \\ F_y &= -kr\sin\theta = -ky \end{aligned}\right\} \tag{3.26}$$

and the equations of motion are

$$\left.\begin{aligned} \ddot{x} + \omega_0^2 x &= 0 \\ \ddot{y} + \omega_0^2 y &= 0 \end{aligned}\right\} \tag{3.27}$$

where, as before, $\omega_0^2 = k/m$. The solutions are

$$\left.\begin{aligned} x(t) &= A\cos(\omega_0 t - \alpha) \\ y(t) &= B\cos(\omega_0 t - \beta) \end{aligned}\right\} \tag{3.28}$$

Thus, the motion is one of simple harmonic oscillation in each of the two directions, both oscillations having the same frequency but possibly differing in amplitude and in phase. We can obtain the equation for the path of the particle by eliminating the time t between the two equations (3.28). First, we write

$$\begin{aligned} y(t) &= B\cos[\omega_0 t - \alpha + (a - \beta)] \\ &= B\cos(\omega_0 t - \alpha)\cos(\alpha - \beta) - B\sin(\omega_0 t - \alpha)\sin(\alpha - \beta) \end{aligned} \tag{3.29}$$

Defining $\delta \equiv \alpha - \beta$ and noting that $\cos(\omega_0 t - \alpha) = x/A$, we have

$$y = \frac{B}{A} x \cos \delta - B\sqrt{1 - \left(\frac{x^2}{A^2}\right)} \sin \delta$$

or,

$$Ay - Bx \cos \delta = -B\sqrt{A^2 - x^2} \sin \delta \tag{3.30}$$

and upon squaring, this becomes

$$A^2y^2 - 2ABxy \cos \delta + B^2x^2 \cos^2 \delta = A^2B^2 \sin^2 \delta - B^2x^2 \sin^2 \delta$$

so that

$$B^2x^2 - 2ABxy \cos \delta + A^2y^2 = A^2B^2 \sin^2 \delta \tag{3.31}$$

If δ is set equal to $\pm\pi/2m$, this equation reduces to the easily recognized equation for an ellipse:

$$\frac{x^2}{A^2} + \frac{y^2}{B^2} = 1; \qquad \delta = \pm\pi/2 \tag{3.32}$$

If the amplitudes are equal, $A = B$, and if $\delta = \pm\pi/2$, we have the special case of circular motion:

$$x^2 + y^2 = A^2; \qquad A = B; \qquad \delta = \pm\pi/2 \tag{3.33}$$

Another special case results if the phase δ vanishes; then we have

$$B^2x^2 - 2ABxy + A^2y^2 = 0; \qquad \delta = 0$$

Factoring,

$$(Bx - Ay)^2 = 0$$

which is the equation of a straight line:

$$y = \frac{B}{A} x; \qquad \delta = 0 \tag{3.34}$$

Similarly, the phase $\delta = \pm\pi$ yields the straight line of opposite slope:

$$y = -\frac{B}{A} x; \qquad \delta = \pm\pi \tag{3.35}$$

The curves of Fig. 3-3 illustrate Eq. (3.31) for the case $A = B$; $\delta = 90°$ or 270° yields a circle and $\delta = 180°$ or 360° (0°) yields a straight line, while all other values of δ yield ellipses.

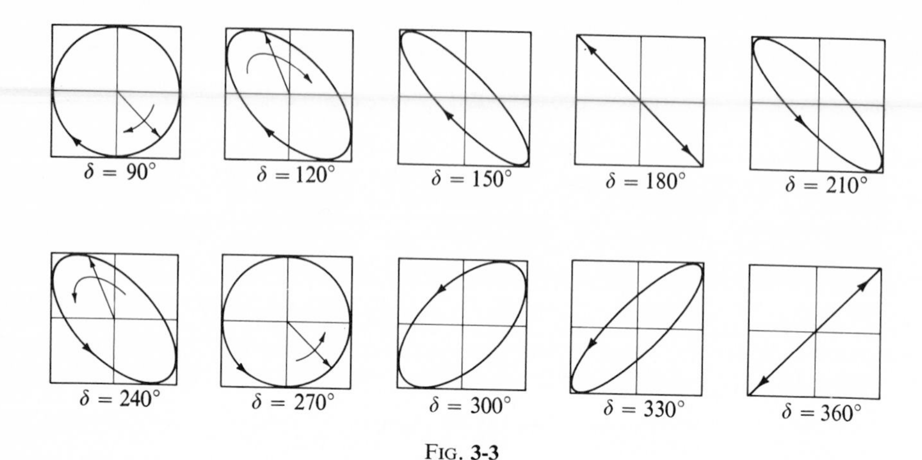

FIG. 3-3

In the *general* case of two-dimensional oscillations, the frequencies for the motions in the x- and y-directions need not be equal, so that Eq. (3.28) becomes

$$\left.\begin{aligned} x(t) &= A\cos(\omega_x t - \alpha) \\ y(t) &= B\cos(\omega_y t - \beta) \end{aligned}\right\} \tag{3.36}$$

The path of the motion is no longer an ellipse, but is a *Lissajous curve*.* Such a curve will be *closed* if the motion repeats itself at regular intervals of time. This will be possible only if the frequencies ω_x and ω_y are *commensurable*, i.e., if ω_x/ω_y is a rational fraction. Such a case is shown in Fig. 3-4, in which $\omega_y = \frac{3}{4}\omega_x$ (also, $A = B$ and $\alpha = \beta$). In the event that the ratio of the

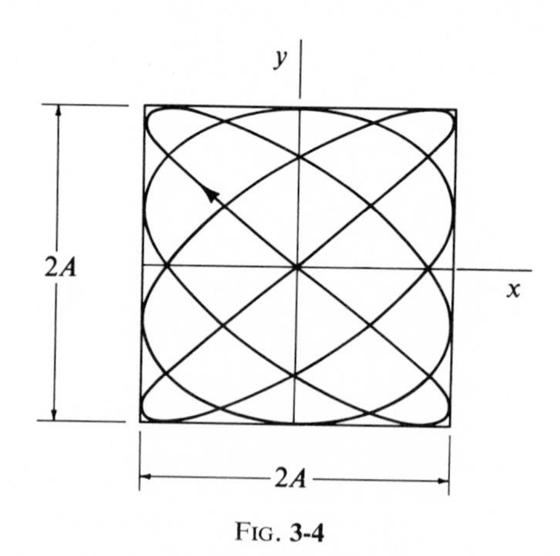

FIG. 3-4

* First demonstrated in 1857 by Jules Lissajous (1822–1880).

frequencies is not a rational fraction, the curve will be *open*; that is, the moving particle will never pass twice through the same point with the same velocity. In such a case it can be shown that, after a sufficiently long time has elapsed, the curve will pass arbitrarily close to any given point lying within the rectangle $2A \times 2B$ and will therefore "fill" the rectangle.*

It is interesting to note that the two-dimensional oscillator is an example of a system in which an infinitesimal change can result in a qualitatively different type of motion. Although the motion will be along a closed path in the event that the two frequencies are commensurable, if the frequency ratio deviates from a rational fraction by even an infinitesimal amount, then the path will no longer be closed and it will "fill" the rectangle. In order for the path to be closed, the frequency ratio must be known to be a rational fraction with infinite precision.

In the event that the frequencies for the motions in the x- and y-directions are different, then the shape of the resulting Lissajous curve† is strongly dependent upon the phase difference $\delta \equiv \alpha - \beta$. Figure 3-5 shows the results for the case $\omega_y = 2\omega_x$ for phase differences of 0, $\pi/3$, and $\pi/2$.

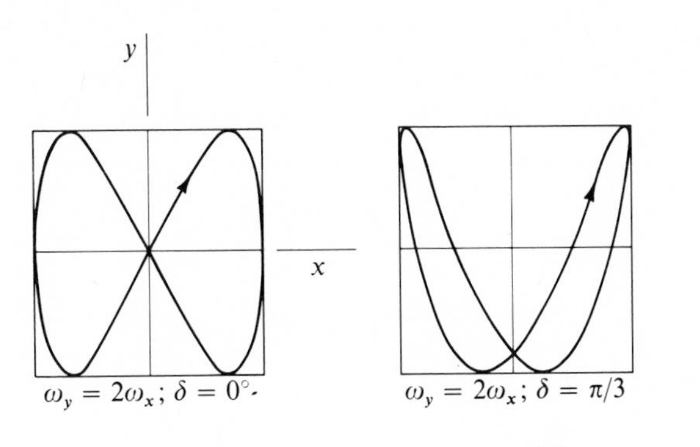

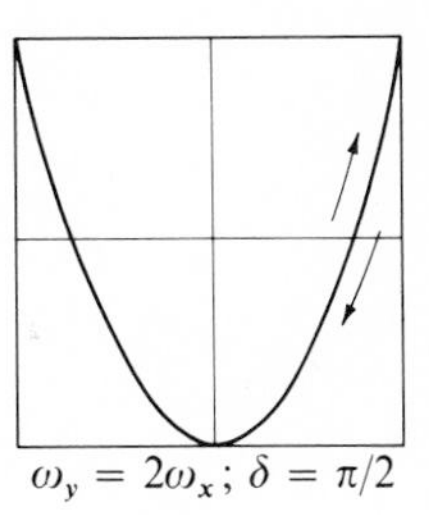

FIG. 3-5

3.5 Damped Oscillations

The motion represented by the simple harmonic oscillator is termed a *free oscillation*; once set into oscillation, the motion would never cease. This is, of course, an oversimplification of the actual physical case in which dissipative or frictional forces would eventually damp the motion to the point that the oscillations would no longer occur. It is possible to analyze the motion

* A proof is given, for example, by Haag (Ha62, p. 36).

† Sketches of the Lissajous curves for several different frequency ratios and phase differences may be found, for example, in Sears (Se58, p. 313).

in such a case by incorporating into the differential equation a term which represents the damping force. Now, it does not seem reasonable that the damping force should in general depend on the displacement, but it could be a function of the velocity or perhaps of some higher time derivative of the displacement. It is frequently assumed that the damping force is a linear function of the velocity,* $\mathbf{F}_d = \alpha\mathbf{v}$. We will consider here only one-dimensional damped oscillations so that we can represent the damping term by $-b\dot{x}$. The parameter b must be *positive* in order that the force indeed be *resisting*. (A force $-b\dot{x}$ with $b < 0$ would be in the *same* direction as the restoring force rather than opposed to it.)

Thus, if a particle of mass m moves under the combined influence of a linear restoring force $-kx$ and a resisting force $-b\dot{x}$ the differential equation which describes the motion is

$$m\ddot{x} + b\dot{x} + kx = 0 \tag{3.37}$$

which we can write as

$$\boxed{\ddot{x} + 2\beta\dot{x} + \omega_0^2 x = 0} \tag{3.38}$$

Here, $\beta \equiv b/2m$ is the *damping parameter* and $\omega_0 = \sqrt{k/m}$ is the characteristic frequency in the absence of damping. The roots of the auxiliary equation are [cf. Eq. (C.8), Appendix C]:

$$\left.\begin{aligned} r_1 &= -\beta + \sqrt{\beta^2 - \omega_0^2} \\ r_2 &= -\beta - \sqrt{\beta^2 - \omega_0^2} \end{aligned}\right\} \tag{3.39}$$

Therefore, the general solution of Eq. (3.38) is

$$\boxed{x(t) = e^{-\beta t}[A_1 \exp(\sqrt{\beta^2 - \omega_0^2}\, t) + A_2 \exp(-\sqrt{\beta^2 - \omega_0^2}\, t)]} \tag{3.40}$$

There are three general cases of interest:

(a) Underdamping: $\omega_0^2 > \beta^2$

(b) Critical damping : $\omega_0^2 = \beta^2$

(c) Overdamping: $\omega_0^2 < \beta^2$

As we will see, only the case of underdamping results in oscillatory motion. These three cases will be discussed separately.

* See Section 2.4 for a discussion of the dependence of resisting forces on the velocity.

(a) UNDERDAMPED MOTION. For this case it is convenient to define

$$\omega_1^2 \equiv \omega_0^2 - \beta^2 \tag{3.41}$$

where $\omega_1^2 > 0$; then the exponents in the brackets of Eq. (3.40) are imaginary and the solution becomes

$$x(t) = e^{-\beta t}[A_1 e^{i\omega_1 t} + A_2 e^{-i\omega_1 t}] \tag{3.42}$$

This can be rewritten as*

$$x(t) = Ae^{-\beta t}\cos(\omega_1 t - \delta) \tag{3.43}$$

We call the quantity ω_1 the *frequency* of the damped oscillator. Strictly speaking, it is not possible to define a frequency when damping is present since the motion is not periodic; i.e., the oscillator never passes twice through a given point with the same velocity. If the damping is small, then

$$\omega_1 = \sqrt{\omega_0^2 - \beta^2} \cong \omega_0$$

so that the term "frequency" may be used, but the meaning is not precise unless $\beta = 0$. Nevertheless, for simplicity, we will refer to ω_1 as the "frequency" of the damped oscillator, and we note that this quantity is *less* than the frequency of the oscillator in the absence of damping.

The maximum amplitude of the motion of the damped oscillator decreases with time because of the factor $\exp(-\beta t)$, where $\beta > 0$, and the envelope of the displacement *versus* time curve is given by

$$x_{en} = \pm Ae^{-\beta t} \tag{3.44}$$

This envelope as well as the displacement curve is shown in Fig. 3-6 for the case $\delta = 0$. Also shown in this figure is the sinusoidal curve for undamped motion ($\beta = 0$). It is evident from a comparison of the two curves that the

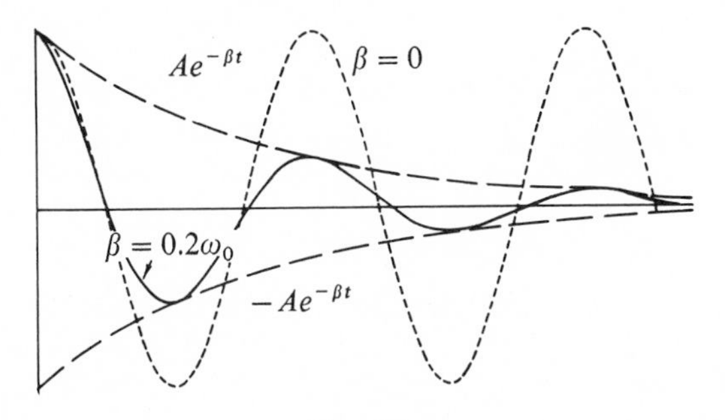

FIG. 3-6

* See Exercise B.6, Appendix B.

frequency for the damped case is *less* (i.e., that the period is *longer*) than that for the undamped case.

The ratio of the amplitudes of the oscillation at two successive maxima is

$$\frac{Ae^{-\beta T}}{Ae^{-\beta(T+\tau_1)}} = e^{\beta\tau_1} \tag{3.45}$$

where the first of any pair of maxima occurs at $t = T$ and where $\tau_1 = 2\pi/\omega_1$. The quantity $\exp(\beta\tau_1)$ is called the *decrement* of the motion; the logarithm of $\exp(\beta\tau_1)$, viz., $\beta\tau_1$, is known as the *logarithmic decrement* of the motion.

In contrast to the simple harmonic oscillator discussed previously, the energy of the damped oscillator is not constant in time. Rather, energy is continually given up to the damping medium and dissipated as heat (or, perhaps, as radiation in the form of fluid waves). The rate of energy loss is proportional to the square of the velocity (see Problem 3-11), so the decrease of

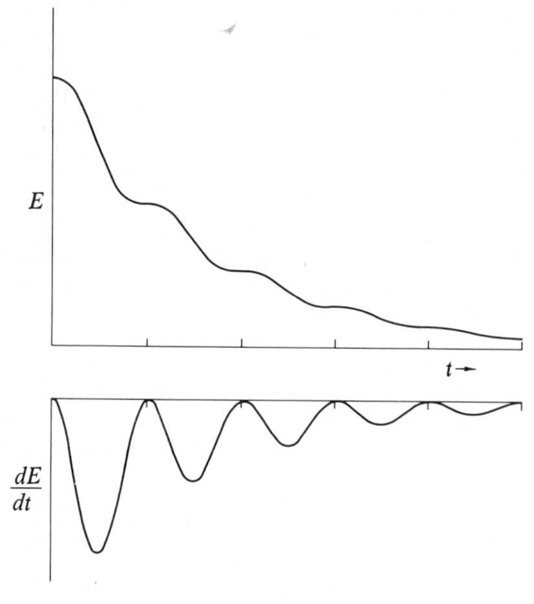

FIG. 3-7

energy does not take place uniformly. The loss rate will be a maximum when the particle attains its maximum velocity near (but not exactly at) the equilibrium position and it will instantaneously vanish when the particle is at maximum amplitude and has zero velocity. The total energy and the rate of energy loss for the damped oscillator are shown in Fig. 3-7.

The phase diagram for the damped oscillator can be constructed as follows. First, we write the expressions for the displacement and the velocity:

$$\left.\begin{aligned} x(t) &= Ae^{-\beta t}\cos(\omega_1 t - \delta) \\ \dot{x}(t) &= -Ae^{-\beta t}[\beta\cos(\omega_1 t - \delta) + \omega_1 \sin(\omega_1 t - \delta)] \end{aligned}\right\} \tag{3.46}$$

These equations can be converted into a more easily recognized form by introducing a change of variables according to the following linear transformations:

$$u = \omega_1 x; \qquad w = \beta x + \dot{x} \tag{3.47}$$

Then,

$$\left.\begin{aligned} u &= \omega_1 A e^{-\beta t}\cos(\omega_1 t - \delta) \\ w &= -\omega_1 A e^{-\beta t}\sin(\omega_1 t - \delta) \end{aligned}\right\} \tag{3.48}$$

If we represent u and w in polar coordinates (see Fig. 3-8), then

$$\rho = \sqrt{u^2 - w^2}; \qquad \varphi = \omega_1 t \tag{3.49}$$

Thus,

$$\rho = \omega_1 A e^{-(\beta/\omega_1)\varphi} \tag{3.50}$$

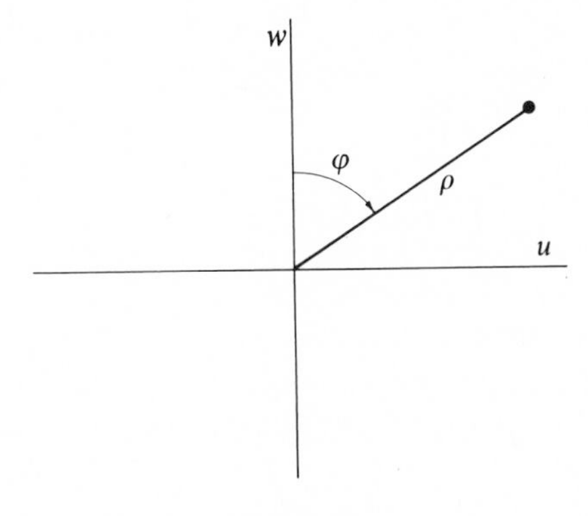

FIG. 3-8

which is the equation of a logarithmic spiral. Since the transformation from x, $\dot{x}$, to u, w is a linear transformation, the phase path has basically the same shape in the phase plane as in the u-w plane. Figure 3-9 shows a spiral phase path of the underdamped oscillator. The continually decreasing magnitude of the radius vector for a representative point in the phase plane always indicates damped motion of the oscillator.

(b) CRITICALLY DAMPED MOTION. In the event that the damping force is sufficiently large (i.e., if $\beta^2 > \omega_0^2$), the system is prevented from undergoing oscillatory motion. If there is zero initial velocity, the displacement decreases monotonically from its initial value to the equilibrium position ($x = 0$).

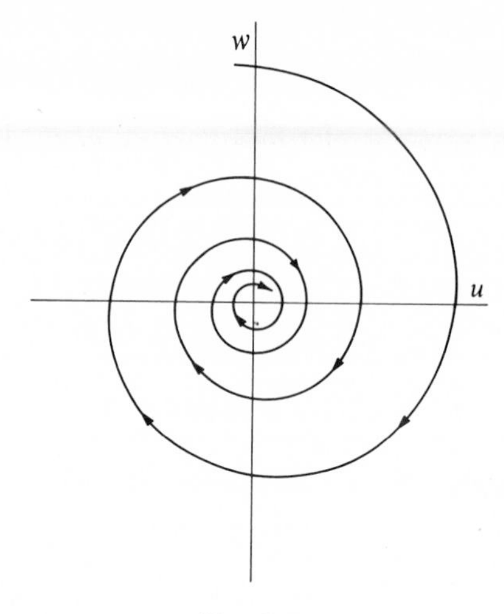

FIG. 3-9

The case of *critical damping* occurs when β^2 is just equal to ω_0^2. The roots of the auxiliary equation are then equal and the function x must be written as [cf. Eq. (C.11), Appendix C]:

$$x(t) = (A + Bt)e^{-\beta t} \tag{3.51}$$

The displacement curve for critical damping is shown in Fig. 3-10 for the case in which the initial velocity is zero. For a given set of initial conditions, a critically damped oscillator will approach equilibrium at a rate more rapid than that for either an overdamped or an underdamped oscillator. This fact is of importance in the design of certain practical oscillatory systems (e.g., galvanometers) when it is desired that the system return to equilibrium as rapidly as possible.

(c) OVERDAMPED MOTION. If the damping parameter β is even larger than ω_0, then overdamping results. Since $\beta^2 > \omega_0^2$, the exponents in the brackets of Eq. (3.28) become real quantities:

$$x(t) = e^{-\beta t}[A_1 e^{\omega_2 t} + A_2 e^{-\omega_2 t}] \tag{3.52}$$

where

$$\omega_2 = \sqrt{\beta^2 - \omega_0^2} \tag{3.53}$$

Note that ω_2 does not represent a "frequency" since the motion is not periodic; the displacement asymptotically approaches the equilibrium position as shown in Fig. 3-10.

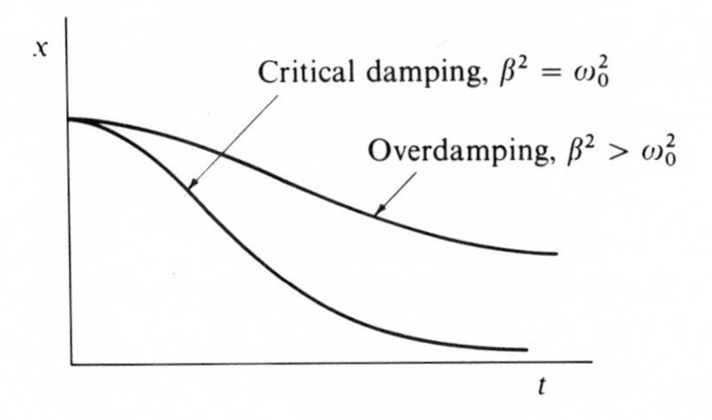

FIG. 3-10

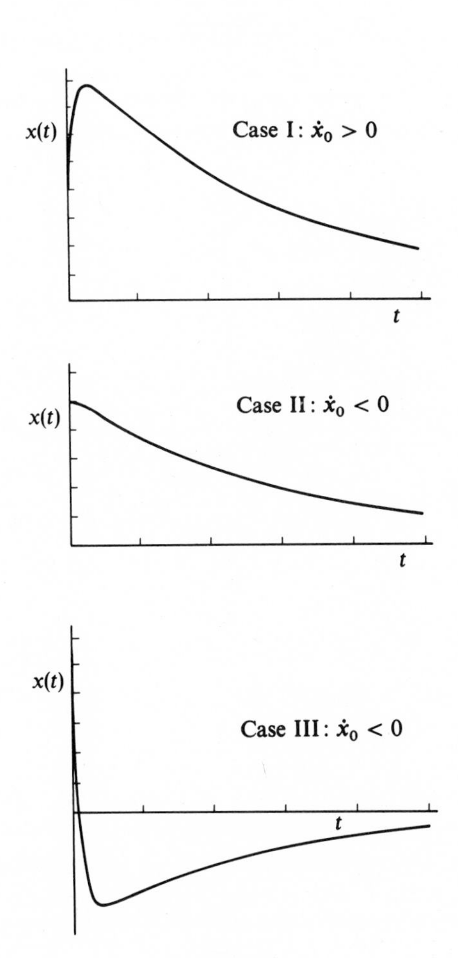

FIG. 3-11

The case of overdamping results in a non-oscillatory asymptotic decrease of the amplitude to zero. But depending on the initial value of the velocity, there may be a change of sign of x before the displacement approaches zero. If we limit our considerations to initial positive displacements, $x(0) \equiv x_0 > 0$, there are three cases of interest for the initial velocity $\dot{x}(0) \equiv \dot{x}_0$:

I. $\dot{x}_0 > 0$, so that $x(t)$ reaches a maximum at some $t > 0$ before approaching zero.

II. $\dot{x}_0 < 0$, with $x(t)$ monotonically approaching zero.

III. $\dot{x}_0 < 0$, but sufficiently large so that $x(t)$ changes sign, reaches a minimum value, and then approaches zero.

These three cases are illustrated in Fig. 3-11. The phase plane is divided into six regions (three for $x_0 > 0$ and three for $x_0 < 0$), depending on which of the above three initial conditions apply for the velocity. Figure 3-12 shows some typical phase paths for each of these regions. It can be shown (see

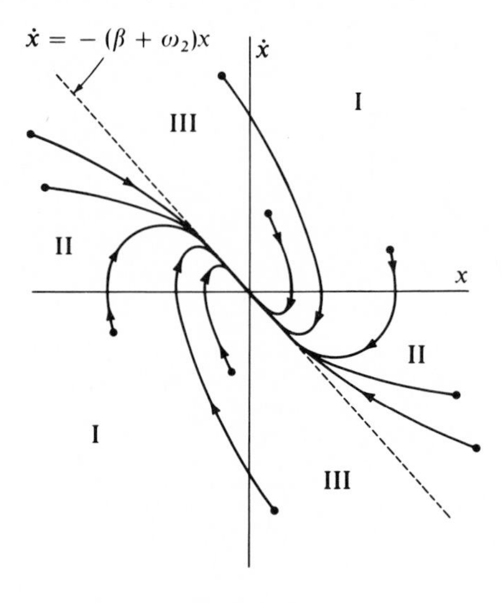

FIG. 3-12

Problem 3-15) that the line $\dot{x} = -(\beta + \omega_2)x$ separates the regions II and III. It is also clear that the phase path for the case of zero initial velocity starts on the x-axis at the appropriate value of x_0 and then proceeds through region II to the origin. The case of critical damping is to be associated with the curves of Fig. 3-12 rather than with the spiral path of Fig. 3-9, since the motion of a critically damped system is not oscillatory.

Phase diagrams for some nonlinear systems are discussed in Chapter 5.

3.6 Electrical Oscillations

The discussion of oscillatory phenomena thus far has been concerned exclusively with the motion of a particle. The results which we have obtained, however, are not restricted to this simple situation. There are many oscillatory *systems* which behave in much the same manner as does a single particle. A prime example is found in the oscillations of an electrical circuit; the equations which describe such oscillations have direct mechanical analogs. Indeed, because of its great practical importance, the electrical case has been so thoroughly investigated that the situation is frequently reversed and mechanical vibrations are analyzed in terms of the "equivalent electrical circuit."

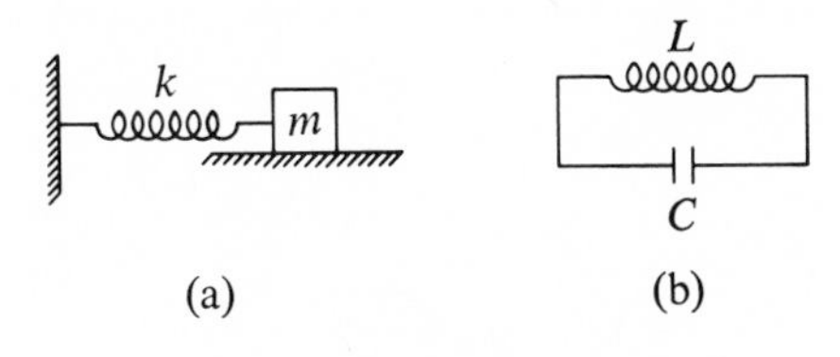

FIG. 3-13

Consider the simple mechanical oscillator shown in Fig. 3-13a, where the mass m slides on a frictionless platform. We know that the equation of motion is

$$m\ddot{x} + kx = 0 \tag{3.54}$$

and that the oscillation frequency is given by

$$\omega_0 = \sqrt{\frac{k}{m}} \tag{3.55}$$

Next, consider the electrical circuit shown in Fig. 3-13b. At some instant of time t the charge on the capacitor C is $q(t)$ and the current flowing through the inductor L is $I(t) = \dot{q}$. Application of Kirchhoff's equation to this circuit gives for the voltage drops around the circuit,

$$L\frac{dI}{dt} + \frac{1}{C}\int I\,dt = 0 \tag{3.56}$$

or, in terms of q,

$$L\ddot{q} + \frac{1}{C}q = 0 \tag{3.57}$$

This equation is of exactly the same form as Eq. (3.54); hence, the solution is

$$q(t) = q_0 \cos \omega_0 t \tag{3.58}$$

where the frequency is

$$\omega_0 = \frac{1}{\sqrt{LC}} \tag{3.59}$$

and where we have set the phase equal to zero by stipulating that $q(t=0) = q_0$ and $I(t=0) = 0$.

By comparing the terms in Eqs. (3.54) and (3.57), we see that the electric analog of mass (or inertia) is *inductance* and that the *compliance* of the spring, represented by the reciprocal of the spring constant k, is to be identified with the *capacitance* C. Altogether, we have

$$\begin{aligned} m &\rightarrow L, \qquad x \rightarrow q \\ \frac{1}{k} &\rightarrow C, \qquad \dot{x} \rightarrow I \end{aligned}$$

Differentiating the expression for $q(t)$, we find

$$\dot{q} = I(t) = -\omega_0 q_0 \sin \omega_0 t \tag{3.60}$$

Upon squaring $q(t)$ and $I(t)$, we can write

$$\frac{1}{2} LI^2 + \frac{1}{2}\frac{q^2}{C} = \frac{1}{2}\frac{q_0^2}{C} = \text{const.} \tag{3.61}$$

The term $\frac{1}{2}LI^2$ represents the energy stored in the inductor (and corresponds to mechanical kinetic energy), whereas the term $\frac{1}{2}(q^2/C)$ represents the energy stored in the capacitor (and corresponds to mechanical potential energy). The sum of these two energies is constant, indicating that the system is conservative. We will see presently that an electrical circuit can be conservative only if it does not contain resistance (an ideal situation and one that is unrealistic from a practical point of view).

The mass-spring combination illustrated in Fig. 3-14a differs from that of Fig. 3-13a by the addition of a constant force due to the weight of the mass: $F_1 = mg$. Without this gravitational force, the equilibrium position would be at $x = 0$; the addition of the force causes an extension of the spring by an amount $h = mg/k$ and displaces the equilibrium position to $x = h$. Therefore, the equation of motion is just Eq. (3.54) with x replaced by $x - h$:

$$m\ddot{x} + kx = kh \tag{3.62}$$

with solution

$$x(t) = h + A \cos \omega_0 t \tag{3.63}$$

where we have chosen the initial conditions, $x(t=0) = h + A$ and $\dot{x}(t=0) = 0$.

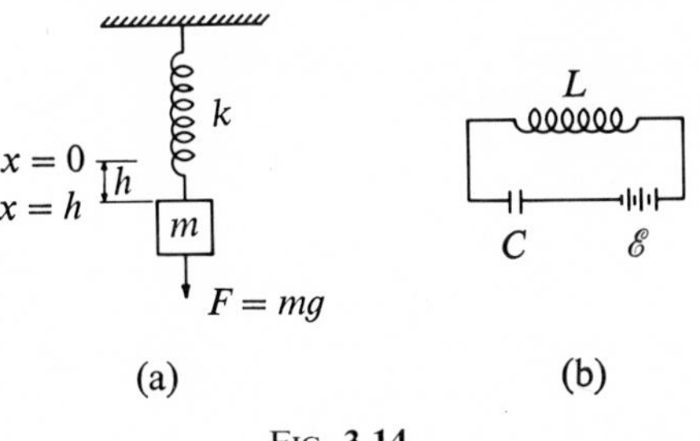

FIG. 3-14

In Fig. 3-14b we have added a battery (with emf $\mathscr{E}$) to the circuit of Fig. 3-13b. Kirchhoff's equation now becomes

$$L\frac{dI}{dt} + \frac{1}{C}\int I\,dt = \mathscr{E} = \frac{q_1}{C} \tag{3.64}$$

where q_1 represents the charge that must be applied to C in order to produce a voltage $\mathscr{E}$. Using $I = \dot{q}$, we have

$$L\ddot{q} + \frac{q}{C} = \frac{q_1}{C} \tag{3.65}$$

If $q = q_0$ and $I = 0$ at $t = 0$, the solution is

$$q(t) = q_1 + (q_0 - q_1)\cos\omega_0 t \tag{3.66}$$

which is the exact electrical analog of Eq. (3.63).

The addition of damping to the mechanical oscillator of Fig. 3-13a can be represented by a "dash pot" containing some viscous fluid, as in Fig. 3-15a. The equation of motion is

$$m\ddot{x} + b\dot{x} + kx = 0. \tag{3.67}$$

Kirchhoff's equation for the analogous electrical circuit of Fig. 3-15b is

$$L\ddot{q} + R\dot{q} + \frac{1}{C}q = 0 \tag{3.68}$$

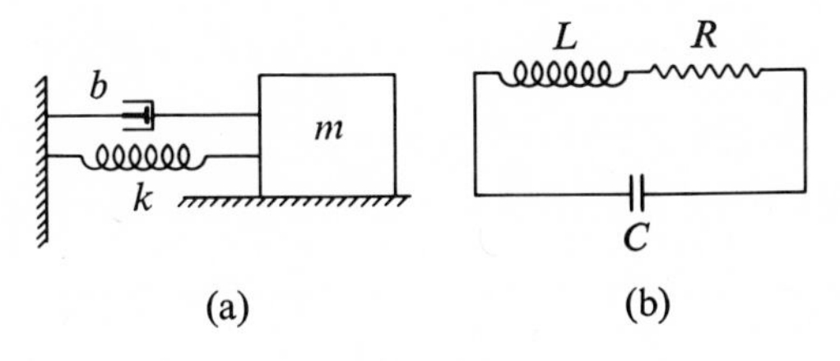

(a) (b)

FIG. 3-15

so that the electrical resistance R corresponds to the mechanical damping resistance b. The analogy between mechanical and electrical quantities can be summarized as in Table 3-1.

Table 3-1

Mechanical		Electrical	
x	Displacement	q	Charge
$\dot{x}$	Velocity	$\dot{q}=I$	Current
m	Mass	L	Inductance
b	Damping resistance	R	Resistance
$1/k$	Mechanical compliance	C	Capacitance
F	Amplitude of impressed force	$\mathscr{E}$	Amplitude of impressed emf

Because of the reciprocal nature of the correspondence between mechanical compliance and electrical resistance, the addition of springs and capacitors to systems must be made in different ways to produce the same effects. For example, consider the mass in Fig. 3-16a to which two springs are attached in tandem. If a force F is applied to the mass, spring 1 will extend an amount $x_1 = F/k_1$, while spring 2 will extend an amount $x_2 = F/k_2$. The total extension will be

$$x = x_1 + x_2 = F\left(\frac{1}{k_1} + \frac{1}{k_2}\right) \tag{3.69}$$

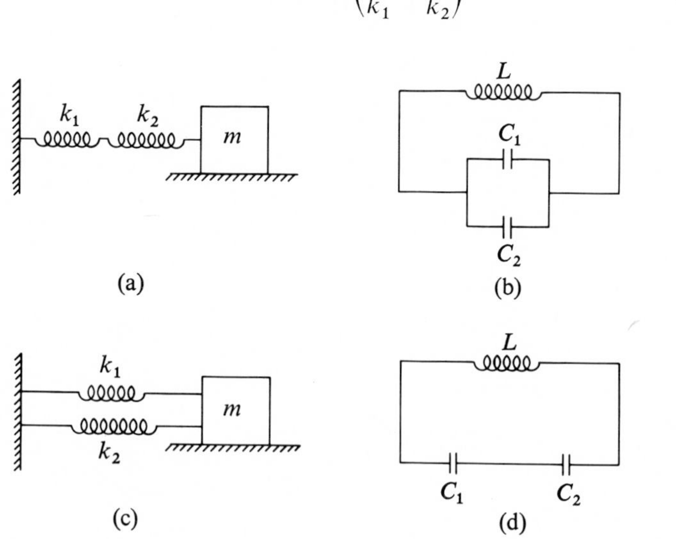

FIG. **3-16**

The electrical analog of this equation is (see Fig. 3-16b)

$$q = \mathscr{E}(C_1 + C_2) \tag{3.70}$$

Thus, springs acting in *series* are equivalent to capacitors acting in *parallel.* Similarly, springs in parallel behave in the same way as capacitors in series (see Fig. 3-16c and d).

Suggested References

The subject of harmonic oscillations is a standard topic and is treated in most texts on mechanics. More or less the same material is discussed at the intermediate level by: Blass (Bl62, Chapter 2), Bradbury (Br68, Chapter 4), Constant (Co54, Vol. 1, Chapter 6), Hauser (Ha65, Chapter 4), Houston (Ho48, Chapter 4), Joos and Freeman (Jo50, Chapter 5), Page (Pa52, Section 27), Slater and Frank (Sl47, Chapter 2), Sommerfeld (So50, Chapter 3), Symon (Sy60, Chapter 2), and Wangsness (Wa63, Chapter 5).

The treatments of Becker (Be54, Chapter 7), Magnus (Ma65c, Chapters 1 and 2,) and Halfman (Ha62a, Vol. 1, Chapter 7) contain somewhat more detail than those listed above.

At the advanced level, Landau and Lifshitz (La60, Chapter 5) give a brief but elegant discussion. A standard work on vibratory phenomena is that of Morse (Mo48, see, particularly, Chapter 2). McLachlan's little book (Mc51) treats a variety of topics (linear, one-dimensional systems are discussed in Chapters 1 and 2). A comprehensive treatment is given in the excellent book by Andronow and Chaikin (An49, Chapter 1).

Problems

3-1. A simple harmonic oscillator consists of a 100-gram mass attached to a spring whose force constant is 10^4 dyne/cm. The mass is displaced a distance of 3 cm and released from rest. Calculate the following quantities: (a) the natural frequency ν_0 and the period τ_0, (b) the total energy, and (c) the maximum velocity.

3-2. In the preceding problem allow the motion to take place in a resisting medium. It is found that after oscillating for 10 sec, the maximum amplitude has decreased to half the initial value. Calculate the following quantities: (a) The damping parameter β, (b) the frequency ν_1 (compare with the undamped frequency ν_0), and (c) the decrement of the motion.

3-3. The oscillator of Problem 3-1 is set into motion by giving it an initial velocity of 1 cm/sec at its equilibrium position. Calculate the following quantities: (a) the maximum displacement and (b) the maximum potential energy.

3-4. Consider a simple harmonic oscillator. Calculate the *time* averages of the kinetic and potential energies over one cycle and show that these quantities are equal. Why is this a reasonable result? Next, calculate the *space* averages of the kinetic and potential energies. Discuss the results.

3-5. Obtain an expression for the fraction of a complete period that a simple harmonic oscillator spends within a small interval Δx at a position x. Sketch curves of this function *versus* x for several different amplitudes. Discuss the physical significance of the results. Comment on the areas under the various curves.

3-6. Two masses, m_1 and m_2, are free to slide on a horizontal frictionless surface and are connected by a spring whose force constant is k. Find the frequency of oscillatory motion for this system.

3-7. A body of uniform cross-sectional area A and of mass density ρ floats in a liquid and, at equilibrium, displaces a volume V. Show that the period of small oscillations about the equilibrium position is given by

$$\tau = 2\pi\sqrt{V/gA}$$

where g is the acceleration due to gravity.

3-8. A pendulum is suspended from the cusp of a cycloid* which is cut in a rigid support, as shown in the figure. The path described by the pendulum

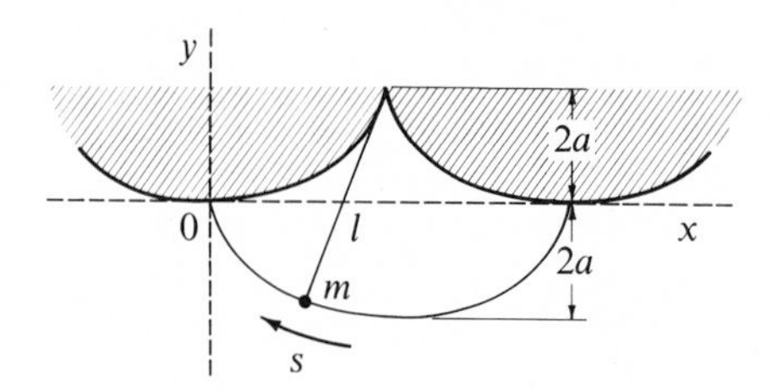

bob is cycloidal and is given by

$$x = a(\varphi - \sin \varphi); \qquad y = a(\cos \varphi - 1)$$

where the length of the pendulum is $l = 4a$, and where φ is the angle of rotation of the circle that generates the cycloid. Show that the oscillations are exactly isochronous with a frequency $\omega_0 = \sqrt{g/l}$, independent of the amplitude.

* The reader unfamiliar with the properties of cycloids should consult a text on analytic geometry.

3-9. A particle of mass m is at rest at the end of a spring (force constant $= k$) which hangs from a fixed support. At $t = 0$ a constant downward force F is applied to the mass and acts for a time t_0. Show that after the force is removed, the displacement of the mass from its equilibrium position ($x = x_0$) is

$$x - x_0 = F\,[\cos \omega(t - t_0) - \cos \omega t]$$

where $\omega^2 = k/m$.

3-10. If the amplitude of a damped oscillator decreases to $1/e$ of its initial value after n periods, show that the frequency of the oscillator must be approximately $[1 - (8\pi^2 n^2)^{-1}]$ times the frequency of the corresponding undamped oscillator.

3-11. Derive the expressions for the energy and energy loss curves shown in Fig. 3-7 for the damped oscillator. Calculate the *average rate* at which the damped oscillator loses energy (i.e., compute a time average over one cycle).

3-12. A simple pendulum consists of a mass m suspended from a fixed point by means of a weightless, extensionless rod of length l. Obtain the equation of motion and, in the approximation that $\sin \theta \cong \theta$, show that the natural frequency is $\omega_0 = \sqrt{g/l}$, where g is the acceleration imparted by gravity. Discuss the motion in the event that the motion takes place in a viscous medium for which the retarding force is $2m\sqrt{gl}\,\dot{\theta}$.

3-13. Show that Eq. (3.51) is indeed the solution for the case of critical damping by assuming a solution of the form $x(t) = y(t)\exp(-\beta t)$ and determining the function $y(t)$.

3-14. Express the displacement $x(t)$ and the velocity $\dot{x}(t)$ for the overdamped oscillator in terms of hyperbolic functions.

3-15. Refer to Fig. 3-12, and show that the equation of the line which separates region II from region III of the phase plane is $\dot{x} = -(\beta + \omega_2)x$. Proceed as follows: For definiteness, assume $x_0 > 0$. Show that if $x(t)$ vanishes for some positive time t_1, then t_1 is a root of the equation

$$e^{(q_2 - q_1)t_1} = 1 - \frac{x_0(q_2 - q_1)}{\dot{x}_0 + q_2 x_0}$$

where $q_1 \equiv \beta - \omega_2$ and $q_2 \equiv +\beta + \omega_2$. Show that the condition for a positive root t_1 is

$$\frac{x_0}{\dot{x}_0 + q_2 x_0} \leq 0$$

which leads to the desired equation of the line. What path is followed by a representative point which has the initial condition $\dot{x}_0 = -(\beta + \omega_2)x_0$? Sketch $x(t)$ for this case.

3-16. Discuss the motion of a particle described by Eq. (3.37) in the event that $b < 0$ (i.e., the damping resistance is *negative*).

3-17. Compute the oscillation frequency for the circuit of Fig. 3-15b in the event that $L = 0.1$ henry, $C = 10$ microfarad, $R = 100$ ohm.

3-18. Show that for an *R-L-C* circuit in which the resistance is small, the logarithmic decrement of the oscillations is approximately $\pi R\sqrt{C/L}$.

3-19. An *R-L-C* circuit (see Fig. 3-15b) contains an inductor of 0.01 henry and a resistor of 100 ohms. The oscillation frequency is 1 kHz. If, at $t = 0$, the voltage across the capacitor is 10 volts and the current is zero, find the current 0.2 milliseconds later.

CHAPTER 4

Driven Oscillations

4.1 Introduction

In the preceding chapter we found that a particle or an electrical circuit undergoing free (i.e., undamped) oscillations would remain in motion forever. In every real system, however, there is always a certain amount of friction (or electrical resistance) that eventually damps the motion to rest.* This damping of the oscillations may be prevented if there exists some mechanism for supplying the system with energy from an external source at a rate equal to that at which it is absorbed by the damping medium. Motions of this type are called *driven oscillations*. Because we have elected to consider only *linear* systems at this stage of the development, the analysis of driven (or *forced*) oscillations is straightforward, and many techniques are available for treating the motion. Several of these methods will be used in this chapter.

* It is therefore clear that a damping term of the form $-b\dot{x}$ cannot be a completely accurate representation of the real situation because such a term will bring the system to rest only after an *infinitely* long time.

4.2 Sinusoidal Driving Forces

The simplest case of driven oscillations is that in which an external driving force varying harmonically with time is applied to the oscillator. The total force on the particle is then

$$F = -kx - b\dot{x} + F_0 \cos \omega t \tag{4.1}$$

where we consider a linear restoring force and a viscous damping force in addition to the driving force. The equation of motion becomes

$$m\ddot{x} + b\dot{x} + kx = F_0 \cos \omega t \tag{4.2}$$

or, using our previous notation,

$$\boxed{\ddot{x} + 2\beta\dot{x} + \omega x_0^2 = A \cos \omega t} \tag{4.3}$$

where $A = F_0/m$ and where ω is the frequency of the driving force. As discussed in Appendix C, the solution of Eq. (4.3) consists of two parts, a *complementary function* $x_c(t)$, which is the solution of Eq. (4.3) with the right-hand side set equal to zero, plus a *particular solution* $x_p(t)$, which reproduces the right-hand side. The complementary solution is clearly the same as that given in Eq. (3.40):

$$x_c(t) = e^{-\beta t}[A_1 \exp(\sqrt{\beta^2 - \omega_0^2}\, t) + A_2 \exp(-\sqrt{\beta^2 - \omega_0^2}\, t)] \tag{4.4}$$

For the particular solution we try

$$x_p(t) = D \cos(\omega t - \delta) \tag{4.5}$$

Substituting $x_p(t)$ in Eq. (4.3) and expanding $\cos(\omega t - \delta)$ and $\sin(\omega t - \delta)$, we obtain

$$\begin{aligned} &\{A - D[(\omega_0^2 - \omega^2)\cos \delta + 2\omega\beta \sin \delta]\}\cos \omega t \\ &\qquad - \{D[(\omega_0^2 - \omega^2)\sin \delta - 2\omega\beta \cos \delta]\}\sin \omega t = 0 \end{aligned} \tag{4.6}$$

Since $\sin \omega t$ and $\cos \omega t$ are linearly independent functions, this equation can be satisfied in general only if the coefficient of each term vanishes identically. From the $\sin \omega t$ term, we have

$$\tan \delta = \frac{2\omega\beta}{\omega_0^2 - \omega^2} \tag{4.7}$$

so that we can write

$$\left.\begin{aligned}\sin\delta &= \frac{2\omega\beta}{\sqrt{(\omega_0^2-\omega^2)^2+4\omega^2\beta^2}}\\ \cos\delta &= \frac{\omega_0^2-\omega^2}{\sqrt{(\omega_0^2-\omega^2)^2+4\omega^2\beta^2}}\end{aligned}\right\} \tag{4.8}$$

And from the coefficient of the cos ωt term, we have

$$\begin{aligned}D &= \frac{A}{(\omega_0^2-\omega^2)\cos\delta+2\omega\beta\sin\delta}\\ &= \frac{A}{\sqrt{(\omega_0^2-\omega^2)^2+4\omega^2\beta^2}}\end{aligned} \tag{4.9}$$

Thus, the particular integral is

$$\boxed{x_p(t) = \frac{A}{\sqrt{(\omega_0^2-\omega^2)^2+4\omega^2\beta^2}}\cos(\omega t-\delta)} \tag{4.10}$$

with

$$\boxed{\delta = \tan^{-1}\left(\frac{2\omega\beta}{\omega_0^2-\omega^2}\right)} \tag{4.11}$$

The general solution is

$$x(t) + x_c(t) + x_p(t) \tag{4.12}$$

But $x_c(t)$ here represents *transient* effects (i.e., effects that depend upon the initial conditions), and the terms contained in this solution damp out with time because of the factor $\exp(-\beta t)$. The term $x_p(t)$ represents the steady-state effects and contains all of the information for t large compared to $1/\beta$. Thus,

$$x(t \gg 1/\beta) = x_p(t)$$

The steady-state solution will be of importance in many of the applications and problems.

The quantity δ [see Eq. (4.11)] represents the phase difference between the driving force and the resultant motion. That is, there is a real delay between the action of the driving force and the response of the system. For a fixed ω_0, as ω increases from 0, the phase increases from $\delta = 0$ at $\omega = 0$ to $\delta = \pi/2$ at $\omega = \omega_0$ and to π as $\omega \to \infty$. The variation of δ with ω is shown in Fig. 4-1.

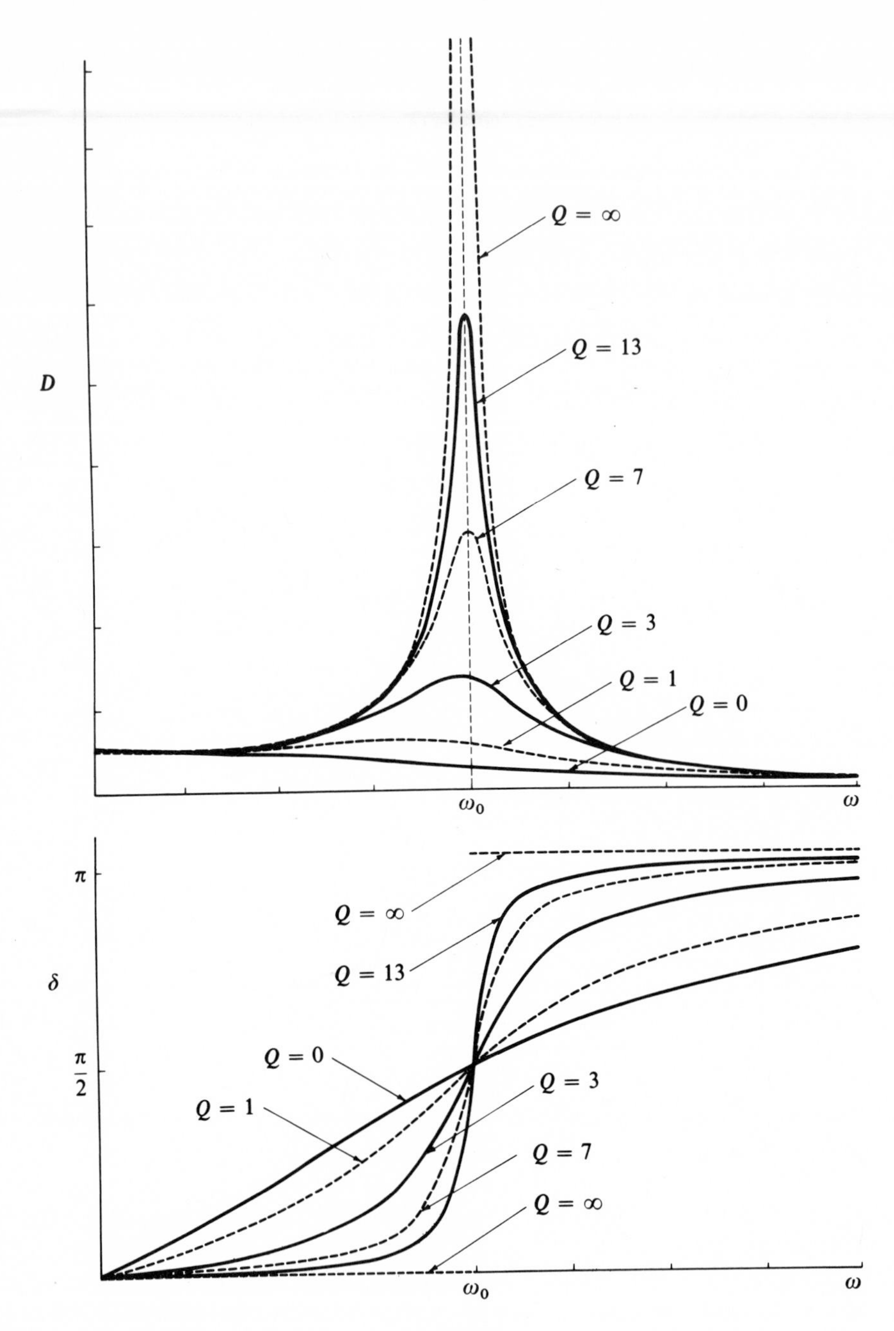

D
$Q = \infty$
$Q = 13$
$Q = 7$
$Q = 3$
$Q = 1$
$Q = 0$
ω_0
ω
π
δ
$\frac{\pi}{2}$
$Q = \infty$
$Q = 13$
$Q = 0$
$Q = 1$
$Q = 3$
$Q = 7$
$Q = \infty$
ω_0
ω

FIG. **4-1**

In order to find the frequency ω_R at which the amplitude D is a maximum (i.e., the *amplitude resonance frequency*), we set

$$\left.\frac{dD}{d\omega}\right|_{\omega=\omega_R} = 0$$

Performing the differentiation, we find

$$\omega_R = \sqrt{\omega_0^2 - 2\beta^2} \tag{4.13}$$

Thus, the resonance frequency ω_R is lowered as the damping coefficient β is increased. There is no resonance, of course, if $\beta^2 > \omega_0^2/2$, for then ω_R is imaginary and D decreases monotonically with increasing ω.

We may now compare the oscillation frequencies for the various cases which we have considered:

(a) Free oscillations, no damping:

$$\omega_0^2 = \frac{k}{m} \qquad \text{[Eq. (3.4)]}$$

(b) Free oscillations, damping:

$$\omega_1^2 = \omega_0^2 - \beta^2 \qquad \text{[Eq. (3.41)]}$$

(c) Driven oscillations, damping:

$$\omega_R^2 = \omega_0^2 - 2\beta^2 \qquad \text{[Eq. (4.13)]}$$

and we note that $\omega_0 > \omega_1 > \omega_R$.

It is customary to describe the degree of damping in an oscillating system in terms of the "quality factor" or the "Q" of the system:

$$Q \equiv \frac{\omega_R}{2\beta} \tag{4.14}$$

If there is little damping, then Q is very large and the shape of the resonance curve approaches that for an undamped oscillator. On the other hand, the resonance can be completely destroyed if the damping is large and Q is very small. Figure 4-1 shows the resonance and phase curves for several different values of Q. These curves indicate the lowering of the resonance frequency with a decrease in Q (i.e., with an increase of the damping coefficient β). The effect is not large, however; the frequency shift is less than 3 percent even for Q as small as 3 and is about 18 percent for $Q = 1$.

For a lightly damped oscillator, it is easy to show that (see Problem 4-3)

$$Q \cong \frac{\omega_0}{\Delta\omega} \tag{4.15}$$

where $\Delta\omega$ represents the frequency interval between the points on the amplitude resonance curve that are $1/\sqrt{2} = 0.707$ of the maximum amplitude.

The values of Q which are found in real physical situations vary greatly. In rather ordinary mechanical systems (e.g., loudspeakers) the values may be in the range from a few to 100 or so. Quartz crystal oscillators may have Q's of 10^4. Highly tuned electrical circuits, including resonant cavities, may have values of 10^4 to 10^5. We may also define Q's for some atomic systems. According to the classical picture, the oscillation of electrons within atoms leads to optical radiation. The sharpness of spectral lines is limited by the damping due to the loss of energy by radiation (*radiation damping**). The minimum width of a line can be calculated classically and is found to be* $\Delta\omega \cong 2 \times 10^{-8}\omega$. The Q of such an oscillator is therefore approximately 5×10^7. Resonances with the largest known Q's are found in the radiation from gas *lasers* (*l*ight *a*mplification by *s*timulated *e*mission of *r*adiation). Measurements with such devices† have yielded Q's of approximately 10^{14}.

Equation (4.13) gives the frequency for *amplitude* resonance. We now calculate the frequency for *kinetic energy* resonance.

The kinetic energy is given by $T = \frac{1}{2}m\dot{x}^2$, and computing $\dot{x}$ from Eq. (4.10) we have

$$\dot{x} = \frac{-A\omega}{\sqrt{(\omega_0^2 - \omega^2)^2 + 4\omega^2\beta^2}} \sin(\omega t - \delta) \tag{4.16}$$

so that the kinetic energy becomes

$$T = \frac{mA^2}{2} \cdot \frac{\omega^2}{(\omega_0^2 - \omega^2)^2 + 4\omega^2\beta^2} \sin^2(\omega t - \delta) \tag{4.17}$$

In order to obtain a value of T which is independent of the time, we compute the average of T over one complete period of oscillation. Thus,

$$\langle T \rangle = \frac{mA^2}{2} \cdot \frac{\omega^2}{(\omega_0^2 - \omega^2)^2 + 4\omega^2\beta^2} \langle \sin^2(\omega t - \delta) \rangle \tag{4.18}$$

Now, the average value of the square of the sine function taken over one period is‡

$$\begin{aligned} \langle \sin^2(\omega t - \delta) \rangle &= \frac{\omega}{2\pi} \int_0^{2\pi/\omega} \sin^2(\omega t - \delta)\, dt \\ &= \tfrac{1}{2} \end{aligned} \tag{4.19}$$

* See Marion (Ma65b, Sections 9.9 and 11.3).

† See, for example, A. Javan, E. A. Ballik, and W. L. Bond, *J. Opt. Soc. Am.* **52**, 96 (1962).

‡ The reader should prove the important result that the average over a complete period of $\sin^2 \omega t$ *or* $\cos^2 \omega t$ is equal to $\frac{1}{2}$: $\langle \sin^2 \omega t \rangle = \langle \cos^2 \omega t \rangle = \frac{1}{2}$.

Therefore,

$$\langle T \rangle = \frac{mA^2}{4} \cdot \frac{\omega^2}{(\omega_0^2 - \omega^2)^2 + 4\omega^2\beta^2} \tag{4.20}$$

The value of ω for $\langle T \rangle$ a maximum is labeled ω_E and is obtained from

$$\left. \frac{d\langle T \rangle}{d\omega} \right|_{\omega = \omega_E} = 0 \tag{4.21}$$

Differentiating Eq. (4.20) and equating the result to zero, we find

$$\omega_E = \omega_0 \tag{4.22}$$

so that the kinetic energy resonance occurs at the natural frequency of the system for undamped oscillations.

We see therefore that the amplitude resonance occurs at a frequency $\sqrt{\omega_0^2 - 2\beta^2}$ whereas the kinetic energy resonance occurs at ω_0. Since the potential energy is proportional to the square of the amplitude, the potential energy resonance must also occur at $\sqrt{\omega_0^2 - 2\beta^2}$. That the kinetic and potential energies resonate at different frequencies is a result of the fact that the damped oscillator is not a conservative system; energy is continually exchanged with the driving mechanism and energy is being transferred to the damping medium.

4.3 Transient Effects

Although we have emphasized the steady-state motion of the driven oscillator in the preceding section, the transient effects are often of considerable importance, especially in certain types of electrical circuits.

The details of the motion during the period of time before the transient effects have disappeared (i.e., $t \lesssim 1/\beta$) are strongly dependent on the conditions of the oscillator at the time that the driving force is first applied and also on the relative magnitudes of the driving frequency ω and the damping frequency $\sqrt{\beta^2 - \omega_0^2}$. Figure 4-2 illustrates the transient motion of an oscillator when driving frequencies less than and greater than $\omega_1 = \sqrt{\beta^2 - \omega_0^2}$ are applied. In the event that $\omega < \omega_1$ (Fig. 4-2a), the transient response of the oscillator greatly distorts the sinusoidal shape of the forcing function during the time interval immediately following the application of the driving force, whereas if $\omega > \omega_1$ (Fig. 4-2b), the effect is a modulation of the forcing function with little distortion of the high-frequency sinusoidal oscillations.

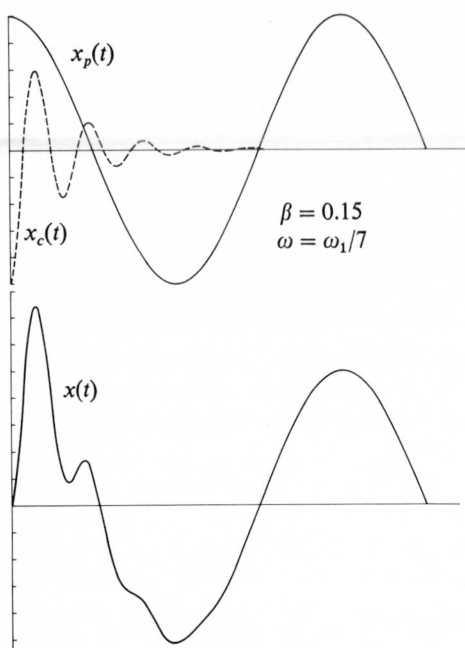

FIG. **4-2a**

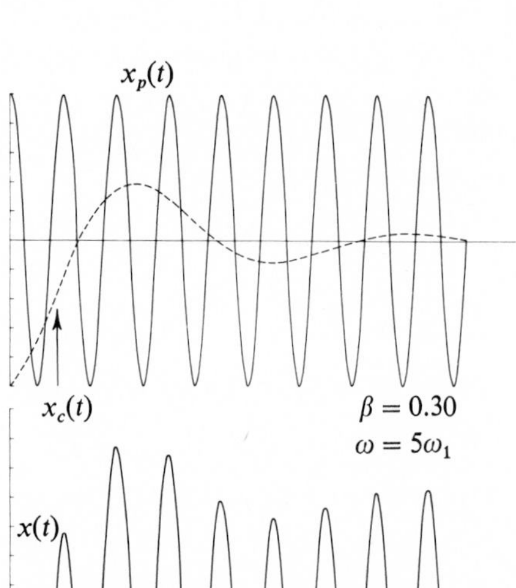

FIG. **4-2b**

4.4 Driven Electrical Oscillations

The electrical equivalent of the oscillator we have been discussing is shown in Fig. 4-3. The circuit equation is

$$L\ddot{q} + R\dot{q} + \frac{1}{C}q = \mathscr{E}_0 \cos \omega t \tag{4.23}$$

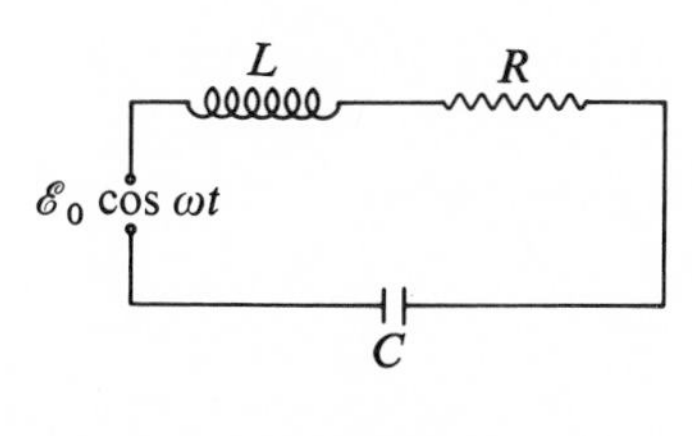

FIG 4-3.

In much of the succeeding material we will find it convenient to employ a mathematical device in order to simplify the analysis. According to Euler's relation [see Appendix B, Eq. (B.13a)], we can express a complex exponential function as

$$e^{i\omega t} = \cos \omega t + i \sin \omega t \tag{4.24}$$

Hence, the real part of $\exp(i\omega t)$ is

$$\text{Re } e^{i\omega t} = \cos \omega t \tag{4.25}$$

Therefore, instead of $\cos \omega t$, we may write $\exp(i\omega t)$ if we understand that only the real part is to be considered when applying the resulting expressions to physical situations. One of the several advantages of this device becomes apparent when we differentiate $\exp(i\omega t)$ with respect to time:

$$\frac{d}{dt} e^{i\omega t} = i\omega \, e^{i\omega t}$$

Hence, the operator d/dt can simply be replaced by $i\omega$.

Equation (4.23) can now be written as

$$\boxed{L\ddot{q} + R\dot{q} + \frac{1}{C} q = \mathscr{E}_0 \, e^{i\omega t}} \tag{4.26}$$

When using electrical circuits, one rarely measures directly the charge on a capacitor; rather, the quantity of interest usually is the *current*. Therefore, differentiating Eq. (4.26) with respect to time and recalling that $\dot{q}(t) = I(t)$, we have

$$L\ddot{I} + R\dot{I} + \frac{1}{C}I = i\omega\mathscr{E}_0 e^{i\omega t} \tag{4.27}$$

Since the applied emf is harmonic, we try a steady-state solution for $I(t)$ that is also harmonic:

$$I(t) = I_0 e^{i\omega t} \tag{4.28}$$

where I_0 may be complex. Substituting this expression into Eq. (4.27) gives

$$\left(-L\omega^2 + iR\omega + \frac{1}{C}\right) I_0 e^{i\omega t} = i\omega\mathscr{E}_0 e^{i\omega t} \tag{4.29}$$

Dividing by $i\omega \exp(i\omega t)$, we obtain

$$\left[R + i\left(\omega L - \frac{1}{\omega C}\right)\right] I_0 = \mathscr{E}_0 \tag{4.30}$$

This equation has the form of Ohm's law, $RI = \mathscr{E}$, if, instead of the normal dc resistance, we identify the coefficient of the current amplitude as the *impedance* Z:

$$ZI_0 = \mathscr{E}_0 \tag{4.31}$$

where

$$\boxed{Z = R + i\left(\omega L - \frac{1}{\omega C}\right)} \tag{4.32}$$

The impedance of a circuit consists of real and imaginary parts. The real part is the resistance R and the imaginary part is the *reactance* X. The reactance, in turn, consists of the *inductive* reactance, $X_L = \omega L$, and the *capacitive* reactance, $X_C = -1/\omega C$. The fact that the impedance is complex means that the current is not generally in phase with the applied emf (see below).

The impedance can be written in terms of a magnitude and a phase factor:

$$Z = |Z|e^{i\phi} \tag{4.33}$$

where

$$|Z| = \left[R^2 + \left(\omega L - \frac{1}{\omega C}\right)^2\right]^{\frac{1}{2}} \tag{4.34a}$$

and

$$\phi = \tan^{-1}\left(\frac{\omega L - \dfrac{1}{\omega C}}{R}\right) \tag{4.34b}$$

The current can now be expressed as

$$I(t) = \frac{\mathscr{E}_0}{|Z|} e^{i(\omega t - \phi)} \tag{4.35}$$

or, since it is only the real part of this expression that is physically meaningful, the actual current is

$$I(t) = \frac{\mathscr{E}_0}{|Z|} \cos(\omega t - \phi) \tag{4.36}$$

From the form of this equation, it is apparent that $I(t)$ will attain its maximum value when $|Z|$ is a minimum, that is, when $\omega L - 1/\omega C = 0$. Hence, the condition for current resonance is

$$\omega_0 = \frac{1}{\sqrt{LC}} \tag{4.37}$$

which is identical with the result we previously obtained for the series *R-L-C* circuit [Eq. (3.59)]. Also, from Eq. (4.34b), we see that $\phi = 0$ at current resonance. Notice that at resonance the reactance vanishes and the impedance becomes purely resistive.

The meaning of the phase angle ϕ is the following:

$\phi > 0$: The current reaches a maximum *later* than does the voltage; i.e., the current *lags* the voltage.
$\phi < 0$: The current reaches a maximum *earlier* than does the voltage; i.e., the current *leads* the voltage.
$\phi = 0$: The current and the voltage are *in phase*; i.e., the circuit is *resonant*.

In an *R-L-C* circuit energy is alternately stored in and released by the inductor and the capacitor, but energy is *dissipated* by the resistor through Joule heating. The rate at which energy is dissipated is the *power* dissipation, and the instantaneous value is given by*

$$P(t) = [\text{Re}\, I(t)][\text{Re}\, \mathscr{E}(t)] \tag{4.38}$$

A more useful quantity is the *average* power dissipation, i.e., the time average of $P(t)$ over one complete cycle of oscillation:

$$\langle P \rangle = \frac{1}{\tau} \int_0^\tau P(t)\, dt \tag{4.39}$$

* In order to calculate a real physical quantity, such as the power, we must always use the real physical components of the expression. That is, $P(t)$ is the product of the *real part* of $I(t)$ and the *real part* of $\mathscr{E}(t)$; $P(t)$ is *not* given by $\text{Re}(I\mathscr{E})$.

Now,

$$\text{Re } \mathscr{E}(t) = |\mathscr{E}_0| \cos \omega t \tag{4.40}$$

Also,

$$\begin{aligned} I(t) &= I_0 e^{i\omega t} \\ &= \frac{\mathscr{E}_0}{|Z|} e^{i(\omega t - \phi)} \end{aligned} \tag{4.41}$$

so that

$$\begin{aligned} \text{Re } I(t) &= \frac{\mathscr{E}_0}{|Z|} \cos(\omega t - \phi) \\ &= |I_0| \cos(\omega t - \phi) \end{aligned} \tag{4.42}$$

Hence,

$$\langle P \rangle = \frac{1}{\tau} \int_0^\tau |I_0| \cdot |\mathscr{E}_0| \cos \omega t \cos(\omega t - \phi)\, dt \tag{4.43}$$

Expanding $\cos(\omega t - \phi)$ and evaluating the two resulting integrals, we find

$$\langle P \rangle = \tfrac{1}{2} |I_0| \cdot |\mathscr{E}_0| \cos \phi \tag{4.44}$$

In a purely resistive circuit, the average power dissipation is equal to $\frac{1}{2}|I_0| \cdot |\mathscr{E}_0|$; the remaining term in the expression for $\langle P \rangle$, viz., $\cos \phi$, is called the *power factor* (p.f.) for the circuit. Explicitly,

$$\text{p.f.} = \cos \phi = \frac{R}{\left[R^2 + \left(\omega L - \dfrac{1}{\omega C}\right)^2\right]^{\frac{1}{2}}} = \frac{R}{|Z|} \tag{4.45}$$

The Q of an electrical circuit is defined in terms of the *current* resonance frequency.* For the series *R-L-C* circuit of Fig. 4-3, we have

$$Q = \frac{\omega_0 L}{R} = \frac{1}{\omega_0 RC} \tag{4.46}$$

since $\omega_0 = 1\sqrt{LC}$. Therefore, the power factor can be expressed as

$$\text{p.f.} = \frac{R}{|Z|} = \left[1 + Q^2 \frac{\omega^2}{\omega_0^2}\left(1 - \frac{\omega_0^2}{\omega^2}\right)^2\right]^{-\frac{1}{2}} \tag{4.47}$$

* Thus, there is a difference between the electrical and mechanical cases. The mechanical Q is defined in terms of the amplitude resonance Eq. (4.14): $Q = \omega_R/2\beta$. Since it is *current* and not *charge* that is important in the electrical case, the interesting resonance frequency is ω_0, not ω_R; the current resonance is the same as the kinetic energy (or velocity) resonance of the mechanical case. For systems with large Q, $\omega_R \cong \omega_0$, and the distinction is not important.

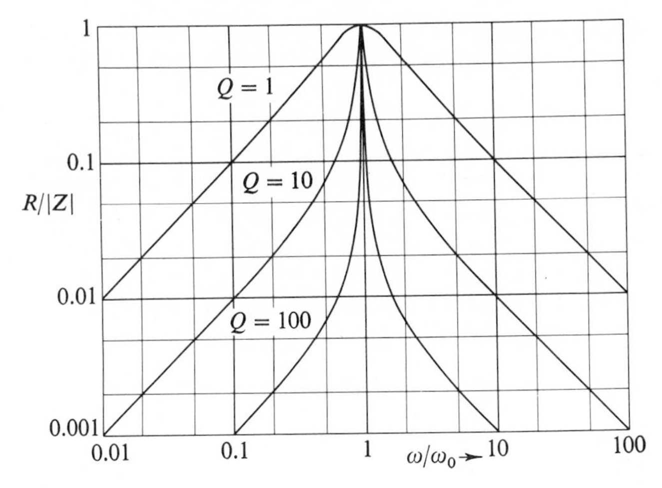

FIG. 4-4

The resonance characteristics of the series R-L-C circuit are best summarized in terms of power factor curves, such as those shown in Fig. 4-4. The full width of each curve between the points at which $R/|Z| = 1/\sqrt{2}$ (the *half-power* points) is given by

$$\Delta\omega = \frac{\omega_0}{Q} \tag{4.48}$$

The *bandwidth* of the circuit is

$$\text{bandwidth} = \Delta\nu = \frac{\omega_0}{2\pi Q} = \frac{R}{2\pi L} \tag{4.49}$$

For arrangements of R, L, and C other than the series connection of Fig. 4-3, the various quantities which we have calculated assume different forms. Some other circuits of interest are examined in the problems.

4.5 The Principle of Superposition—Fourier Series

The oscillations which we have been discussing obey a differential equation of the form

$$\left(\frac{d^2}{dt^2} + a\frac{d}{dt} + b\right)x(t) = A\cos\omega t \tag{4.50}$$

The quantity in parentheses on the left-hand side is a *linear operator* which we may represent by $\mathbf{L}$. If we generalize the time-dependent forcing function on the right-hand side, we can write the equation of motion as

$$\mathbf{L}x = F(t) \tag{4.51}$$

An important property of linear operators is that they obey the *principle of superposition*. This property is a result of the fact that linear operators are distributive. That is,

$$\mathbf{L}(x_1 + x_2) = \mathbf{L}(x_1) + \mathbf{L}(x_2) \tag{4.52}$$

Therefore, if we have two solutions, $x_1(t)$ and $x_2(t)$, for two different forcing functions, $F_1(t)$ and $F_2(t)$,

$$\mathbf{L}x_1 = F_1(t), \qquad \mathbf{L}x_2 = F_2(t) \tag{4.53}$$

we can add these equations (multiplied by arbitrary constants α_1 and α_2) and obtain

$$\mathbf{L}(\alpha_1 x_1 + \alpha_2 x_2) = \alpha_1 F_1(t) + \alpha_2 F_2(t) \tag{4.54}$$

Clearly, we can extend this argument to a set of solutions $x_n(t)$, each of which is appropriate for a given $F_n(t)$:

$$\mathbf{L}\left(\sum_{n=1}^{N} \alpha_n x_n(t)\right) = \sum_{n=1}^{N} \alpha_n F_n(t) \tag{4.55}$$

This equation is just Eq. (4.51) if we identify the linear combinations as

$$\begin{aligned} x(t) &= \sum_{n=1}^{N} \alpha_n x_n(t) \\ F(t) &= \sum_{n=1}^{N} \alpha_n F_n(t) \end{aligned} \tag{4.56}$$

If each of the individual functions $F_n(t)$ has a simple harmonic dependence on time, such as $\cos \omega_n t$, we know that the corresponding solution $x_n(t)$ is given by Eq. (4.10). Thus, if $F(t)$ has the form

$$F(t) = \sum_n \alpha_n \cos(\omega_n t - \phi_n) \tag{4.57}$$

the steady-state solution is

$$x(t) = \frac{1}{m} \sum_n \frac{\alpha_n}{\sqrt{(\omega_0^2 - \omega_n^2)^2 + 4\omega_n^2 \beta^2}} \cos(\omega_n t - \phi_n - \delta_n) \tag{4.58}$$

where

$$\delta_n = \tan^{-1}\left(\frac{2\omega_n \beta}{\omega_0^2 - \omega_n^2}\right) \tag{4.59}$$

We can obviously write down similar solutions for the case in which $F(t)$ is represented by a series of terms, $\sin(\omega_n t - \phi_n)$. We therefore arrive at the important conclusion that if some arbitrary forcing function $F(t)$ can be expressed as a series (finite or infinite) of harmonic terms, the complete solution can also be written as a similar series of harmonic terms. This is an extremely useful result since, according to Fourier's theorem, *any* arbitrary periodic function (subject to certain conditions that are not very restrictive) can be represented by a series of harmonic terms.* Thus, in the usual physical case in which $F(t)$ is periodic with period $\tau = 2\pi/\omega$,

$$F(t + \tau) = F(t) \tag{4.60}$$

we then have

$$F(t) = \tfrac{1}{2}a_0 + \sum_{n=1}^{\infty} (a_n \cos n\omega t + b_n \sin n\omega t) \tag{4.61}$$

where

$$\left.\begin{aligned} a_n &= \frac{2}{\tau}\int_0^{\tau} F(t')\cos n\omega t'\, dt' \\ b_n &= \frac{2}{\tau}\int_0^{\tau} F(t')\sin n\omega t'\, dt' \end{aligned}\right\} \tag{4.62}$$

or, since $F(t)$ has a period τ, we can replace the integral limits 0 and τ by the limits $-\frac{1}{2}\tau = -\pi/\omega$ and $+\frac{1}{2}\tau = +\pi/\omega$:

$$\left.\begin{aligned} a_n &= \frac{\omega}{\pi}\int_{-\pi/\omega}^{+\pi/\omega} F(t')\cos n\omega t'\, dt' \\ b_n &= \frac{\omega}{\pi}\int_{-\pi/\omega}^{+\pi/\omega} F(t')\sin n\omega t'\, dt' \end{aligned}\right\} \tag{4.62a}$$

Before we discuss the response of damped systems to arbitrary forcing functions (in the following section), we give two examples of the Fourier representation of periodic functions.

▶ Example 4.5(a) Sawtooth Driving Force

A sawtooth function is shown in Fig. 4-5. In this case $F(t)$ is an *odd* function, $F(-t) = -F(t)$, and is expressed by

$$F(t) = A \cdot \frac{t}{\tau} = \frac{\omega A}{2\pi} t, \qquad -\tau/2 < t < \tau/2 \tag{1}$$

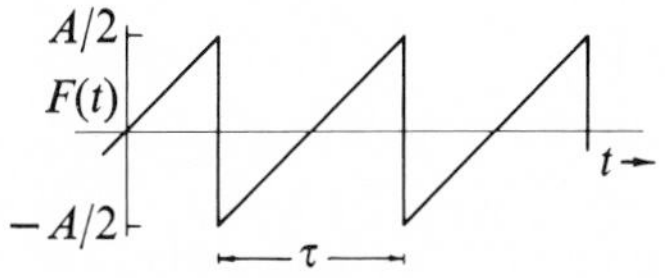

FIG. 4-5

* Fourier series are discussed further in Section 14.8.

Since $F(t)$ is odd, the coefficients a_n all vanish identically. The b_n are given by

$$b_n = \frac{\omega^2 A}{2\pi^2} \int_{-\pi/\omega}^{+\pi/\omega} t' \sin n\omega t' \, dt'$$

$$= \frac{\omega^2 A}{2\pi^2} \left[-\frac{t' \cos n\omega t'}{n\omega} + \frac{\sin n\omega t'}{n^2\omega^2} \right]_{-\pi/\omega}^{+\pi/\omega}$$

$$= \frac{\omega^2 A}{2\pi^2} \cdot \frac{2\pi}{n\omega^2} \cdot (-1)^{n+1} = \frac{A}{n\pi}(-1)^{n+1} \tag{2}$$

where the term $(-1)^{n+1}$ takes account of the fact that

$$-\cos n\pi = \begin{cases} +1, & n \text{ odd} \\ -1, & n \text{ even} \end{cases} \tag{3}$$

Therefore, we have

$$F(t) = \frac{A}{\pi}\left[\sin \omega t - \frac{1}{2} \sin 2\omega t + \frac{1}{3} \sin 3\omega t - \cdots \right] \tag{4}$$

Figure 4-6 shows the results for 2 terms, 5 terms, and 8 terms of this expansion. It is clear that the convergence toward the sawtooth function is none too rapid.

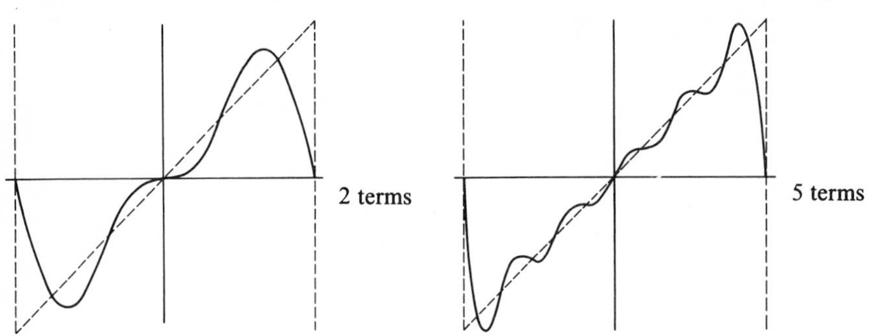

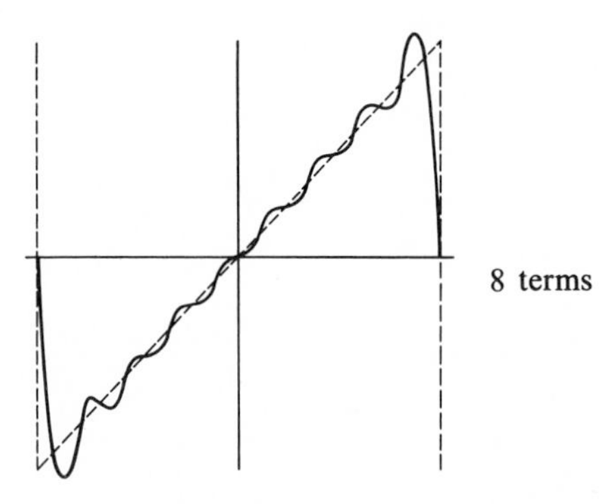

FIG. 4-6

Two features of the expansion should be noted. At the points of discontinuity ($t = \pm\tau/2$) the series yields the mean value (*zero*), and in the region immediately adjacent to the points of discontinuity, the expansion "overshoots" the original function. This latter effect is known as the *Gibb's phenomenon** and occurs in all orders of approximation. The Gibb's overshoot amounts to about 9% on each side of *any* discontinuity even in the limit of an infinite series.

▶ Example 4.5(b) Half-Wave Driving Force

Consider the function shown in Fig. 4-7 which consists of the positive portions of a sine function. (Such a function represents, for example, the output of a half-wave rectifying circuit.) Therefore, we have for the first complete period

$$F(t) = \begin{cases} \sin \omega t, & 0 < t < \pi/\omega \\ 0, & \pi/\omega < t < 2\pi/\omega \end{cases} \tag{1}$$

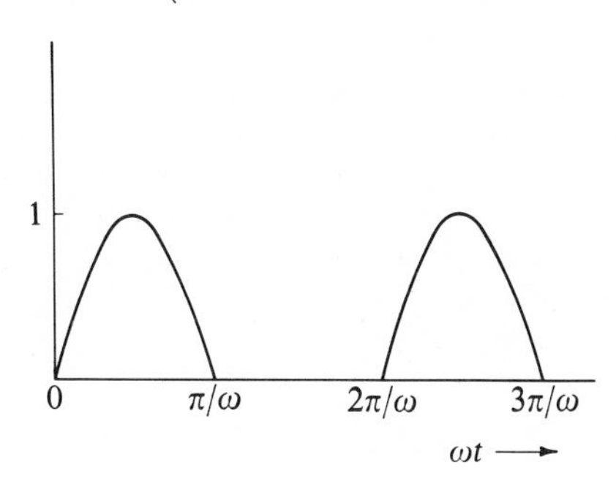

FIG. 4-7

The evaluation of the Fourier coefficients yields the following results:

$$a_r = \begin{cases} -\dfrac{2}{\pi(r^2-1)}, & r \text{ even (or 0)} \\ 0, & r \text{ odd} \end{cases}$$

$$b_1 = \tfrac{1}{2}$$

$$b_r = 0, \qquad r \geq 2$$

The solid curve in Fig. 4-8 shows the function,

$$\begin{aligned} F(t) &= \frac{1}{2}a_0 + b_1 \sin \omega t + a_2 \cos 2\omega t + a_4 \cos 4\omega t \\ &= \frac{1}{\pi} + \frac{1}{2} \sin \omega t - \frac{2}{3\pi} \cos 2\omega t - \frac{2}{15\pi} \cos 4\omega t \end{aligned} \tag{2}$$

* Josiah Willard Gibbs (1839–1903) discovered this effect empirically in 1898. A detailed discussion is given, e.g., by Davis (Da63, pp. 113–118). The amount of overshoot is actually 8.9490 ··· %.

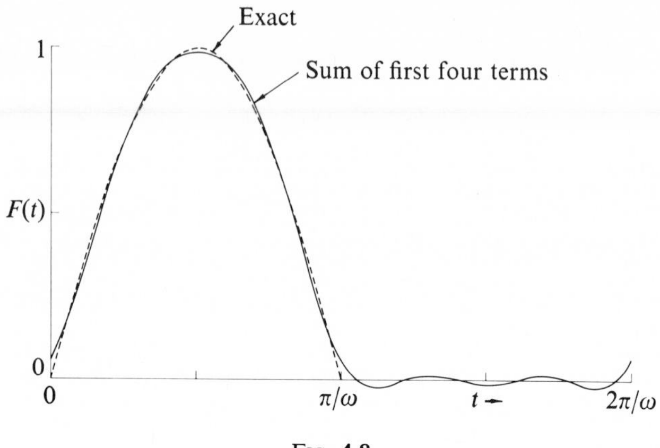

FIG. 4-8

and the dashed curve is the exact function. It is apparent that even the first few terms of the expansion give a fairly close approximation to the function. The original function in this case is "smoother" than the sawtooth function of the preceding example, so the convergence of the Fourier series is much more rapid. This is a general result: A highly discontinuous function can be approximated with reasonable accuracy by a Fourier series only if a large number of terms is used. That is, the representation of an irregular function contains many harmonics of the fundamental frequency.

4.6 The Response of Linear Oscillators to Impulsive Forcing Functions

In the previous discussions we have mainly considered steady-state oscillations. For many types of physical problems (particularly those involving oscillating electrical circuits), the transient effects are quite important; indeed, the transient solution may be of dominating interest in such cases. In this section we shall investigate the transient behavior of a linear oscillator which is subjected to a driving force that acts discontinuously. Of course, a "discontinuous" force is an idealization since it always takes a finite time to apply a force. But if the application time is small compared to the natural period of the oscillator, the result of the ideal case is a close approximation to the actual physical situation.

The differential equation which describes the motion of a damped oscillator is

$$\ddot{x} + 2\beta\dot{x} + \omega_0^2 x = \frac{F(t)}{m} \tag{4.63}$$

The general solution is composed of the complementary and particular solutions:

$$x(t) = x_c(t) + x_p(t) \tag{4.64}$$

We can write the complementary solution as

$$x_c(t) = e^{-\beta t}(A_1 \cos \omega_1 t + A_2 \sin \omega_1 t) \tag{4.65}$$

where

$$\omega_1 \equiv \sqrt{\omega_0^2 - \beta^2} \tag{4.66}$$

The particular solution $x_p(t)$ will, of course, depend on the nature of the forcing function $F(t)$.

Two types of idealized discontinuous forcing functions are of considerable interest; these are the *step function* (or *Heaviside function*) and the *impulse*

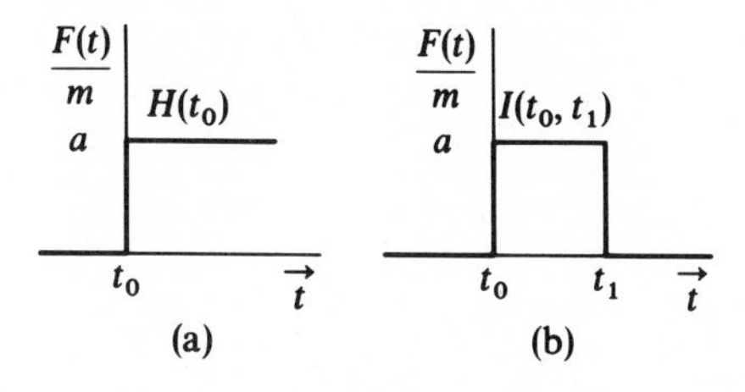

FIG. 4-9

function, shown in Figs. 4-9a and 4-9b, respectively. The step function H is given by

$$H(t_0) = \begin{cases} 0, & t < t_0 \\ a, & t > t_0 \end{cases} \tag{4.67}$$

where a is a constant with the dimensions of acceleration, and where the argument t_0 indicates that the time of application of the force is $t = t_0$.

The impulse function I may be considered to be a positive step function applied at $t = t_0$ followed by a negative step function applied at some later time t_1. Thus,

$$I(t_0, t_1) = H(t_0) - H(t_1)$$

$$I(t_0, t_1) = \begin{cases} 0, & t < t_0 \\ a, & t_0 < t < t_1 \\ 0, & t > t_1 \end{cases} \tag{4.68}$$

Although we write the Heaviside and impulse functions as $H(t_0)$ and $I(t_0, t_1)$ for simplicity, these functions, of course, depend on the time t and are more properly written as $H(t; t_0)$ and $I(t; t_0, t_1)$.

(a) RESPONSE TO A STEP FUNCTION. For this case the differential equation which describes the motion is

$$\ddot{x} + 2\beta\dot{x} + \omega_0^2 x = a, \qquad t > t_0 \tag{4.69}$$

We consider the initial conditions to be $x(t_0) = 0$ and $\dot{x}(t_0) = 0$. The particular solution is just a constant, and examination of Eq. (4.69) shows that it must be a/ω_0^2. Thus, the general solution for $t > t_0$ is

$$x(t) = e^{-\beta(t-t_0)}[A_1 \cos \omega_1(t - t_0) + A_2 \sin \omega_1(t - t_0)] + \frac{a}{\omega_0^2} \tag{4.70}$$

Application of the initial conditions yields

$$A_1 = -\frac{a}{\omega_0^2}; \qquad A_2 = -\frac{\beta a}{\omega_1 \omega_0^2} \tag{4.71}$$

Therefore, for $t > t_0$, we have

$$x(t) = \frac{a}{\omega_0^2}\left[1 - e^{-\beta(t-t_0)} \cos \omega_1(t - t_0) - \frac{\beta e^{-\beta(t-t_0)}}{\omega_1} \sin \omega_1(t - t_0)\right] \tag{4.72}$$

and, of course, $x(t) = 0$ for $t < t_0$.

If, for simplicity, we take $t_0 = 0$, the solution can be expressed as

$$x(t) = \frac{H(0)}{\omega_0^2}\left[1 - e^{-\beta t} \cos \omega_1 t - \frac{\beta e^{-\beta t}}{\omega_1} \sin \omega_1 t\right] \tag{4.72a}$$

This response function is shown in Fig. 4-10 for the case $\beta = 0.2\omega_0$. It is clear that the ultimate condition of the oscillator (i.e., the steady-state condition) is simply a displacement by an amount a/ω_0^2.

In the event that there is no damping, $\beta = 0$ and $\omega_1 = \omega_0$. Then, for $t_0 = 0$ we have

$$x(t) = \frac{H(0)}{\omega_0^2}[1 - \cos \omega_0 t], \qquad \beta = 0 \tag{4.73}$$

so that the oscillation is sinusoidal with amplitude extremes $x = 0$ and $x = 2a/\omega_0^2$ (see Fig. 4-10).

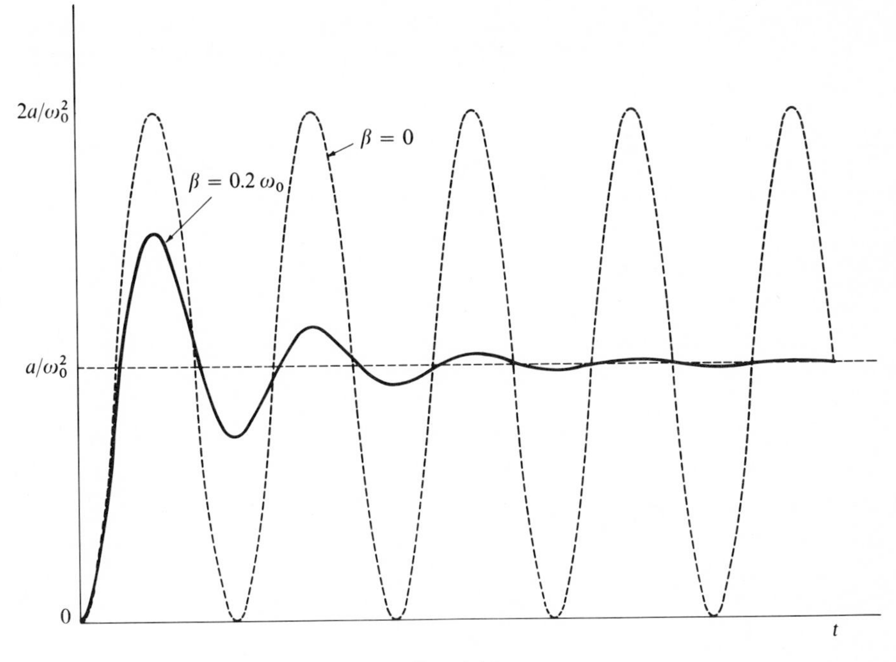

FIG. **4-10**

(b) RESPONSE TO AN IMPULSE FUNCTION. If we consider the impulse function as the difference between two step functions separated by a time $t_1 - t_0 = \tau$, then since the system is linear, the general solution for $t > t_1$ is given by the superposition of the solutions [Eq. (4.72)] for the two step functions taken individually:

$$\begin{aligned}
x(t) = \frac{a}{\omega_0^2}\left[1 - e^{-\beta(t-t_0)}\cos\omega_1(t-t_0) - \frac{\beta e^{-\beta(t-t_0)}}{\omega_1}\sin\omega_1(t-t_0)\right] \\
- \frac{a}{\omega_0^2}\left[1 - e^{-\beta(t-t_0-\tau)}\cos\omega_1(t-t_0-\tau)\right. \\
\left. - \frac{\beta e^{-\beta(t-t_0-\tau)}}{\omega_1}\sin\omega_1(t-t_0-\tau)\right] \\
= \frac{ae^{-\beta(t-t_0)}}{\omega_0^2}\left[e^{\beta\tau}\cos\omega_1(t-t_0-\tau) - \cos\omega_1(t-t_0)\right. \\
\left. + \frac{\beta e^{\beta\tau}}{\omega_1}\sin\omega_1(t-t_0-\tau) - \frac{\beta}{\omega_1}\sin\omega_1(t-t_0)\right], \quad t > t_1
\end{aligned} \tag{4.74}$$

The *total* response [i.e., Eqs. (4.72) and (4.74)] to an impulse function of duration $\tau = 5 \times 2\pi/\omega_1$, which is applied at $t = t_0$, is shown in Fig. 4-11 for the case $\beta = 0.2\omega_0$.

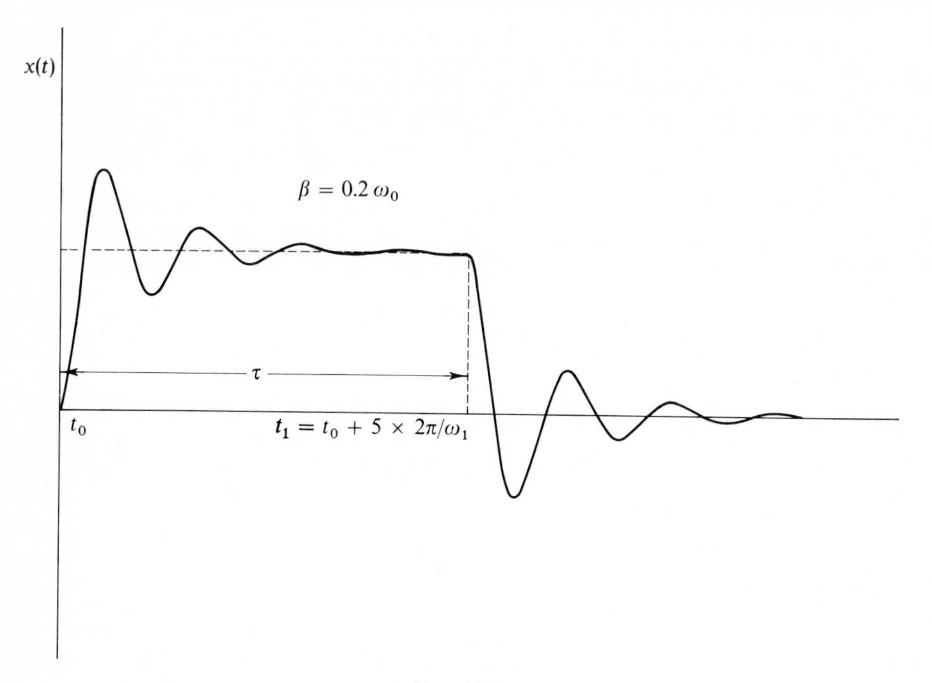

FIG. **4-11**

If the duration t of the impulse function is allowed to approach zero, then the response function will become vanishingly small. But if we allow $a \to \infty$ as $\tau \to 0$ in such a way that the product $a\tau$ is constant, then the response will be finite. This particular limiting case is of considerable importance since it approximates the application of a driving force which is a "spike" at $t = t_0$ (i.e., $\tau \ll 2\pi/\omega_1$).* If we expand Eq. (4.74) and allow $\tau \to 0$ but with $a\tau = b$, we obtain

$$x(t) = \frac{b}{\omega_1} e^{-\beta(t-t_0)} \sin \omega_1(t - t_0), \qquad t > t_0 \tag{4.75}$$

* A "spike" of this type is usually termed a *delta function* and is written $\delta(t - t_0)$. The delta function has the property that $\delta(t) = 0$ for $t \neq 0$ and $\delta(0) = \infty$, but

$$\int_{-\infty}^{+\infty} \delta(t - t_0)\, dt = 1$$

Therefore, this is not a proper function in the mathematical sense, but it can be defined as the limit of a well-behaved and highly local function (such as a Gaussian function) as the width parameter approaches zero. See also Marion (Ma65b, Section 1.11).

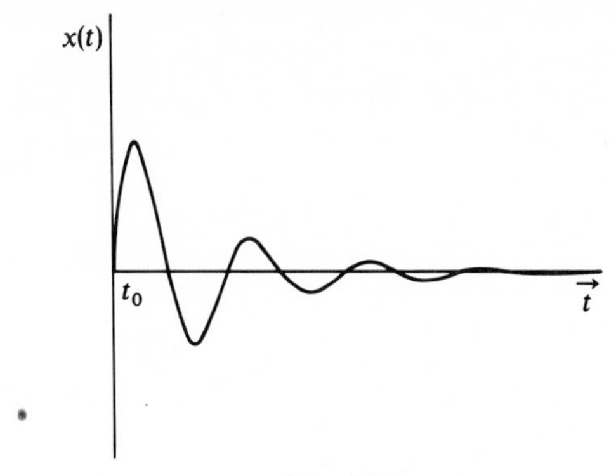

FIG. **4-12**

This response function is shown in Fig. 4-12 for the case $\beta = 0.2\omega_0$. It is evident that as t becomes large, the oscillator returns to its original position of equilibrium.

The fact that the response of a linear oscillator to an impulsive driving force can be represented in the simple manner of Eq. (4.75) leads to a powerful technique for dealing with general forcing functions, which was developed by George Green.* Green's method is based upon representing an arbitrary forcing function as a series of impulses, as shown schematically in Fig. 4-13.

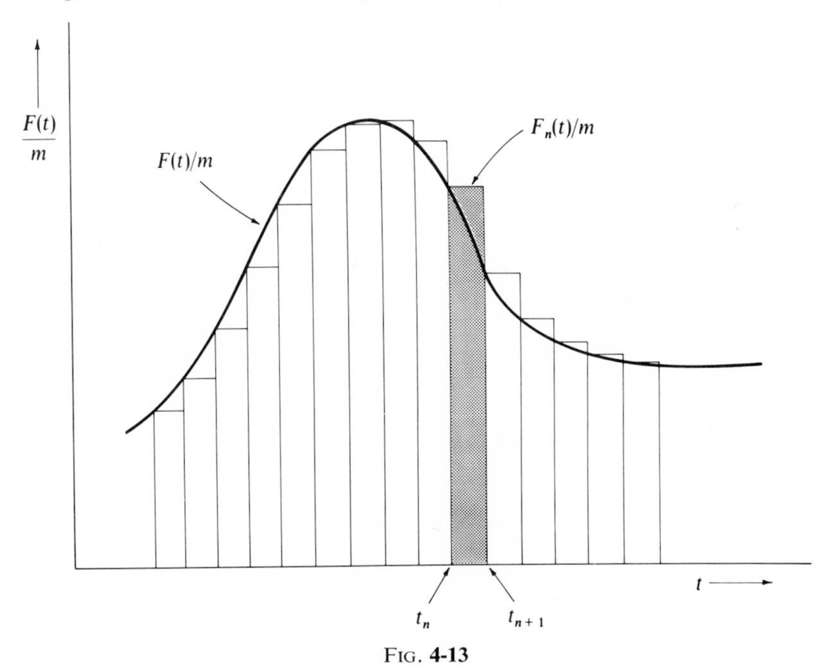

FIG. **4-13**

* George Green (1793–1841), a self-educated English mathematician.

If the driven system is linear, the principle of superposition is valid and it is permissible to express the inhomogeneous part of the differential equation as the sum of individual forcing functions $F_n(t)/m$, which in Green's method are impulse functions:

$$\ddot{x} + 2\beta\dot{x} + \omega_0^2 x = \sum_{n=-\infty}^{\infty} \frac{F_n(t)}{m}$$

$$= \sum_{n=-\infty}^{\infty} I_n(t) \tag{4.76}$$

where

$$I_n(t) = I(t_n, t_{n+1})$$

$$= \begin{cases} a_n(t_n), & t_n < t < t_{n+1} \\ 0, & \text{otherwise} \end{cases} \tag{4.77}$$

The interval of time over which I_n acts is $t_{n+1} - t_n = \tau$, and $\tau \ll 2\pi/\omega_1$. The solution for the nth impulse is, according to Eq. (4.75),

$$x_n(t) = \frac{a_n(t_n)\tau}{\omega_1} e^{-\beta(t-t_n)} \sin \omega_1(t - t_n), \qquad t > t_n + \tau \tag{4.78}$$

and the solution for all of the impulses up to and including the Nth impulse is

$$x(t) = \sum_{n=-\infty}^{N} \frac{a_n(t_n)\tau}{\omega_1} e^{-\beta(t-t_n)} \sin \omega_1(t - t_n), \qquad t_N < t < t_{N+1} \tag{4.79}$$

If we allow the interval τ to approach zero and write t_n as t', then the sum becomes an integral:

$$x(t) = \int_{-\infty}^{t} \frac{a(t')}{\omega_1} e^{-\beta(t-t')} \sin \omega_1(t - t')\, dt' \tag{4.80}$$

We define

$$G(t, t') \equiv \begin{cases} \dfrac{1}{m\omega_1} e^{-\beta(t-t')} \sin \omega_1(t - t'), & t \geq t' \\ 0, & t < t' \end{cases} \tag{4.81}$$

Then, since

$$ma(t') = F(t') \tag{4.82}$$

we have

$$\boxed{x(t) = \int_{-\infty}^{t} F(t')G(t, t')\, dt'} \tag{4.83}$$

The function $G(t, t')$ is known as the *Green's function* for the linear oscillator equation (4.63). The solution expressed by Eq. (4.82) is valid only for an oscillator initially at rest in its equilibrium position since the solution which we used for a single impulse [Eq. (4.75)] was obtained for just such an initial condition. For other initial conditions the general solution may be obtained in an analogous manner.

Green's method is generally useful for the solution of linear, inhomogeneous differential equations. The main advantage of the method lies in the fact that the Green's function $G(t, t')$, which is the solution of the equation for an infinitesimal element of the inhomogeneous part, *already contains the initial conditions*, so that the general solution, expressed by the integral of $F(t')G(t, t')$, automatically contains the initial conditions also.

▶ Example 4.6 An Application of Green's Method

Consider an exponentially-decaying forcing function which begins at $t = 0$:

$$F(t) = F_0 e^{-\gamma t}, \qquad t > 0 \tag{1}$$

The solution for $x(t)$ according to Green's method is

$$x(t) = \frac{F_0}{m\omega_1}\int_0^t e^{-\gamma t'} e^{-\beta(t-t')} \sin \omega_1(t - t')\, dt' \tag{2}$$

Making a change of variable to $z = \omega_1(t - t')$, we find

$$\begin{aligned} x(t) &= \frac{F_0}{m\omega_1^2}\int_0^{\omega_1 t} e^{-\gamma t}\, e^{[(\gamma-\beta)/\omega_1]z} \sin z\, dz \\ &= \frac{F_0/m}{(\gamma-\beta)^2 + \omega_1^2}\left[e^{-\gamma t} - e^{-\beta t}\left(\cos \omega_1 t - \frac{\gamma-\beta}{\omega_1} \sin \omega_1 t\right)\right] \end{aligned} \tag{3}$$

This response function is illustrated in Fig. 4-14 for three different combinations of the damping parameters, β and γ. When γ is large compared to β, and if both are small compared to ω_0, then the response approaches that for a "spike"; compare Fig. 4-12 with the upper curve in Fig. 4-14. When γ is small compared to β, the response approaches the shape of the forcing function itself, i.e., an initial increase followed by an exponential decay. The lower curve in Fig. 4-14 shows a decaying amplitude on which is superimposed a residual oscillation. When β and γ are equal, Eq. (3) becomes

$$x(t) = \frac{F_0}{m\omega_1^2} e^{-\beta t}(1 - \cos \omega_1 t), \qquad \beta = \gamma \tag{4}$$

Thus, the response is oscillatory with a "period" equal to $2\pi/\omega_1$ but with an exponentially decaying amplitude, as shown in the middle curve of Fig. 4-14.

A response of the type given by Eq. (3) could result, for example, if a quiescent, but intrinsically oscillatory electronic circuit were suddenly driven by the decaying voltage on a capacitor.

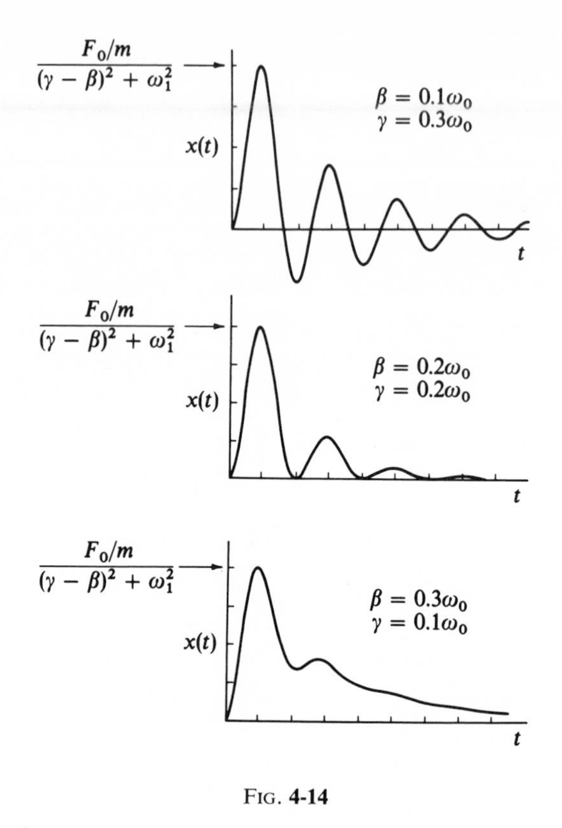

FIG. **4-14**

4.7 The Laplace Transform Method

In the preceding sections we have mainly used straightforward methods in solving the differential equations that describe oscillatory motion. The procedure has been to obtain a general solution and then to impose the initial conditions in order to obtain the desired particular solution. Green's method (Section 4.6) is a technique for automatically including initial conditions. Another related procedure that also yields a particular solution is the *Laplace transform method.* This technique, which is generally useful for obtaining solutions to linear differential equations, allows the reduction of a *differential* equation to an *algebraic* equation, with an obvious consequent decrease in complexity. This is accomplished by defining the *Laplace transform* $f(p)$ of a function $F(t)$ according to

$$f(p) = \int_0^\infty e^{-pt} F(t)\, dt \tag{4.84}$$

The Laplace transform of a function $F(t)$ exists if $F(t)$ is sectionally continuous in every finite interval, $0 < t < \infty$, and if $F(t)$ increases at a rate less than exponential as t becomes infinitely large. In general, the parameter p may be complex, but we shall not have occasion to consider such cases here. The Laplace transform of a function $F(t)$ will be denoted by $\mathscr{L}[F(t)] = f(p)$.

Most of the interesting Laplace transforms can be evaluated by using elementary methods of integration. For example, if $F(t) = 1$,

$$\mathscr{L}[1] = \int_0^\infty e^{-pt}\,dt = -\frac{1}{p}e^{-pt}\Big|_0^\infty = \frac{1}{p}, \qquad p > 0$$

Similarly, for $F(t) = e^{-at}$,

$$\mathscr{L}[e^{-at}] = \int_0^\infty e^{-pt}e^{-at}\,dt = -\frac{1}{p+a}e^{-(p+a)t}\Big|_0^\infty$$

$$= \frac{1}{p+a}, \qquad p + a > 0$$

The list of transforms appearing in Table 4.1 can be verified by similar calculations.*

Table 4.1

SOME IMPORTANT LAPLACE TRANSFORMS

$\mathscr{L}[1] = \dfrac{1}{p}$,	$p > 0$
$\mathscr{L}[e^{at}] = \dfrac{1}{p-a}$,	$p > a$
$\mathscr{L}[t^n] = \dfrac{n!}{p^{n+1}}$,	$p > 0$, $n = 1, 2, 3, \ldots$
$\mathscr{L}[t^n e^{at}] = \dfrac{n!}{(p-a)^{n+1}}$,	$p > 0$, $n = 1, 2, 3, \ldots$
$\mathscr{L}[\sin \omega t] = \dfrac{\omega}{p^2 + \omega^2}$,	$p > 0$
$\mathscr{L}[\cos \omega t] = \dfrac{p}{p^2 + \omega^2}$	$p > 0$
$\mathscr{L}[e^{-at} \sin \omega t] = \dfrac{\omega}{(p+a)^2 + \omega^2}$,	$p + a > 0$
$\mathscr{L}[e^{-at} \cos \omega t] = \dfrac{p+a}{(p+a)^2 + \omega^2}$,	$p + a > 0$

* A useful tabulation of Laplace transforms is given, for example, by Churchill (Ch44). An extensive list is included in Erdélyi (Er54).

Some important properties of Laplace transforms are the following:

I. The Laplace transformation is linear. If α and β are constants, then

$$\mathscr{L}[\alpha F(t) + \beta G(t)] = \alpha\mathscr{L}[F(t)] + \beta\mathscr{L}[G(t)] \tag{4.85}$$

II. The Laplace transform of the derivative of $F(t)$ is given by

$$\mathscr{L}[\dot{F}(t)] = p\mathscr{L}[F(t)] - F(0) \tag{4.86}$$

This is easily proved by an integration by parts:

$$\begin{aligned}\mathscr{L}[\dot{F}(t)] &= \int_0^\infty e^{-pt}\frac{dF}{dt}\,dt \\ &= F(t)e^{-pt}\Big|_0^\infty + p\int_0^\infty e^{-pt}F(t)\,dt \\ &= -F(0) + p\mathscr{L}[F(t)]\end{aligned}$$

The transforms of higher derivatives can be calculated similarly; for example,

$$\mathscr{L}[\ddot{F}(t)] = p^2\mathscr{L}[F(t)] - pF(0) - \dot{F}(0) \tag{4.87}$$

III. If $F(t)$ is zero until a certain time $\tau > 0$, it is frequently advantageous to transform the origin of the time scale to the instant $t = \tau$. This may be effected as follows. Since $F(t - \tau) = 0$ for $t < \tau$, we define a new variable $t' = t - \tau$ so that

$$\begin{aligned}\mathscr{L}[F(t-\tau)] &= \int_0^\infty e^{-pt}F(t-\tau)\,dt = \int_0^\infty e^{-p(t'+\tau)}F(t')\,dt' \\ &= e^{-p\tau}f(p) = e^{-p\tau}\mathscr{L}[F(t)]\end{aligned} \tag{4.88}$$

IV. The substitution of $p + a$ for the parameter p in the transform corresponds to multiplying $F(t)$ by e^{-at}. For example,

$$\mathscr{L}[\cos\omega t] = \frac{p}{p^2+\omega^2}, \qquad p > 0$$

so that

$$\mathscr{L}[e^{-at}\cos\omega t] = \frac{p+a}{(p+a)^2+\omega^2}, \qquad p + a > 0$$

▶ Example 4.7 Response of a Damped Linear Circuit to a Step Function

The equation of motion is just Eq. (4.69),

$$\ddot{x} + 2\beta\dot{x} + \omega_0^2 x = a, \qquad t > 0 \tag{1}$$

where the initial conditions are $x(0) = 0$, $\dot{x}(0) = 0$. We multiply Eq. (1) by $\exp(-pt)$ and integrate:

$$\int_0^\infty e^{-pt}\ddot{x}\,dt + 2\beta\int_0^\infty e^{-pt}\dot{x}\,dt + \omega_0^2\int_0^\infty e^{-pt}x\,dt = a\int_0^\infty e^{-pt}\,dt \tag{2}$$

In terms of Laplace transforms, we have

$$p^2 f(p) + 2\beta p f(p) + \omega_0^2 f(p) = a/p \tag{3}$$

from which

$$f(p) = \frac{a}{p(p^2 + 2\beta p + \omega_0^2)} \tag{4}$$

Using the method of partial fractions, this result can be expressed as

$$f(p) = \frac{a}{\omega_0^2}\left[\frac{1}{p} - \frac{p + 2\beta}{p^2 + 2\beta p + \omega_0^2}\right] \tag{5}$$

Completing the square in the denominator of the second term gives

$$f(p) = \frac{a}{\omega_0^2}\left[\frac{1}{p} - \frac{p + 2\beta}{(p+\beta)^2 + \omega_0^2 - \beta^2}\right] \tag{6}$$

This can be rewritten in standard form as

$$f(p) = \frac{a}{\omega_0^2}\left[\frac{1}{p} - \frac{p + \beta}{(p+\beta)^2 + \omega_1^2} - \frac{\beta}{\omega_1}\,\frac{\omega_1}{(p+\beta)^2 + \omega_1^2}\right] \tag{7}$$

where, as usual,

$$\omega_1^2 \equiv \omega_0^2 - \beta^2 \tag{8}$$

The transforms of Table 4.1 can now be used to identify the various terms in Eq. (7). We find

$$F(t) = x(t) = \frac{a}{\omega_0^2}\left[1 - e^{-\beta t}\cos\omega_1 t - \frac{\beta e^{-\beta t}}{\omega_1}\sin\omega_1 t\right], \qquad t > 0 \tag{9}$$

which is the same as Eq. (4.72a).

Suggested References

The books listed as suggested references for Chapter 3 all contain material on driven oscillations. Of particular interest is the discussion of Halfman (Ha62a, Vol. 1, Chapter 7) where response functions are treated with a leaning toward engineering applications. Many details also are given by Magnus (Ma65c, Chapter 5).

Intermediate texts on electricity and magnetism frequently contain material on electrical oscillations in connection with ac circuit analysis; see, for example, Peck (Pe53, Chapter 11), Reitz and Milford (Re60, Chapter 13), and Scott (Sc59, Chapter 9). Electrical engineering texts may be consulted for even greater detail.

Green's functions are discussed, for example, by Churchill (Ch58, Chapter 9) in connection with a general treatment of operational methods for the solution of differential equations. A brief introduction is given by Arfken (Ar66, Section 16.5). An extensive discussion is given by Dettman (De62, Chapter 4).

Problems

4-1. For a damped, driven oscillator, show that the average kinetic energy is the same at a frequency which is a given number of octaves* above resonance as at a frequency which is the same number of octaves below resonance.

4-2. Show that if a driven oscillator is only lightly damped, the Q of the system is approximately

$$Q \cong 2\pi \times \left(\frac{\text{Total energy}}{\text{Energy loss during one period}} \right)$$

4-3. For a lightly damped oscillator, show that $Q \cong \omega_0/\Delta\omega$ [Eq. (4.15)].

4-4. Plot a *velocity* resonance curve for a driven, damped oscillator with $Q = 6$ and show that the full width of the curve between the points corresponding to $\dot{x}_{max}/\sqrt{2}$ is approximately equal to $\omega_0/6$.

4-5. An electrical circuit consists of a resistor R and a capacitor C connected in series to a source of alternating emf. Find the expression for the current as a function of time and show that it decreases to zero as the frequency of the alternating emf approaches zero.

4-6. In a series R-L-C circuit, show that the voltage amplitude across the inductor, as a function of the frequency of the impressed emf, reaches a maximum at a frequency different from the resonance frequency $1/\sqrt{LC}$. Obtain an explicit formula for this frequency.

4-7. A source of alternating emf is connected to two impedances, Z_1 and Z_2, in series. Obtain an expression for the power dissipated in Z_2.

4-8. Calculate the impedance of the circuit shown in the figure by considering the two branches (with impedance Z_1 and Z_2) to be connected in parallel, so that

$$\frac{1}{Z} = \frac{1}{Z_1} + \frac{1}{Z_2}$$

* An *octave* is a frequency interval in which the highest frequency is just twice the lowest frequency.

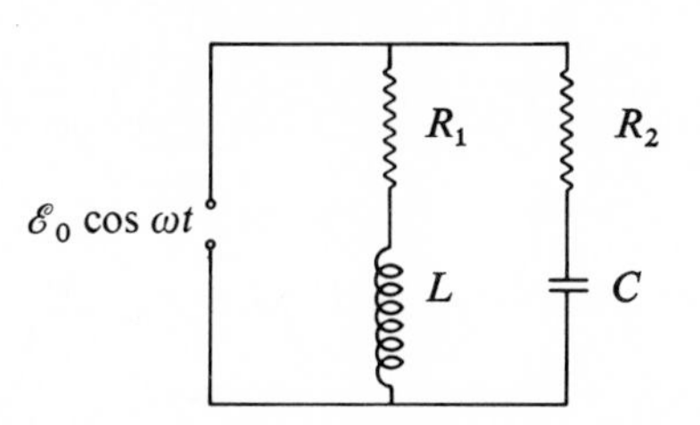

Obtain an expression for the total current when it is *in phase* with the impressed emf (i.e., when the circuit is in resonance). Show that the current vanishes if $R_1 = R_2 = 0$. Interpret this result physically.

4-9. Consider the parallel *R-L-C* circuit shown in the figure. Calculate the impedance of this circuit and show that for $\omega = \omega_0 = 1/\sqrt{LC}$, the current

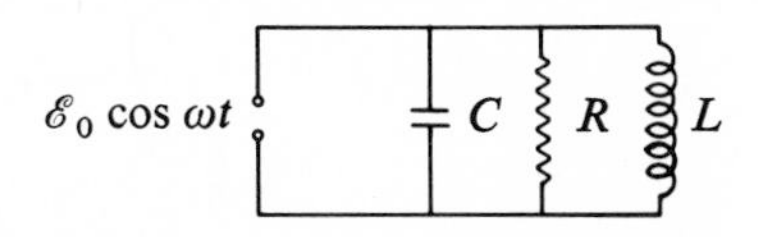

flow is a *minimum* (i.e., there is current *anti*resonance). What is the Q of the circuit? Plot $|Z|/R$ linearly *versus* a logarithmic frequency scale. Show that the curve is then *symmetrical.*

4-10. The figure illustrates a mass m_1 driven by a sinusoidal force whose frequency is ω. The mass m_1 is attached to a rigid support by means of a spring of force constant k and slides on a second mass m_2. The frictional force

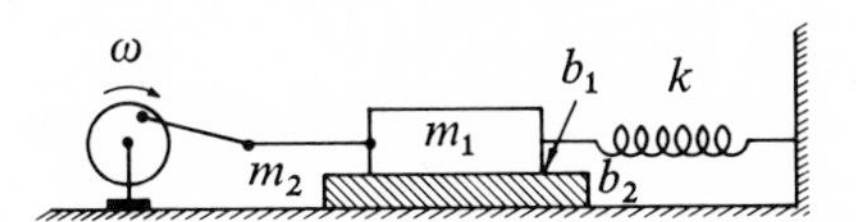

between m_1 and m_2 is represented by the damping parameter b_1 and the friction force between m_2 and the support is represented by b_2. Construct the electrical analog of this system and calculate the impedance.

4-11. Show that the Fourier series of Eq. (4.61) can be expressed as

$$F(t) = \tfrac{1}{2}a_0 + \sum_{n=1}^{\infty} c_n \cos(n\omega t - \phi_n)$$

Relate the coefficients c_n to the a_n and b_n of Eqs. (4.62).

4-12. Obtain the Fourier expansion of the function

$$F(t) = \begin{cases} -1, & -\pi/\omega < t < 0 \\ +1, & 0 < t < \pi/\omega \end{cases}$$

in the interval $-\pi/\omega < t < \pi/\omega$. Calculate and plot the sums of the first two terms, the first three terms, and the first four terms to demonstrate the convergence of the series.

4-13. Obtain the Fourier series which represents the function

$$F(t) = \begin{cases} 0, & -2\pi/\omega < t < 0 \\ \sin \omega t, & 0 < t < 2\pi/\omega \end{cases}$$

4-14. Obtain the Fourier representation of the output of a full-wave rectifier. Plot the first three terms of the expansion and compare with the exact function.

4-15. A damped linear oscillator, originally at rest in its equilibrium position, is subjected to a forcing function given by

$$\frac{F(t)}{m} = \begin{cases} 0, & t < 0 \\ a \times (t/\tau), & 0 < t < \tau \\ a, & t > \tau \end{cases}$$

Find the response function. Allow $\tau \to 0$ and show that the solution becomes that for a step function.

4-16. Obtain the response of a linear oscillator to a step function and to an impulse function (in the limit $\tau \to 0$) for the case of overdamping. Sketch the response functions.

4-17. Calculate the maximum values of the amplitudes of the response functions shown in Figs. 4-10 and 4-12. Obtain numerical values (in units of a/ω_0^2) for $\beta = 0.2\omega_0$.

4-18. Consider an undamped linear oscillator with a natural frequency ω_0. Calculate and sketch the response function for an impulse forcing function which acts for a time $\tau = 2\pi/\omega_0$. Give a physical interpretation of the results.

4-19. Obtain the response of a linear oscillator to the forcing function

$$\frac{F(t)}{m} = \begin{cases} 0, & t < 0 \\ a \sin \omega t, & 0 < t < \pi/\omega \\ 0, & t > \pi/\omega \end{cases}$$

4-20. Derive an expression for the displacement of a linear oscillator analogous to Eq. (4.75) but for the initial conditions $x(t_0) = x_0$ and $\dot{x}(t_0) = \dot{x}_0$.

4-21. Derive the Green's method solution for the response due to an arbitrary forcing function by considering the function to consist of a series of step functions; i.e., start from Eq. (4.72) rather than from Eq. (4.75).

4-22. Use Green's method to obtain the response of a damped oscillator to a forcing function of the form

$$F(t) = \begin{cases} 0, & t < 0 \\ F_0 e^{-\gamma t} \sin \omega t, & t > 0 \end{cases}$$

4-23. Verify by direct integration the Laplace transforms given in the third through the sixth entries in Table 4.1.

4-24. Use the Laplace transform method to obtain the response of a series R-L-C circuit to which is applied a step-function voltage at $t = 0$.

4-25. Calculate the Laplace transform of the trapezoidal pulse illustrated in the figure.

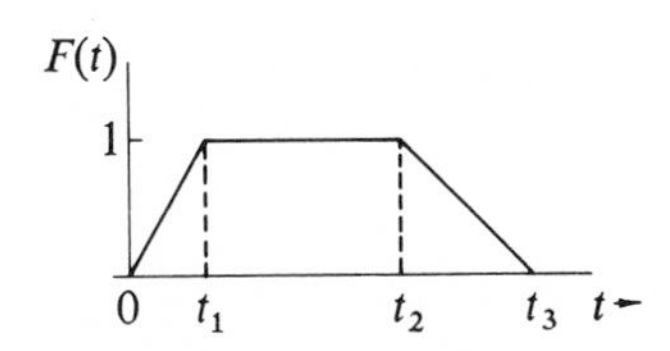

4-26. Work Example 4.7 using the Laplace transform method.

CHAPTER 5

Nonlinear Oscillations

5.1 Introduction

The extensive discussion of linear oscillatory systems given in the preceding two chapters is warranted by the great importance of oscillatory phenomena in many areas of physics and engineering. It is frequently permissible to use the linear approximation in the analysis of such systems. The usefulness of these analyses is due in large measure to the fact that *analytical* methods can usually be employed. When pressed to divulge greater detail, however, Nature insists on being *non*linear in general.

In making the transition from linear to nonlinear systems, we must exercise some caution since many of the more useful results of linear analysis are not applicable for nonlinear systems. Most important is the fact that the principle of superposition does not apply for nonlinear differential equations. Therefore, if we find the response of a nonlinear system to two separate forcing functions, the application of a combination of the forcing functions will not in general result in a response which is the sum of the individual responses. This fact makes difficult the generalization of solutions of nonlinear equations; essentially every problem must be treated as a special case. Therefore, in order to contend with these equations, a large number of techniques have been developed. In this chapter we shall discuss several of the more useful methods of attack. We begin with qualitative considerations which can be expressed graphically in terms of *phase diagrams* and then proceed to some of the more quantitative methods. Certain types of nonlinear equations can be solved exactly in terms of *elliptic integrals* but more often one must resort to various approximation procedures. Some of these methods are discussed in detail.

5.2 Oscillations for General Potential Functions

Consider the motion of a particle in some arbitrary conservative force field for which the potential energy of the particle as a function of its displacement may be represented by a curve such as that in Fig. 5-1. Now, in general, for a conservative system,

$$E = T + U = \text{const.}$$

or,

$$E = \tfrac{1}{2}m\dot{x}^2 + U(x) \tag{5.1}$$

from which

$$\boxed{\dot{x} = \sqrt{\frac{2}{m}[E - U(x)]}} \tag{5.2}$$

If the particle has a total energy E as indicated by the horizontal dashed line in Fig. 5-1, then we can distinguish several possible cases for the motion:

(a) $x < x_1$. In this region the potential energy exceeds the total energy, so that, according to Eq. (5.2), the velocity is imaginary. Hence, the particle is excluded from this region.

(b) $x_1 < x < x_2$. In this region the total energy exceeds the potential energy so that the particle is permitted to move within the limits $x = x_1$ and

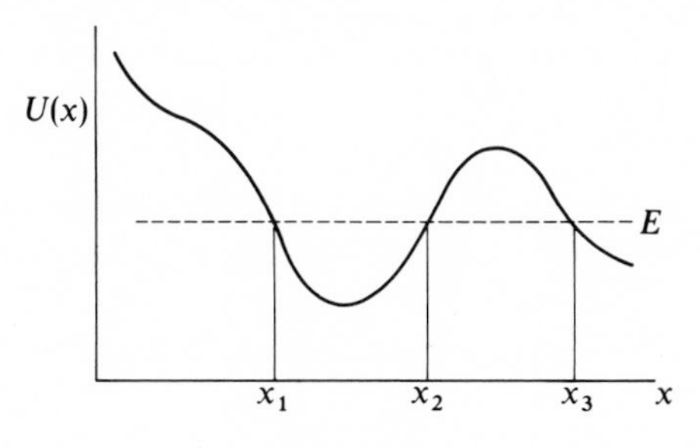

FIG. 5-1

$x = x_2$. At these two extremes, we have $E = U$ and hence the velocity is zero. These points are called the *turning points* of the motion, since the particle must reverse its motion upon reaching the *potential barrier* at either extreme. The motion is therefore oscillatory within the *potential well* defined by $U(x)$ between $x = x_1$ and $x = x_2$.

(c) $x_2 < x < x_3$. This region is again excluded since $E < U$.

(d) $x > x_3$. If $U(x)$ is always less than its value at x_3, the entire region from $x = x_3$ to $x \to \infty$ is a permissible region. If the particle approaches $x = x_3$ from larger x, then when the potential barrier is encountered the particle is *reflected* and proceeds unimpeded out again to infinitely large x.

The period of motion in a potential well may be obtained by integrating Eq. (5.2):

$$\tau = \int dt = \int \frac{dx}{\sqrt{\frac{2}{m}(E - U)}} \tag{5.3}$$

If $E = \text{const.}$ is considered the parameter of the motion, then, since the motion is symmetrical, the period is equal to twice the transit time from $x = x_1$ to $x = x_2$:

$$\boxed{\tau(E) = \sqrt{2m} \int_{x_1(E)}^{x_2(E)} \frac{dx}{\sqrt{E - U(x)}}} \tag{5.4}$$

where the turning points $x = x_1$ and $x = x_2$ are given by the roots of Eq. (5.2) for $\dot{x} = 0$, i.e., for $U(x) = E$. Note that this is an *improper integral* (the integrand becomes infinite at the limits), but on physical grounds the integral must *exist* since the motion lies entirely within the potential well.

If the potential energy is of the form

$$U(x) = \tfrac{1}{2}kx^2 \tag{5.5a}$$

then the corresponding force is

$$F(x) = -kx \tag{5.5b}$$

This is just the case of simple harmonic motion discussed in Section 3.2. Now, if a particle moves in a potential well which is some arbitrary function of distance, then in the vicinity of the minimum of the well, the potential can usually be quite well approximated by a parabola. Therefore, if the energy of the particle is only slightly greater than U_{min}, only small amplitudes are possible and the motion is approximately simple harmonic. If the energy is appreciably greater than U_{min} so that the amplitude of the motion cannot be considered small, then it may no longer be sufficiently accurate to make the approximation $U(x) \cong \frac{1}{2}kx^2$ and we must deal with a *nonlinear* force.

In many physical situations the deviation of the force from linearity is *symmetrical* about the equilibrium position (which we take to be at $x = 0$). That is, the *magnitude* of the force exerted on a particle is the same at $-x$ as at x; the *direction* of the force is, of course, opposite in the two cases. Therefore, in a symmetric situation, the first correction to a linear force must be a term proportional to x^3; hence,

$$F(x) \cong -kx + \varepsilon x^3 \tag{5.6a}$$

where ε is usually a small quantity. The potential corresponding to such a force is

$$U(x) = \tfrac{1}{2}kx^2 - \tfrac{1}{4}\varepsilon x^4 \tag{5.6b}$$

Depending upon the sign of the quantity ε, the force may either be greater or less than the linear approximation. If $\varepsilon > 0$, then the force is less than the linear term alone and the system is said to be *soft*; if $\varepsilon < 0$, then the force is greater and the system is *hard*. Figure 5-2 shows the form of the force and the potential for a *soft* and a *hard* system.

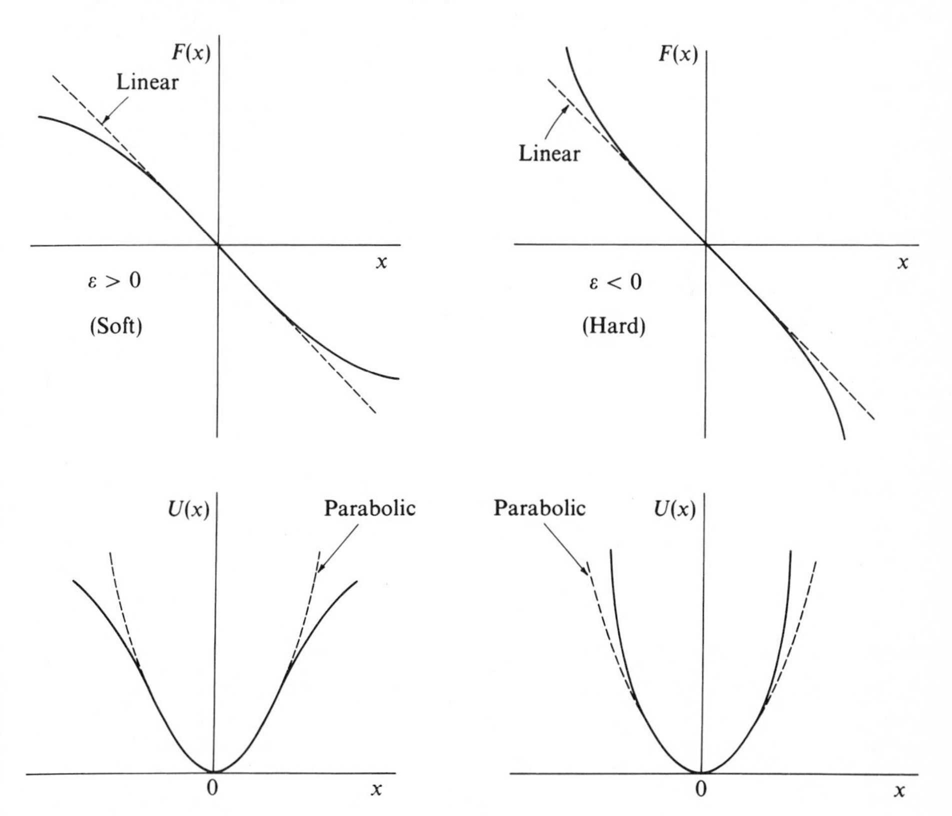

FIG. 5-2

▶ Example 5.2 An Intrinsically Nonlinear Spring System

Consider a particle of mass m which is suspended between two identical springs, as in Fig. 5-3a. If both springs are in their unextended conditions (i.e., there is no tension, and therefore no potential energy, in either spring) when the particle is

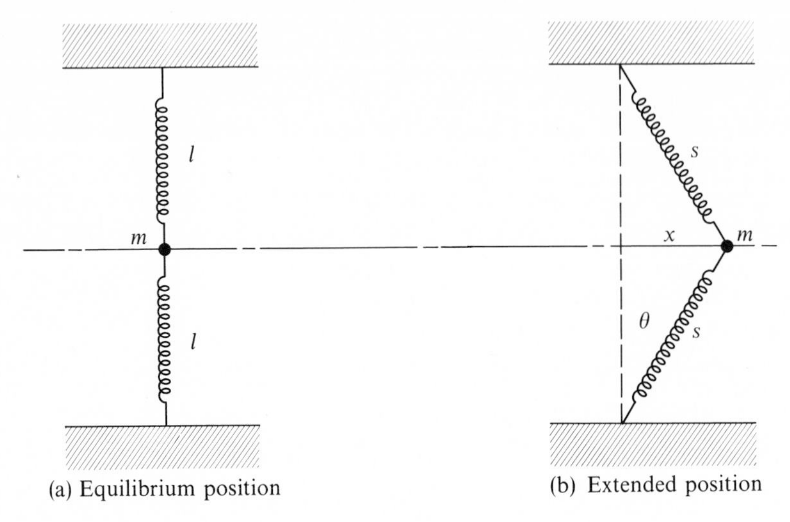

(a) Equilibrium position (b) Extended position

FIG. **5-3**

in its equilibrium position, and if we neglect gravitational forces, then when the particle is displaced from equilibrium (Fig. 5-3b), each spring exerts a force $-k(s - l)$ on the particle (k is the force constant of each spring). The net (horizontal) force on the particle is

$$F = -2k(s - l)\sin\theta \tag{1}$$

Now,

$$s = \sqrt{l^2 + x^2} \tag{2}$$

so that

$$\sin\theta = \frac{x}{s} = \frac{x}{\sqrt{l^2 + x^2}} \tag{3}$$

Hence,

$$\begin{aligned} F &= -\frac{2kx}{\sqrt{l^2 + x^2}} \cdot (\sqrt{l^2 + x^2} - l) \\ &= -2kx\left(1 - \frac{1}{\sqrt{1 + (x/l)^2}}\right) \end{aligned} \tag{4}$$

If we consider x/l to be a small quantity and expand the radical, we find

$$F = -kl\left(\frac{x}{l}\right)^3\left[1-\frac{3}{4}\left(\frac{x}{l}\right)^2+\cdots\right] \tag{5}$$

If we neglect all terms except the leading term, we have, approximately,

$$F(x) \cong -(k/l^2)x^3 \tag{6}$$

Therefore, even if the amplitude of the motion is sufficiently restricted so that x/l is a small quantity, we still have the result that the force is proportional to x^3. The system is therefore *intrinsically nonlinear*. On the other hand, if it had been necessary to stretch each spring a distance d in order to attach it to the mass when at the equilibrium position, then we would find for the force (see Problem 5-2)

$$F(x) \cong -2(kd/l)x - [k(l-d)/l^3]x^3 \tag{7}$$

and a linear term is introduced.

From Eq. (7) we identify

$$\varepsilon = -k(l-d)/l^3 < 0 \tag{8}$$

Thus, the system is *hard*, and for oscillations with small amplitude the motion is approximately simple harmonic.

In real physical situations we are often concerned with symmetric forces and potentials, but there also exist cases which have asymmetric forms; for example,

$$F(x) = -kx + \lambda x^2 \tag{5.7a}$$

the potential for which is

$$U(x) = \tfrac{1}{2}kx^2 - \tfrac{1}{3}\lambda x^3 \tag{5.7b}$$

This case is illustrated in Fig. 5-4, in which the system is *hard* for $x > 0$ and *soft* for $x < 0$.

In Section 5.5 we shall investigate some details of the motion in an asymmetric potential.

5.3 Phase Diagrams for Nonlinear Systems

The construction of a phase diagram for a nonlinear system may be accomplished by using Eq. (5.2); thus,

$$\dot{x}(x) \propto \sqrt{E - U(x)}$$

Since $U(x)$ is in general a complicated function, it is possible to obtain an analytic expression for $\dot{x}(x)$ only in rare instances; it is usually necessary to

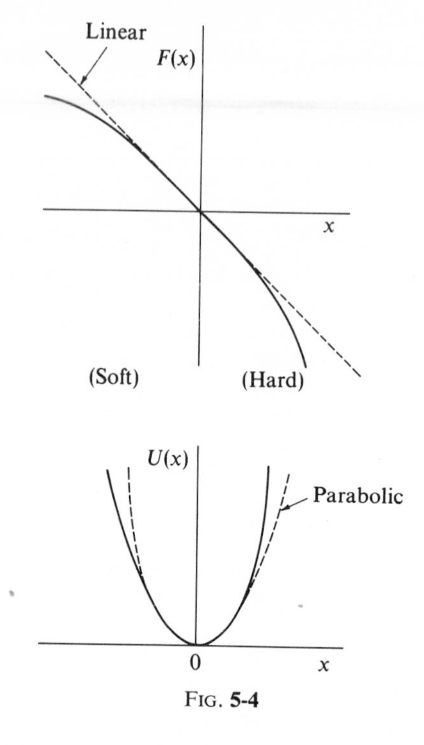

FIG. **5-4**

resort to various approximation procedures. On the other hand, it is relatively easy to obtain a qualitative picture of the phase diagram for the motion of a particle in an arbitary potential. For example, consider the asymmetric potential shown in the upper portion of Fig. 5-5, which represents a system that is *soft* for $x < 0$ and is *hard* for $x > 0$. If there is no damping present, then since $\dot{x}$ is proportional to $\sqrt{E - U(x)}$, it is clear that the phase diagram must be of the form shown in the lower portion of the figure. Three of the oval phase paths are drawn, corresponding to the three values of the total energy indicated by the dotted lines in the potential diagram. For a total energy only slightly greater than that of the minimum of the potential, the oval phase paths approach ellipses. If the system is damped, then the oscillating particle will "spiral down the potential well" and eventually come to rest at the equilibrium position, $x = 0$.

For the case shown in Fig. 5-5, if the total energy E of the particle is less than the height to which the potential rises on either side of $x = 0$, then the particle is "trapped" in the potential well (cf. the region $x_1 < x < x_2$ in Fig. 5-1). The point $x = 0$ is a position of *stable* equilibrium. We shall consider some details regarding the question of *stability* in a later chapter. For the present it suffices to state that a position of equilibrium is said to be stable if

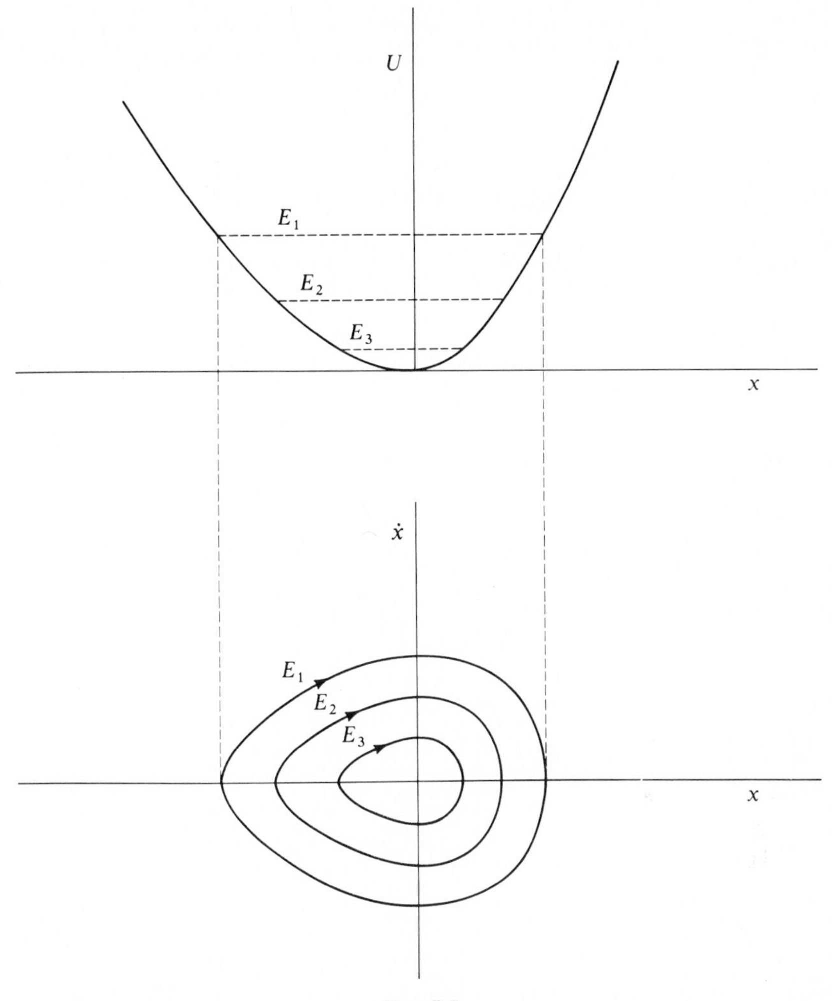

FIG. **5-5**

a small disturbance results in locally *bounded* motion; this is clearly the case for the potential of Fig. 5-5.

In the vicinity of the *maximum* of a potential, a qualitatively different type of motion occurs, as shown in Fig. 5-6. Here, the point $x = 0$ is one of *unstable* equilibrium, since, if a particle is at rest at this point, then a slight disturbance will result in locally *unbounded* motion.*

* It is clear that the definition of instability must be in terms of *locally* unbounded motion, for if there are other maxima of the potential which are greater than the one shown at $x = 0$, the motion will be bounded by these other potential barriers.

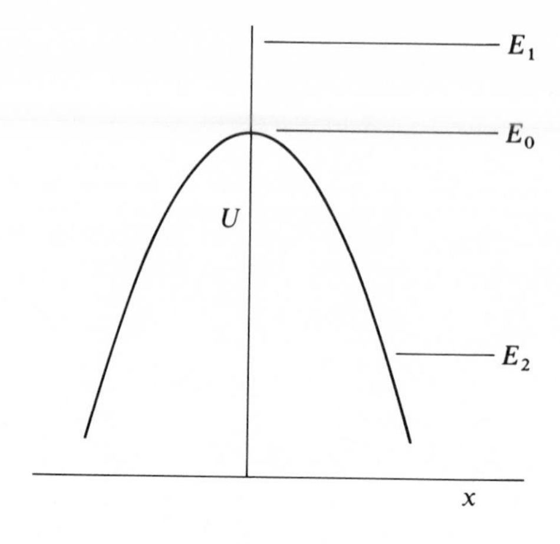

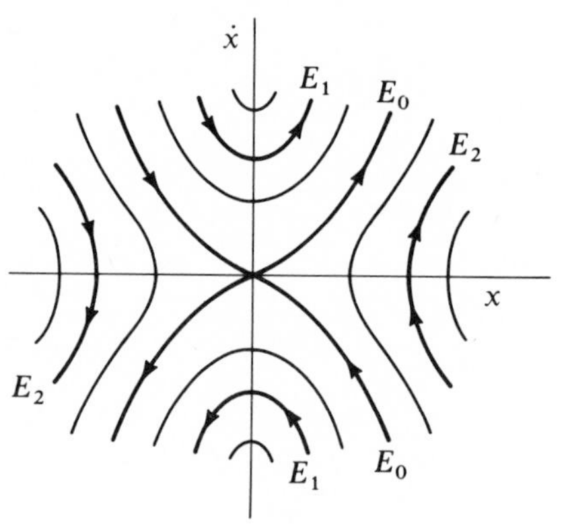

FIG. **5-6**

If the potential in Fig. 5-6 were parabolic, i.e., if $U(x) = -\frac{1}{2}kx^2$, then the phase paths corresponding to the energy E_0 would be straight lines and those corresponding to the energies E_1 and E_2 would be hyberbolas. This is therefore the limit to which the phase paths of Fig. 5-6 would approach if the nonlinear term in the potential were made to decrease in magnitude.

By referring to the phase paths for the potentials shown in Figs. 5-5 and 5-6, it is possible to rapidly construct a phase diagram for any arbitrary potential (such as that in Fig. 5-1).

An important type of nonlinear equation was extensively studied by van der Pol in connection with an investigation of oscillations in vacuum tube circuits.* This equation has the form

* B. van der Pol, *Phil. Mag.* **2**, 978 (1926). Extensive treatments of van der Pol's equation may be found, for example, in Minorsky (Mi47) or in Andronow and Chaikin (An49); brief discussions are given by Lindsay (Li51, pp. 64–66) and by Pipes (Pi46, pp. 606–610).

$$\ddot{x} - \mu(x_0^2 - x^2)\dot{x} + \omega_0^2 x = 0 \tag{5.8}$$

where μ is a small, positive parameter. A system which is described by van der Pol's equation has the following interesting property. If the amplitude $|x|$ exceeds the critical value $|x_0|$, then the coefficient of $\dot{x}$ is positive and the system is damped. On the other hand, if $|x| < |x_0|$, then there is *negative damping*, i.e., the amplitude of the motion *increases*. It follows that there must be some amplitude for which the motion neither increases nor decreases with time. Such a curve in the phase plane is called the *limit cycle** (see Fig. 5-7). Phase paths which lie outside the limit cycle spiral *inward* while those

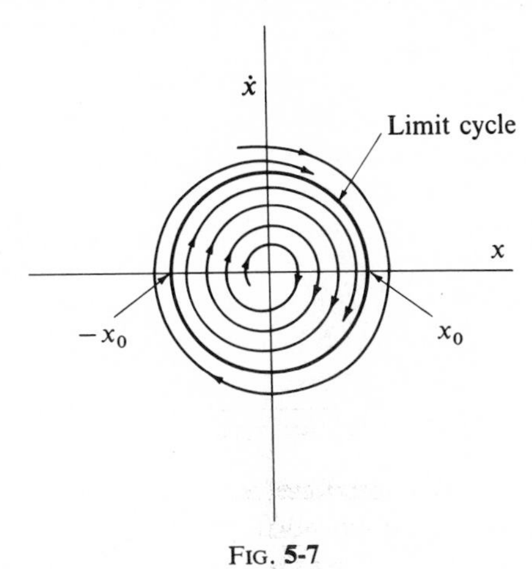

FIG. 5-7

which lie inside the limit cycle spiral *outward*. Inasmuch as the limit cycle defines locally *bounded* motion, we may refer to the situation which it represents as being *stable*.

A system which is described by van der Pol's equation has the property that it is *self-limiting*. That is, once set into motion under conditions that lead to an increasing amplitude, the amplitude is automatically prevented from growing without bound. The system has this property whether the initial amplitude is greater or smaller than the critical (limiting) amplitude, x_0.

5.4 The Plane Pendulum

The solutions of certain types of nonlinear oscillation problems can be expressed in closed form by means of elliptic integrals.† An example of this type is the *plane pendulum*. Consider a particle of mass m that is constrained

* The term was introduced by Poincaré and is often called the *Poincaré limit cycle*.
† See Appendix E.4 for a list of some elliptic integrals.

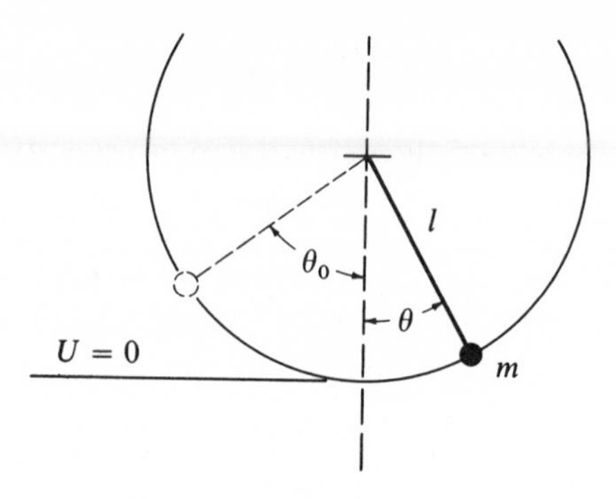

FIG. 5-8

by a weightless, extensionless rod to move in a vertical circle of radius l, as in Fig. 5-8. The gravitational force acts, of course, in the downward direction, but the component of this force which influences the motion is that which is *perpendicular* to the support rod. This force component, shown in Fig. 5-9, is simply $F(\theta) = -mg \sin \theta$. The plane pendulum is clearly a nonlinear system with a symmetric restoring force. It is only for small angular deviations that a linear approximation may be used.

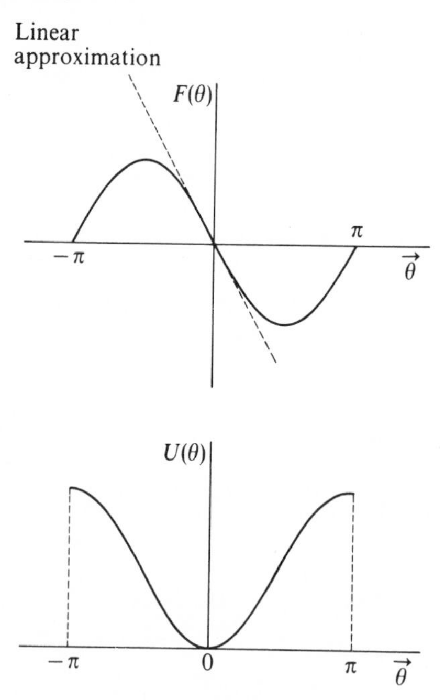

FIG. 5-9

The equation of motion for the plane pendulum is most easily obtained by equating the torque about the support axis to the product of the angular acceleration and the moment of inertia about the same axis:

$$I\ddot{\theta} = lF \tag{5.9}$$

or, since $I = ml^2$,

$$\boxed{\ddot{\theta} + \omega_0^2 \sin\theta = 0} \tag{5.10}$$

where

$$\omega_0^2 \equiv \frac{g}{l} \tag{5.11}$$

If the amplitude of the motion is small, we may approximate $\sin\theta \cong \theta$, and the equation of motion becomes identical with that for the simple harmonic oscillator:

$$\ddot{\theta} + \omega^2\theta = 0 \tag{5.12}$$

In this approximation, the period is given by the familiar expression

$$\tau \cong 2\pi\sqrt{\frac{l}{g}} \tag{5.13}$$

If we wish to obtain the general result for the period in the event that the amplitude is finite, we may begin with Eq. (5.10). However, since the system is *conservative*, we can use the fact that

$$T + U = E = \text{const.} \tag{5.14}$$

to obtain a solution by considering the energy of the system rather than by solving the equation of motion.

If the zero of potential energy is taken to be the lowest point on the circular path described by the pendulum bob (i.e., $\theta = 0$; see Fig. 5-9), the kinetic and potential energies can be expressed as

$$\left.\begin{aligned} T &= \tfrac{1}{2}I\omega^2 = \tfrac{1}{2}ml^2\dot{\theta}^2 \\ U &= mgl(1 - \cos\theta) \end{aligned}\right\} \tag{5.15}$$

If we let $\theta = \theta_0$ at the highest point of the motion, then

$$\left.\begin{aligned} T(\theta = \theta_0) &= 0 \\ U(\theta = \theta_0) &= E = mgl(1 - \cos\theta_0) \end{aligned}\right\} \tag{5.16}$$

Using the trigonometric identity

$$\cos \theta = 1 - 2 \sin^2(\theta/2)$$

we have

$$E = 2mgl \sin^2(\theta_0/2) \tag{5.17a}$$

and

$$U = 2mgl \sin^2(\theta/2) \tag{5.17b}$$

Expressing the kinetic energy as the difference between the total energy and the potential energy, we have

$$\tfrac{1}{2}ml^2\dot{\theta}^2 = 2mgl[\sin^2(\theta_0/2) - \sin^2(\theta/2)]$$

or,

$$\dot{\theta} = 2\sqrt{\frac{g}{l}}\,[\sin^2(\theta_0/2) - \sin^2(\theta/2)]^{\frac{1}{2}} \tag{5.18}$$

from which

$$dt = \frac{1}{2}\sqrt{\frac{l}{g}}\,[\sin^2(\theta_0/2) - \sin^2(\theta/2)]^{-\frac{1}{2}}\,d\theta \tag{5.19}$$

This equation may be integrated to obtain the period τ. Since the motion is symmetrical, the integral over θ from $\theta = 0$ to $\theta = \theta_0$ will yield $\tau/4$; hence,

$$\tau = 2\sqrt{\frac{l}{g}}\int_0^{\theta_0} [\sin^2(\theta_0/2) - \sin^2(\theta/2)]^{-\frac{1}{2}}\,d\theta \tag{5.20}$$

That this is actually an *elliptic integral of the first kind** may be seen more clearly by making the substitutions

$$z = \frac{\sin(\theta/2)}{\sin(\theta_0/2}; \qquad k = \sin(\theta_0/2) \tag{5.21}$$

Then,

$$dz = \frac{\cos(\theta/2)}{2\sin(\theta_0/2)}\,d\theta = \frac{\sqrt{1 - k^2z^2}}{2k}\,d\theta \tag{5.22}$$

from which

$$\tau = 4\sqrt{\frac{l}{g}}\int_0^1 [(1 - z^2)(1 - k^2z^2)]^{-\frac{1}{2}}\,dz \tag{5.23}$$

Numerical values for integrals of this type can be found in various tables.

* Refer to Eqs. (E.28a) and (E.28b), Appendix E.

In order for oscillatory motion to result, $\theta_0 < \pi$, or, equivalently, $\sin(\theta_0/2) = k < 1$. For this case, the integral in Eq. (5.23) can be evaluated by expanding $(1 - k^2z^2)^{-\frac{1}{2}}$ in a power series:

$$(1 - k^2z^2)^{-\frac{1}{2}} = 1 + \frac{k^2z^2}{2} + \frac{3k^4z^4}{8} + \cdots \tag{5.24}$$

Then the expression for the period becomes

$$\begin{aligned}\tau &= 4\sqrt{\frac{l}{g}}\int_0^1 \frac{dz}{(1-z^2)^{\frac{1}{2}}}\left[1 + \frac{k^2z^2}{2} + \frac{3k^4z^4}{8} + \cdots\right]\\ &= 4\sqrt{\frac{l}{g}}\cdot\left[\frac{\pi}{2} + \frac{k^2}{2}\cdot\frac{1}{2}\cdot\frac{\pi}{2} + \frac{3k^4}{8}\cdot\frac{3}{8}\cdot\frac{\pi}{2} + \cdots\right]\\ &= 2\pi\sqrt{\frac{l}{g}}\left[1 + \frac{k^2}{4} + \frac{9k^4}{64} + \cdots\right]\end{aligned} \tag{5.25}$$

If k is large (i.e., near 1), then many terms will be necessary to produce a reasonably accurate result. However, for small k, the expansion converges rapidly, and since $k = \sin(\theta_0/2)$, then $k \cong (\theta_0/2) - (\theta_0^3/48)$, and the result, correct to the fourth order, is

$$\tau \cong 2\pi\sqrt{\frac{l}{g}}\left[1 + \frac{1}{16}\theta_0^2 + \frac{11}{3072}\theta_0^4\right] \tag{5.26}$$

We therefore see that although the plane pendulum is not isochronous, it is very nearly so for small amplitudes of oscillation.*

The phase diagram for the plane pendulum (Fig. 5-10) is easy to construct since Eq. (5.18) provides the necessary relationship, $\dot{\theta} = \dot{\theta}(\theta)$. The parameter θ_0 specifies the total energy through Eq. (5.17a). If θ and θ_0 are small angles, then Eq. (5.18) can be written as

$$\left(\sqrt{\frac{l}{g}}\,\dot{\theta}\right)^2 + \theta^2 \cong \theta_0^2 \tag{5.27}$$

That is, if the coordinates of the phase plane are θ and $\dot{\theta}/\sqrt{g/l}$, then the phase paths near $\theta = 0$ are approximately circles. This result is expected, since for small θ_0 the motion is approximately simple harmonic.

For $-\pi < \theta < \pi$ and $E < 2mgl \equiv E_0$, the situation is equivalent to a particle bound in the potential well $U(\theta) = mgl(1 - \cos\theta)$ (see Fig. 5-9).

* This fact was discovered by Galileo in the cathedral at Pisa in 1581. The expression for the period of small oscillations was given by Huygens in 1673. Finite oscillations were first treated by Euler in 1736.

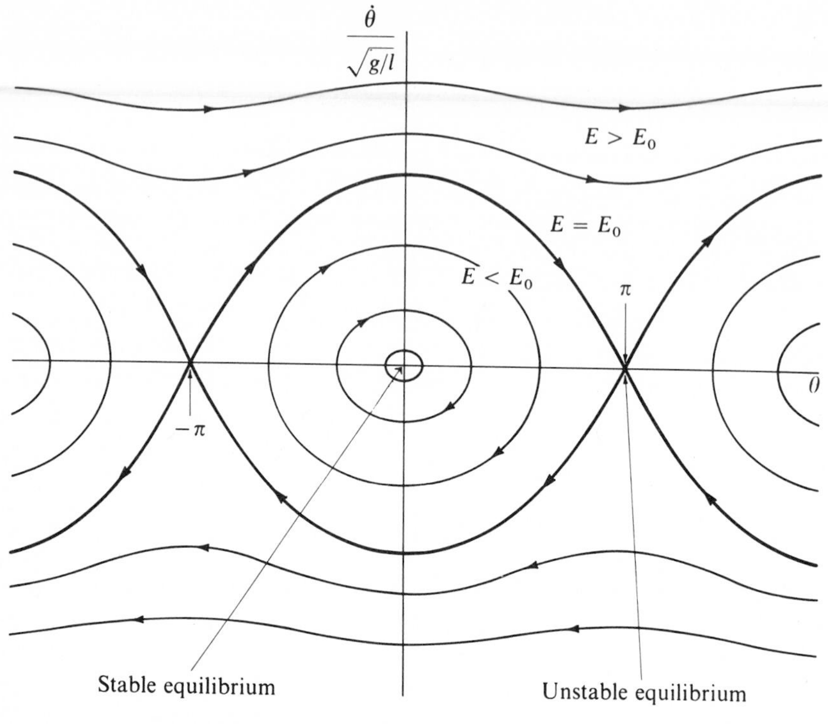

FIG. **5-10**

The phase paths are therefore closed curves for this region and are given by Eq. (5.18). Since the potential is periodic in θ, exactly the same phase paths exist for the regions $\pi < \theta < 3\pi$, $-3\pi < \theta < -\pi$, etc. The points $\theta = \cdots$, -2π, 0, 2π, $\cdots$ along the θ-axis are positions of *stable* equilibrium.

For values of the total energy that exceed E_0, the motion is no longer oscillatory, although it is still periodic. This situation corresponds to the pendulum executing complete revolutions about its support axis.

If the total energy equals E_0, then Eq. (5.17a) shows that $\theta_0 = \pi$. In this case Eq. (5.18) reduces to

$$\dot{\theta} = \pm 2\sqrt{\frac{g}{l}}\cos(\theta/2)$$

so that the phase paths for $E = E_0$ are just cosine functions (see the heavy curves in Fig. 5-10). There are two branches, depending on the direction of motion.

The phase paths for $E = E_0$ do not actually represent possible continuous motions of the pendulum. If the pendulum were at rest at, say $\theta = \pi$ (which is a point on the $E = E_0$ phase paths), then any small disturbance would cause the motion to follow *closely but not exactly on* one of the phase paths which diverge from $\theta = \pi$ since the total energy would be $E = E_0 + \delta$, where δ is a small but *nonzero* quantity. If the motion were along one of the $E = E_0$ phase paths, the pendulum would reach one of the points $\theta = n\pi$ with exactly zero velocity, but only after an infinite time! (This may be verified by evaluating Eq. (5.20) for $\theta_0 = \pi$; the result is $\tau \to \infty$.)

A phase path which separates locally bounded motion from locally unbounded motion (such as the path for $E = E_0$ in Fig. 5-10) is called a *separatrix*. A *separatrix* always passes through a point of unstable equilibrium.

■ 5.5 Nonlinear Oscillations in an Asymmetric Potential—The Method of Perturbations

We now consider a case in which the potential can be represented as

$$U(x) = \tfrac{1}{2}kx^2 - \tfrac{1}{3}m\lambda x^3 \tag{5.28}$$

with the force given by

$$F(x) = -kx + m\lambda x^2 \tag{5.29}$$

The differential equation which describes free oscillations without damping for a particle moving under the influence of such a force is

$$\ddot{x} + \omega_0^2 x - \lambda x^2 = 0 \tag{5.30}$$

We shall obtain an approximate solution to this equation by a *perturbation* calculation. Now, λ is by hypothesis a small quantity, so that the motion of the particle must be close to that of a simple harmonic oscillator.* Therefore, we can express the difference between the true motion and simple harmonic motion as a power series in λ. Thus, the solution may be written as

$$x(t) = x_0 + \lambda x_1 + \lambda^2 x_2 + \cdots \tag{5.31}$$

where x_0 is the solution for the simple oscillator and $x_1, x_2, \ldots,$ are functions of the time that are to be determined. We shall limit ourselves to a first-order calculation; i.e., we assume that λ is sufficiently small so that the approximation

$$x(t) \cong x_0 + \lambda x_1 \tag{5.32}$$

* One must be chary in making such statements concerning nonlinear equations. Reasoning which is based on experience with linear equations is not necessarily valid when dealing with nonlinear equations. We will simply *assume* that a perturbation expansion of this type can be made. A fundamental difficulty with this approach will be discussed in the following section.

is adequate. Substituting this expression into Eq. (5.30), we find

$$\ddot{x}_0 + \lambda \ddot{x}_1 + \omega_0^2 x_0 + \omega_0^2 \lambda x_1 - \lambda x_0^2 - 2\lambda^2 x_0 x_1 - \lambda^3 x_1^2 = 0 \tag{5.33}$$

But, λ is a small quantity, so that the terms in λ^2 and λ^3 can be neglected. [This is the same order of approximation as the neglect of such terms in Eq. (5.31) to obtain Eq. (5.32).] Then, collecting terms,

$$(\ddot{x}_0 + \omega_0^2 x_0) + \lambda(\ddot{x}_1 + \omega_0^2 x_1 - x_0^2) = 0 \tag{5.34}$$

This equation must be valid for arbitrary λ (as long as λ is small); hence, each term in parentheses must vanish separately:

$$\ddot{x}_0 + \omega_0^2 x_0 = 0 \tag{5.35a}$$

$$\ddot{x}_1 + \omega_0^2 x_1 = x_0^2 \tag{5.35b}$$

The procedure is to obtain a solution for $x_0(t)$ from Eq. (5.35a) and then to substitute the square into Eq. (5.35b) and solve for $x_1(t)$. First, we have

$$x_0(t) = A \cos \omega_0 t \tag{5.36}$$

where, for simplicity, we neglect the phase δ. Then,

$$\begin{aligned} x_0^2 &= A^2 \cos^2 \omega_0 t \\ &= \frac{A^2}{2} + \frac{A^2}{2} \cos 2\omega_0 t \end{aligned} \tag{5.37}$$

Substitution of x^2 into Eq. (5.35b) gives

$$\ddot{x}_1 + \omega_0^2 x_1 = \frac{A^2}{2} + \frac{A^2}{2} \cos 2\omega_0 t \tag{5.38}$$

Since the complementary function for Eq. (5.38) already occurs in the solution for $x_0(t)$ and since the final result for $x(t)$ will involve the sum of $x_0(t)$ and $x_1(t)$, we need only consider a particular integral for $x_1(t)$. We choose

$$x_1(t) = B \cos 2\omega_0 t + C \tag{5.39}$$

Then,

$$-3\omega_0^2 B \cos 2\omega_0 t + \omega_0^2 C = \frac{A^2}{2} + \frac{A^2}{2} \cos 2\omega_0 t \tag{5.40}$$

Hence,

$$B = -\frac{A^2}{6\omega_0^2}; \qquad C = \frac{A^2}{2\omega_0^2} \tag{5.41}$$

Thus, the first-order function becomes

$$x_1(t) = -\frac{A^2}{6\omega_0^2}\cos 2\omega_0 t + \frac{A^2}{2\omega_0^2} \tag{5.42}$$

and the solution for $x(t)$, correct to first order in λ, is

$$x(t) = A\cos\omega_0 t - \lambda\cdot\frac{A^2}{6\omega_0^2}(\cos 2\omega_0 t - 3) \tag{5.43}$$

Therefore, we see that the inclusion of a term $m\lambda x^2$ in the expression for the force introduces into the motion a frequency of $2\omega_0$, i.e., a second harmonic of the fundamental or natural frequency ω_0. If terms involving higher powers of λ had been allowed in Eq. (5.32), then the series of harmonics $2\omega_0, 3\omega_0, 4\omega_0, \ldots$, would have been generated.

5.6 The Problem of Secular Terms in the Approximate Solutions

The perturbation procedure for dealing with nonlinear equations that was described in the preceding section was originated by Poisson and extended by Poincaré. The development given is correct *as far as it goes*, but if one attempts to obtain the solution in the next order of approximation (i.e., x_2), it is immediately evident that the Poisson–Poincaré procedure is subject to a fundamental difficulty. The problem is the occurrence in the solution of a term proportional to the time (called a *secular* term, one that always increases with time). Now, the bound motion of a particle in a conservative one-dimensional potential is clearly *periodic* and so no term in the solution $x(t)$ that is proportional to t (or some power of t) is physically meaningful. The reason for the occurrence of such a term is easy to understand. When the nonlinear term is zero, the solution is periodic with period $2\pi/\omega$. However, the inclusion of the nonlinear term may render the solution periodic but with a period different from $2\pi/\omega$. For example, if the frequency of the unperturbed solution is ω_0 and if the nonlinearity of the system alters this frequency by a small amount δ, the expansion of $\sin(\omega_0 + \delta)t$ results in

$$\sin(\omega_0 + \delta)t = \sin\omega_0 t + t\delta\cos\omega_0 t - \frac{t^2\delta^2}{2}\sin\omega_0 t - \cdots$$

Therefore, if the approximation procedure is terminated after a finite number of terms, secular terms will inevitably appear.

In the example discussed in Section 5.5 the secular term appears only if the solution is carried to the second order of approximation (see Problem

5-10), but in the case of the simple symmetric potential, the difficulty arises even in *first* order. We can show this explicitly as follows. The equation of motion is

$$\ddot{x} + \omega_0^2 x = \varepsilon x^3 \tag{5.44}$$

and we attempt a perturbation solution by writing (to first order)

$$x(t) \cong x_0 + \varepsilon x_1 \tag{5.45}$$

The equation of motion then separates into [cf. Eqs. (5.35)]

$$\ddot{x}_0 + \omega_0^2 x_0 = 0 \tag{5.46a}$$

$$\ddot{x}_1 + \omega_0^2 x_1 = x_0^3 \tag{5.46b}$$

We choose the zero-order solution to be

$$x_0(t) = A \cos \omega_0 t \tag{5.47}$$

so that Eq. (5.46b) becomes

$$\begin{aligned} \ddot{x}_1 + \omega_0^2 x_1 &= A^3 \cos^3 \omega_0 t \\ &= \tfrac{1}{4} A^3 (3 \cos \omega_0 t + \cos 3\omega_0 t) \end{aligned} \tag{5.48}$$

The solution to this equation is

$$x_1(t) = \frac{A^3}{32\omega_0^2} (12\omega_0 t \sin \omega_0 t - \cos 3\omega_0 t) \tag{5.49}$$

and the secular term arises in the first-order solution. This term has no physical significance and must be eliminated from the solution. General procedures for dealing with the spurious secular terms in nonlinear equations are discussed, for example, by Andronow and Chaikin (An49) and by Minorsky (Mi47).

We will illustrate one method for eliminating the secular terms by applying to Eq. (5.44) the procedure of Kryloff and Bogoliuboff which is derived from the methods of Gylden and Lindstedt used in celestial mechanics. Again, we expand $x(t)$:

$$x(t) = x_0 + \varepsilon x_1 + \varepsilon^2 x_2 + \cdots \tag{5.50a}$$

and we also expand (the square of) the frequency:

$$\omega_0^2 = \omega^2 + \alpha_1 \varepsilon + \alpha_2 \varepsilon^2 + \cdots \tag{5.50b}$$

The constants α_n are to be determined so that the secular terms are eliminated. Substituting Eqs. (5.50a) and (5.50b) into Eq. (5.44) and equating to zero the coefficients of like powers of ε, we have

$$\ddot{x}_0 + \omega^2 x_0 = 0 \tag{5.51a}$$

$$\ddot{x}_1 + \omega^2 x_1 = x_0^3 - \alpha_1 x_0 \tag{5.51b}$$

$$\ddot{x}_2 + \omega^2 x_2 = 3x_0^2 x_1 - \alpha_1 x_1 - \alpha_2 x_0 \tag{5.51c}$$

The solution obtained from these equations will be correct to second order in ε; higher approximations may be obtained by continuing the procedure.

The initial conditions are

$$x(0) = A; \qquad \dot{x}(0) = 0$$

which, in view of the expansion, Eq. (5.50a), lead to

$$x_0(0) = A, \qquad \dot{x}_0(0) = 0$$

$$x_i(0) = 0, \qquad \dot{x}_i(0) = 0; \qquad i = 1, 2, 3, \ldots$$

From Eq. (5.51a) we have the zero-order (or *generating*) solution

$$x_0(t) = A \cos \omega t \tag{5.52}$$

which satisfies the initial conditions. Substituting for x_0 in Eq. (5.51b) we have

$$\begin{aligned} \ddot{x}_1 + \omega^2 x_1 &= A^3 \cos^3 \omega t - \alpha_1 A \cos \omega t \\ &= (\tfrac{3}{4}A^3 - \alpha_1 A) \cos \omega t + \tfrac{1}{4}A^3 \cos 3\omega t \end{aligned} \tag{5.53}$$

In this equation it is the term proportional to $\cos \omega t$ that is the culprit. The solution to an inhomogeneous equation of this type is composed of two parts, a complementary solution (i.e., a solution of the homogeneous equation) and a particular integral (i.e., a term or terms that reproduce the right-hand side). Ordinarily, a term proportional to $\cos \omega t$ would be required in the particular integral, but for Eq. (5.53) such a term already occurs in the complementary solution. Therefore,* it is necessary to include a term proportional to $t \cos \omega t$ in the particular integral. This is the secular term and it can be eliminated by setting equal to zero the coefficient of $\cos \omega t$ in Eq. (5.53). Hence, α_1 is required to be

$$\alpha_1 = \tfrac{3}{4}A^2$$

Equation (5.53) then reduces to

$$\ddot{x}_1 + \omega^2 x_1 = \tfrac{1}{4}A^3 \cos 3\omega t \tag{5.54}$$

* See Section C.2, Appendix C.

The solution satisfying the initial conditions is

$$x_1(t) = \frac{A^3}{32\omega^2}(\cos \omega t - \cos 3\omega t) \tag{5.55}$$

Therefore, correct to first order, the complete solution is

$$x(t) = \left(1 + \frac{\varepsilon A^2}{32\omega^2}\right) A \cos \omega t - \frac{\varepsilon A^3}{32\omega^2}\cos 3\omega t \tag{5.56}$$

Also, correct to first order, the frequency is obtained by substituting for α_1 in Eq. (5.50b):

$$\begin{aligned}\omega_0 &= (\omega^2 + \alpha_1 \varepsilon)^{\frac{1}{2}} \\ &= (\omega^2 + \tfrac{3}{4}A^2\varepsilon)^{\frac{1}{2}}\end{aligned}$$

or,

$$\omega \cong \omega_0 \left(1 + \frac{3A^2}{4\omega_0^2}\varepsilon\right)^{-\frac{1}{2}}$$

where ω_0 has been substituted for ω in the radical. Thus, expanding the radical,

$$\omega \cong \omega_0 \left(1 - \frac{3A^2}{8\omega_0^2}\varepsilon\right) \tag{5.57}$$

Next, for the second-order solution we substitute x_0 and x_1 from Eqs. (5.52) and (5.54) into Eq. (5.51c):

$$\begin{aligned}\ddot{x}_2 + \omega^2 x_2 &= \frac{3A^5}{32\omega^2}\cos^2 \omega t\,(\cos \omega t - \cos 3\omega t) \\ &= -\frac{3A^5}{128\omega^2}(\cos \omega t - \cos 3\omega t) - \alpha_2 A \cos \omega t\end{aligned} \tag{5.58}$$

Using the identities

$$\left.\begin{aligned}\cos^3 \omega t &= \tfrac{3}{4}\cos \omega t + \tfrac{1}{4}\cos 3\omega t \\ \cos^2 \omega t \cos 3\omega t &= \tfrac{1}{4}\cos \omega t + \tfrac{1}{2}\cos 3\omega t + \tfrac{1}{4}\cos 5\omega t\end{aligned}\right\} \tag{5.59}$$

the terms in Eq. (5.58) proportional to cos $3\omega t$ cancel, and we have

$$\ddot{x}_2 + \omega^2 x_2 = \left(\frac{3A^5}{128\omega^2} - \alpha_2 A\right)\cos \omega t - \frac{3A^5}{128\omega^2}\cos 5\omega t \tag{5.60}$$

Again, the coefficient of the cos ωt term must vanish, giving

$$\alpha_2 = -\frac{3A^4}{128\omega^2}$$

The solution of Eq. (5.60) which satisfies the initial conditions is then

$$x_2(t) = -\frac{A^5}{1024\omega^4}(\cos \omega t - \cos 5\omega t) \tag{5.61}$$

The complete solution, correct to second order, is

$$x(t) = \left(1 + \frac{\varepsilon A^2}{32\omega^2} - \frac{\varepsilon^2 A^4}{1024\omega^4}\right) A \cos \omega t + \left(-\frac{\varepsilon A^2}{32\omega^2}\right) A \cos 3\omega t + \left(\frac{\varepsilon^2 A^4}{1024\omega^4}\right) A \cos 5\omega t \tag{5.62}$$

The frequency, to the same approximation, is

$$\omega_0 = \left(\omega^2 + \frac{3}{4}A^2\varepsilon - \frac{3A^4}{128\omega^2}\varepsilon^2\right)^{\frac{1}{2}}$$

or,

$$\omega \cong \omega_0\left(1 - \frac{3A^2}{8\omega_0^2}\varepsilon + \frac{3A^4}{256\omega_0^4}\varepsilon^2\right) \tag{5.63}$$

This procedure can obviously be extended to any desired order of approximation.

5.7 The Occurrence of Subharmonics

The analyses in the preceding sections have shown that nonlinear terms in the expression for the force acting on an oscillating particle give rise to vibration frequencies that are multiples of the driving frequency of the system. Under some conditions it is possible to obtain, in addition to the harmonics, frequencies that are rational fractions of the fundamental, the *subharmonics*. We shall demonstrate this phenomenon by considering the conditions for the occurrence of the subharmonic of order 3 (i.e., frequency $\omega/3$). We choose a symmetric, nonlinear force so that the equation of motion is

$$\ddot{x} + \omega_0^2 x - \varepsilon x^3 = B \cos \omega t \tag{5.64}$$

For the trial function we use

$$x_1(t) = C \cos \omega t + D \cos(\omega/3)t \tag{5.65}$$

If we substitute this expression for $x_1(t)$ into Eq. (5.64) and use the appropriate trigonometric identities to simplify the result, then the requirement that the coefficient of $\cos(\omega/3)t$ must vanish yields

$$D\left[\left(\omega_0^2 - \frac{\omega^2}{9}\right) - \frac{3}{4}\varepsilon(D^2 + CD + 2C^2)\right] = 0 \tag{5.66}$$

If this equation is to be valid for $D \neq 0$, we must have

$$\omega_0^2 - \frac{\omega^2}{9} = \frac{3}{4}\varepsilon(D^2 + CD + 2C^2) \tag{5.67}$$

This is a quadratic in D, the solution for which is

$$D = -\frac{C}{2} \pm \frac{1}{2}\left[\frac{16}{27\varepsilon}(9\omega_0^2 - \omega^2) - 7C^2\right]^{\frac{1}{2}} \tag{5.68}$$

Since D must be real, the radicand in this expression must be positive or zero, so that

$$\frac{16}{27\varepsilon}(9\omega_0^2 - \omega^2) \geq 7C^2 \tag{5.69}$$

from which the restriction on ω is

$$\omega \leq 3\sqrt{\omega_0^2 - \frac{21\varepsilon C^2}{16}}, \qquad \varepsilon > 0 \tag{5.70a}$$

or,

$$\omega \geq 3\sqrt{\omega_0^2 + \frac{21\varepsilon C^2}{16}}, \qquad \varepsilon < 0 \tag{5.70b}$$

Under certain conditions the subharmonic alone will be present (see Problem 5-13).

■ 5.8 Intermodulation and Combination Tones

A linear oscillator obeys the principle of superposition. That is, if the response of such an oscillator is known for two forcing functions taken separately, the response to both forcing functions applied simultaneously is just the sum of the individual responses. Nonlinear systems do not exhibit this property. The human ear, for example, is known to be nonlinear in its response. If two tones are sounded together, the two original tones plus additional tones, which arise from the nonlinearity of the ear, will be audible. These latter tones are called *combination tones* and result from the *intermodulation* of the impressed tones.

The response of the human ear is described reasonably well by including in the oscillator equation a nonlinear term proportional to the square of the displacement. We therefore can analyze the situation described above by seeking a solution to the equation

$$\ddot{x} + \omega_0^2 x + \beta x^2 = A \cos \omega_1 t + B \cos \omega_2 t \tag{5.71}$$

where A and B are the amplitudes of the impressed tones of frequencies ω_1 and ω_2. Since β is assumed to be small, we use as a trial solution that solution which would result for $\beta = 0$. Thus, for the steady-state situation, we have

$$x_1(t) = \frac{A}{\omega_0^2 - \omega_1^2} \cos \omega_1 t + \frac{B}{\omega_0^2 - \omega_2^2} \cos \omega_2 t \tag{5.72}$$

Substituting this expression into the βx^2 term in Eq. (5.71) and expanding the squares of the cosines, we find

$$\begin{aligned} \ddot{x} + \omega_0^2 x = {} & A \cos \omega_1 t + B \cos \omega_2 t \\ & - \beta \left\{ \frac{A^2}{2(\omega_0^2 - \omega_1^2)^2} (\cos 2\omega_1 t + 1) + \frac{B^2}{2(\omega_0^2 - \omega_2^2)^2} (\cos 2\omega_2 t + 1) \right. \\ & \left. + \frac{AB}{(\omega_0^2 - \omega_1^2)(\omega_0^2 - \omega_2^2)} [\cos(\omega_1 + \omega_2)t - \cos(\omega_1 - \omega_2)t] \right\} \end{aligned} \tag{5.73}$$

Without writing down the solution of this equation, it is clear that the solution will contain terms with the following frequencies: ω_1 and ω_2, the original (or *generator*) frequencies; $2\omega_1$ and $2\omega_2$, the second harmonics; and $\omega_1 \pm \omega_2$, the combination tones. We could continue the successive approximation procedure and generate solutions with additional frequencies but the new frequencies brought out in the higher approximations would be of smaller amplitude.

Although two combination tones and two harmonics of second order are produced in the ear by two impressed tones, only the difference tone, of frequency $\omega_1 - \omega_2$, can be lower in pitch than either of the original tones. The difference tone is therefore easier to perceive and identify and thus is the most important of the additional tones. When the generator tones are greater than one octave apart, however, the difference tone is no longer of a pitch lower than either generator; under this condition the difference tone is difficult to hear. The tones with frequencies $\omega_1 + \omega_2$, $2\omega_1$, and $2\omega_2$ are always difficult to perceive.

The transmission of audio signals by amplitude-modulated radio waves makes use of the production of combination frequencies in nonlinear devices. The carrier signal (of some radio frequency ω_1) is modulated by the audio signal (of frequency ω_2) and the two components are mixed in a nonlinear vacuum-tube or transistor circuit. This signal is applied to a narrow-band amplifier which passes frequencies in the vicinity of ω_1. The output of the transmitter then consists of the carrier frequency ω_1 and the *sidebands* with frequencies $\omega_1 \pm \omega_2$. The procedure is reversed in the receiver and an audio signal is delivered to the listener.

Suggested References

The subject of nonlinear oscillations is one of the more difficult problem areas in classical physics and in modern engineering. Much of the literature of this field is quite complicated. Some of the treatments that are not overly complex are those of Andronow and Chaikin (An49, Chapter 2), Hayashi (Ha64, Chapters 1 and 2), Ku (Ku58, Chapters 1–4), McCusky (Mc59, Chapter 6), and Stoker (St50, Chapters 2, 4, and 5); much use is made of phase diagrams. Another standard work on the subject is that of Minorsky (Mi47).

Of special interest are the original papers of van der Pol, a pioneer in the field (va60, especially papers 15, 27, 40, 58, and 76).

At the intermediate level, brief discussions are given by Constant (Co54, Chapter 6), Sharman (Sh63, Chapter 3), and Halfman (Ha62, Chapter 7). Several different methods of solution of nonlinear problems including Laplace transforms, are discussed by Pipes (Pi46, Chapter 22).

Modern approaches to nonlinear problems that emphasize the mathematician's viewpoint (but are still quite readable) are those of Bellman (Be64) and Hochstadt (Ho64, Chapters 6 and 7).

Problems

5-1. Show that the period of oscillation of a particle of mass m in a potential $U = A|x|^n$, is given by

$$\tau = \frac{2}{n}\sqrt{\frac{2\pi m}{E}} \cdot \left(\frac{E}{A}\right)^{1/n} \cdot \frac{\Gamma\left(\frac{1}{n}\right)}{\Gamma\left(\frac{1}{2} + \frac{1}{n}\right)}$$

Take $n = 2$, evaluate the gamma functions,* and thus show that τ reduces to the normal expression for a parabolic potential.

5-2. Refer to Example 5.2. If each of the springs must be stretched a distance d in order to attach the particle at the equilibrium position (i.e., in its equi-

* Gamma functions are defined in Appendix E.3.

forces of magnitude kd), then show that the potential in which the particle moves is approximately

$$U(x) \cong (kd/l)x^2 + [k(l-d)/4l^3]x^4$$

5-3. Construct a phase diagram for the potential in Fig. 5-1.

5-4. Construct a phase diagram for the potential $U(x) = -(\lambda/3)x^3$.

5-5. Lord Rayleigh used the equation

$$\ddot{x} - (a - b\dot{x}^2)\dot{x} + \omega_0^2 x = 0$$

in his discussion of nonlinear effects in acoustic phenomena.* Show that by differentiating this equation with respect to the time and making the substitution $y = y_0\sqrt{3b/a}\,\dot{x}$, van der Pol's equation results:

$$\ddot{y} - \frac{a}{y_0^2}(y_0^2 - y^2)\dot{y} + \omega_0^2 y = 0$$

5-6. Derive the expression for the phase paths of the plane pendulum in the event that the total energy is $E > 2mgl$. Note that this is just the case of a particle moving in a periodic potential $U(\theta) = mgl(1 - \cos\theta)$.

5-7. Solve by a successive approximation procedure and obtain a result accurate to four significant figures:

(a) $x + x^2 + 1 = \tan x, \qquad 0 \le x \le \pi/2$
(b) $x(x + 3) = 10 \sin x, \qquad x > 0$
(c) $1 + x + \cos x = e^x, \qquad x > 0$

(It may be profitable to make a crude graph in order to choose a reasonable first approximation.)

5-8. Consider the free motion of a plane pendulum for the case that the amplitude is not small. Show that the *horizontal component* of the motion may be represented by the approximate expression

$$\ddot{x} + \omega_0^2 x - \varepsilon x^3 = 0$$

where $\omega_0^2 = g/l$ and $\varepsilon = g/2l^3$, with l equal to the length of the suspension.

* J. W. S. Rayleigh, *Phil. Mag.* **15** (April, 1883); see also Ra94, Section 68a.

Use a successive approximation procedure to solve this nonlinear equation and show how the frequency of the motion is related to the amplitude.

5-9. Investigate the motion of an undamped particle subject to a force of the form

$$F(x) = \begin{cases} -kx, & |x| < a \\ -(k+\delta)x + \delta a, & |x| > a \end{cases}$$

where k and δ are positive constants.

5-10. Obtain a solution to

$$\ddot{x} + \omega_0^2 x - \lambda x^2 = 0$$

correct to second order by writing

$$x(t) = x_0 + \lambda x_1 + \lambda^2 x_2$$

and using the method of perturbations. Show that a secular term arises in the second-order solution.

5-11. Consider a damped oscillator whose motion is described by

$$\ddot{x} + \lambda \dot{x}|\dot{x}| + \omega_0^2 x = 0$$

The initial conditions are $x(0) = a$, $\dot{x}(0) = 0$. Use the method of perturbations described in Section 5.6 and show that to an accuracy of order λ^2 the amplitude after any half cycle is smaller than that for the previous half cycle in the ratio $a_{n+1}/a_n = 1 - 4\lambda a_n/3 + 16\lambda^2 a_n^2/9$.

5-12. Consider the equation (in simplified notation)

$$\ddot{x} + x^3 = A \cos t$$

Use a perturbation procedure and calculate the amplitudes of the fundamental and third-harmonic components of the motion for $A = 0.2$.

[There are three kinds of harmonic solutions in this case; the exact results for the first two terms (to four decimal places) are

$$x_1(t) = -0.2066 \cos t - 0.0003 \cos 3t$$

$$x_2(t) = 1.2103 \cos t + 0.0658 \cos 3t$$

$$x_3(t) = -1.0161 \cos t - 0.0352 \cos 3t$$

Satisfactory agreement with these exact solutions will be obtained in second approximation.]

5-13. Refer to Section 5.7. Find the conditions under which the subharmonic of frequency $\omega/3$ is the only oscillation present. Show that there are three allowed amplitudes which differ only in phase.

5-14. Investigate the combination tones produced when the term βx^2 in Eq. (5.71) is replaced by βx^3. Show that there are six combination tones of second order.

CHAPTER 6

Some Methods in the Calculus of Variations

6.1 Introduction

In the following chapter we shall restate the Newtonian equations of motion in a more general form called *Lagrange's equations.* We shall also show that these equations can be derived from a variational principle* called *Hamilton's Principle.* It is not necessary to use Hamilton's variational principle to obtain Lagrange's equations, but the statement of dynamics in these terms allows an elegant unification of mechanics and provides a powerful method for extending the theory to include field phenomena. This latter topic is beyond the scope of this book, but it seems appropriate to lay the groundwork at this stage for more general treatments of dynamics. Therefore, even though we shall use variational techniques in only a small part of the succeeding development, these methods are of sufficient importance in physics that a brief discussion of the principles seems in order here.

* The development of the calculus of variations was begun by Newton (1686), and was extended by Johann and Jakob Bernoulli (1696) and by Euler (1744). Legendre (1786), Lagrange (1788), Hamilton (1833), and Jacobi (1837) all made important contributions. The names of Dirichlet and Weierstrass are particularly associated with the establishment of a rigorous mathematical foundation for the subject.

In the following sections we will consider only the *necessary* conditions in treating variational problems; the *sufficient* conditions are quite involved and the reader is referred to texts on the calculus of variations for details. The discussion in this chapter will therefore be limited to those aspects of the theory of variations which have a direct bearing on classical systems.

6.2 Statement of the Problem

The basic problem of the calculus of variations is to determine the function $y(x)$ such that the integral

$$J = \int_{x_1}^{x_2} f\{y(x), y'(x); x\}\, dx \tag{6.1}$$

shall be an *extremum* (i.e., either a maximum or a minimum). In Eq. (6.1), $y'(x) \equiv dy/dx$, and the semicolon in f separates the independent variable x from the dependent variable $y(x)$ and its derivative $y'(x)$. The functional* f is considered as given and the limits of integration are fixed†; the function $y'(x)$ is then to be varied until an extreme value of J is found. By this it is meant that if a function $y = y(x)$ gives to the integral J a minimum value, then any *neighboring function*, no matter how close to $y(x)$, must make J increase. The definition of a neighboring function may be made as follows. We give to all possible functions y a parametric representation $y = y(\alpha, x)$, such that for $\alpha = 0$, $y = y(0, x) = y(x)$ is the function which yields an extremum for J. We can then write

$$y(\alpha, x) = y(0, x) + \alpha\eta(x) \tag{6.2}$$

where $\eta(x)$ is some function of x which has a continuous first derivative and which vanishes at x_1 and x_2, since the varied function $y(\alpha, x)$ must be identical with $y(x)$ at the end points of the path: $\eta(x_1) = \eta(x_2) = 0$. The situation is depicted schematically in Fig. 6-1.

If functions of the type given by Eq. (6.2) are considered, the integral J becomes a function of the parameter α:

$$J(\alpha) = \int_{x_1}^{x_2} f\{y(\alpha, x), y'(\alpha, x); x\}\, dx \tag{6.3}$$

* The quantity f depends on the functional form of the dependent variable $y(x)$ and is called a *functional.*

† It is not necessary that the limits of integration be considered fixed. If they are allowed to vary, then the problem increases to finding not only $y(x)$ but also x_1 and x_2 such that J is an extremum.

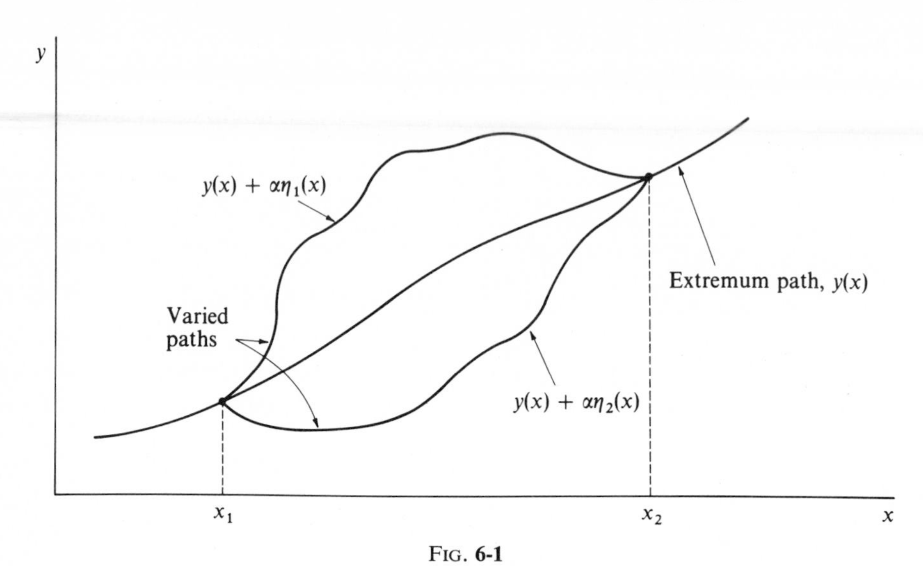

FIG. 6-1

The condition that the integral have a *stationary value* (i.e., that an extremum result) is that J be independent of α in first order, or, equivalently, that

$$\left.\frac{\partial J}{\partial \alpha}\right|_{\alpha=0} = 0 \tag{6.4}$$

for all functions $\eta(x)$. This is only a *necessary* condition; it is not sufficient. But as stated earlier, we shall not pursue here the details of the sufficient conditions.

▶ Example 6.2 Two Variational Problems

(a) As a first example of varied paths and the extremum condition, let us consider the simple function

$$y(x) = x \tag{1}$$

We may construct neighboring paths by adding to $y(x)$ a sinusoidal variation:

$$y(\alpha, x) = x + \alpha \sin x \tag{2}$$

These paths are illustrated in Fig. 6-2 for $\alpha = 0$ and for two different nonvanishing values of α. We now inquire as to the value of the integral of $f = (dy/dx)^2$ between the limits $x = 0$ and $x = 2\pi$. Clearly, the function $\eta(x) = \sin x$ obeys the end-point conditions, viz., $\eta(0) = \eta(2\pi) = 0$. Therefore, we have

$$\frac{dy}{dx} = 1 + \alpha \cos x \tag{3}$$

and the integral becomes

$$J(\alpha) = \int_0^{2\pi} (1 + 2\alpha \cos x + \alpha^2 \cos^2 x)\, dx$$

$$= 2\pi + \alpha^2 \pi \tag{4}$$

Thus, we see that the value of $J(\alpha)$ is always greater than $J(0)$, no matter what value (positive or negative) is chosen for α. This may be expressed by the relation

$$\left.\frac{\partial J}{\partial \alpha}\right|_{\alpha=0} = 0 \tag{5}$$

which is just the condition of Eq. (6.4).

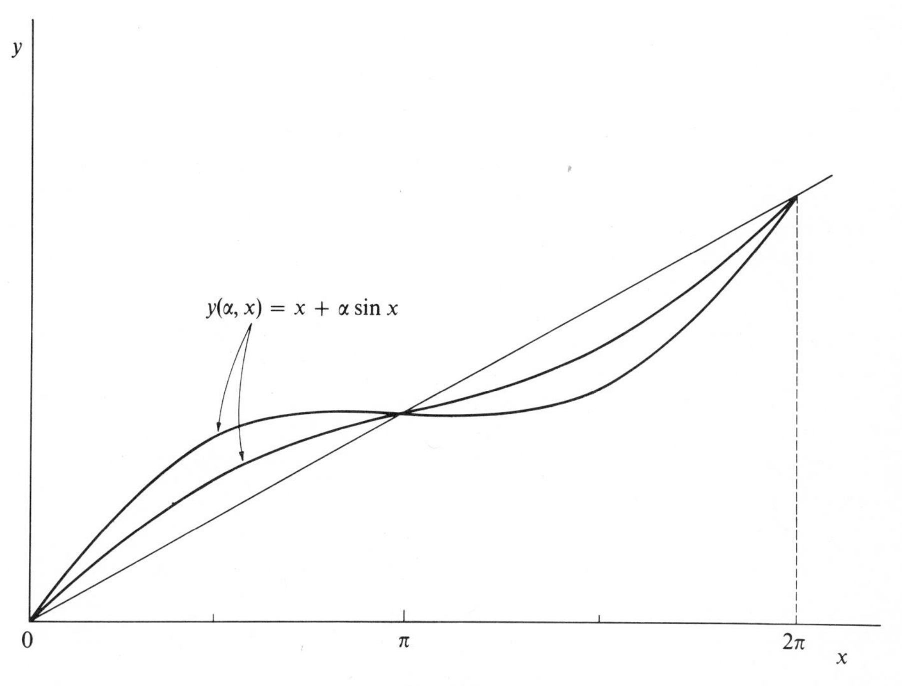

FIG. 6-2

(b) Next, consider the equation of the line which yields the shortest distance between the points $(x_1, y_1) = (0, 0)$ and $(x_2, y_2) = (1, 0)$. Clearly, this equation is

$$y(x) = 0 \tag{1}$$

i.e., just the x-axis. Now, let the path be varied in the following manner:

$$y(\alpha, x) = y(0, x) + \alpha\eta(x)$$

$$= 0 + \alpha(x^2 - x) \tag{2}$$

Thus, the original function $y=0$, which we know from the outset will yield a minimum path length, is to be varied by $\alpha(x^2-x)$. The function $\eta(x)=(x^2-x)$ satisfies the requirement that $\eta(x_1)=\eta(0)=0$ and $\eta(x_2)=\eta(1)=0$.

The increment of distance ds is

$$ds=\sqrt{dx^2+dy^2}=\sqrt{1+(dy/dx)^2}\,dx \tag{3}$$

and the total length of the path is

$$s=\int_0^1 \sqrt{1+y'^2}\,dx \tag{4}$$

Now,

$$y'=\frac{dy}{dx}=\alpha(2x-1) \tag{5}$$

so that

$$s=\int_0^1 [(4\alpha^2)x^2+(-4\alpha^2)x+(\alpha^2+1)]^{\frac{1}{2}}\,dx \tag{6}$$

This integral is of the form

$$\int \sqrt{ax^2+bx+c}\,dx$$

and can be evaluated by using Eqs. (E.11) and (E.8b) (see Appendix E):

$$s=\tfrac{1}{2}\sqrt{\alpha^2+1}+\frac{1}{2\alpha}\sinh^{-1}\alpha \tag{7}$$

Since we are considering only neighboring paths, α is a small quantity. Therefore, both the radical and $\sinh^{-1}\alpha$ may be expanded:

$$\begin{aligned} s&=\frac{1}{2}\left(1+\frac{\alpha^2}{2}+\cdots\right)+\frac{1}{2\alpha}\left(\alpha-\frac{1}{6}\alpha^3+\cdots\right) \\ &=1+\frac{\alpha^2}{6}+O(\alpha^4) \end{aligned} \tag{8}$$

where $O(\alpha^4)$ means "terms of order α^4 and higher."

The quantity $s(\alpha)$ above is identified with $J(\alpha)$ [see Eq. (6.3)], and its minimum value must be found:

$$\frac{\partial J}{\partial\alpha}=\frac{\partial s}{\partial\alpha}=\frac{\alpha}{3}+O(\alpha^3) \tag{9}$$

which implies that $\alpha=0$ will make $\partial J/\partial\alpha$ vanish and therefore yield a minimum for J (or s). Then,

$$J(\alpha=0)=s(\alpha=0)=1 \tag{10}$$

which it must, since this is just the distance from (0, 0) to (1, 0). Thus, Eq. (6.4) is verified for this example: the condition $(\partial J/\partial\alpha)|_{\alpha=0} = 0$ produces an extremum for J.

6.3 Euler's Equation

In order to determine the result of the condition (6.4), we perform the indicated differentiation in Eq. (6.3). Thus,

$$\frac{\partial J}{\partial \alpha} = \frac{\partial}{\partial \alpha}\int_{x_1}^{x_2} f\{y, y'; x\}\, dx$$

Since the limits of integration are fixed, the differential operation affects only the integrand. Hence,

$$\frac{\partial J}{\partial \alpha} = \int_{x_1}^{x_2}\left(\frac{\partial f}{\partial y}\frac{\partial y}{\partial \alpha} + \frac{\partial f}{\partial y'}\frac{\partial y'}{\partial \alpha}\right) dx \tag{6.5}$$

From Eq. (6.2) we have

$$\frac{\partial y}{\partial \alpha} = \eta(x); \qquad \frac{\partial y'}{\partial \alpha} = \frac{d\eta}{dx} \tag{6.6}$$

Then, Eq. (6.5) becomes

$$\frac{\partial J}{\partial \alpha} = \int_{x_1}^{x_2}\left(\frac{\partial f}{\partial y}\eta(x) + \frac{\partial f}{\partial y'}\frac{d\eta}{dx}\right) dx \tag{6.7}$$

The second term in the integrand can be integrated by parts:

$$\int_{x_1}^{x_2}\frac{\partial f}{\partial y'}\frac{d\eta}{dx}\,dx = \frac{\partial f}{\partial y'}\eta(x)\bigg|_{x_1}^{x_2} - \int_{x_1}^{x_2}\frac{d}{dx}\left(\frac{\partial f}{\partial y'}\right)\eta(x)\,dx$$

The integrated term vanishes because $\eta(x_1) = \eta(x_2) = 0$. Therefore, Eq. (6.5) becomes

$$\begin{aligned}\frac{\partial J}{\partial \alpha} &= \int_{x_1}^{x_2}\left(\frac{\partial f}{\partial y}\frac{\partial y}{\partial \alpha} - \frac{d}{dx}\left(\frac{\partial f}{\partial y'}\right)\frac{\partial y}{\partial \alpha}\right) dx \\ &= \int_{x_1}^{x_2}\left(\frac{\partial f}{\partial y} - \frac{d}{dx}\frac{\partial f}{\partial y'}\right)\eta(x)\,dx\end{aligned} \tag{6.8}$$

The integral in Eq. (6.8) now gives the appearance of being independent of α. However, the functions y and y' with respect to which the derivatives of f are taken, are still functions of α. When $\alpha = 0$, however, $y(\alpha, x) = y(0, x) = y(x)$, and the dependence on α disappears.

Since $(\partial J/\partial\alpha)|_{\alpha=0}$ must vanish, and since $\eta(x)$ is an arbitrary function (subject to the conditions stated above), the integrand in Eq. (6.8) must itself vanish for $\alpha = 0$:

$$\boxed{\frac{\partial f}{\partial y} - \frac{d}{dx}\frac{\partial f}{\partial y'} = 0} \tag{6.9}$$

where now y and y' are the original functions, independent of α. This result is known as *Euler's equation** and it is a necessary condition that J have an extreme value.

6.4 The Brachistochrone Problem

▶ **Example 6.4** One of the classic problems the solution to which is most easily accomplished by using the calculus of variations is that of the *brachistrochrone*.† Consider a particle which moves in a constant force field starting at rest from some point (x_1, y_1) to some lower point (x_2, y_2). The problem is to find the path which allows the particle to accomplish the transit in the least possible time. The coordinate system may be chosen so that the point (x_1, y_1) is at the origin. Further, let the force field be directed along the positive x-axis as in Fig. 6-3.

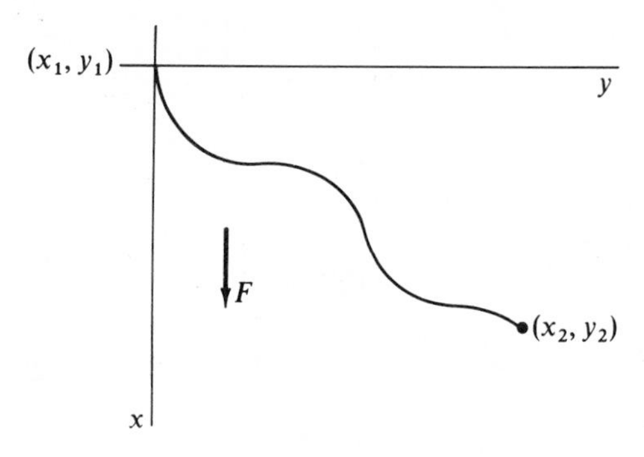

FIG. **6-3**

Since the force on the particle is constant and if we ignore the possibility of friction, the field is conservative and the total energy of the particle is $T + U = \text{const.}$ If we choose to measure the potential from the point $x = 0$ [i.e., $U(x = 0) = 0$], then since the particle starts from rest, $T + U = 0$. The kinetic energy is $T = \frac{1}{2}mv^2$, and

* Derived first by Euler in 1744. When applied to mechanical systems, this is known as the *Euler–Lagrange equation.*

† First solved by Johann Bernoulli, 1696.

the potential energy is $U = -Fx = -mgx$, where g is the acceleration imparted by the force. Thus,

$$v = \sqrt{2gx} \tag{1}$$

The time required for the particle to make the transit from the origin to (x_2, y_2) is

$$t = \int_{(x_1,y_1)}^{(x_2,y_2)} \frac{ds}{v} = \int \frac{(dx^2 + dy)^{\frac{1}{2}}}{(2gx)^{\frac{1}{2}}}$$

$$= \int_{x_1=0}^{x_2} \left(\frac{1+y'^2}{2gx}\right)^{\frac{1}{2}} dx \tag{2}$$

The time of transit is the quantity for which a minimum is desired. Since the constant $(2g)^{-\frac{1}{2}}$ does not affect the final equation, the functional f may be identified as

$$f = \left(\frac{1+y'^2}{x}\right)^{\frac{1}{2}} \tag{3}$$

And, since $\partial f/\partial y = 0$, the Euler equation becomes

$$\frac{d}{dx}\frac{\partial f}{\partial y'} = 0$$

or,

$$\frac{\partial f}{\partial y'} = \text{const.} \equiv (2a)^{-\frac{1}{2}} \tag{4}$$

Performing the differentiation and squaring the result, we have

$$\frac{y'^2}{x(1+y'^2)} = \frac{1}{2a} \tag{5}$$

This may be put in the form

$$y = \int \frac{x\,dx}{(2ax - x^2)^{\frac{1}{2}}} \tag{6}$$

Now make the following change of variable:

$$\begin{aligned} x &= a(1 - \cos\theta) \\ dx &= a \sin\theta \, d\theta \end{aligned} \tag{7}$$

The integral then becomes

$$y = \int a(1 - \cos\theta)\, d\theta$$

and,

$$y = a(\theta - \sin\theta) + \text{const.} \tag{8}$$

Now, the parametric equations for a *cycloid* passing through the origin are

$$\left.\begin{aligned} x &= a(1 - \cos\theta) \\ y &= a(\theta - \sin\theta) \end{aligned}\right\} \tag{9}$$

which is just the solution found; therefore, the constant of integration vanishes. The path is then as shown in Fig. 6-4, and the constant a must be adjusted to allow the cycloid to pass through the specified point (x_2, y_2). The solution to the problem

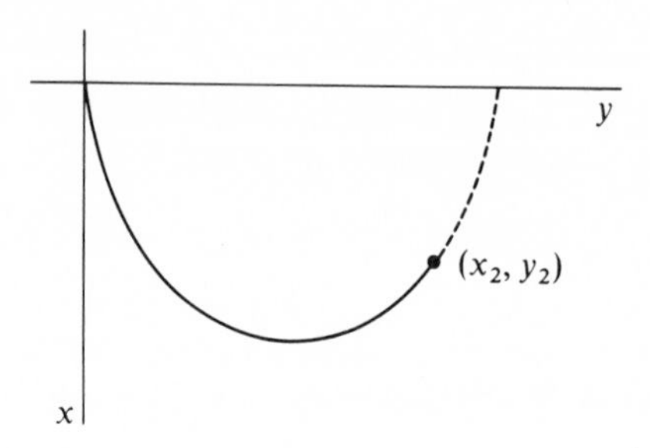

FIG. 6-4

of the brachistochrone does indeed yield a path which the particle traverses in a *minimum* time. It must be noted, however, that the procedures of variational calculus are designed only to produce an extremum—either a minimum or a maximum. It is almost always the case in dynamics that we desire (and find) a minimum for the problem.

6.5 The "Second Form" of Euler's Equation

We first note that

$$\frac{df}{dx} = \frac{d}{dx} f\{y, y'; x\} = \frac{\partial f}{\partial x} + \frac{\partial f}{\partial y}\frac{dy}{dx} + \frac{\partial f}{dy'}\frac{dy'}{dx}$$

$$= \frac{\partial f}{\partial x} + y'\frac{\partial f}{\partial y} + y''\frac{\partial f}{\partial y'} \tag{6.10}$$

Now,

$$\frac{d}{dx}\left(y'\frac{\partial f}{\partial y'}\right) = y''\frac{\partial f}{\partial y'} + y'\frac{d}{dx}\frac{\partial f}{\partial y'}$$

or, substituting from Eq. (6.10) for $y''(\partial f/\partial y')$,

$$\frac{d}{dx}\left(y'\frac{\partial f}{\partial y'}\right) = \frac{df}{dx} - \frac{\partial f}{\partial x} - y'\frac{\partial f}{\partial y} + y'\frac{d}{dx}\frac{\partial f}{\partial y'} \tag{6.11}$$

The last two terms in Eq. (6.11) may be written as

$$y'\left(\frac{d}{dx}\frac{\partial f}{\partial y'} - \frac{\partial f}{\partial y}\right)$$

which vanishes in view of the Euler equation [Eq. (6.9)]. Therefore,

$$\boxed{\frac{\partial f}{\partial x} - \frac{d}{dx}\left(f - y' \frac{\partial f}{\partial y'}\right) = 0} \tag{6.12}$$

This so-called "second form" of the Euler equation is of use in cases in which f does not depend explicitly on x, and $\partial f/\partial x = 0$. Then,

$$f - y' \frac{\partial f}{\partial y'} = \text{const.} \qquad \left(\text{for } \frac{\partial f}{\partial x} = 0\right) \tag{6.13}$$

▶ **Example 6.5 Geodesic on a Sphere**

A "geodesic" is a line which represents the shortest path between any two points, when the path is restricted to lie on some surface. The element of length on the surface of a sphere of radius ρ is given by [see Eq. (1.104) with $dr = 0$]:

$$ds = \rho(d\theta^2 + \sin^2 \theta \, d\varphi^2)^{1/2} \tag{1}$$

Therefore, the distance s between points 1 and 2 is

$$s = \rho \int_1^2 \left[\left(\frac{d\theta}{d\varphi}\right)^2 + \sin^2 \theta\right]^{1/2} d\varphi \tag{2}$$

and, if s is to be a minimum, f is identified as

$$f = (\theta_\varphi^2 + \sin^2 \theta)^{1/2} \tag{3}$$

where $\theta_\varphi \equiv d\theta/d\varphi$. Since $\partial f/\partial \varphi = 0$, we may use the second form of the Euler equation [Eq. (6.13)] which yields

$$(\theta_\varphi^2 + \sin^2 \theta)^{1/2} - \theta_\varphi \cdot \frac{\partial}{\partial \theta_\varphi}(\theta_\varphi^2 + \sin^2 \theta)^{1/2} = \text{const.} \equiv a \tag{4}$$

Differentiating and multiplying through by f, we have

$$\sin^2 \theta = a(\theta_\varphi^2 + \sin^2 \theta)^{1/2} \tag{5}$$

This may be solved for $d\varphi/d\theta = \theta_\varphi^{-1}$, with the result

$$\frac{d\varphi}{d\theta} = \frac{a \csc^2 \theta}{(1 - a^2 \csc^2 \theta)^{1/2}} \tag{6}$$

Solving for φ, we obtain

$$\varphi = \sin^{-1}\left(\frac{\cot \theta}{\beta}\right) + \alpha \tag{7}$$

where α is the constant of integration and $\beta^2 \equiv (1 - a^2)/a^2$. Upon rewriting, Eq. (7) becomes

$$\cot \theta = \beta \sin(\varphi - \alpha) \tag{8}$$

In order to interpret this result, we convert the equation to rectangular coordinates by multiplying through by $\rho \sin \theta$, to obtain, upon expanding $\sin(\varphi - \alpha)$,

$$(\beta \cos \alpha)\rho \sin \theta \sin \varphi - (\beta \sin \alpha)\rho \sin \theta \cos \varphi = \rho \cos \theta \tag{9}$$

Since α and β are constants, we may write them as

$$\beta \cos \alpha \equiv A; \qquad \beta \sin \alpha \equiv B \tag{10}$$

Then Eq. (9) becomes

$$A(\rho \sin \theta \sin \varphi) - B(\rho \sin \theta \cos \varphi) = (\rho \cos \theta) \tag{11}$$

The quantities in the parentheses are just the expressions for y, x, and z, respectively, in spherical coordinates (see Fig. F-3, Appendix F); therefore, Eq. (11) may be written as

$$Ay - Bx = z \tag{12}$$

which is the equation of a plane passing through the center of the sphere. Hence, the geodesic on a sphere is the path which that plane forms at the intersection with the surface of the sphere, i.e., a great circle. Note that the great circle is the *maximum* as well as the *minimum* "straight-line" distance between two points on the surface of a sphere.

6.6 Functions with Several Dependent Variables

The Euler equation derived above was the solution of the variational problem in which it was desired to find the single function $y(x)$ such that the integral of the functional f was an extremum. The case, more commonly encountered in mechanics, is that in which f is a functional of several dependent variables:

$$f = f\{y_1(x), y_1'(x), y_2(x), y_2'(x), \ldots; x\} \tag{6.14}$$

or simply,

$$f = f\{y_i(x), y_i'(x); x\}, \qquad i = 1, 2, \ldots, n \tag{6.14a}$$

Then, in analogy with Eq. (6.2), we write

$$y_i(\alpha, x) = y_i(0, x) + \alpha\eta_i(x) \tag{6.15}$$

The development proceeds in an exactly analogous manner, with the result [cf. Eq. (6.8)]:

$$\frac{\partial J}{\partial \alpha} = \int_{x_1}^{x_2} \sum_i \left(\frac{\partial f}{\partial y_i} - \frac{d}{dx}\frac{\partial f}{\partial y_i'}\right)\eta_i(x)\, dx \tag{6.16}$$

Since the individual variations, i.e., the $\eta_i(x)$, are all independent, the vanishing of Eq. (6.16) when evaluated at $\alpha = 0$ requires the separate vanishing of *each* expression in the brackets:

$$\boxed{\frac{\partial f}{\partial y_i} - \frac{d}{dx}\frac{\partial f}{\partial y_i'} = 0, \qquad i = 1, 2, \ldots, n} \tag{6.17}$$

6.7 The Euler Equations When Auxiliary Conditions Are Imposed

If it is desired to find, for example, the shortest path between two points on a surface, then, in addition to the conditions discussed above, there is the condition that the path must satisfy the equation of the surface, say, $g\{y_i; x\} = 0$. The use of such an equation was implicit in the solution of the problem of the geodesic on a sphere where the condition was

$$g = \sum_i x_i^2 - \rho^2 = 0$$

that is,

$$r = \rho = \text{const.}$$

In the general case, however, explicit use of the auxiliary equation or equations must be made. These equations are also called *equations of constraint.* Consider the case in which

$$f = f\{y_i, y_i'; x\} = f\{y, y', z, z'; x\} \tag{6.18}$$

Then the equation corresponding to Eq. (6.8) for the case of *two* variables is

$$\frac{\partial J}{\partial \alpha} = \int_{x_1}^{x_2} \left[\left(\frac{\partial f}{\partial y} - \frac{d}{dx}\frac{\partial f}{\partial y'} \right) \frac{\partial y}{\partial \alpha} + \left(\frac{\partial f}{\partial z} - \frac{d}{dx}\frac{\partial f}{\partial z'} \right) \frac{\partial z}{\partial \alpha} \right] dx \tag{6.19}$$

But now, there also exists an equation of constraint of the form

$$g\{y_i; x\} = g\{y, z; x\} = 0 \tag{6.20}$$

and the variations $\partial y/\partial \alpha$ and $\partial z/\partial \alpha$ are no longer independent, so that the expressions in parentheses in Eq. (6.19) do not separately vanish at $\alpha = 0$.

Differentiating g from Eq. (6.20), we have

$$dg = \left(\frac{\partial g}{\partial y}\frac{\partial y}{\partial \alpha} + \frac{\partial g}{\partial z}\frac{\partial z}{\partial \alpha} \right) d\alpha = 0 \tag{6.21}$$

where no term in x appears since $\partial x/\partial\alpha = 0$. Now,

$$\left.\begin{aligned} y(\alpha, x) &= y(x) + \alpha\eta_1(x) \\ z(\alpha, x) &= z(x) + \alpha\eta_2(x) \end{aligned}\right\} \tag{6.22}$$

Therefore,

$$\frac{\partial g}{\partial y}\eta_1(x) = -\frac{\partial g}{\partial z}\eta_2(x) \tag{6.23}$$

And Eq. (6.19) becomes

$$\frac{\partial J}{\partial\alpha} = \int_{x_1}^{x_2}\left[\left(\frac{\partial f}{\partial y} - \frac{d}{dx}\frac{\partial f}{\partial y'}\right)\eta_1(x) + \left(\frac{\partial f}{\partial z} - \frac{d}{dx}\frac{\partial f}{\partial z'}\right)\eta_2(x)\right] dx$$

Factoring $\eta_1(x)$ out of the square brackets and writing Eq. (6.23) as

$$\frac{\eta_2(x)}{\eta_1(x)} = -\frac{\partial g/\partial y}{\partial g/\partial z}$$

we have

$$\frac{\partial J}{\partial\alpha} = \int_{x_1}^{x_2}\left[\left(\frac{\partial f}{\partial y} - \frac{d}{dx}\frac{\partial f}{\partial y'}\right) - \left(\frac{\partial f}{\partial z} - \frac{d}{dx}\frac{\partial f}{\partial z'}\right)\left(\frac{\partial g/\partial y}{\partial g/\partial z}\right)\right]\eta_1(x)\, dx \tag{6.24}$$

This latter equation now contains the single arbitrary function $\eta_1(x)$, which is not in any way restricted by Eqs. (6.22), and upon setting $\alpha = 0$, the expression in the brackets must vanish. Thus, we have

$$\left(\frac{\partial f}{\partial y} - \frac{d}{dx}\frac{\partial f}{\partial y'}\right)\left(\frac{\partial g}{\partial y}\right)^{-1} = \left(\frac{\partial f}{\partial z} - \frac{d}{dx}\frac{\partial f}{\partial z'}\right)\left(\frac{\partial g}{\partial z}\right)^{-1} \tag{6.25}$$

Now, the left-hand side of this equation involves only derivatives of f and g with respect to y and y', while the right-hand side involves only derivatives with respect to z and z'. Because y and z are both functions of x, the two sides of Eq. (6.25) may be set equal to a function of x, which we will write as $\lambda(x)$:

$$\left.\begin{aligned} \frac{\partial f}{\partial y} - \frac{d}{dx}\frac{\partial f}{\partial y'} + \lambda(x)\frac{\partial g}{\partial y} &= 0 \\ \frac{\partial f}{\partial z} - \frac{d}{dx}\frac{\partial f}{\partial z'} + \lambda(x)\frac{\partial g}{\partial z} &= 0 \end{aligned}\right\} \tag{6.26}$$

The complete solution to the problem now depends upon finding *three* functions: $y(x)$, $z(x)$, and $\lambda(x)$. But there are *three* relations which may be used: the two equations (6.26) and the equation of constraint, Eq. (6.20).

Thus, there is a sufficient number of relations to allow a complete solution. Note that here $\lambda(x)$ is considered to be *undetermined** and is obtained as a part of the solution.

For the general case of several dependent variables and several auxiliary conditions, we have the following set of equations:

$$\frac{\partial f}{\partial y_i} - \frac{d}{dx}\frac{\partial f}{\partial y_i'} + \sum_j \lambda_j(x)\frac{\partial g_j}{\partial y_i} = 0 \tag{6.27}$$

$$g_j\{y_i\,;x\} = 0 \tag{6.28}$$

If $i = 1, 2, \ldots, m$ and $j = 1, 2, \ldots, n$, Eq. (6.27) represents m equations in $m + n$ unknowns, but there are also the n equations of constraint (6.28). Thus, there are $m + n$ equations in $m + n$ unknowns and the system is soluble.

Equation (6.28) is, of course, equivalent to the set of n differential equations

$$\sum_i \frac{\partial g_j}{\partial y}\,dy_i = 0, \qquad \begin{cases} i = 1, 2, \ldots, m \\ j = 1, 2, \ldots, n \end{cases} \tag{6.29}$$

In problems in mechanics the constraint equations are frequently differential equations rather than algebraic equations. Therefore, equations of the type (6.29) are sometimes more useful than the equations represented by (6.28). [See Section 7.5 for an amplification of this point.]

▶ **Example 6.7 Disk Rolling on an Inclined Plane**

Consider a disk rolling without slipping on an inclined plane, as in Fig. 6-5. The configuration of this system can be uniquely described in terms of the "coordinates"† y and θ. The relation between these coordinates is

$$y = R\theta \tag{1}$$

where R is the radius of the disk. Hence, the equation of constraint is

$$g(y, \theta) = y - R\theta = 0 \tag{2}$$

and

$$\frac{\partial g}{\partial y} = 1; \qquad \frac{\partial g}{\partial \theta} = -R \tag{3}$$

are the quantities associated with λ, the single undetermined multiplier for this case.

* The function $\lambda(x)$ is known as a *Lagrange undetermined multiplier*, and was introduced in Lagrange's *Mechanique Analytique*, 1788.

† These are actually the *generalized coordinates* discussed in Section 7.3; see also Example 7.5(b).

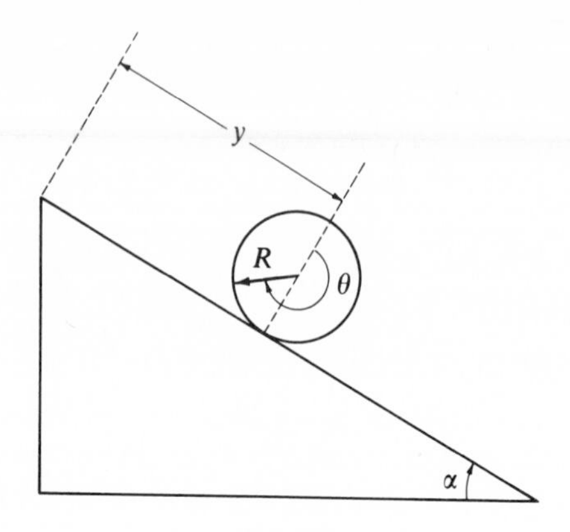

FIG. **6-5**

6.8 The δ Notation

In analyses which make use of the calculus of variations it is customary to use a shorthand notation to represent the variation. Thus, Eq. (6.8), which can be written as

$$\frac{\partial J}{\partial \alpha} d\alpha = \int_{x_1}^{x_2} \left(\frac{\partial f}{\partial y} - \frac{d}{dx} \frac{\partial f}{\partial y'} \right) \frac{\partial y}{\partial \alpha} \, d\alpha \, dx \tag{6.30}$$

may be expressed as

$$\delta J = \int_{x_1}^{x_2} \left(\frac{\partial f}{\partial y} - \frac{d}{dx} \frac{\partial f}{\partial y'} \right) \delta y \, dx \tag{6.31}$$

where

$$\left. \begin{aligned} \frac{\partial J}{\partial \alpha} d\alpha &\equiv \delta J \\ \frac{\partial y}{\partial \alpha} d\alpha &\equiv \delta y \end{aligned} \right\} \tag{6.32}$$

The condition of extremum then becomes

$$\delta J = \delta \int_{x_1}^{x_2} f\{y, y'; x\} \, dx = 0 \tag{6.33}$$

Taking the variation symbol δ inside the integral (since, by hypothesis, the limits of integration are not affected by the variation), we have

$$\begin{aligned} \delta J &= \int_{x_1}^{x_2} \delta f \, dx \\ &= \int_{x_1}^{x_2} \left(\frac{\partial f}{\partial y} \delta y + \frac{\partial f}{\partial y'} \delta y' \right) dx \end{aligned} \tag{6.34}$$

But,

$$\delta y' = \delta\left(\frac{dy}{dx}\right) = \frac{d}{dx}(\delta y) \tag{6.35}$$

so that

$$\delta J = \int_{x_1}^{x_2} \left(\frac{\partial f}{\partial y}\,\delta y + \frac{\partial f}{\partial y'}\frac{d}{dx}\,\delta y\right) dx \tag{6.36}$$

Upon integrating the second term by parts as before, we find

$$\delta J = \int_{x_1}^{x_2} \left(\frac{\partial f}{\partial y} - \frac{d}{dx}\frac{\partial f}{\partial y'}\right) \delta y\, dx \tag{6.37}$$

Since the variation δy is arbitrary, the extremum condition, $\delta J = 0$, requires the integrand to vanish, thereby yielding the Euler equation (6.9).

Although the δ notation is frequently used, it is important to realize that it is only a shorthand expression of the more precise differential quantities.

Suggested References

Accounts of the calculus of variations at the intermediate level which are pointed toward the application in Hamilton's Principle are to be found, for example, in Bradbury (Br68, Chapter 11), Goldstein (Go50, Chapter 2), Houston (Ho48, Chapter 5), and Margenau and Murphy (Ma43, Chapter 6).

One of the most useful books dealing with variational techniques in mechanics is that by Lanczos (La49).

The treatment of Halfman (Ha62a, Vol. II, Chapters 10 and 11) includes a discussion of performance optimization and several applications in the field of engineering physics.

A useful account at the intermediate-to-advanced level which gives applications of several variational methods is that of Irving and Mullineux (Ir59, Chapter 7). A book at the same level with many applications is that of Weinstock (We52). See also Arfken (Ar66, Chapter 17).

A brief mathematical treatment is given by Dettman (De62, Chapter 2). One of the standard works, with a mathematical viewpoint is, that of Bliss (Bl46). The discussion by Courant and Hilbert (Co53, Chapter 4) is advanced but directed toward applications in physics.

Problems

6-1. Consider the line connecting $(x_1, y_1) = (0, 0)$ and $(x_2, y_2) = (1, 1)$. Show explicitly that the function $y(x) = x$ produces a minimum path length by using the varied function $y(\alpha, x) = x + \alpha \sin \pi(1 - x)$. Use the first few terms in the expansion of the resulting elliptic integral to show the equivalent of Eq. (6.4).

6-2. Show that the shortest distance between two points on a plane is a straight line.

6-3. Show that the shortest distance between two points in (three-dimensional) space is a straight line.

6-4. Show that the geodesic on the surface of a right circular cylinder is a helix.

6-5. Consider the surface generated by revolving a line connecting two fixed points, (x_1, y_1) and (x_2, y_2), about an axis which is coplanar with the two points. Find the equation of the line which connects the points such that the surface area generated by the revolution (i.e., the area of the surface of revolution) is a minimum. Obtain the solution by using Eq. (6.12). (The result is a *catenary*.)

6-6. Solve Problem 6-5 by using Eq. (6.8).

6-7. Reexamine the problem of the brachistochrone (Example 6.4) and show that the time required for a particle to move (frictionlessly) to the final point (x_2, y_2) is $(\pi/2)\sqrt{a/g}$, *independent* of the starting point.

6-8. Find the dimensions of the parallelepiped of maximum volume that is circumscribed by (a) a sphere of radius R, and (b) an ellipsoid with semi-axes a, b, c.

6-9. Find an expression involving the function $\phi(x_1, x_2, x_3)$ which has a minimum average value of the square of its gradient within a certain volume V of space.

6-10. Find the ratio of the radius R to the height H of a right-circular cylinder of fixed volume V that will minimize the surface area A.

6-11. A disk of radius R rolls without slipping inside the parabola $y = ax^2$. Find the equation of constraint. Express the condition which allows the disk to roll so that it contacts the parabola at one and only one point, independent of its position.

CHAPTER 7

Hamilton's Principle—Lagrangian and Hamiltonian Dynamics

7.1 Introduction

Experience has shown that, if relativistic effects can be neglected, the motion of a particle in an inertial reference frame is correctly described by the Newtonian equation $\mathbf{F} = \dot{\mathbf{p}}$. In the event that the particle is not required to move in some complicated manner and if rectangular coordinates are used to describe the motion, then usually the equations of motion are relatively simple. However, if either of these restrictions is removed, the equations can become quite complex and difficult to manipulate. For example, if a particle is constrained to move on the surface of a sphere, the equations of motion result from the projection of the Newtonian vector equation onto that surface. The representation of the acceleration vector in spherical coordinates is a formidable expression, as the reader who has worked Problem 1-19 can readily testify.

Moreover, if a particle is constrained to move on a given surface, there must exist certain forces (called *forces of constraint*) which maintain the particle in contact with the specified surface. For the case in which a particle moves on a smooth horizontal surface, the force of constraint is simply $\mathbf{F}_c = -m\mathbf{g}$. But, if the particle is, say, a bead sliding down a curved wire, the force of constraint can be quite complicated. Indeed, in particular situations it may be difficult or even impossible to obtain explicit expressions for the forces of constraint. In solving a problem by using the Newtonian procedure, however, it is necessary to know *all* of the forces since the quantity **F** which appears in the fundamental equation is the *total* force acting on a body.

In order to circumvent some of the practical difficulties which arise in attempts to apply Newton's equations to particular problems, alternative procedures may be developed. All such approaches are in essence *a posteriori* since it is known beforehand that a result equivalent to the Newtonian equations must be obtained. Thus, in order to effect a simplification it is not required to formulate a *new* theory of mechanics—the Newtonian theory is quite correct—but only to devise an alternative method of dealing with complicated problems in a general manner. Such a method is contained in *Hamilton's Principle* and the equations of motion which result from the application of this principle are called *Lagrange's equations.*

If Lagrange's equations are to constitute a proper description of the dynamics of particles, then they must be equivalent to Newton's equations. On the other hand, Hamilton's Principle can be applied to a wide range of physical phenomena (particularly those involving *fields*) with which Newton's equations are not usually associated. To be sure, each of the results than can be obtained from Hamilton's Principle was *first* obtained, as were Newton's equations, by the correlation of experimental facts. Hamilton's Principle has not provided us with any new physical theories, but it has allowed a satisfying unification of many individual theories by means of a single basic postulate. This is not an idle exercise in hindsight, since it is the goal of physical theory not only to give precise mathematical formulation to observed phenomena but also to describe these effects with an economy of fundamental postulates and in the most unified manner possible. Indeed, Hamilton's Principle is one of the most elegant and far-reaching principles of physical theory.

In view of the wide range of applicability that Hamilton's Principle has been found to possess (even though this is an after-the-fact discovery), it is not unreasonable to assert that Hamilton's Principle is more "fundamental" than are Newton's equations. Therefore, we shall proceed by first postulating Hamilton's Principle; we shall then obtain Lagrange's equations and show that these are equivalent to Newton's equations.

Since we have already discussed (in Chapters 2 through 5) dissipative phenomena at some length, we shall henceforth confine our attention to *conservative* systems. Consequently, we shall not discuss the more general set of Lagrange's equations which take into account the effects of non-conservative forces. The reader is referred to the literature for these details.*

7.2 Hamilton's Principle

Minimal principles in physics have a long and interesting history. The search for such principles is predicated on the notion that Nature always acts in such a way that certain important quantities are minimized when a physical process takes place. The first such minimum principles were developed in the field of optics. Hero of Alexandria, in the second century B.C., found that the law governing the reflection of light could be obtained by asserting that a light ray, traveling from one point to another by a reflection from a plane mirror, always takes the shortest possible path. A simple geometrical construction will verify that this minimum principle does indeed lead to the equality of the angles of incidence and reflection for a light ray reflected from a plane mirror. Hero's principle of the *shortest path* cannot, however, yield a correct law for *refraction*. In 1657 Fermat† reformulated the principle by postulating that a light ray always travels from one point to another in a medium by a path that requires the least time. Fermat's principle of *least time* leads immediately, not only to the correct law of reflection, but also to Snell's law of refraction.‡

Minimum principles continued to be sought, and in the latter part of the seventeenth century the beginnings of the calculus of variations were developed by Newton, Leibniz, and the Bernoullis when such problems as the brachistochrone (see Section 6.4) and the shape of a hanging chain (a *catenary*) were solved.

The first application of a general minimum principle in mechanics was made in 1747 by Maupertuis,§ who asserted that dynamical motion takes place with minimum action. Maupertuis' principle of *least action* was based on theological grounds (action is minimized through the "wisdom of God"),

*See for example, Goldstein (Go50, Chapter 2), or for a comprehensive discussion, Whittaker (Wh37, Chapter 8).

† Pierre de Fermat (1601–1665), a French lawyer, linguist, and amateur mathematician.

‡ In 1661 Fermat correctly deduced the law of refraction which had been discovered experimentally in about 1621 by Willebrord Snell (1591–1626), a Dutch mathematical prodigy. For an excerpt from Fermat's paper, see Magie (Ma35, pp. 278–280); the derivation is also given by Lindsay and Margenau (Li36, p. 135).

§ Pierre-Louise-Moreau de Maupertuis (1698–1759), French mathematician and astronomer. The first use to which Maupertuis put the principle of least action was to restate Fermat's derivation of the law of refraction (1744).

and his concept of "action" was rather vague. (Recall that *action* is a quantity with the dimensions of *length* × *momentum* or *energy* × *time*.) Only later was a firm mathematical foundation of the principle given by Lagrange (1760). Although it is a useful form from which to make the transition from classical mechanics to optics and to quantum mechanics, the principle of least action is less general than Hamilton's Principle and, indeed, can be derived from it. We forego a detailed discussion here.*

In 1828 Gauss developed a method of treating mechanics by his principle of *least constraint*; a modification was later made by Hertz and embodied in his principle of *least curvature*. These principles† are closely related to Hamilton's Principle and add nothing to the content of Hamilton's more general formulation; their mention only serves to emphasize the continual concern with minimal principles in physics.

In two papers published in 1834 and 1835, Hamilton‡ announced the dynamical principle upon which it is possible to base all of mechanics and, indeed, most of classical physics. Hamilton's Principle may be stated as follows§:

> Of all the possible paths along which a dynamical system may move from one point to another within a specified time interval (consistent with any constraints), the actual path followed is that which minimizes the time integral of the difference between the kinetic and potential energies.

In terms of the calculus of variations, Hamilton's Principle becomes

$$\delta \int_{t_1}^{t_2} (T - U)\, dt = 0 \tag{7.1}$$

This variational statement of the principle requires only that $T - U$ be an *extremum*, not necessarily a *minimum*, but in almost all applications of importance in dynamics the minimum condition obtains.

Now, the kinetic energy of a particle expressed in fixed, rectangular coordinates is a function only of the $\dot{x}_i$ and if the particle moves in a conservative force field, the potential energy is a function only of the x_i:

$$T = T(\dot{x}_i); \qquad U = U(x_i)$$

* See, for example, Goldstein (Go50, pp. 228–235) or Sommerfeld (So50, pp. 204–209).

† See, for example, Lindsay and Margenau (Li36, pp. 112–120) or Sommerfeld (So50, pp. 210–214).

‡ Sir William Rowan Hamilton (1805–1865), Scottish mathematician and astronomer, and later, Irish Astronomer Royal.

§ The general meaning of "the path of a system" will be made clear in Section 7.3.

If we define the difference of these quantities to be

$$\begin{aligned} L &\equiv T - U \\ &= L(x_i, \dot{x}_i) \end{aligned} \tag{7.2}$$

then Eq. (7.1) becomes

$$\boxed{\delta \int_{t_1}^{t_2} L(x_i, \dot{x}_i)\, dt = 0} \tag{7.3}$$

The function L appearing in this expression may be identified with the functional f of the variational integral (see Section 6.6),

$$\delta \int_{x_1}^{x_2} f\{y_i(x), y_i'(x); x\}\, dx$$

if we make the transformations

$$\begin{aligned} x &\to t \\ y_i(x) &\to x_i(t) \\ y_i'(x) &\to \dot{x}_i(t) \\ f\{y_i(x), y_i'(x); x\} &\to L(x_i, \dot{x}_i) \end{aligned}$$

Therefore, the Euler–Lagrange equations (6.17) corresponding to Eq. (7.3) are

$$\boxed{\frac{\partial L}{\partial x_i} - \frac{d}{dt}\frac{\partial L}{\partial \dot{x}_i} = 0, \qquad i = 1, 2, 3} \tag{7.4}$$

These are the *Lagrange equations of motion* for the particle and the quantity L is called the *Lagrange function* or *Lagrangian* for the particle.

By way of example, let us obtain the Lagrange equation of motion for the one-dimensional harmonic oscillator. With the usual expressions for the kinetic and potential energies, we have

$$L = T - U = \tfrac{1}{2}m\dot{x}^2 - \tfrac{1}{2}kx^2$$

Applying Eq. (7.4) there results

$$m\ddot{x} + kx = 0$$

which is identical with the Newtonian equation of motion.

It may seem that the Lagrangian procedure is a rather complicated one if it can only duplicate the simple results of Newtonian theory. However, let us continue illustrating the method by considering the case of the plane pendulum (see Section 7.4). Using Eqs. (7.15) for T and U we have for the Lagrangian function

$$L = \tfrac{1}{2}ml^2\dot{\theta}^2 - mgl(1 - \cos\theta)$$

We now treat θ *as if it were a rectangular coordinate* and apply the operations specified in Eq. (7.4); we obtain

$$\ddot{\theta} + \frac{g}{l}\sin\theta = 0$$

which again is identical with the Newtonian equation. This is a remarkable result; it has been obtained by calculating the kinetic and potential energies in terms of θ rather than x and then applying a set of operations designed for use with rectangular rather than angular coordinates. We are therefore led to suspect that the Lagrange equations are more general than the form of Eq. (7.4) would indicate. We shall pursue this matter in Section 7.4.

Another important characteristic of the method employed in the two simple examples above is the fact that nowhere in the calculations did there enter any statement regarding *force*. Thus, the equations of motion were obtained only by specifying certain properties associated *with the particle* (the kinetic and potential energies), and without the necessity of explicitly taking into account the fact that there was an external agency acting *on the particle* (the force). Therefore, insofar as *energy* can be defined independently of Newtonian concepts, Hamilton's Principle allows the calculation of the equations of motion of a body completely without recourse to Newtonian theory. We shall return to this important point in Sections 7.5 and 7.7.

7.3 Generalized Coordinates

We now seek to take advantage of the flexibility in the specification of coordinates which the two examples of the preceding section have suggested is inherent in Lagrange's equations.

We shall consider a general mechanical system which consists of a collection of n discrete, point particles, some of which may be connected together to form rigid bodies. In order to specify the state of such a system at a given time, it is necessary to use n radius vectors. Since each radius vector consists of a triple of numbers (e.g., the rectangular coordinates), $3n$ quantities must be specified in order to describe the positions of all the particles. If there exist equations of constraint which relate some of these coordinates to others (as would be the case, for example, if some of the particles were connected to form rigid bodies or if the motion were constrained to lie along some path or on some surface), then not all of the $3n$ coordinates are independent. In fact, if there are m equations of constraint, then $3n - m$ coordinates are independent, and the system is said to possess $3n - m$ *degrees of freedom*.

It is important to note that if $s = 3n - m$ coordinates are required in a given case, it is not necessary to choose s rectangular coordinates, or even s curvilinear coordinates (spherical, cylindrical, etc.); it is possible to choose *any* s parameters, as long as they completely specify the state of the system. These s quantities need not even have the dimensions of length. Depending on the problem at hand, it may prove more convenient to choose some of the parameters with dimensions of *energy*, some with dimensions of $(length)^2$, some which are *dimensionless*, etc. In Example 6.7, a disk rolling down an inclined plane was described in terms of one coordinate which was a length and one which was an angle. The name *generalized coordinates* is given to any set of quantities which completely specifies the state of a system. The generalized coordinates are customarily written as $q_1, q_2, \ldots$ or simply as the q_j. A set of independent generalized coordinates whose number equals the number s of degrees of freedom of the system and which are not restricted by the constraints will be called a *proper* set of generalized coordinates. In certain instances it may be advantageous to use generalized coordinates whose number exceeds the number of degrees of freedom and to explicitly take into account the constraint relations through the use of the Lagrange undetermined multipliers. Such would be the case, for example, if it were desired to calculate the forces of constraint [see Example 7.5(b)].

The choice of a set of generalized coordinates for the description of a system is not unique; there are in general many sets of quantities (in fact, an *infinite* number!) which will completely specify the state of a given system. For example, in the problem of the disk rolling down the inclined plane, we might choose as coordinates the height of the center of mass of the disk above some reference level and the distance through which some point on the rim has traveled since the start of the motion. The ultimate test of the "suitability" of a particular set of generalized coordinates is whether the resulting equations of motion are sufficiently simple to allow a straightforward interpretation. Unfortunately, no general rules can be stated for the selection of the "most suitable" set of generalized coordinates for a given problem; a certain skill must be developed through experience.

In addition to the generalized coordinates, we may define a set of quantities which consists of the time derivatives of the q_j: $\dot{q}_1, \dot{q}_2, \ldots$ or simply the $\dot{q}_j$. In analogy with the nomenclature for rectangular coordinates, we call the $\dot{q}_j$ the *generalized velocities*.

If we allow for the possibility that the equations connecting the $x_{\alpha,i}$ and the q_j explicitly contain the time, then the set of transformation equations is given by*

$$x_{\alpha,i} = x_{\alpha,i}(q_1, q_2, \ldots, q_s, t), \qquad \begin{cases} \alpha = 1, 2, \ldots, n \\ i = 1, 2, 3 \end{cases}$$

$$= x_{\alpha,i}(q_j, t), \qquad j = 1, 2, \ldots, s \tag{7.5a}$$

In general, the rectangular components of the velocities will depend on the generalized coordinates, the generalized velocities, and the time:

$$\dot{x}_{\alpha,i} = \dot{x}_{\alpha,i}(q_j, \dot{q}_j, t) \tag{7.5b}$$

We may also write the inverse transformations as

$$q_j = q_j(x_{\alpha,i}, t) \tag{7.5c}$$

$$\dot{q}_j = \dot{q}_j(x_{\alpha,i}, \dot{x}_{\alpha,i}, t) \tag{7.5d}$$

In addition, there are also $m = 3n - s$ equations of constraint of the form

$$f_k = f_k(x_{\alpha,i}, t), \qquad k = 1, 2, \ldots, m \tag{7.6}$$

▶ **Example 7.3 Particle Moving on a Hemispherical Surface**

Consider a point particle which moves on the surface of a hemisphere of radius R whose center is at the origin. Thus, the motion always takes place on the surface

$$x^2 + y^2 + z^2 - R^2 = 0, \qquad z \geq 0 \tag{1}$$

Let us choose as our generalized coordinates the cosines of the angles between the x-, y-, and z-axes and the line which connects the particle with the origin. Therefore,

$$q_1 = \frac{x}{R}; \qquad q_2 = \frac{y}{R}; \qquad q_3 = \frac{z}{R} \tag{2}$$

But the sum of the squares of the direction cosines of a line equals unity. Hence,

$$q_1^2 + q_2^2 + q_3^2 = 1 \tag{3}$$

Therefore, this set of q_j does not constitute a proper set of generalized coordinates since we can write q_3 as a function of q_1 and q_2:

$$q_3 = \sqrt{1 - q_1^2 + q_2^2} \tag{4}$$

We may, however, choose $q_1 = x/R$ and $q_2 = y/R$ as proper generalized coordinates and these quantities, together with the equation of constraint, Eq. (4), or

$$z = \sqrt{R^2 - x^2 - y^2}$$

will be sufficient to uniquely specify the position of the particle. This is an obvious result since it is clear that only two coordinates (e.g., latitude and longitude) are necessary to specify a point on the surface of a sphere, but the example serves to

* In this chapter we attempt to simplify the notation by reserving the subscript i to designate rectangular axes; therefore, we always have $i = 1, 2, 3$.

illustrate the fact that the equations of constraint can always be used to reduce a trial set of coordinates to a proper set of generalized coordinates.

The state of a system that consists of n particles and which is subject to m constraints that connect some of the $3n$ rectangular coordinates is completely specified by $s = 3n - m$ generalized coordinates. Therefore, we may represent the state of such a system by a point in an s-dimensional space called *configuration space*. Each dimension of this space corresponds to one of the q_j. The time history of a system may be represented by a curve in configuration space, each point specifying the *configuration* of the system at a particular instant. Through each such point there passes an infinity of curves representing possible motions of the system; each curve corresponds to a particular set of initial conditions. We may therefore speak of the "path" of a system as it "moves" through configuration space, but we must be careful not to confuse this terminology with that applied to the motion of a particle along a path in ordinary three-dimensional space.

It is also to be noted that a dynamical path in a configuration space consisting of proper generalized coordinates is automatically consistent with the constraints on the system since the coordinates are chosen to correspond only to realizable motions of the system.

7.4 Lagrange's Equations of Motion in Generalized Coordinates

In view of the definitions in the preceding sections, we may now restate Hamilton's Principle as follows:

> Of all the possible paths along which a dynamical system may move from one point to another in configuration space within a specified time interval, the actual path followed is that which minimizes the time integral of the Lagrangian function for the system.

In order to set up the variational form of Hamilton's Principle in generalized coordinates, we may take advantage of an important property of the Lagrangian which we have not so far emphasized. The Lagrangian for a system is defined to be the difference between the kinetic and potential energies. But *energy* is a scalar quantity and so *the Lagrangian is a scalar function*. Hence the Lagrangian must be *invariant with respect to coordinate transformations*. We are therefore assured that no matter what generalized coordinates are chosen for the description of a system, the Lagrangian will have the same value for a given condition of the system.* Although the

* A "given condition of the system" means that each constituent particle is at a specified position and has a specified velocity.

Lagrangian will be expressed by means of different *functions* depending on the generalized coordinates used, the *value* of the Lagrangian is unique* for a given condition. It is therefore immaterial whether we express the Lagrangian in terms of the $x_{\alpha,i}$ and $\dot{x}_{\alpha,i}$ or the q_j and $\dot{q}_j$:

$$\begin{aligned} L &= T(\dot{x}_{\alpha,i}) - U(x_{\alpha,i}) \\ &= T(q_j, \dot{q}_j, t) - U(q_j, t) \end{aligned} \tag{7.7a}$$

That is,

$$\begin{aligned} L &= L(q_1, q_2, \ldots, q_s; \dot{q}_1, \dot{q}_2, \ldots, \dot{q}_s; t) \\ &= L(q_j, \dot{q}_j, t) \end{aligned} \tag{7.7b}$$

Thus, Hamilton's Principle becomes

$$\boxed{\delta \int_{t_1}^{t_2} L(q_j, \dot{q}_j, t)\, dt = 0} \tag{7.8}$$

If we refer to the definitions of the quantities in Section 6.6 and make the identifications

$$\begin{aligned} x &\to t \\ y_i(x) &\to q_j(t) \\ y_i'(x) &\to \dot{q}_j(t) \\ f\{y_i, y_i'; x\} &\to L(q_j, \dot{q}_j, t) \end{aligned}$$

then the Euler equations (6.17) corresponding to the variational problem stated in Eq. (7.8) becomes

$$\boxed{\frac{\partial L}{\partial q_j} - \frac{d}{dt}\frac{\partial L}{\partial \dot{q}_j} = 0, \qquad j = 1, 2, \ldots, s} \tag{7.9}$$

These are the Euler–Lagrange equations of motion for the system (usually called simply *Lagrange's equations*†). There are s of these equations, and together with the m equations of constraint and the initial conditions that

* Since the potential energy U is undetermined to within an additive constant, the Lagrangian is always similarly indefinite.

† First derived for a mechanical system (although not, of course, by using Hamilton's Principle) by Joseph Louis Lagrange (1736–1813) and presented in his famous treatise *Mechanique Analytique* in 1788. In this monumental work, which encompasses all phases of mechanics (statics, dynamics, hydrostatics, and hydrodynamics), Lagrange succeeded in placing the subject on a firm and unified mathematical foundation. The treatise is mathematical rather than physical in nature; Lagrange was quite proud of the fact that the entire work contains not a single diagram.

are imposed,* they provide a complete description of the motion of the system.

It is important to realize that the validity of Lagrange's equations requires the following two conditions:

(1) the forces acting on the system (apart from any forces of constraint) must be derivable from a potential (or several potentials); and

(2) the equations of constraint must be relations that connect the *coordinates* of the particles and may be functions of the time; i.e., we must have constraint relations of the form given by Eq. (7.6). If the constraints can be expressed in this manner, they are termed *holonomic* constraints. If the equations do not explicitly contain the time, the constraints are said to be *fixed* or *scleronomic*; moving constraints are *rheonomic*.

As stated earlier, we shall consider only the motion of systems subject to conservative forces. Such forces can always be derived from potential functions, so that condition (1) is satisfied. This is not a necessary restriction on either Hamilton's Principle or Lagrange's equations; the theory can readily be extended to include nonconservative forces. Similarly, Hamilton's Principle can also be formulated to include certain types of nonholonomic constraints, but the treatment here will be confined to holonomic systems.†

▶ Example 7.4 Particle on the Surface of a Cone

Consider the motion of a particle of mass m which is constrained to move on the surface of a cone of half-angle α and which is subject to a gravitational force. Let the axis of the cone correspond to the z-axis and let the apex of the cone be located at the origin, as in Fig. 7-1. Since the problem possesses cylindrical symmetry, we choose r, θ, and z as the generalized coordinates. We have, however, the equation of constraint

$$z = r \cot \alpha \tag{1}$$

so that there are only two degrees of freedom for the system and therefore there are only two proper generalized coordinates. We may use Eq. (1) to eliminate either the coordinate z or r; we choose to do the former. Then, the square of the velocity is [see Eqs. (1.101)]:

$$\begin{aligned} v^2 &= \dot{r}^2 + r^2\dot{\theta}^2 + \dot{z}^2 \\ &= \dot{r}^2 + r^2\dot{\theta}^2 + \dot{r}^2 \cot^2 \alpha \\ &= \dot{r}^2 \csc^2 \alpha + r^2\dot{\theta}^2 \end{aligned} \tag{2}$$

* Since there are s second-order differential equations, $2s$ initial conditions must be supplied in order to determine the motion uniquely.

† For details concerning these topics see the "advanced" Suggested References at the end of the chapter.

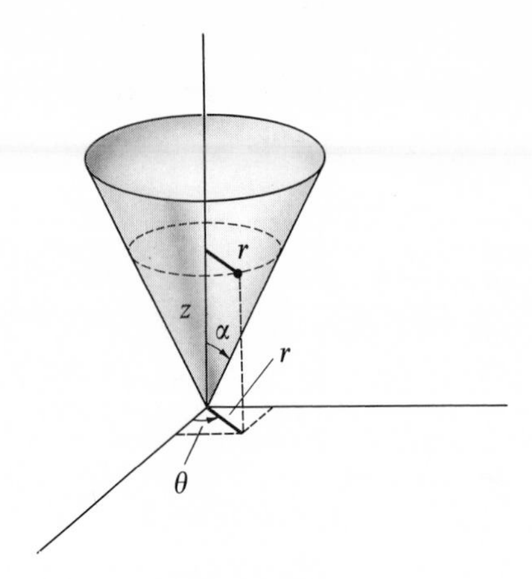

FIG. 7-1

The potential energy is [if we choose $U = 0$ at $z = 0$]:

$$U = mgz = mgr \cot \alpha \tag{3}$$

so that the Lagrangian is

$$L = \tfrac{1}{2}m(\dot{r}^2 \csc^2 \alpha + r^2\dot{\theta}^2) - mgr \cot \alpha \tag{4}$$

We note first that L does not explicitly contain θ. Therefore $\partial L/\partial\theta = 0$, and the Lagrange equation for the coordinate θ is

$$\frac{d}{dt}\frac{\partial L}{\partial \dot{\theta}} = 0 \tag{5}$$

Hence,

$$\frac{\partial L}{\partial \dot{\theta}} = mr^2\dot{\theta} = \text{const.} \tag{6}$$

But, $mr^2\dot{\theta} = mr^2\omega$ is just the angular momentum about the z-axis. Therefore, Eq. (6) expresses the conservation of angular momentum about the axis of symmetry of the system.

The Lagrange equation for r is

$$\frac{\partial L}{\partial r} - \frac{d}{dt}\frac{\partial L}{\partial \dot{r}} = 0 \tag{7}$$

Calculating the derivatives, we find

$$\ddot{r} - r\dot{\theta}^2 \sin^2 \alpha + g \sin \alpha \cos \alpha = 0 \tag{8}$$

which is the equation of motion for the coordinate r.

We shall return to this example in Section 8.11 and examine the motion in more detail.

■ 7.5 Lagrange's Equations with Undetermined Multipliers

Constraints which can be expressed as algebraic relations among the coordinates are holonomic constraints. If a system is subject only to such constrain s, a proper set of generalized coordinates can always be found in terms of which the equations of motion are free from explicit reference to the constraints.

Any constraints that must be expressed in terms of the *velocities* of the particles in the system are of the form

$$f_l(x_{\alpha,i}, \dot{x}_{\alpha,i}, t) = 0 \tag{7.10}$$

and constitute nonholonomic constraints *unless* the equations can be integrated to yield relations among the coordinates.*

▶ Example 7.5(a) A Holonomic Constraint

Consider a constraint relation of the form

$$\sum_i A_i \dot{x}_i + B = 0, \qquad i = 1, 2, 3 \tag{1}$$

In general, this equation is nonintegrable and therefore the constraint is nonholonomic. But if A_i and B have the forms

$$A_i = \frac{\partial f}{\partial x_i}; \qquad B = \frac{\partial f}{\partial t}; \qquad f = f(x_i, t) \tag{2}$$

then Eq. (1) may be written as

$$\sum_i \frac{\partial f}{\partial x_i}\frac{\partial x_i}{\partial t} + \frac{\partial f}{\partial t} = 0 \tag{3}$$

But this is just

$$\frac{df}{dt} = 0 \tag{4}$$

which can be integrated to yield

$$f(x_i, t) - \text{const.} = 0 \tag{5}$$

so that the constraint is actually holonomic.

* Such constraints are sometimes called *semiholonomic.*

From the above example, we conclude that constraints expressible in differential form as

$$\sum_j \frac{\partial f_k}{\partial q_j} dq_j + \frac{\partial f_k}{\partial t} dt = 0 \tag{7.11}$$

are entirely equivalent to those which have the form of Eq. (7.6).

If the constraint relations for a problem are given in differential form rather than as algebraic expressions, it is possible to incorporate them directly into Lagrange's equations by means of the Lagrange undetermined multipliers (see Section 6.7) without the necessity of first performing the integrations. That is, for constraints expressible as in Eq. (6.29),

$$\sum_j \frac{\partial f_k}{\partial q_j} dq_j = 0 \qquad \begin{cases} j = 1, 2, \ldots, s \\ k = 1, 2, \ldots, m \end{cases} \tag{7.12}$$

the Lagrange equations are [Eq. (6.27)]:

$$\boxed{\frac{\partial L}{\partial q_j} - \frac{d}{dt}\frac{\partial L}{\partial \dot{q}_j} + \sum_k \lambda_k(t) \frac{\partial f_k}{\partial q_j} = 0} \tag{7.13}$$

In fact, since the variation process involved in Hamilton's Principle holds the time constant at the end points, we could add to Eq. (7.12) a term $(\partial f_k/\partial t)\, dt$ without affecting the equations of motion. Thus, constraints expressed by Eq. (7.11) also lead to the Lagrange equations given in Eq. (7.13).

The great advantage of the Lagrangian formulation of mechanics is that the explicit inclusion of the forces of constraint is not necessary. That is, the emphasis is placed on the dynamics of the system rather than the calculation of the forces acting on each component of the system. In certain instances, however, it may be desired to know the forces of constraint. It is therefore worth pointing out that in Lagrange's equations expressed as in Eq. (7.13), the undetermined multipliers $\lambda_s(t)$ are just these forces of constraint.*

▶ Example 7.5(b) Disk Rolling Down an Inclined Plane

Let us consider again the case of the disk rolling down an inclined plane (see Example 6.7 and Fig. 6-5). The kinetic energy may be separated into translational

* See, for example, Goldstein (Go50, p. 42). Explicit calculations of the forces of constraint in some specific problems are carried out by Becker (Be54, Chapters 11 and 13) and by Symon (Sy60, p. 374 ff.)

and rotational terms*:

$$T = \tfrac{1}{2}M\dot{y}^2 + \tfrac{1}{2}I\dot{\theta}^2$$
$$= \tfrac{1}{2}M\dot{y}^2 + \tfrac{1}{4}MR^2\dot{\theta}^2 \tag{1}$$

where M is the mass of the disk and R is the radius; $I = \tfrac{1}{2}MR^2$ is the moment of inertia of the disk about a central axis. The potential energy is

$$U = Mg(l - y) \sin \alpha \tag{2}$$

where l is the length of the inclined surface of the plane and where the disk is assumed to have zero potential energy at the bottom of the plane. The Lagrangian is therefore

$$L = T - U$$
$$= \tfrac{1}{2}M\dot{y}^2 + \tfrac{1}{4}MR^2\dot{\theta}^2 + Mg(y - l) \sin \alpha \tag{3}$$

The equation of constraint is

$$f(y, \theta) = y - R\theta = 0 \tag{4}$$

Now, the system has only one degree of freedom if we insist that the rolling takes place without slipping. Therefore, we may choose either y or θ as the proper coordinate and use Eq. (4) to eliminate the other. Alternatively, we may continue to consider *both* y and θ as generalized coordinates and use the method of undetermined multipliers. The Lagrange equations in this case are

$$\left.\begin{aligned} \frac{\partial L}{\partial y} - \frac{d}{dt}\frac{\partial L}{\partial \dot{y}} + \lambda \frac{\partial f}{\partial y} = 0 \\ \frac{\partial L}{\partial \theta} - \frac{d}{dt}\frac{\partial L}{\partial \dot{\theta}} + \lambda \frac{\partial f}{\partial \theta} = 0 \end{aligned}\right\} \tag{5}$$

Performing the differentiations, we obtain

$$Mg \sin \alpha - M\ddot{y} + \lambda = 0 \tag{6a}$$

$$-\tfrac{1}{2}MR^2\ddot{\theta} - \lambda R = 0 \tag{6b}$$

Also, from the constraint equation, we have

$$y = R\theta \tag{6c}$$

The three Eqs. (6) constitute a soluble system for the three unknowns y, θ, λ. Differentiating the equation of constraint [Eq (6c)], we obtain

$$\ddot{\theta} = \ddot{y}/R \tag{7}$$

Hence, combining Eqs. (6b) and (7), we find

$$\lambda = -\tfrac{1}{2}M\ddot{y} \tag{8}$$

* We anticipate here a well-known result from rigid-body dynamics that is discussed in Chapter 12.

and then using this expression in Eq. (6a), there results

$$\ddot{y} = \frac{2g \sin \alpha}{3} \tag{9}$$

with

$$\lambda = -\frac{Mg \sin \alpha}{3} \tag{10}$$

so that Eq. (6b) yields

$$\ddot{\theta} = \frac{2g \sin \alpha}{3R} \tag{11}$$

Thus, we have three equations for the quantities $\ddot{y}$, $\ddot{\theta}$, and λ which can be immediately integrated.

We note that if the disk were to slide without friction down the plane, then we would have $\ddot{y} = g \sin \alpha$. Therefore, the rolling constraint reduces the acceleration to $\frac{2}{3}$ of the value for frictionless sliding. The magnitude of the force of friction which gives rise to the constraint is just λ, i.e. $(Mg/3) \sin \alpha$.

Note that we may eliminate $\dot{\theta}$ from the Lagrangian by substituting $\dot{\theta} = \dot{y}/R$ from the equation of constraint:

$$L = \tfrac{3}{4}M\dot{y}^2 + Mg(y - l) \sin \alpha \tag{12}$$

The Lagrangian is then expressed in terms of only one proper coordinate and the single equation of motion is immediately obtained from Eq. (7.9):

$$Mg \sin \alpha - \tfrac{3}{2}M\ddot{y} = 0 \tag{13}$$

which is the same as Eq. (9). Although this procedure is simpler, it cannot be used to obtain the force of constraint.

7.6 The Equivalence of Lagrange's and Newton's Equations

As we have emphasized from the outset, the Lagrangian and Newtonian formulations of mechanics are equivalent: the viewpoint is different but the content is the same. We now explicitly demonstrate this equivalence by showing that the two sets of equations of motion are in fact the same.

In Eq. (7.9) let us choose the generalized coordinates to be the rectangular coordinates. Then, Lagrange's equations (for a single particle) become

$$\frac{\partial L}{\partial x_i} - \frac{d}{dt}\frac{\partial L}{\partial \dot{x}_i} = 0, \qquad i = 1, 2, 3 \tag{7.14}$$

or,

$$\frac{\partial(T-U)}{\partial x_i} - \frac{d}{dt}\frac{\partial(T-U)}{\partial \dot{x}_i} = 0$$

But, in rectangular coordinates and for a conservative system, we have $T = T(\dot{x}_i)$ and $U = U(x_i)$, so that

$$\frac{\partial T}{\partial x_i} = 0 \qquad \text{and} \qquad \frac{\partial U}{\partial \dot{x}_i} = 0$$

Therefore, Lagrange's equations become

$$-\frac{\partial U}{\partial x_i} = \frac{d}{dt}\frac{\partial T}{\partial \dot{x}_i} \tag{7.15}$$

We also have (for a conservative system)

$$-\frac{\partial U}{\partial x_i} = F_i$$

and

$$\frac{d}{dt}\frac{\partial T}{\partial \dot{x}_i} = \frac{d}{dt}\frac{\partial}{\partial \dot{x}_i}\left(\sum_{j=1}^{3} \tfrac{1}{2} m\dot{x}_j^2\right) = \frac{d}{dt}(m\dot{x}_i) = \dot{p}_i$$

so that Eq. (7.15) yields the Newtonian equations, as required:

$$F_i = \dot{p}_i \tag{7.16}$$

Thus, the Lagrangian and Newtonian equations are identical in the event that the generalized coordinates are the rectangular coordinates.

7.7 The Essence of Lagrangian Dynamics

In the preceding sections several general and important statements were made concerning the Lagrangian formulation of mechanics. Before proceeding further, it seems worthwhile to summarize these points in order to emphasize the differences between the Lagrangian and Newtonian viewpoints.

Historically, the Lagrange equations of motion expressed in generalized coordinates were derived prior to the statement of Hamilton's Principle.* We have elected to deduce Lagrange's equations by postulating Hamilton's

* Lagrange's equations, 1788; Hamilton's Principle, 1834.

Principle because this is the most straightforward approach and is also the formal method by which the unification of classical dynamics is possible.

First and foremost, it must be reiterated that Lagrangian dynamics does not constitute a *new* theory in any sense of the word. The results of a Lagrangian analysis or a Newtonian analysis must be the same for any given mechanical system; it is only the method used to obtain these results that is different.

Whereas the Newtonian approach places the emphasis on an outside agency acting *on* a body (the *force*), the Lagrangian method deals only with quantities which are associated *with* the body (the kinetic and potential *energies*). In fact, nowhere in the Lagrangian formulation does the concept of *force* enter. This is a particularly important property of the method for a variety of reasons. First, since energy is a scalar quantity, the Lagrangian function for a system is invariant to coordinate transformations. Indeed, such transformations are not restricted to be between various orthogonal coordinate systems in ordinary space; they may also be transformations between *ordinary* coordinates and *generalized* coordinates. Thus, it is possible to pass from ordinary space (in which the equations of motion may be quite complicated) to a configuration space which can be chosen to yield maximum simplification for a particular problem. We are accustomed to thinking of mechanical systems in terms of *vector* quantities such as force, velocity, angular momentum, torque, etc., but in the Lagrangian formulation, the equations of motion are obtained entirely in terms of *scalar* operations in configuration space.

Another important aspect of the force-versus-energy viewpoint is that in certain situations it may not even be possible to state explicitly all the forces acting on a body (as is sometimes the case for forces of constraint), whereas it is still possible to give expressions for the kinetic and potential energies. It is just this fact that makes Hamilton's Principle useful for quantum-mechanical systems in which the forces are sometimes not known but the energies are known.

The differential statement of mechanics contained in Newton's equations or the integral statement embodied in Hamilton's Principle (and the resulting Lagrangian equations) have been shown to be entirely equivalent. Hence, there can be no distinction between these viewpoints which are based on the description of *physical effects*. From a philosophical standpoint, however, it is possible to make a distinction. In the Newtonian formulation, a certain force on a body is considered to produce a definite motion; that is, a definite *effect* is always associated with a certain *cause*. According to Hamilton's Principle, however, the motion of a body may be considered to result from the attempt of Nature to achieve a certain *purpose*, namely, to minimize the time integral of the difference between the kinetic and potential energies.

Clearly, the operational solving of problems in mechanics is not dependent on adopting one or the other of these views, but historically such considerations had a profound influence on the development of dynamics (as, for example, in Maupertuis' principle, mentioned in Section 7.2). The interested reader is referred to Margenau's excellent book (Ma50, Chapter 19) for a discussion of these matters.

7.8 A Theorem Concerning the Kinetic Energy

If the kinetic energy is expressed in fixed, rectangular coordinates, the result is a homogeneous, quadratic function of the $\dot{x}_{\alpha,i}$:

$$T = \frac{1}{2}\sum_{\alpha=1}^{n}\sum_{i=1}^{3} m_\alpha \dot{x}_{\alpha,i}^2 \tag{7.17}$$

We now wish to determine the dependence of T on the generalized coordinates and velocities. The equations connecting the rectangular coordinates and the generalized coordinates are [see Eq. (7.5a)]

$$x_{\alpha,i} = x_{\alpha,i}(q_j, t), \qquad j = 1, 2 \ldots, s \tag{7.18}$$

Hence,

$$\dot{x}_{\alpha,i} = \sum_{j=1}^{s} \frac{\partial x_{\alpha,i}}{\partial q_j}\dot{q}_j + \frac{\partial x_{\alpha,i}}{\partial t} \tag{7.19}$$

Evaluating the square of $\dot{x}_{\alpha,i}$, we obtain

$$\dot{x}_{\alpha,i}^2 = \sum_{j,k} \frac{\partial x_{\alpha,i}}{\partial q_j}\frac{\partial x_{\alpha,i}}{\partial q_k}\dot{q}_j\dot{q}_k + 2\sum_j \frac{\partial x_{\alpha,i}}{\partial q_j}\frac{\partial x_{\alpha,i}}{\partial t}\dot{q}_j + \left(\frac{\partial x_{\alpha,i}}{\partial t}\right)^2 \tag{7.20}$$

And the kinetic energy becomes

$$T = \sum_\alpha \sum_{i,j,k} \frac{1}{2}m_\alpha \frac{\partial x_{\alpha,i}}{\partial q_j}\frac{\partial x_{\alpha,i}}{\partial q_k}\dot{q}_j\dot{q}_k + \sum_\alpha\sum_{i,j} m_\alpha \frac{\partial x_{\alpha,i}}{\partial q_j}\frac{\partial x_{\alpha,i}}{\partial t}\dot{q}_j + \sum_\alpha\sum_i \frac{1}{2}m_\alpha\left(\frac{\partial x_{\alpha,i}}{\partial t}\right)^2 \tag{7.21}$$

Thus, with obvious notation, we have the general result

$$T = \sum_{j,k} a_{jk}\dot{q}_j\dot{q}_k + \sum_j b_j\dot{q}_j + c \tag{7.22}$$

Now, a case of particular importance is that in which the system is *scleronomic* so that the time does not appear explicitly in the equations of transformation [Eq. (7.18)]; then the partial time derivatives vanish:

$$\frac{\partial x_{\alpha,i}}{\partial t} = 0; \qquad b_j = 0; \qquad c = 0$$

Therefore, under these conditions, the kinetic energy is a *homogeneous quadratic function* of the generalized velocities:

$$\boxed{T = \sum_{j,k} a_{jk} \dot{q}_j \dot{q}_k} \tag{7.23}$$

Next, we differentiate Eq. (7.23) with respect to $\dot{q}_l$:

$$\frac{\partial T}{\partial \dot{q}_l} = \sum_k a_{lk} \dot{q}_k + \sum_j a_{jl} \dot{q}_j$$

Multiplying this equation by $\dot{q}_l$ and summing over l, we have

$$\sum_l \dot{q}_l \frac{\partial T}{\partial \dot{q}_l} = \sum_{k,l} a_{lk} \dot{q}_k \dot{q}_l + \sum_{j,l} a_{jl} \dot{q}_j \dot{q}_l$$

Now, *all* of the indices are dummies, so that both terms on the right-hand side are identical:

$$\boxed{\sum_l \dot{q}_l \frac{\partial T}{\partial \dot{q}_l} = 2 \sum_{j,k} a_{jk} \dot{q}_j \dot{q}_k = 2T} \tag{7.24}$$

This important result is a special case of *Euler's Theorem* which states that if $f(y_k)$ is a homogeneous function of the y_k which is of degree n, then

$$\sum_k y_k \frac{\partial f}{\partial y_k} = nf \tag{7.25}$$

7.9 The Conservation of Energy

According to our previous arguments,* *time* is homogeneous within an inertial reference frame. Therefore, the Lagrangian that describes a *closed system* (i.e., a system which does not interact with anything outside the system) cannot depend explicitly on the time.† That is,

$$\frac{\partial L}{\partial t} = 0$$

so that the total derivative of the Lagrangian becomes

$$\frac{dL}{dt} = \sum_j \frac{\partial L}{\partial q_j} \dot{q}_j + \sum_j \frac{\partial L}{\partial \dot{q}_j} \ddot{q}_j \tag{7.26}$$

* See Section 2.3.

† The Lagrangian will likewise be independent of the time if the system exists in a uniform force field.

where the usual term, $\partial L/\partial t$, does not now appear. But Lagrange's equations are

$$\frac{\partial L}{\partial q_j} = \frac{d}{dt}\frac{\partial L}{\partial \dot{q}_j} \tag{7.27}$$

Using Eq. (7.27) to substitute for $\partial L/\partial q_j$ in Eq. (7.26), we have

$$\frac{dL}{dt} = \sum_j \dot{q}_j \frac{d}{dt}\frac{\partial L}{\partial \dot{q}_j} + \sum_j \frac{\partial L}{\partial \dot{q}_j}\ddot{q}_j$$

or,

$$\frac{dL}{dt} - \sum_j \frac{d}{dt}\left(\dot{q}_j \frac{\partial L}{\partial \dot{q}_j}\right) = 0$$

so that

$$\frac{d}{dt}\left(L - \sum_j \dot{q}_j \frac{\partial L}{\partial \dot{q}_j}\right) = 0 \tag{7.28}$$

Therefore, the quantity in the parentheses is constant in time; denote this constant by $-H$:

$$L - \sum_j \dot{q}_j \frac{\partial L}{\partial \dot{q}_j} = -H = \text{const.} \tag{7.29}$$

If the potential energy U does not depend explicitly on the velocities $\dot{x}_{\alpha,i}$ or the time t, then $U = U(x_{\alpha,i})$. Now, the relations which connect the rectangular coordinates and the generalized coordinates are of the form $x_{\alpha,i} = x_{\alpha,i}(q_j)$ or $q_j = q_j(x_{\alpha,i})$, where we exclude the possibility of an explicit time dependence in the transformation equations. Therefore, $U = U(q_j)$, and $\partial U/\partial \dot{q}_j = 0$. Thus,

$$\frac{\partial L}{\partial \dot{q}_j} = \frac{\partial(T-U)}{\partial \dot{q}_j} = \frac{\partial T}{\partial \dot{q}_j}$$

Equation (7.29) can then be written as

$$(T - U) - \sum_j \dot{q}_j \frac{\partial T}{\partial \dot{q}_j} = -H \tag{7.30}$$

And, using Eq. (7.24), we have

$$(T - U) - 2T = -H$$

or,

$$T + U = E = H = \text{const.} \tag{7.31}$$

And the total energy E is a constant of the motion for this case.

The function H is called the *Hamiltonian* of the system and may be defined as in Eq. (7.29) (but see also the comments in Section 7.12). It is important to note that the Hamiltonian H is equal to the total energy E only if the following conditions are met:

(a) The equations of transformation connecting the rectangular and generalized coordinates [Eq. (7.18)] must be independent of the time, thus insuring that the kinetic energy is a homogeneous quadratic function of the $\dot{q}_j$.
(b) The potential energy must be velocity-independent, thus allowing the elimination of the terms $\partial U/\partial \dot{q}_j$ from the equation for H [Eq. (7.30)].

Therefore, the questions *Does $H = E$ for the system?* and *Is energy conserved for the system?* pertain to two *different* aspects of the problem and each question must be examined separately. We may, for example, have cases in which the Hamiltonian does not equal the total energy, but, nevertheless, the energy is conserved. Thus, consider a conservative system and let the description be made in terms of generalized coordinates that are in motion with respect to fixed, rectangular axes. The transformation equations then contain the time and the kinetic energy is *not* a homogeneous, quadratic function of the generalized velocities. Clearly, the choice of a mathematically convenient set of generalized coordinates cannot alter the physical fact that energy is conserved, but in the moving coordinate system the Hamiltonian is no longer equal to the total energy.

7.10 The Conservation of Linear Momentum

Since space is *homogeneous* in an inertial reference frame, the Lagrangian of a closed system will be unaffected by a translation of the entire system in space. Consider an infinitesimal translation of every radius vector $\mathbf{r}_\alpha$ such that $\mathbf{r}_\alpha \rightarrow \mathbf{r}_\alpha + \delta\mathbf{r}$; this amounts to translating the entire system by $\delta\mathbf{r}$. For simplicity, let us examine a system consisting of only a single particle (by including a summation over α we could consider an n-particle system in an entirely equivalent manner), and let us write the Lagrangian in terms of rectangular coordinates, $L = L(x_i, \dot{x}_i)$. The change in L due to the infinitesimal displacement $\delta\mathbf{r} = \sum_i \delta x_i \mathbf{e}_i$ is

$$\delta L = \sum_i \frac{\partial L}{\partial x_i} \partial x_i + \sum_i \frac{\partial L}{\partial \dot{x}_i} \delta \dot{x}_i = 0 \tag{7.32}$$

Now, we consider only a *displacement*, so that the δx_i are not explicit or implicit functions of the time. Thus,

$$\delta \dot{x}_i = \delta \frac{dx_i}{dt} = \frac{d}{dt} \delta x_i \equiv 0 \tag{7.33}$$

Therefore, δL becomes

$$\delta L = \sum_i \frac{\partial L}{\partial x_i} \delta x_i = 0 \tag{7.34}$$

Since each of the δx_i is an independent displacement, δL will vanish identically only if each of the partial derivatives of L vanishes:

$$\frac{\partial L}{\partial x_i} = 0 \tag{7.35}$$

Then, according to Lagrange's equations,

$$\frac{d}{dt} \frac{\partial L}{\partial \dot{x}_i} = 0 \tag{7.36}$$

and,

$$\frac{\partial L}{\partial \dot{x}_i} = \text{const.} \tag{7.37}$$

or,

$$\begin{aligned} \frac{\partial (T - U)}{\partial \dot{x}_i} = \frac{\partial T}{\partial \dot{x}_i} &= \frac{\partial}{\partial \dot{x}_i} \left(\frac{1}{2} m \sum_j \dot{x}_j^2 \right) \\ &= m\dot{x}_i = p_i = \text{const.} \end{aligned} \tag{7.38}$$

Thus, the homogeneity of space implies that the linear momentum **p** of a closed system is constant in time.

This result may also be interpreted according to the following statement: If the Lagrangian of a system (not necessarily *closed*) is invariant with respect to translation in a certain direction, then the linear momentum of the system in that direction is constant in time.

7.11 The Conservation of Angular Momentum

It was stated in Section 2.3 that one characteristic of an inertial reference frame is that space is *isotropic*; i.e., the mechanical properties of a closed system are unaffected by the orientation of the system. In particular, the

Lagrangian of a closed system will not change if the system is rotated through an infinitesimal angle.*

If a system is rotated about a certain axis by an infinitesimal angle $\delta\theta$ (see Fig. 7-2), the radius vector $\mathbf{r}$ to a given point will change to $\mathbf{r} + \delta\mathbf{r}$, where [see Eq. (1.106)]

$$\delta\mathbf{r} = \delta\boldsymbol{\theta} \times \mathbf{r} \tag{7.39}$$

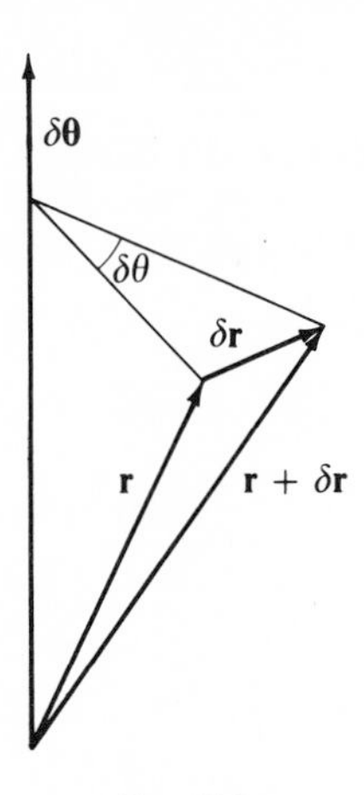

FIG. 7-2

Now, the velocity vectors will also change upon rotation of the system, and since the transformation equation for all vectors is the same, we have

$$\delta\dot{\mathbf{r}} = \delta\boldsymbol{\theta} \times \dot{\mathbf{r}} \tag{7.40}$$

As in Section 7.10, we consider only a single particle and express the Lagrangian in rectangular coordinates. The change in L due to the infinitesimal rotation is

$$\delta L = \sum_i \frac{\partial L}{\partial x_i}\,\delta x_i + \sum_i \frac{\partial L}{\partial \dot{x}_i}\,\delta \dot{x}_i = 0 \tag{7.41}$$

Equations (7.37) and (7.38) show that the rectangular components of the momentum vector are given by

$$p_i = \frac{\partial L}{\partial \dot{x}_i} \tag{7.42}$$

* We limit the rotation to an infinitesimal angle since we wish to be able to represent the rotation by a vector; see Section 1.15.

Lagrange's equations may then be expressed by

$$\dot{p}_i = \frac{\partial L}{\partial x_i} \tag{7.43}$$

Hence, Eq. (7.41) becomes

$$\delta L = \sum_i \dot{p}_i \, \delta x_i + \sum_i p_i \, \delta \dot{x}_i = 0 \tag{7.44}$$

or,

$$\dot{\mathbf{p}} \cdot \delta \mathbf{r} + \mathbf{p} \cdot \delta \dot{\mathbf{r}} = 0 \tag{7.45}$$

And using Eqs. (7.39) and (7.40), this equation may be written as

$$\dot{\mathbf{p}} \cdot (\delta \boldsymbol{\theta} \times \mathbf{r}) + \mathbf{p} \cdot (\delta \boldsymbol{\theta} \times \dot{\mathbf{r}}) = 0 \tag{7.46}$$

We may permute in cyclic order the factors of a triple scalar product without altering the value. Thus,

$$\delta \boldsymbol{\theta} \cdot (\mathbf{r} \times \dot{\mathbf{p}}) + \delta \boldsymbol{\theta} \cdot (\dot{\mathbf{r}} \times \mathbf{p}) = 0$$

or,

$$\delta \boldsymbol{\theta} \cdot [(\mathbf{r} \times \dot{\mathbf{p}}) + (\dot{\mathbf{r}} \times \mathbf{p})] = 0 \tag{7.47}$$

The terms in the brackets are just the factors which result from the differentiation with respect to time of $\mathbf{r} \times \mathbf{p}$:

$$\delta \boldsymbol{\theta} \cdot \frac{d}{dt}(\mathbf{r} \times \mathbf{p}) = 0 \tag{7.48}$$

Since $\delta \boldsymbol{\theta}$ is arbitrary, we must have

$$\frac{d}{dt}(\mathbf{r} \times \mathbf{p}) = 0$$

so that

$$\mathbf{r} \times \mathbf{p} = \text{const.} \tag{7.49}$$

But $\mathbf{r} \times \mathbf{p} = \mathbf{L}$; therefore, the angular momentum of the particle is constant in time.

An important corollary of this theorem is the following. Consider a system in an external force field. If the field possesses an axis of symmetry, then the Lagrangian of the system is invariant with respect to rotations about the symmetry axis. Hence, the angular momentum of the system about the axis of symmetry is constant in time. This is exactly the case discussed in Example 7.4: the vertical direction was an axis of symmetry of the system and the angular momentum about that axis was found to be conserved.

The importance of the connection between *symmetry* properties and the *invariance* of physical quantities can hardly be overemphasized. The association goes beyond momentum conservation, indeed beyond classical systems, and finds wide application in modern theories of field phenomena and elementary particles.

We have been able to derive the conservation theorems for a closed system simply by considering the properties of an inertial reference frame. The results can be summarized as in Table 7.1.

Table 7.1

Characteristic of inertial frame	Property of Lagrangian	Conserved quantity
Time homogeneous	Not explicit function of time	Total energy
Space homogeneous	Invariant to translation	Linear momentum
Space isotropic	Invariant to rotation	Angular momentum

Thus, there are seven constants (or integrals) of the motion for a closed system: total energy, linear momentum (three components), and angular momentum (three components). These and only these seven integrals have the property that they are *additive* for the particles composing the system; they possess this property whether or not there is an interaction among the particles.

7.12 The Canonical Equations of Motion—Hamiltonian Dynamics

In Section 7.10 we found that if the potential energy of a system is velocity-independent, then the linear momentum components in rectangular coordinates are given by

$$p_i = \frac{\partial L}{\partial \dot{x}_i} \tag{7.50}$$

By analogy we extend this result to the case in which the Lagrangian is expressed in generalized coordinates and define the *generalized momenta** according to

$$p_j = \frac{\partial L}{\partial \dot{q}_j} \tag{7.51}$$

(Unfortunately, the customary notation for ordinary momentum and generalized momentum is the same, even though the two quantities may be quite different.) The Lagrange equations of motion are then expressed by

$$\dot{p}_j = \frac{\partial L}{\partial q_j} \tag{7.52}$$

Using the definition of the generalized momenta, Eq. (7.29) for the Hamiltonian may be written as

$$H = \sum_j p_j \dot{q}_j - L \tag{7.53}$$

Now, the Lagrangian is considered to be a function of the generalized coordinates, the generalized velocities, and possibly the time. The dependence of L on the time may arise either if the constraints are time dependent or if the transformation equations connecting the rectangular and generalized coordinates explicitly contain the time. (Recall that we do not consider time-dependent potentials.) We may solve Eq. (7.51) for the generalized velocities and express them as

$$\dot{q}_j = \dot{q}_j(q_k, p_k, t) \tag{7.54}$$

Thus, in Eq. (7.53) we may make a change of variables from the $(q_j, \dot{q}_j, t)$ set to the (q_j, p_j, t) set† and express the Hamiltonian as

$$H(q_k, p_k, t) = \sum_j p_j \dot{q}_j - L(q_k, \dot{q}_k, t) \tag{7.55}$$

This equation is written in a manner which stresses the fact that *the Hamiltonian is always considered as a function of the* (q_k, p_k, t) *set, whereas the Lagrangian is a function of the* $(q_k, \dot{q}_k, t)$ *set*:

* The terms "generalized coordinates," "generalized velocities," and "generalized momenta" were introduced in 1867 by Sir William Thomson (later, Lord Kelvin) and P. G. Tait in their famous treatise *Natural Philosophy*.

† This change of variables is similar to that frequently encountered in thermodynamics and falls in the general class of the so-called *Legendre transformations* (used first by Euler and perhaps even by Leibniz). A general discussion of Legendre transformations with emphasis on their importance in mechanics is given by Lanczos (La49, Chapter 6).

$$\boxed{H = H(q_k, p_k, t); \qquad L = L(q_k, \dot{q}_k, t)} \tag{7.56}$$

Therefore, the total differential of H is

$$dH = \sum_k \left(\frac{\partial H}{\partial q_k} dq_k + \frac{\partial H}{\partial p_k} dp_k \right) + \frac{\partial H}{\partial t} dt \tag{7.57}$$

According to Eq. (7.55) we can also write

$$dH = \sum_k \left(\dot{q}_k \, dp_k + p_k \, d\dot{q}_k - \frac{\partial L}{\partial q_k} dq_k - \frac{\partial L}{\partial \dot{q}_k} d\dot{q}_k \right) - \frac{\partial L}{\partial t} dt \tag{7.58}$$

Using Eqs. (7.51) and (7.52) to substitute for $\partial L/\partial q_k$ and $\partial L/\partial \dot{q}_k$, the second and fourth terms in the parentheses in Eq. (7.58) cancel and there remains

$$dH = \sum_k (\dot{q}_k \, dp_k - \dot{p}_k \, dq_k) - \frac{\partial L}{\partial t} dt \tag{7.59}$$

If we identify the coefficients* of dq_k, dp_k, and dt between Eqs. (7.57) and (7.59), we find

$$\dot{q}_k = \frac{\partial H}{\partial p_k} \tag{7.60}$$

$$-\dot{p}_k = \frac{\partial H}{\partial q_k} \tag{7.61}$$

and

$$-\frac{\partial L}{\partial t} = \frac{\partial H}{\partial t} \tag{7.62}$$

Furthermore, using Eqs. (7.60) and (7.61) in Eq. (7.57), each term in the parentheses vanishes, and it follows that

$$\frac{dH}{dt} = \frac{\partial H}{\partial t} \tag{7.63}$$

Equations (7.60) and (7.61) are *Hamilton's equations of* motion†; because

* The assumptions implicitly contained in this procedure are examined in the following section.

† This set of equations was first obtained by Lagrange in 1809 and Poisson also derived similar equations in the same year. But neither of these individuals recognized the equations as a basic set of equations of motion; this point was first realized by Cauchy in 1831. Hamilton first derived the equations in 1834 from a fundamental variational principle and made them the basis for a far-reaching theory of dynamics. Thus, the designation *Hamilton's* equations is fully deserved.

of their symmetrical appearance, they are also known as the *canonical equations of motion*. The description of motion by means of these equations is termed *Hamiltonian dynamics*.

Equation (7.63) expresses the fact that if H does not explicitly contain the time, then the Hamiltonian is a conserved quantity. As we have seen previously (Section 7.9), the Hamiltonian will equal the total energy $T + U$ in the event that the potential energy is velocity-independent and the transformation equations between the $x_{\alpha,i}$ and the q_j do not explicitly contain the time. Under these conditions, and if $\partial H/\partial t = 0$, then $H = E = \text{const.}$

There are $2s$ canonical equations and they replace the s Lagrange equations. (Recall that $s = 3n - m$ is the number of degrees of freedom of the system.) But the canonical equations are *first-order* differential equations, whereas the Lagrange equations are of *second-order*.* In order to use the canonical equations in solving a problem, the Hamiltonian must first be constructed as a function of the generalized coordinates and momenta. It may be possible in some instances to do this directly; in more complicated cases it may be necessary first to set up the Lagrangian and then to calculate the generalized momenta according to Eq. (7.51). The equations of motion are then given by the canonical equations.

▶ Example 7.12 Particle Moving on a Cylindrical Surface

As an illustration of the formulation of a problem using the Hamiltonian method, let us consider a particle of mass m that is constrained to move on the surface of a cylinder, the defining equation of which is

$$x^2 + y^2 = R^2 \tag{1}$$

The particle is to be subject to a force directed toward the origin and proportional to the distance of the particle from the origin:

$$\mathbf{F} = -k\mathbf{r} \tag{2}$$

The situation is illustrated in Fig. 7-3. The potential corresponding to the force $\mathbf{F}$ is

$$\begin{aligned} U &= \tfrac{1}{2}kr^2 = \tfrac{1}{2}k(x^2 + y^2 + z^2) \\ &= \tfrac{1}{2}k(R^2 + z^2) \end{aligned} \tag{3}$$

In cylindrical coordinates the square of the velocity is [Eq. (1.101)]

$$v^2 = \dot{r}^2 + r^2\dot{\theta}^2 + \dot{z}^2 \tag{4}$$

But here $r = R$ is a constant so that the kinetic energy is

$$T = \tfrac{1}{2}m(R^2\dot{\theta}^2 + \dot{z}^2) \tag{5}$$

* This is not a special result; any set of s second-order equations can always be replaced by a set of $2s$ first-order equations.

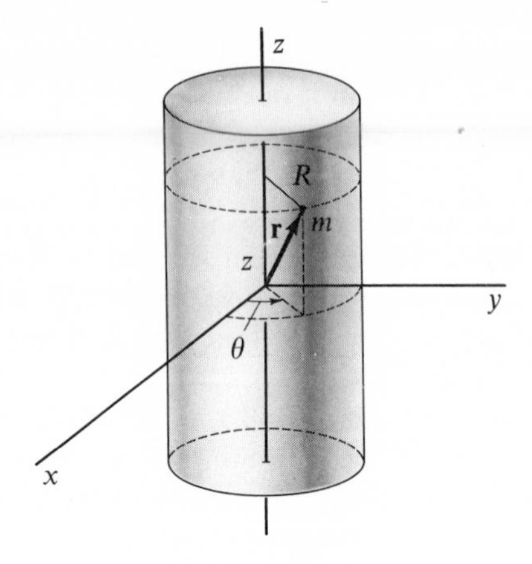

FIG. 7-3

We may now write the Lagrangian as

$$L = T - U = \tfrac{1}{2}m(R^2\dot{\theta}^2 + \dot{z}^2) - \tfrac{1}{2}k(R^2 + z^2) \tag{6}$$

The generalized coordinates are θ and z, and the generalized momenta are

$$p_\theta = \frac{\partial L}{\partial \dot{\theta}} = mR^2\dot{\theta} \tag{7a}$$

$$p_z = \frac{\partial L}{\partial \dot{z}} = m\dot{z} \tag{7b}$$

Since the system is conservative and since the equations of transformation between rectangular and cylindrical coordinates do not explicitly involve the time, the Hamiltonian H is just the total energy expressed in terms of the variables θ, p_θ, z, and p_z but θ does not occur explicitly, so that

$$\begin{aligned} H(z, p_\theta, p_z) &= T + U \\ &= \frac{p_\theta^2}{2mR^2} + \frac{p_z^2}{2m} + \tfrac{1}{2}kz^2 \end{aligned} \tag{8}$$

where the constant term $\tfrac{1}{2}kR^2$ has been suppressed. The equations of motion are therefore found from the canonical equations:

$$\dot{p}_\theta = -\frac{\partial H}{\partial \theta} = 0 \tag{9a}$$

$$\dot{p}_z = -\frac{\partial H}{\partial z} = -kz \tag{9b}$$

$$\dot{\theta} = \frac{\partial H}{\partial p_\theta} = \frac{p_\theta}{mR^2} \tag{9c}$$

$$\dot{z} = \frac{\partial H}{\partial p_z} = \frac{p_z}{m} \tag{9d}$$

Equations (9c) and (9d) just duplicate Eqs. (7a) and (7b). Equation (9a) gives

$$p_\theta = mR^2\dot{\theta} = \text{const.} \tag{10}$$

so that the angular momentum about the z-axis is a constant of the motion; this result is assured since the z-axis is the symmetry axis of the problem. Combining Eqs. (7b) and (9b), we find

$$\ddot{z} + \omega_0^2 z = 0 \tag{11}$$

where

$$\omega_0^2 \equiv k/m \tag{12}$$

Therefore, the motion in the z-direction is simple harmonic.

The same equations of motion for the above problem can be found by the Lagrangian method using the function L defined by Eq. (6). In this case the Lagrange equations of motion are easier to obtain than are the canonical equations. In fact, it is quite often true that the Lagrangian method leads more readily to the equations of motion than does the Hamiltonian method. However, because of the fact that there is greater freedom in choosing the variable in the Hamiltonian formulation of a problem (the q_k and the p_k are independent, whereas the q_k and the $\dot{q}_k$ are not), there is sometimes a certain practical advantage to be gained by using the Hamiltonian method. For example, in celestial mechanics, particularly in the event that the motions are subject to perturbations caused by the influence of other bodies, it proves convenient to formulate the problem in terms of Hamiltonian dynamics. Generally speaking, however, the great power of the Hamiltonian approach to dynamics does not manifest itself in simplifying the solutions to mechanics problems; rather, it does so in providing the base from which extensions may be made to other fields.

The generalized coordinate q_k and the generalized momentum p_k are said to be *canonically conjugate* quantities. According to Eqs. (7.60) and (7.61), if q_k does not appear in the Hamiltonian, then $\dot{p}_k = 0$ and the conjugate momentum p_k is a constant of the motion. Coordinates which do not appear explicitly in the expressions for T and U are said to be *cyclic*. A coordinate which is cyclic in H is also cyclic in L. However, even if q_k does not appear in L, the generalized velocity $\dot{q}_k$ related to this coordinate will in general still be present. Thus,

$$L = L(q, \ldots, q_{k-1}, q_{k+1}, \ldots, q_s, \dot{q}_1, \ldots, \dot{q}_s, t)$$

and no reduction in the number of degrees of freedom of the system has been accomplished even though one coordinate is cyclic; there are still s

second-order equations to be solved. On the other hand, in the canonical formulation, if q_k is cyclic, p_k is constant, $p_k = \alpha_k$, and

$$H = H(q_1, \ldots, q_{k-1}, q_{k+1}, \ldots, q_s, p_1 \ldots, p_{k-1}, \alpha_k, p_{k+1}, \ldots, p_s, t)$$

Thus, there are $2s - 2$ first-order equations to be solved and the problem has in fact been reduced in complexity; there are in effect only $s - 1$ degrees of freedom remaining. The coordinate q_k is completely separated and it is *ignorable* as far as the remainder of the problem is concerned. The constant α_k is calculated by applying the initial conditions, and the equation of motion for the cyclic coordinate is

$$\dot{q}_k = \frac{\partial H}{\partial \alpha_k} \equiv \omega_k \tag{7.64}$$

which can be immediately integrated to yield

$$q_k(t) = \int \omega_k \, dt \tag{7.65}$$

The solution for a cyclic coordinate is therefore trivial to reduce to quadrature. Consequently, the canonical formulation of Hamilton is particularly well suited for dealing with problems in which one or more of the coordinates is cyclic. It is clear that the simplest possible solution to a problem would result if the problem could be formulated in such a way that *all* the coordinates were cyclic. Then each coordinate would be described in a trivial manner as in Eq. (7.65). It is in fact possible to find transformations which render all of the coordinates cyclic* and these procedures lead naturally to a formulation of dynamics that is particularly useful in the construction of modern theories of matter. The general discussion of these topics, however, is beyond the scope of this book.†

7.13 Some Comments Regarding Dynamical Variables and Variational Calculations in Physics

We originally obtained Lagrange's equations of motion by stating Hamilton's Principle as a variational integral and then making use of the results of the preceding chapter on the calculus of variations. Since the method and the application were thereby separated, it is perhaps worthwhile to restate the argument in an orderly but abbreviated way.

* Transformations of this type were devised by Carl Gustav Jacob Jacobi (1804–1851). Jacobi's investigations greatly extended the usefulness of Hamilton's methods and these developments are known as *Hamilton–Jacobi theory*.

† See, for example, Goldstein (Go50, Chapter 9).

Hamilton's Principle is expressed by

$$\delta \int_{t_1}^{t_2} L(q_j, \dot{q}_j, t)\, dt = 0 \tag{7.66}$$

Applying the variational procedure specified in Section 6.8, we have

$$\int_{t_1}^{t_2} \left(\frac{\partial L}{\partial q_j} \delta q_j + \frac{\partial L}{\partial \dot{q}_j} \delta \dot{q}_j \right) dt = 0 \tag{7.67}$$

Next, we assert that the δq_j and the $\delta \dot{q}_j$ are *not* independent so that the variation operation and the time differentiation can be interchanged:

$$\delta \dot{q}_j = \delta \left(\frac{dq_j}{dt} \right) = \frac{d}{dt} \delta q_j \tag{7.68}$$

Then, the varied integral becomes (after the integration by parts in which the δq_j are set equal to zero at the end points)

$$\int_{t_1}^{t_2} \left(\frac{\partial L}{\partial q_j} - \frac{d}{dt} \frac{\partial L}{\partial \dot{q}_j} \right) \delta q_j\, dt = 0 \tag{7.69}$$

The requirement that the δq_j be independent variations leads immediately to Lagrange's equations.

In Hamilton's Principle, expressed by the variational integral in Eq. (7.66), the Lagrangian is a function of the generalized coordinates and the generalized velocities. But only the q_j are considered as *independent* variables; the generalized velocities are simply the time derivatives of the q_j. When the integral is reduced to the form given by Eq. (7.69), we state that the δq_j are *independent* variations; thus the integrand must vanish identically and Lagrange's equations result. We may therefore pose the question: since the dynamical motion of the system is completely determined by the initial conditions, what is the meaning of the variations δq_j? Perhaps a sufficient answer is that the variations are to be considered as those which are *geometrically feasible* within the limits of the given constraints, although they are not *dynamically possible.* That is, when using a variational procedure to obtain Lagrange's equations, it is convenient to ignore temporarily the fact that we are dealing with a *physical system* whose motion is completely determined and subject to no variation, and to consider instead only a certain abstract *mathematical* problem. Indeed, this is the spirit in which any variational calculation relating to a physical process must be carried out. In adopting such a viewpoint, we must not be overly concerned with the fact that the variation procedure may be contrary to certain known physical properties of the system. (For example, energy is in general not conserved in passing from the true path to the varied path.) A variational calculation simply tests various *possible* solutions to a problem and prescribes a method for selecting the *correct* solution.

The canonical equations of motion can also be obtained directly from a variational calculation which is based on the so-called *modified Hamilton's Principle.* The Lagrangian function can be expressed as [see Eq. (7.53)]:

$$L = \sum_j p_j \dot{q}_j - H(q_j, p_j, t) \tag{7.70}$$

so that the statement of Hamilton's Principle contained in Eq. (7.66) can be modified to read

$$\boxed{\delta \int_{t_1}^{t_2} \left(\sum_j p_j \dot{q}_j - H \right) dt = 0} \tag{7.71}$$

Carrying out the variation in the standard manner, we obtain

$$\int_{t_1}^{t_2} \sum_j \left(p_j \, \delta\dot{q}_j + \dot{q}_j \, \delta p_j - \frac{\partial H}{\partial p_j} \, \delta q_j - \frac{\partial H}{\partial p_j} \, \delta p_j \right) dt = 0 \tag{7.72}$$

Now, in the Hamiltonian formulation, the q_j and the p_j are considered to be independent. The $\dot{q}_j$ are again *not* independent of the q_j, so Eq. (7.68) can be used to express the first term in Eq. (7.72) as

$$\int_{t_1}^{t_2} \sum_j p_j \, \delta\dot{q}_j \, dt = \int_{t_1}^{t_2} \sum_j p_j \frac{d}{dt} \, \delta q_j \, dt \tag{7.73}$$

Upon integrating by parts, the integrated term vanishes and we have

$$\int_{t_1}^{t_2} \sum_j p_j \, \delta\dot{q}_j \, dt = - \int_{t_1}^{t_2} \sum_j \dot{p}_j \, \delta q_j \, dt \tag{7.74}$$

Equation (7.72) then becomes

$$\int_{t_1}^{t_2} \sum_j \left\{ \left(\dot{q}_j - \frac{\partial H}{\partial p_j} \right) \delta p_j - \left(\dot{p}_j + \frac{\partial H}{\partial q_j} \right) \delta q_j \right\} dt = 0 \tag{7.75}$$

If the δq_j and the δp_j represent *independent variations*, then the terms in the parentheses must separately vanish and Hamilton's canonical equations result.

In the preceding section we obtained the canonical equations by writing two different expressions for the total differential of the Hamiltonian [Eqs. (7.57) and (7.59)] and then equating the coefficients of dq_j and dp_j. Such a procedure is valid if the q_j and the p_j are independent variables. Therefore, both in the previous derivation and in the variational calculation above, the canonical equations were obtained by exploring the independent nature of the generalized coordinates and the generalized momenta.

Now, it is of course true that the coordinates and momenta are not "independent" in the ultimate sense of the word. For, if the time dependence of each of the coordinates is known, $q_j = q_j(t)$, then the problem is completely solved. The generalized velocities can be calculated from

$$\dot{q}_j(t) = \frac{d}{dt} q_j(t)$$

and the generalized momenta are

$$p_j = \frac{\partial}{\partial \dot{q}_j} L(q_j, \dot{q}_j, t)$$

The essential point is that, whereas the q_j and the $\dot{q}_j$ are related by a simple time derivative *independent of the manner in which the system behaves*, the connection between the q_j and the p_j are *the equations of motion themselves*. Therefore, finding the relations that connect the q_j and the p_j (and thereby eliminating the assumed independence of these quantities) is tantamount to solving the problem.

■ 7.14 Phase Space and Liouville's Theorem

As pointed out previously, the generalized coordinates q_j can be used to define an s-dimensional *configuration space* in which every point represents a certain state of the system. Similarly, the *generalized momenta* p_j define an s-dimensional *momentum space* in which every point represents a certain condition of motion of the system. A given point in configuration space specifies only the *position* of each of the particles in the system; nothing can be inferred regarding the *motion* of the particles. The reverse is true, of course, for momentum space. In Chapters 3, 4, and 5 we found it profitable to represent geometrically the dynamics of simple oscillatory systems by means of *phase diagrams*. If we take over this concept for use with more complicated dynamical systems, then a $2s$-dimensional space consisting of the q_j and the p_j will allow the representation of both the positions *and* the momenta of all of the particles. This generalization is called *Hamiltonian phase space* or simply *phase space*.*

▶ Example 7.14 A Phase Space Diagram

The phase diagram for the motion of the particle in Example 7.12(a) may be constructed as follows. The particle has two degrees of freedom (θ, z), so the phase

* Previously, we plotted in the phase diagrams the position versus a quantity proportional to the velocity. In Hamiltonian phase space this latter quantity becomes the generalized momentum.

space for this example is actually four-dimensional: θ, p_θ, z, p_z. But p_θ is constant and therefore may be suppressed. In the z-direction the motion is simple harmonic and so the projection onto the z-p_z plane of the phase path for any total energy H is just an ellipse. Since $\theta = \text{const.}$, the phase path must represent motion increasing uniformly with θ. Thus the phase path on any surface $H = \text{const.}$ is a *uniform elliptic spiral*, as shown in Fig. 7-4.

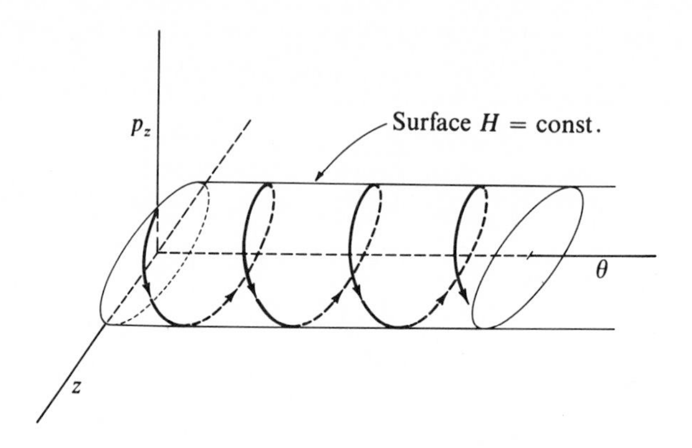

FIG. 7-4

If, at a given instant of time, the positions and momenta of all of the particles in a system are known, then with these quantities as initial conditions the subsequent motion of the system is completely determined. That is, starting from a point $q_j(0)$, $p_j(0)$ in phase space, the representative point which describes the system moves along a unique phase path. In principle, this procedure can always be followed and a solution obtained, but if the number of degrees of freedom of the system is large, then the set of equations of motion may be too complicated to solve in a reasonable time. Moreover, for complex systems, such as a quantity of gas, it is clearly a practical impossibility to determine the initial conditions for each constituent molecule. Since we cannot identify any particular point in phase space as representing the actual conditions at any given time, we must devise some alternative approach to study the dynamics of such systems. We therefore arrive at the point of departure of *statistical mechanics*. The Hamiltonian formulation of dynamics is ideal for the statistical study of complex systems and we demonstrate this in part by now proving a theorem which is fundamental for such investigations.

For a large collection of particles, say, gas molecules, we are unable to identify the particular point in phase space that correctly represents the system. However, we may fill the phase space with a collection of points, each of which represents a *possible* condition of the system. That is, we imagine a large number of systems (each consistent with the known constraints), any of which could conceivably be the actual system. Since we are unable to discuss the details of the motion of the particles in the actual system, we substitute a discussion of an *ensemble* of equivalent systems. Each representative point in phase space corresponds to a single system of the ensemble and the motion of a particular point represents the *independent* motion of that system. Thus, no two of the phase paths may ever intersect.

We may consider the representative points to be sufficiently numerous that we can define a *density in phase* ρ. Of course, the volume elements of the phase space which we use to define the density must be sufficiently large to contain a large number of representative points, but they must also be sufficiently small so that the density may be considered to vary in a continuous manner. The number N of systems whose representative points lie within a volume dv of phase space is

$$N = \rho\, dv$$

where

$$dv = dq_1\, dq_2 \cdots dq_s\, dp_1\, dp_2 \cdots dp_s$$

As before, s is the number of degrees of freedom of each system in the ensemble.

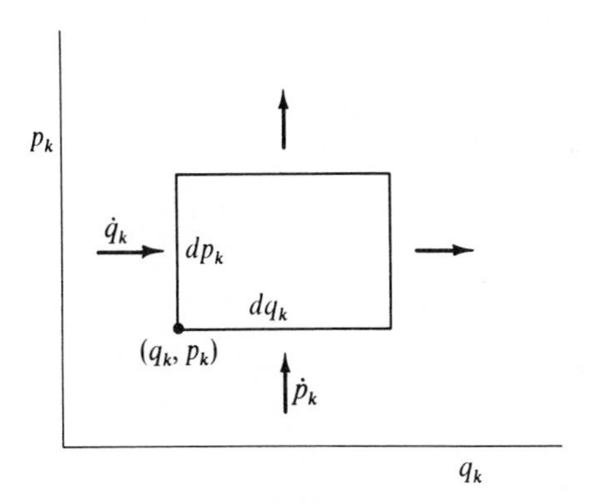

FIG. **7-5**

Consider an element of area in the q_k-p_k plane in phase space as in Fig. 7-5. The number of representative points moving across the left-hand edge into the area per unit time is

$$\rho \frac{dq_k}{dt} dp_k = \rho \dot{q}_k\, dp_k$$

and the number moving across the lower edge into the area per unit time is

$$\rho \frac{dp_k}{dt} dq_k = \rho \dot{p}_k \, dq_k$$

so that the total number of representative points moving *into* the area $dq_k \, dp_k$ per unit time is

$$\rho(\dot{q}_k \, dp_k + \dot{p}_k \, dq_k)$$

By a Taylor series expansion, the number of representative points moving *out* of the area per unit unit time is (approximately)

$$\left(\rho \dot{q}_k + \frac{\partial}{\partial q_k}(\rho \dot{q}_k) \, dq_k\right) dp_k + \left(\rho \dot{p}_k + \frac{\partial}{\partial p_k}(\rho \dot{q}_k) \, dp_k\right) dq_k$$

Hence, the total increase in density in $dq_k \, dp_k$ per unit time is

$$\frac{\partial \rho}{\partial t} dq_k \, dp_k = -\left(\frac{\partial}{\partial q_k}(\rho \dot{q}_k) + \frac{\partial}{\partial p_k}(\rho \dot{p}_k)\right) dq_k \, dp_k$$

Summing this expression over all possible values of k, we find

$$\frac{\partial \rho}{\partial t} + \sum_{k=1}^{s} \left(\frac{\partial \rho}{\partial q_k} \dot{q}_k + \rho \frac{\partial \dot{q}_k}{\partial q_k} + \frac{\partial \rho}{\partial p_k} \dot{p}_k + \rho \frac{\partial \dot{p}_k}{\partial p_k}\right) = 0 \tag{7.76}$$

Now, from Hamilton's equations (7.60), (7.61), we have (if the second partial derivatives of H are continuous)

$$\frac{\partial \dot{q}_k}{\partial q_k} + \frac{\partial \dot{p}_k}{\partial p_k} = 0 \tag{7.77}$$

so that Eq. (7.76) becomes

$$\frac{\partial \rho}{\partial t} = \sum_k \left(\frac{\partial \rho}{\partial q_k} \frac{\partial q_k}{\partial t} + \frac{\partial \rho}{\partial p_k} \frac{\partial p_k}{\partial t}\right) = 0 \tag{7.78}$$

But this is just the total time derivative of ρ, so we conclude that

$$\boxed{\frac{d\rho}{dt} = 0} \tag{7.79}$$

This important result is known as *Liouville's theorem** and states that the density of representative points in phase space corresponding to the motion of a system of particles remains constant during the motion. It must be emphasized that we have been able to establish the invariance of the density

* Published in 1838 by Joseph Liouville (1809–1882).

ρ only because the problem was formulated in *phase space*; an equivalent theorem for configuration space does not exist. Thus, it is necessary to use Hamiltonian dynamics (rather than Lagrangian dynamics) for the discussion of ensembles in statistical mechanics.

According to Liouville's theorem, if we imagine two surfaces (actually *hypersurfaces* in the $2s$-dimensional space) which correspond to energies E and $E + dE$, and consider the density of representative points in some small region between these surfaces,* then as the system moves, this density remains constant in time. Figure 7-6 shows such a pair of surfaces and a

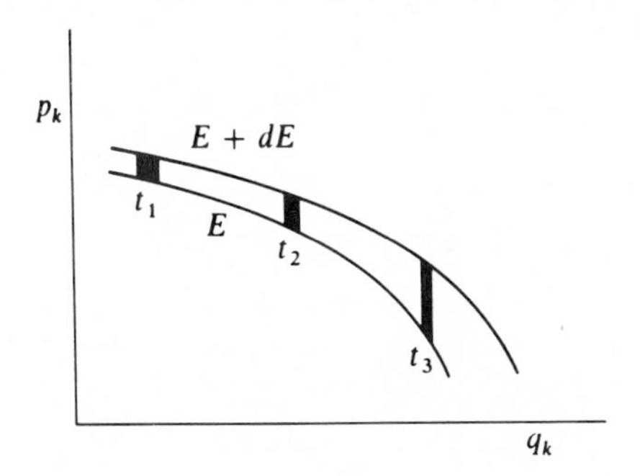

FIG. 7-6

region which has an extent $\Delta q_k \, \Delta p_k$ at a certain time t_1. At some later times t_2 and t_3, the relative magnitudes of Δq_k and Δp_k must change as indicated in order to maintain a constant phase density. This fact can be immediately applied, for example, to the focussing of beams of electrons or other charged particles (see Problem 7-27).

Liouville's theorem is important not only for aggregates of microscopic particles, as in the statistical mechanics of gaseous systems and the focussing properties of charged-particle accelerators, but also in certain macroscopic systems. For example, in stellar dynamics the problem is inverted and by studying the distribution function ρ of stars in the galaxy, the potential U of the galactic gravitational field may be inferred.

7.15 The Virial Theorem

Another important result of a statistical nature is worthy of mention. Consider a collection of particles whose position vectors $\mathbf{r}_\alpha$ and whose

* An ensemble for which the phase density is zero except in the region between E and $E + dE$ was called by Gibbs a *microcanonical ensemble.*

momenta $\mathbf{p}_\alpha$ are both *bounded* (i.e., remain finite for all values of the time). Define a quantity

$$S \equiv \sum_\alpha \mathbf{p}_\alpha \cdot \mathbf{r}_\alpha \tag{7.80}$$

The time derivative of S is

$$\frac{dS}{dt} = \sum_\alpha (\mathbf{p}_\alpha \cdot \dot{\mathbf{r}}_\alpha + \dot{\mathbf{p}}_\alpha \cdot \mathbf{r}_\alpha) \tag{7.81}$$

If we calculate the average value of dS/dt over a time interval τ, we find

$$\left\langle \frac{dS}{dt} \right\rangle = \frac{1}{\tau} \int_0^\tau \frac{dS}{dt}\, dt = \frac{S(\tau) - S(0)}{\tau} \tag{7.82}$$

In the event that the motion of the system is periodic, and if τ is some integer multiple of the period, then $S(\tau) = S(0)$, and $\langle \dot{S} \rangle$ vanishes. However, even if the system does not exhibit any periodicity, then since S is by hypothesis a bounded function, we can make $\langle \dot{S} \rangle$ as small as desired by allowing the time τ to become sufficiently long. Therefore, the time average of the right-hand side of Eq. (7.81) can always be made to vanish (or at least to *approach* zero). Thus, in this limit, we have

$$\left\langle \sum_\alpha \mathbf{p}_\alpha \cdot \dot{\mathbf{r}}_\alpha \right\rangle = -\left\langle \sum_\alpha \dot{\mathbf{p}}_\alpha \cdot \mathbf{r}_\alpha \right\rangle \tag{7.83}$$

On the left-hand side of this equation, $\mathbf{p}_\alpha \cdot \dot{\mathbf{r}}_\alpha$ is twice the kinetic energy; on the right-hand side, $\dot{\mathbf{p}}_\alpha$ is just the force $\mathbf{F}_\alpha$ on the αth particle. Hence,

$$\left\langle 2 \sum_\alpha T_\alpha \right\rangle = -\left\langle \sum_\alpha \mathbf{F}_\alpha \cdot \mathbf{r}_\alpha \right\rangle \tag{7.84}$$

The sum over T_α is the total kinetic energy T of the system, so we have the general result

$$\boxed{\langle T \rangle = -\tfrac{1}{2}\left\langle \sum_\alpha \mathbf{F}_\alpha \cdot \mathbf{r}_\alpha \right\rangle} \tag{7.85}$$

The right-hand side of this equation was called by Clausius* the *virial* of the system, and the *virial theorem* is stated thusly: the average kinetic energy of a system of particles is equal to its virial.

* Rudolph Julius Emmanuel Clausius (1822–1888), a German physicist and one of the founders of thermodynamics.

The virial theorem is particularly useful in the kinetic theory of gases; the equation of state of a perfect gas (Boyle's law) can be derived with only a few additional considerations.*

If the forces $\mathbf{F}_\alpha$ can be derived from potentials U_α, then Eq. (7.85) may be rewritten as

$$\langle T\rangle = \tfrac{1}{2}\left\langle \sum_\alpha \mathbf{r}_\alpha \cdot \mathbf{grad}\, U_\alpha \right\rangle \tag{7.85a}$$

Of particular interest is the case of two particles which interact according to a central, power-law force: $F \propto r^n$. Then, the potential is of the form

$$U = kr^{n+1} \tag{7.86}$$

Therefore,

$$\mathbf{r} \cdot \mathbf{grad}\, U = r\frac{dU}{dr} = k(n+1)r^{n+1} = (n+1)U \tag{7.87}$$

and the virial theorem becomes

$$\boxed{\langle T\rangle = \frac{n+1}{2}\langle U\rangle} \tag{7.88}$$

If the particles have a gravitational interaction, then $n = -2$, and

$$\langle T\rangle = -\tfrac{1}{2}\langle U\rangle, \qquad n = -2$$

This relation is useful in calculating, for example, the energetics in planetary motion.

Suggested References

The reader is again referred to Lindsay and Margenau (Li36, Chapter 3) for a discussion of the foundations of mechanics; Sections 3.12–3.14 present the minimal principles from which Lagrangian and Hamiltonian dynamics may be obtained. A complete and detailed (although quite readable) study of variational principles in mechanics is given by Lanczos (La49); see also Yourgrau and Mandelstam (Yo60, Chapters 1–5).

Generalized coordinates in Lagrangian mechanics is given an interesting treatment by Byerly (By13, Chapter 2).

The Lagrangian function is "derived" in an interesting way by Landau and Lifshitz (La60, Chapter 1).

Lagrange's equations can be obtained in a variety of ways. One of the more frequently used methods involves the application of *d'Alembert's principle*; see, for example, Goldstein (Go50, Chapter 1) or Joos and Freeman (Jo50, Chapter 6). *Virtual work* derivations are given by Slater and Frank (Sl47, Chapter 4) and by Symon (Sy60, Chapter 9); see also Konopinski (Ko69, Chapter 7).

* See, example, Lindsay (Li61, pp. 164–167).

The standard treatise on advanced dynamics is that of Whittaker (Wh37); see, particularly, Chapters 1–4 for Lagrange's method and for various applications.

Hamiltonian mechanics is discussed at the intermediate-to-advanced level by Corben and Stehle (Co60, Chapter 10), Goldstein (Go50, Chapter 7), Konopinski (Ko69, Chapter 8), and McCuskey (Mc59, Chapter 6); Tolman's account (To38, Chapter 2) is particularly clear.

Introductory accounts of classical statistical mechanics (ensembles, Liouville's theorem, kinetic theory, etc.) may be found in Crawford (Cr63, Chapter 16), Houston (Ho48, Chapter 11), Joos and Freeman (Jo50, Chapters 33 and 34), Lindsay (Li61, Chapter 7), and Morse (Mo62, Chapters 16–23). The first few chapters of Tolman (To38) are a clear exposition of the extension of Hamiltonian mechanics to statistical mechanics.

Problems

7-1. A disk rolls without slipping across a horizontal plane. The plane of the disk remains vertical, but it is free to rotate about a vertical axis. What generalized coordinates may be used to describe the motion? Write a differential equation which describes the rolling constraint. Is this equation integrable? Justify your answer by a physical argument. Is the constraint holonomic?

7-2. A body is released from a height of 64 ft and 2 sec later it strikes the ground. The equation for the distance of fall s during a time t could conceivably have any of the forms (where g has different units in the three expressions)

$$s = gt; \qquad s = \tfrac{1}{2}gt^2; \qquad s = \tfrac{1}{4}gt^3$$

all of which yield $s = 64$ ft for $t = 2$ sec. Show that the correct form leads to a minimum for the integral in Hamilton's Principle.

7-3. A sphere of radius ρ is constrained to roll without slipping on the lower half of the inner surface of a hollow cylinder of inside radius R. Determine the Lagrangian function, the equation of constraint, and Lagrange's equations of motion. Find the frequency of small oscillations.

7-4. A particle moves in a plane under the influence of a force $f = -Ar^{\alpha-1}$ directed toward the origin; A and α ($\neq 0$ or 1) are constants. Choose appropriate generalized coordinates and let the potential energy be zero at the origin. Find the Lagrangian equations of motion. Is the angular momentum about the origin conserved? Is the total energy conserved?

7-5. Consider a vertical plane in a gravitational field. Let the origin of a coordinate system be located at some point in this plane. A particle of mass m moves in the plane under the influence of gravity and under the influence of an additional force $f = -Ar^{\alpha-1}$ which is directed toward the origin. (r is the distance from the origin; A and α ($\neq 0$ or 1) are constants.) Choose appropriate generalized coordinates and let the gravitational potential energy be zero along a horizontal line through the origin. Find the Lagrangian equations of motion. Is the angular momentum about the origin conserved? Explain.

7-6. A hoop of mass m and radius R rolls without slipping down an inclined plane of mass M which makes an angle α with the horizontal. Find the Lagrange equations and the integrals of the motion for the case in which the plane can slide without friction along a horizontal surface.

7-7. A double pendulum consists of two simple pendula, with one pendulum suspended from the bob of the other. If the two pendula have equal lengths and have bobs of equal mass and if both pendula are confined to move in the same plane, find Lagrange's equations of motion for the system.

7-8. Consider a region of space which is divided into two parts by a plane. The potential energy of a particle in region 1 is U_1 and in region 2 it is U_2. If a particle of mass m and with velocity v_1 in region 1 passes from region 1 into region 2 such that its path in region 1 makes an angle θ_1 with the normal to the plane of separation and an angle θ_2 with the normal when in region 2, show that

$$\frac{\sin \theta_1}{\sin \theta_2} = \left[1 + \frac{U_1 - U_2}{T_1}\right]^{\frac{1}{2}}$$

where $T_1 = \frac{1}{2}mv_1^2$. What is the optical analog of this problem?

7-9. A disk of mass M and radius R rolls without slipping down a plane which is inclined from the horizontal by an angle α. The disk has a short, weightless axle of negligible radius. From this axis is suspended a simple pendulum of length $l < R$ and whose bob has a mass m. Consider that the motion of the pendulum takes place in the plane of the disk and find Lagrange's equations for the system.

7-10. Two blocks, each of mass M, are connected by an extensionless string of length l. One block is placed on a smooth horizontal surface while the other block hangs over the side, the string passing over a frictionless pulley. Describe the motion of the system when (a) the mass of the string is negligible and when (b) the string has a mass m.

7-11. A particle of mass m is constrained to move on a circle of ıadius R. The circle rotates in space about one point on the circle which is fixed. The rotation takes place in the plane of the circle and with constant angular velocity ω. In the absence of a gravitational force, show that the motion of the particle about one end of a diameter which passes through the pivot point and the center of the circle is the same as that of a plane pendulum in a uniform gravitational field. Explain why this is a reasonable result.

7-12. A point particle slides frictionlessly down a sphere, starting from rest at the top. Show that the particle leaves the sphere after the motion has gone through a polar angle equal to $\cos^{-1}(\frac{2}{3})$.

7-13. *A double Atwood machine** is shown in the figure. Pick a proper set of generalized coordinates and obtain the Lagrange equations for the

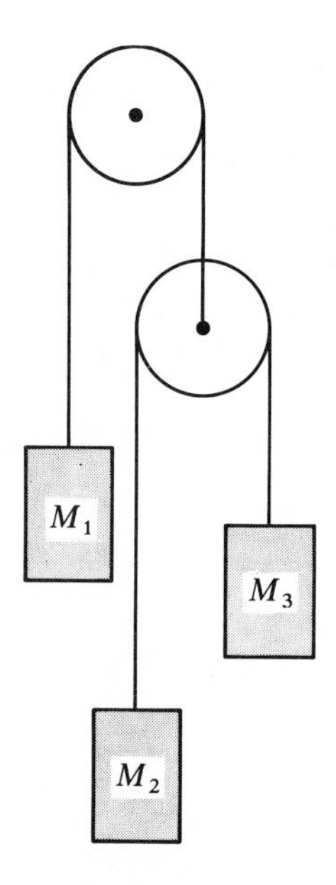

system. Solve for the accelerations of the masses. (The masses of the pulleys and friction are to be neglected.) Also, set up the problem with Lagrange undetermined multipliers and find the tensions in the strings.

7-14. A particle of mass m can slide freely along a wire AB whose perpendicular distance to the origin O is h (see figure). The line OC rotates about

* Devised by George Atwood (1746–1807) in 1784 for the purpose of making accurate measurements of the acceleration due to gravity.

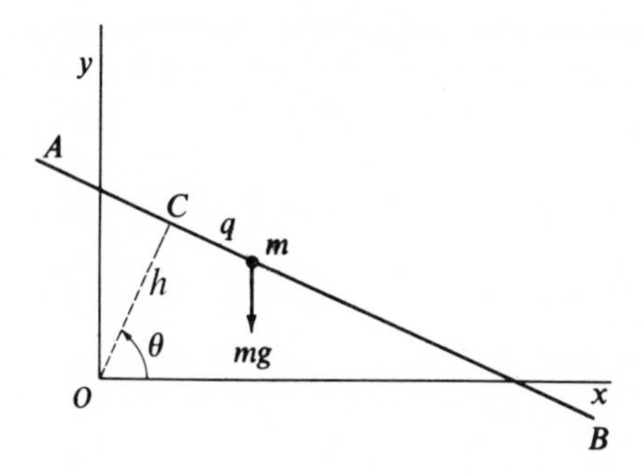

the origin at a constant angular velocity $\dot{\theta} = \omega$. The position of the particle can be described in terms of the angle θ and the distance q to the point C. If the particle is subject to a gravitational force and if the initial conditions are

$$\theta(0) = 0; \qquad q(0) = 0; \qquad \dot{q}(0) = 0$$

show that the time dependence of the coordinate q is

$$q(t) = \frac{g}{2\omega^2}(\cosh \omega t - \cos \omega t)$$

Sketch this result. Compute the Hamiltonian for the system and compare with the total energy. Is the total energy conserved?

7-15. A pendulum is constructed by attaching a mass m to an extensionless string of length l. The upper end of the string is connected to the uppermost point on a vertical disk of radius R ($R < l/\pi$), as in the figure. Obtain the

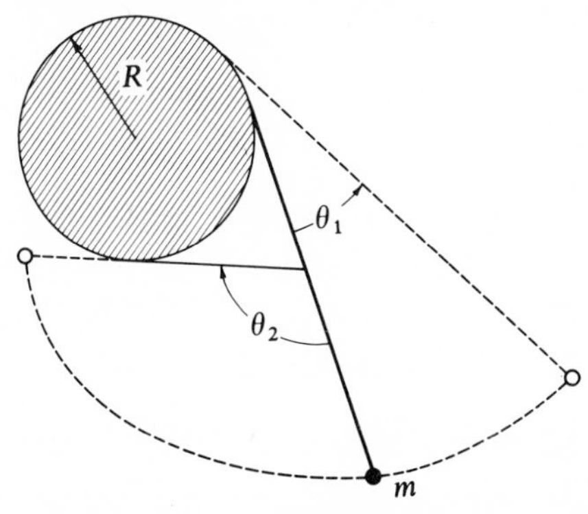

equation of motion of the pendulum and find the frequency of small oscillations. Also, find the line about which the angular motion extends equally in either direction (i.e., $\theta_1 = \theta_2$).

7-16. Two masses m_1 and m_2 $(m_1 \neq m_2)$ are connected by a rigid rod of length d and of negligible mass. An extensionless string of length l_1 is attached to m_1 and connected to a fixed point of support P. Similarly, a string of length l_2 $(l_1 \neq l_2)$ connects m_2 and P. Obtain the equation which describes the motion in the plane of m_1, m_2, and P, and find the frequency of small oscillations around the equilibrium position.

7-17. A circular hoop is suspended in a horizontal plane by three strings, each of length l, which are attached symmetrically to the hoop and are connected to fixed points lying in a plane above the hoop. At equilibrium each string is vertical. Show that the frequency of small rotational oscillations about the vertical through the center of the hoop is the same as that for a simple pendulum of length l.

7-18. A particle is constrained to move (without fiiction) on a circular wire which rotates with constant angular velocity ω about a vertical diameter. Find the equilibrium position of the particle and calculate the frequency of small oscillations around this position. Find, and interpret physically, a *critical angular velocity*, $\omega = \omega_c$, which divides the motion of the particle into two distinct types.

7-19. A particle of mass m moves in one dimension under the influence of a force

$$F(x, t) = \frac{k}{x^2} e^{-(t/\tau)}$$

where k and τ are positive constants. Compute the Lagrangian and Hamiltonian functions. Compare the Hamiltonian and the total energy, and discuss the conservation of energy for the system.

7-20. Consider a particle of mass m which moves freely in a conservative force field whose potential function is U. Find the Hamiltonian function and show that the canonical equations of motion reduce to Newton's equations. (Use rectangular coordinates.)

7-21. Consider a simple, plane pendulum which consists of a mass m attached to a string of length l. After the pendulum is set into motion, the length of the string is shortened at a constant rate

$$\frac{dl}{dt} = -\alpha = \text{const.}$$

The suspension point remains fixed. Compute the Lagrangian and Hamiltonian functions. Compare the Hamiltonian and the total energy, and discuss the conservation of energy for the system.

7-22. A particle of mass m moves under the influence of gravity along the spiral $z = k\theta$, $r = \text{const.}$, where k is a constant and z is vertical. Obtain the Hamiltonian equations of motion.

7-23. Consider any two continuous functions of the generalized coordinates and momenta, $g(q_k, p_k)$ and $h(q_k, p_k)$. The so-called *Poisson brackets* are defined by

$$[g, h] \equiv \sum_k \left(\frac{\partial g}{\partial q_k}\frac{\partial h}{\partial p_k} - \frac{\partial g}{\partial p_k}\frac{\partial h}{\partial q_k}\right)$$

Verify the following properties of the Poisson brackets:

(a) $\dfrac{dg}{dt} = [g, H] + \dfrac{\partial g}{\partial t}$

(b) $\dot{q}_j = [q_j, H]; \qquad \dot{p}_j = [p_j, H]$

(c) $[p_k, p_j] = 0; \qquad [q_k, q_j] = 0$

(d) $[q_k, p_j] = \delta_{kj}$

where H is the Hamiltonian. If the Poisson bracket of two quantities vanishes, the quantities are said to *commute*. If the Poisson bracket of two quantities equals unity, the quantities are said to be *canonically conjugate*. Show that any quantity which does not depend explicitly on the time and which commutes with the Hamiltonian is a constant of the motion of the system. Poisson-bracket formalism is of considerable importance in quantum mechanics.

7-24. A *spherical pendulum* consists of a bob of mass m attached to a weightless, extensionless rod of length l. The end of the rod opposite the bob is allowed to pivot freely (in *all* directions) about some fixed point. Set up the Hamiltonian function in spherical coordinates. (If $p_\varphi = 0$, then the result is the same as that for the plane pendulum.) Combine the term which depends on p_φ with the ordinary potential energy term to define an *effective* potential $V(\theta, p_\varphi)$. Sketch V as a function of θ for several values of p_φ, including $p_\varphi = 0$. Discuss the features of the motion, pointing out the differences between $p_\varphi = 0$ and $p_\varphi \neq 0$. Discuss also the limiting case of the conical pendulum (for which $\theta = \text{const.}$) with reference to the V-θ diagram.

7-25. A particle moves in a spherically symmetric force field in which the potential energy is given by $U(r) = -k/r$. Calculate the Hamiltonian function in spherical coordinates and obtain the canonical equations of motion. Sketch the path that a representative point for the system would follow on a surface $H = \text{const.}$ in phase space. Begin by showing that the motion must lie in a plane so that the phase space is four-dimensional (r, θ, p_r, p_θ, but only the first three are nontrivial). Calculate the projection of the phase path on the phase path on the r-p_r plane, then take into account the variation with θ.

7-26. Four particles are directed upward in a uniform gravitational field with the following initial conditions:

(1) $z(0) = z_0;$ $\quad p_z(0) = p_0$

(2) $z(0) = z_0 + \Delta z_0;$ $\quad p_z(0) = p_0$

(3) $z(0) = z_0;$ $\quad p_z(0) = p_0 + \Delta p_0$

(4) $z(0) = z_0 + \Delta z_0;$ $\quad p_z(0) = p_0 + \Delta p_0$

Show by direct calculation that the representative points corresponding to these particles always define an area in phase space equal to $\Delta z_0 \, \Delta p_0$. Sketch the phase paths and show for several times $t > 0$ the shape of the region whose area remains constant.

7-27. Discuss the implications of Liouville's theorem on the focussing of beams of charged particles by considering the following simple case. An electron beam of circular cross section (radius R_0) is directed along the z-axis. The density of electrons across the beam is constant, but the momentum components transverse to the beam (p_x and p_y) are distributed uniformly over a circle of radius p_0 in momentum space. If some focussing system is used to reduce the beam radius from R_0 to R_1, find the resulting distribution of the transverse momentum components. What is the physical meaning of this result? (Consider the angular divergence of the beam.)

CHAPTER 8

Central-Force Motion

8.1 Introduction

The motion of a system consisting of two bodies which act under the influence of a force directed along the line connecting the centers of the two bodies (i.e., a *central force*) is an extremely important physical problem, the solution to which can be determined completely. The importance of such a problem lies in large measure in two quite different realms of physics—the motion of celestial bodies, such as planets, moons, comets, double stars, etc. and in certain two-body nuclear interactions, such as the scattering of α particles by nuclei. Also, in the pre-quantum-mechanics days, the hydrogen atom was described in terms of a classical two-body central-force picture, and although such a description is still useful in a qualitative sense, the quantum theoretical approach must be used for a detailed description. In addition to some general considerations regarding motion in central-force fields, we shall discuss in this and in the following chapter several of the problems of two bodies that are encountered in celestial mechanics and in nuclear physics. We shall also give a brief discussion of the problem of the motion of *three* interacting bodies.

8.2 The Reduced Mass

The description of a system consisting of two particles requires the specification of six quantities, e.g., the three components of each of the two radius vectors $\mathbf{r}_1$ and $\mathbf{r}_2$ for the particles.* Alternatively, we may choose the three components of the center-of-mass vector $\mathbf{R}$ and the three components of $\mathbf{r} \equiv \mathbf{r}_1 - \mathbf{r}_2$ (see Fig. 8-1a). We shall restrict our attention to systems without

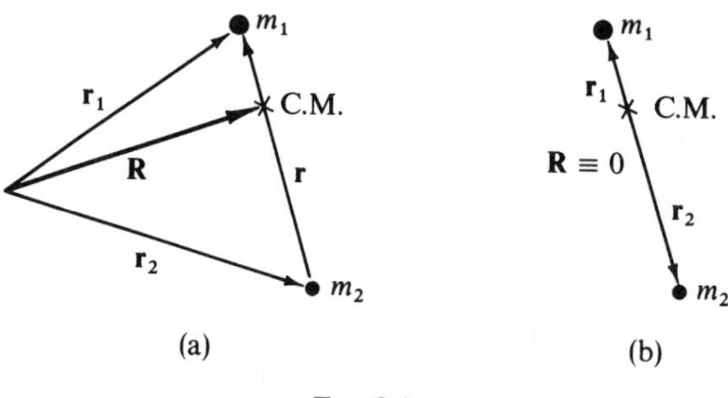

FIG. **8-1**

frictional losses and for which the potential energy is a function only of $r = |\mathbf{r}_1 - \mathbf{r}_2|$. The Lagrangian for such a system may be written as

$$L = \tfrac{1}{2}m_1|\dot{\mathbf{r}}_1|^2 + \tfrac{1}{2}m_2|\dot{\mathbf{r}}_2|^2 - U(r) \tag{8.1}$$

Since translational motion of the system as a whole is uninteresting from the standpoint of the particle orbits with respect to one another, we may choose the origin for the coordinate system to be the center of mass of the particles; i.e., $\mathbf{R} \equiv 0$ (see Fig. 8-1b). Then [cf. Eq. (2.25)],

$$m_1\mathbf{r}_1 + m_2\mathbf{r}_2 = 0 \tag{8.2}$$

This equation, combined with $\mathbf{r} = \mathbf{r}_1 - \mathbf{r}_2$, yields

$$\left.\begin{aligned} \mathbf{r}_1 &= \frac{m_1}{m_1 + m_2}\mathbf{r} \\ \mathbf{r}_2 &= -\frac{m_1}{m_1 + m_2}\mathbf{r} \end{aligned}\right\} \tag{8.3}$$

* The *orientation* of the particles is assumed to be unimportant; i.e., they are spherically symmetrical (or are *point* particles).

Substitution of Eqs. (8.3) into the expression for the Lagrangian gives

$$\boxed{L = \tfrac{1}{2}\mu|\dot{\mathbf{r}}|^2 - U(r)} \tag{8.4}$$

where μ is the *reduced mass*,

$$\boxed{\mu \equiv \frac{m_1 m_2}{m_1 + m_2}} \tag{8.5}$$

We have therefore formally reduced the problem of the motion of two bodies to an *equivalent one-body problem* in which we must determine only the motion of a "particle" of mass μ in the central field described by the potential function $U(r)$. Once the solution for $\mathbf{r}(t)$ is obtained by applying the Lagrange equations to Eq. (8.4), the individual motions of the particles, $\mathbf{r}_1(t)$ and $\mathbf{r}_2(t)$, may be found (if desired) by using Eqs. (8.3). This latter step is not necessary if only the orbits relative to one another are required.

8.3 Conservation Theorems—First Integrals of the Motion

The system which we wish to discuss may be considered to consist of a particle of mass μ which moves in a central-force field described by the potential function $U(r)$. Since the potential energy depends only upon the distance of the particle from the force center and not upon the orientation, the system possesses *spherical symmetry*. That is, the rotation of the system about any fixed axis through the center of force cannot affect the equations of motion. We have already shown (see Section 7.11) that under such conditions the angular momentum of the system is conserved:

$$\mathbf{L} = \mathbf{r} \times \mathbf{p} = \text{const.} \tag{8.6}$$

From this relation it is clear that both the radius vector and the linear momentum vector of the particle lie always in a plane normal to the angular momentum vector $\mathbf{L}$ which is fixed in space (see Fig. 8-2). Therefore, we have only a two-dimensional problem, and the Lagrangian may then be conveniently expressed in plane polar coordinates:

$$\boxed{L = \tfrac{1}{2}\mu(\dot{r}^2 + r^2\dot{\theta}^2) - U(r)} \tag{8.7}$$

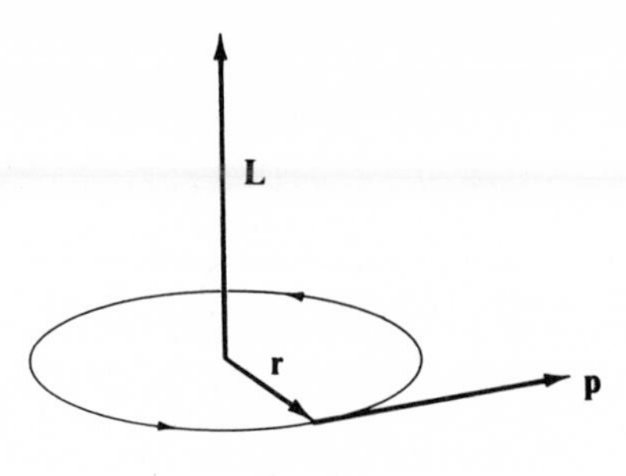

FIG. 8-2

Since the Lagrangian is cyclic in θ, the angular momentum conjugate to the coordinate θ is conserved:

$$\dot{p}_\theta = \frac{\partial L}{\partial \theta} = 0 = \frac{d}{dt}\frac{\partial L}{\partial \dot{\theta}} \tag{8.8}$$

or,

$$p_\theta \equiv \frac{\partial L}{\partial \dot{\theta}} = \mu r^2 \dot{\theta} = \text{const.} \tag{8.9}$$

The symmetry of the system has therefore permitted us to integrate immediately one of the equations of motion. The quantity p_θ is a *first integral* of the motion and we denote its constant value by the symbol l:

$$\boxed{l \equiv \mu r^2 \dot{\theta} = \text{const.}} \tag{8.10}$$

The fact that l is constant has a simple geometrical interpretation. Referring to Fig. 8-3, we see that in describing the path $\mathbf{r}(t)$, the radius vector sweeps out an area $\frac{1}{2}r^2\, d\theta$ in a time interval dt:

$$dA = \tfrac{1}{2}r^2\, d\theta \tag{8.11}$$

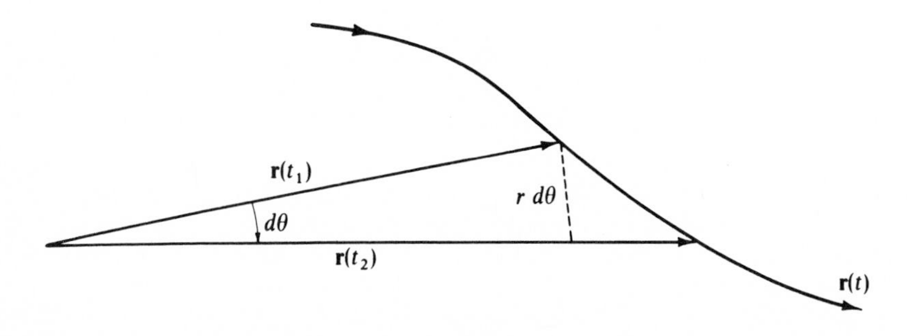

FIG. 8-3

and, upon dividing by the time interval, the *areal velocity* is

$$\frac{dA}{dt} = \tfrac{1}{2}r^2 \frac{d\theta}{dt} = \tfrac{1}{2}r^2\dot{\theta}$$

$$= \frac{l}{2\mu} = \text{const.} \tag{8.12}$$

from which it is seen that the areal velocity is constant in time. This result was obtained empirically by Kepler for the case of planetary motion and is known as *Kepler's Second Law*.* It is important to note that the conservation of the areal velocity is not limited to the case of an inverse-square-law force (the case for planetary motion), but is a general result for central-force motion.

Since we have eliminated from consideration the uninteresting uniform motion of the center of mass of the system, the conservation of linear momentum adds nothing new to the description of the motion. Therefore, the conservation of energy is the only remaining first integral of the problem. The conservation of the total energy E is automatically assured since we have limited the discussion to non-dissipative systems. Thus,

$$T + U = E = \text{const.} \tag{8.13}$$

and,

$$E = \tfrac{1}{2}\mu(\dot{r}^2 + r^2\dot{\theta}^2) + U(r)$$

or,

$$\boxed{E = \tfrac{1}{2}\mu\dot{r}^2 + \tfrac{1}{2}\frac{l^2}{\mu r^2} + U(r)} \tag{8.14}$$

8.4 Equations of Motion

When $U(r)$ is specified, Eq. (8.14) completely describes the system and the integration of this equation gives the general solution of the problem in terms of the parameters E and l. Solving Eq. (8.14) for $\dot{r}$, we have

$$\dot{r} = \frac{dr}{dt} = \sqrt{\frac{2}{\mu}(E - U) - \frac{l^2}{\mu^2 r^2}} \tag{8.15}$$

* Published by Johannes Kepler (1571–1630) in 1609 after an exhaustive study of the compilations made by Tycho Brahe (1546–1601) of the positions of the planet Mars. Kepler's *First Law* deals with the shape of planetary orbits (see Section 8.7).

This equation can be solved for dt and integrated to yield the solution $t = t(r)$. An inversion of this result will then give the equation of motion in the standard form, $r = r(t)$. At present, however, we are more interested in the equation of the path in terms of r and θ. We can write

$$d\theta = \frac{d\theta}{dt}\frac{dt}{dr}\,dr = \frac{\dot{\theta}}{\dot{r}}\,dr \tag{8.16}$$

Into this relation we can substitute $\dot{\theta} = l/\mu r^2$ [Eq. (8.10)] and the expression for $\dot{r}$ from Eq. (8.15). Upon integrating, we have

$$\theta(r) = \int \frac{(l/r^2)\,dr}{\sqrt{2\mu\left(E - U - \dfrac{l^2}{2\mu r^2}\right)}} \tag{8.17}$$

Furthermore, since l is constant in time, $\dot{\theta}$ cannot change sign and therefore $\theta(t)$ must increase monotonically with time.

Although we have reduced the problem to the formal evaluation of an integral, the actual solution can be easily obtained only for certain specific forms of the force law. If the force is proportional to some power of the radial distance, $F(r) \propto r^n$, then the solution can be expressed in terms of elliptic integrals for certain integer and fractional values of n. Only for $n = 1$, -2, and -3 are the solutions expressible in terms of circular functions.* The case $n = 1$ is just that of the harmonic oscillator (see Chapter 3), and the case $n = -2$ is the important inverse-square-law force which is treated in Sections 8.6 and 8.7. These two cases, $n = 1$, -2, are the ones of prime importance in physical situations; details of some other cases of interest will be found in the Problems.

We have therefore solved the problem in a formal way by combining the equations which express the conservation of energy and angular momentum into a single result which gives the equation of the orbit, $\theta = \theta(r)$. We can also attack the problem via Lagrange's equation for the coordinate r:

$$\frac{\partial L}{\partial r} - \frac{d}{dt}\frac{\partial L}{\partial \dot{r}} = 0$$

Using Eq. (8.7) for L, we find

$$\mu(\ddot{r} - r\dot{\theta}^2) = -\frac{\partial U}{\partial r} = F(r) \tag{8.18}$$

* See, for example, Goldstein (Go50, pp. 73–75).

Equation (8.18) can be cast in a form more suitable for certain types of calculations by making a simple change of variable:

$$u \equiv \frac{1}{r}$$

First, we compute

$$\frac{du}{d\theta} = -\frac{1}{r^2}\frac{dr}{d\theta} = -\frac{1}{r^2}\frac{dr}{dt}\frac{dt}{d\theta} = -\frac{1}{r^2}\frac{\dot{r}}{\dot{\theta}}$$

But, from Eq. (8.10), $\dot{\theta} = l/\mu r^2$, so that

$$\frac{du}{d\theta} = -\frac{\mu}{l}\dot{r}$$

Next, we write

$$\frac{d^2u}{d\theta^2} = \frac{d}{d\theta}\left(-\frac{\mu}{l}\dot{r}\right) = \frac{dt}{d\theta}\frac{d}{dt}\left(-\frac{\mu}{l}\dot{r}\right) = -\frac{\mu}{l\dot{\theta}}\ddot{r}$$

And with the same substitution for $\dot{\theta}$ we have

$$\frac{d^2u}{d\theta^2} = -\frac{\mu^2}{l^2}r^2\ddot{r}$$

Therefore, solving for $\ddot{r}$ and $r\dot{\theta}^2$ in terms of u, we find

$$\left.\begin{aligned} \ddot{r} &= -\frac{l^2}{\mu^2}u^2\frac{d^2u}{d\theta^2} \\ r\dot{\theta}^2 &= \frac{l^2}{\mu^2}u^3 \end{aligned}\right\} \tag{8.19}$$

Substituting Eqs. (8.19) into Eq. (8.18), we obtain the transformed equation of motion:

$$\frac{d^2u}{d\theta^2} + u = -\frac{\mu}{l^2}\frac{1}{u^2}F(u) \tag{8.20}$$

which we may also write as

$$\boxed{\frac{d^2}{d\theta^2}\left(\frac{1}{r}\right) + \frac{1}{r} = -\frac{\mu r^2}{l^2}F(r)} \tag{8.20a}$$

This form of the equation of motion is particularly useful if we wish to find the force law which gives a particular known orbit $r = r(\theta)$.

8.5 Orbits in a Central Field

The radial velocity of a particle moving in a central field is given by Eq. (8.15). This equation indicates that $\dot{r}$ will vanish at the roots of the radical, i.e., at points for which

$$E - U(r) - \frac{l^2}{2\mu r^2} = 0 \tag{8.21}$$

The vanishing of $\dot{r}$ implies that a *turning point* in the motion has been reached (cf. the discussion in Section 5.2). In general, Eq. (8.21) possesses two roots: r_{max} and r_{min}. The motion of the particle is therefore confined to the annular region specified by $r_{max} \geq r \geq r_{min}$. Certain combinations of the potential function $U(r)$ and the parameters E and l will produce only a single root for Eq. (8.21). In such a case, $\dot{r} = 0$ for all values of the time; hence, $r =$ const., and the orbit is circular.

If the motion of a particle in the potential $U(r)$ is periodic, then the orbit is *closed*. That is, after a finite number of excursions between the radial limits r_{min} and r_{max}, the motion exactly repeats itself. On the other hand, if the orbit does not close upon itself after a finite number of oscillations, the orbit is said to be *open*. Such a case is shown in Fig. 8-4. Now, from Eq. (8.17)

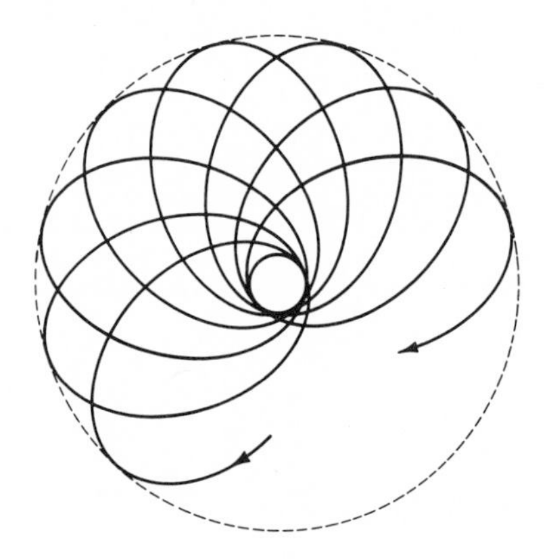

FIG. **8-4**

we can compute the change in the angle θ which results from one complete transit of r from r_{min} to r_{max} and back to r_{min}. Since the motion is symmetrical in time, this angular change is twice that which would result from the passage from r_{min} to r_{max}; thus,

$$\Delta\theta = 2\int_{r_{\min}}^{r_{\max}} \frac{(l/r^2)\,dr}{\sqrt{2\mu\left(E - U - \dfrac{l^2}{2\mu r^2}\right)}} \tag{8.22}$$

Now, the path will be closed only if $\Delta\theta$ is a rational fraction of 2π, i.e., if $\Delta\theta = 2\pi \cdot (a/b)$, where a and b are integers. Under these conditions, after b periods the radius vector of the particle will have made a complete revolutions and will have returned to its original position. It can be shown (see Problem 8-28) that if the potential varies with some integer power of the radial distance, $U(r) \propto r^{n+1}$, then a closed noncircular path can result *only** if $n = -2$ or $+1$. The case $n = -2$ corresponds to an inverse-square-law force, e.g., the gravitational or electrostatic force, while the second case ($n = +1$) corresponds to the harmonic oscillator potential, the two-dimensional case for which was discussed in Section 3.4 and where a closed path for the motion was found to result in the event that the ratio of the angular frequencies for the x and y motions is a rational fraction.

8.6 Centrifugal Energy and the Effective Potential

In the expressions above for $\dot{r}$, $\Delta\theta$, etc., a common term is the radical

$$\sqrt{E - U - \frac{l^2}{2\mu r^2}}$$

The last term in the radical has the dimensions of energy, and, according to Eq. (8.14), can also be written as

$$\frac{l^2}{2\mu r^2} = \tfrac{1}{2}\mu r^2\dot{\theta}^2$$

If we interpret this quantity as a "potential energy,"

$$U_c \equiv \frac{l^2}{2\mu r^2} \tag{8.23}$$

then the "force" that must be associated with U_c is

$$F_c = -\frac{\partial U_c}{\partial r} = \frac{l^2}{\mu r^3} = \mu r\dot{\theta}^2 \tag{8.24}$$

This quantity is traditionally called the *centrifugal force*,† although it is not

* Certain fractional values of n will also lead to closed orbits, but in general these cases are uninteresting from a physical standpoint.

† The expression is more readily recognized in the form $F_c = mr\omega^2$. The first real appreciation of centrifugal force was by Christian Huygens (1629–1695), who made a detailed examination in connection with his study of the conical pendulum in 1659.

a "force" in the ordinary sense of the word.* We shall, however, continue to use this unfortunate terminology since it is customary and convenient.

Thus, we see that the term $l^2/2\mu r^2$ can be interpreted as the *centrifugal potential energy* of the particle, and as such, can be included with $U(r)$ in an *effective potential* defined by

$$\boxed{V(r) \equiv U(r) + \frac{l^2}{2\mu r^2}} \tag{8.25}$$

$V(r)$ is therefore a *fictitious* potential which combines the real potential function $U(r)$ with the energy term associated with the angular motion about the center of force. For the case of inverse-square-law central-force motion, the force is given by

$$F(r) = -\frac{k}{r^2} \tag{8.26}$$

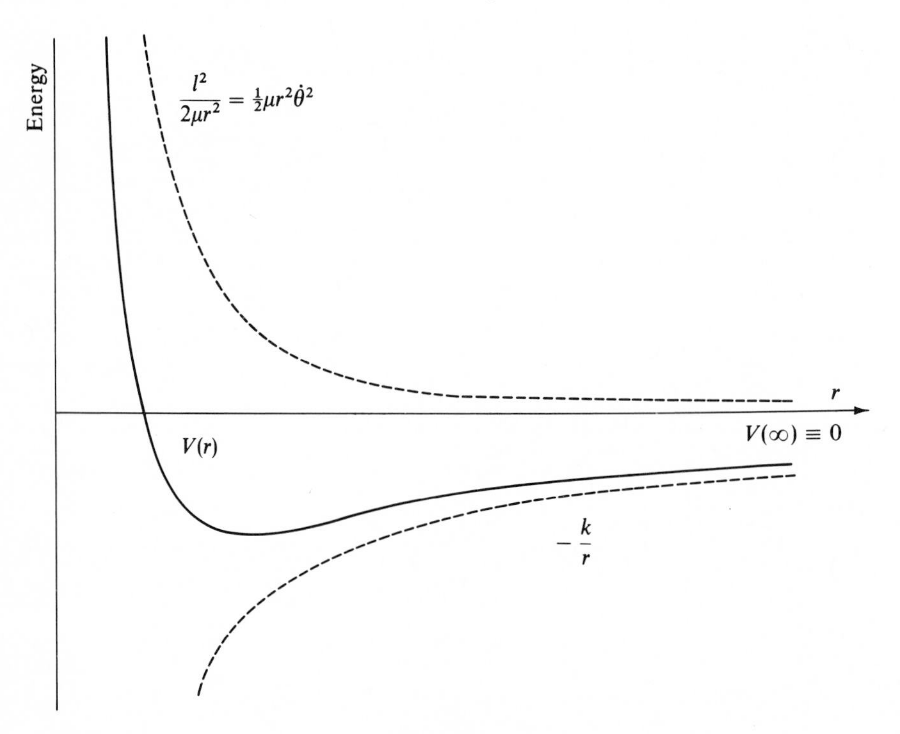

FIG. **8-5**

* See Section 11.3 for a more critical discussion of centrifugal force.

from which

$$U(r) = -\int F(r)\,dr = -\frac{k}{r} \tag{8.27}$$

Therefore, the effective potential function for gravitational attraction is

$$V(r) = -\frac{k}{r} + \frac{l^2}{2\mu r^2} \tag{8.28}$$

This effective potential and its components are shown in Fig. 8-5. The value of the potential is arbitrarily taken to be zero at $r = \infty$. [This is implicit in Eq. (8.27) where the constant of integration was omitted.]

We may now draw conclusions similar to those in Section 5.2 where the motion of a particle in an arbitrary potential well was discussed. If we plot the total energy E of the particle on a diagram similar to Fig. 8-5, we may identify three regions of interest (see Fig. 8-6). If the total energy is positive

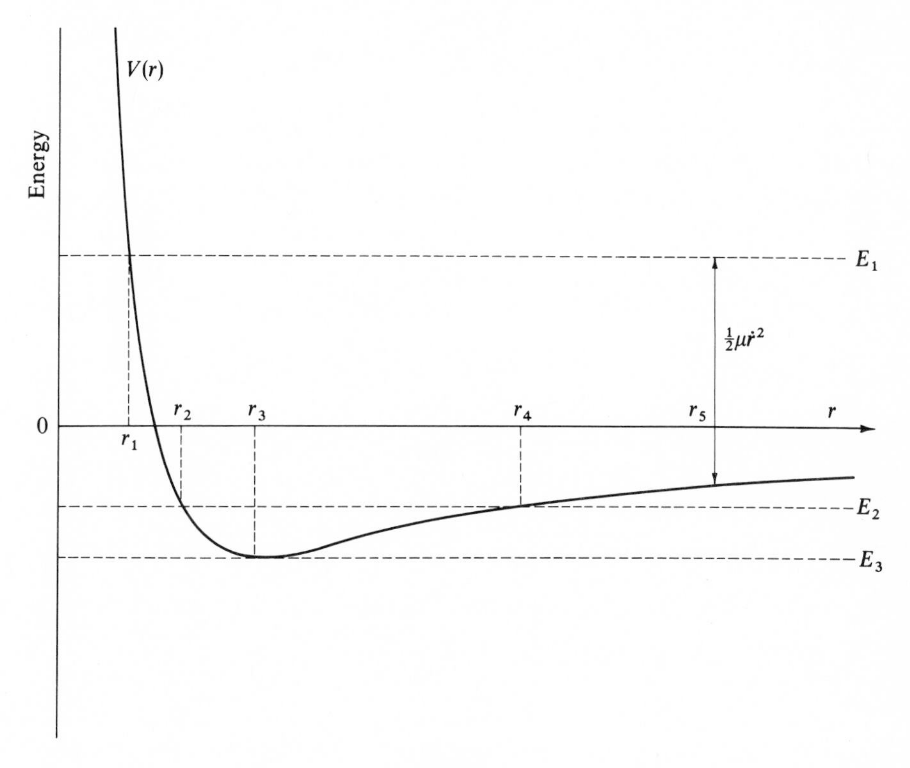

FIG. 8-6

or zero (e.g., $E_1 \geq 0$), then the motion is unbounded; the particle moves toward the force center (located at $r = 0$) from infinitely far away until it "strikes" the potential barrier at the *turning point* $r = r_1$ and is reflected back towards infinitely large r. Note that the height of the constant total energy line above $V(r)$ at any r, such as r_5 in Fig. 8-6, is equal to $\frac{1}{2}\mu\dot{r}^2$. Thus, the radial velocity $\dot{r}$ vanishes and changes sign at the turning point (or points).

If the total energy is negative* and lies between zero and the minimum value of $V(r)$, as does E_2, then the motion is bounded, with $r_2 \leq r \leq r_4$. The values r_2 and r_4 are the turning points, or the *apsidal distances*, of the orbit. If E equals the minimum value of the effective potential energy (see E_3 in Fig. 8-6), then the radius of the particle's path is limited to the single value r_3, and then $\dot{r} = 0$ for all values of the time; hence, the motion is circular.

Values of E which are less than $V_{\min} = -(\mu k^2/2l^2)$ do not result in physically real motion, since for such cases $\dot{r}^2 < 0$ and the velocity would be imaginary.

8.7 Planetary Motion—Kepler's Problem

The equation for the path of a particle moving under the influence of a central force whose magnitude is inversely proportional to the square of the distance between the particle and the force center can be obtained from [cf. Eq. (8.17)]:

$$\theta(r) = \int \frac{(l/r^2)\,dr}{\sqrt{2\mu\left(E + \dfrac{k}{r} - \dfrac{l^2}{2\mu r^2}\right)}} + \text{const.} \tag{8.29}$$

The integral can be easily evaluated if the variable is changed to $u \equiv 1/r$ (see Problem 8-2). If the origin of θ is defined so that the integration constant is zero, we find

$$\cos\theta = \frac{\dfrac{l^2}{\mu k}\cdot\dfrac{1}{r} - 1}{\sqrt{1 + \dfrac{2El^2}{\mu k^2}}} \tag{8.30}$$

* Note that negative values of the total energy arise only because of the arbitrary choice of $V(r) = 0$ at $r = \infty$.

Let us now define the following constants:

$$\left.\begin{aligned} \alpha &\equiv \frac{l^2}{\mu k} \\ \varepsilon &\equiv \sqrt{1 + \frac{2El^2}{\mu k^2}} \end{aligned}\right\} \tag{8.31}$$

Then, Eq. (8.30) can be written as

$$\boxed{\frac{\alpha}{r} = 1 + \varepsilon \cos \theta} \tag{8.32}$$

This is the equation of a conic section with one focus at the origin*; the quantity ε is called the *eccentricity* and 2α is termed the *latus rectum* of the orbit.

The minimum value for r occurs when $\cos \theta$ is a maximum, i.e., for $\theta = 0$. Thus, the choice of *zero* for the constant in Eq. (8.29) corresponds to measuring θ from r_{min}, which position is called the *pericenter*; r_{max} corresponds to the *apocenter*.† The general term for turning points is *apsides*.

Various values of the eccentricity (and, hence, of the energy E) classify the orbits according to different conic sections (see Fig. 8-7):

$\varepsilon > 1$;	$E > 0$	(hyperbola)
$\varepsilon = 1$;	$E = 0$	(parabola)
$0 < \varepsilon < 1$;	$V_{min} < E < 0$	(ellipse)
$\varepsilon = 0$;	$E = V_{min}$	(circle)
$\varepsilon < 0$;	$E < V_{min}$	(not allowed)

For the case of planetary motion, the orbits are ellipses with major and minor axes (a and b, respectively) given by

$$a = \frac{\alpha}{1 - \varepsilon^2} = \frac{k}{2|E|} \tag{8.33a}$$

$$b = \frac{\alpha}{\sqrt{1 - \varepsilon^2}} = \frac{l}{\sqrt{2\mu|E|}} \tag{8.33b}$$

* Johann Bernoulli (1667–1748) appears to have been the first to prove that *all* possible orbits of a body moving in a potential proportional to $1/r$ are conic sections (1710).

† The corresponding terms for motion about the sun are *perihelion* and *aphelion*, and for motion about the Earth, *perigee* and *apogee*.

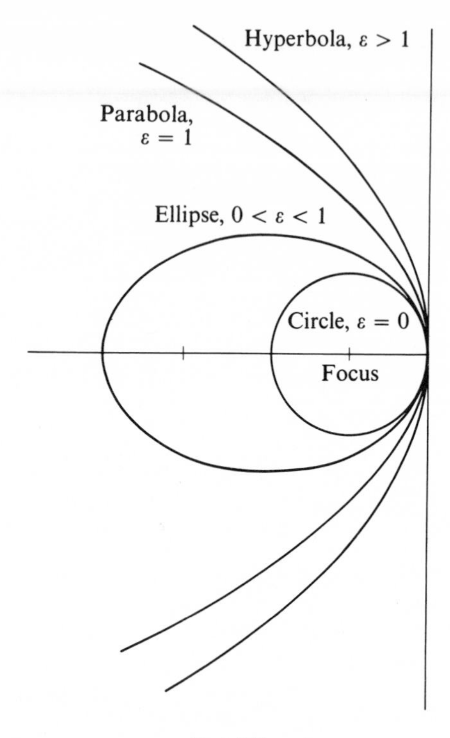

FIG. 8-7

Thus, the major axis depends only on the energy of the particle, whereas the minor axis is a function of both first integrals of the motion, E and l. The geometry of elliptic orbits in terms of the parameters α, ε, a, and b is shown in Fig. 8-8; P and P' are the foci. From this diagram we see that the apsidal distances ($r_{\min}$ and $r_{\max}$ as measured from the foci to the orbit) are

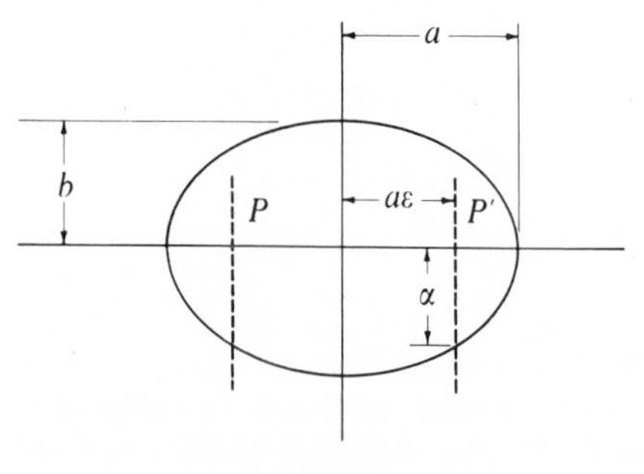

FIG. 8-8

given by:

$$\left.\begin{aligned} r_{\min} &= a(1-\varepsilon) = \frac{\alpha}{1+\varepsilon} \\ r_{\max} &= a(1+\varepsilon) = \frac{\alpha}{1-\varepsilon} \end{aligned}\right\} \tag{8.34}$$

In order to find the period for elliptic motion, we rewrite Eq. (8.12) for the areal velocity as

$$dt = \frac{2\mu}{l}\,dA$$

Since the entire area A of the ellipse is swept out in one complete period τ,

$$\int_0^\tau dt = \frac{2\mu}{l}\int_0^A dA$$

$$\tau = \frac{2\mu}{l}\,A \tag{8.35}$$

Now, the area of an ellipse is given by $A = \pi ab$, and using a and b from Eqs. (8.33), we find

$$\tau = \frac{2\mu}{l}\cdot \pi ab = \frac{2\mu}{l}\cdot \pi \cdot \frac{k}{2|E|}\cdot \frac{l}{\sqrt{2\mu|E|}}$$

$$= \pi k\sqrt{\frac{\mu}{2}}\cdot |E|^{-3/2} \tag{8.36}$$

We also note from Eqs. (8.33) that the minor axis can be written as

$$b = \sqrt{\alpha a} \tag{8.37}$$

Therefore, since $\alpha = l^2/\mu k$, the period τ can also be expressed as

$$\boxed{\tau^2 = \frac{4\pi^2\mu}{k}\,a^3} \tag{8.38}$$

This result, that the square of the period is proportional to the cube of the major axis of the elliptic orbit, is known as *Kepler's Third Law*.* Note that

* Published by Kepler in 1619. Kepler's *Second Law* was stated in Section 8.3. The *First Law* (1609) expresses the fact that the planets move in elliptical orbits with the Sun at one focus. It should be noted that Kepler's work preceded by almost 80 years Newton's enunciation of his general laws of motion. Indeed, Newton's conclusions were based to a great extent on Kepler's pioneering studies (and upon those of Galileo and Huygens).

this result is concerned with the equivalent one-body problem, so that account must be taken of the fact that it is the *reduced* mass μ which occurs in Eq. (8.38). Kepler's actual statement of his conclusion was that the squares of the periods of the planets were proportional to the cubes of the major axes of their orbits, with the same proportionality constant for all planets. In this sense, the statement is only approximately correct, since the reduced mass is different for each planet. In particular, since the gravitational force is given by

$$F(r) = -\gamma \frac{m_1 m_2}{r^2} = -\frac{k}{r^2}$$

we identify $k = \gamma m_1 m_2$. Therefore, the expression for the square of the period becomes*

$$\tau^2 = \frac{4\pi^2 a^3}{\gamma(m_1 + m_2)} \cong \frac{4\pi^2 a^3}{\gamma m_2}, \qquad m_1 \ll m_2$$

so that Kepler's statement is valid only if the mass m_1 of a planet can be neglected with respect to the mass m_2 of the Sun. (But note, for example, that the mass of Jupiter is about 1/1000 of the mass of the Sun, so that the departure from the approximate law is not difficult to observe in this case.)

■ 8.8 Kepler's Equation

We have found in Eq. (8.32) the relationship between the coordinates r and θ which describes the motion of a particle attracted toward a center by a force that varies inversely with r^2. For the purposes of astronomical calculations, however, it is not $r(\theta)$ that is desired, but the function $\theta(t)$ so that the direction of the body (a planet, comet, etc.) may be found at any time. Furthermore, we wish to have an expression giving θ (called the *true anomaly*†) as a function of time which involves as parameters only the two fundamental observable constants of the orbit, viz., the period τ and the

* In astronomical calculations, the gravitational constant γ is usually replaced by a quantity k^2, which is the square of the so-called *Gaussian gravitational constant*. If the unit of mass is the solar mass and if the unit of length is the value of a for the Earth's orbit (the *astronomical unit*), then $k = 0.017\,202\,098\,95$. This is the value calculated by Gauss in 1809 and was adopted as the standard by the International Astronomical Union in 1938. In order to maintain k fixed as better measurements are made, the value of the astronomical unit is adjusted.

† The historical term "anomaly" is not used in the sense of "strange," but rather as "deviation," since θ measures the angular deviation from some fixed point, which in astronomical calculations is usually the perihelion.

eccentricity ε. We can make such a calculation in the following way. Since it requires a time τ for the radius vector of the body to sweep out the entire area πab of an elliptic orbit, and since the areal velocity is a constant of the motion, then in a time t an area $(\pi ab/\tau)t$ will be swept out. We can equate this expression to the integral of the area by writing

$$\frac{\pi ab}{\tau} t = \int dA \tag{8.39}$$

According to Eq. (8.11), if we take $\theta = 0$ at $t = 0$, we have

$$\frac{\pi ab}{\tau} t = \frac{1}{2}\int_0^\theta r^2\, d\theta \tag{8.39a}$$

Equation (8.32) can be written as

$$r = \frac{\alpha}{1 + \varepsilon \cos \theta} \tag{8.40}$$

Therefore [see Eq. (E.16), Appendix E]:

$$\begin{aligned}\frac{\pi ab}{\tau} t &= \frac{\alpha^2}{2}\int_0^\theta \frac{d\theta}{(1 + \varepsilon\cos\theta)^2}\\ &= \frac{\alpha^2}{2(1-\varepsilon^2)}\left[\frac{2}{\sqrt{1-\varepsilon^2}}\tan^{-1}\left(\frac{(1-\varepsilon)\tan(\theta/2)}{\sqrt{1-\varepsilon^2}}\right) - \frac{\varepsilon\sin\theta}{1+\varepsilon\cos\theta}\right]\end{aligned}$$

Noting that $ab = \alpha^2(1-\varepsilon^2)^{-3/2}$, we can simplify this expression to

$$\frac{2\pi t}{\tau} = 2\tan^{-1}\left(\sqrt{\frac{1-\varepsilon}{1+\varepsilon}}\tan\frac{\theta}{2}\right) - \frac{\varepsilon\sqrt{1-\varepsilon^2}\sin\theta}{1+\varepsilon\cos\theta} \tag{8.41}$$

This is a formidable equation and one that is certainly not easy to use. To make matters worse, it is not $t(\theta)$ that is necessary, but rather $\theta(t)$; that is, the equation must be inverted. Clearly, this cannot be done in any simple way; only a series expansion is possible:

$$\begin{aligned}\theta(t) = \frac{2\pi t}{\tau} &+ 2\varepsilon\sin\frac{2\pi t}{\tau} + \frac{5}{4}\varepsilon^2\sin\frac{4\pi t}{\tau}\\ &+ \frac{1}{12}\varepsilon^3\left(13\sin\frac{6\pi t}{\tau} - 3\sin\frac{2\pi t}{\tau}\right) + \cdots\end{aligned} \tag{8.42}$$

If ε is a sufficiently small quantity, then the terms in ε^2 and higher powers can be neglected and an easily handled expression results. For planetary studies, however, such an approximation is not permissible since the eccentricities of most of the planets are greater than 0.04. (Table 8.1 gives some

Table 8.1

SOME PROPERTIES OF THE PRINCIPAL OBJECTS IN THE SOLAR SYSTEM

Name	Semimajor axis of orbit (in astronomical units[a])	Period (years)	Eccentricity	Mass (in units of the mass of the Earth[b])
Sun	—	—	—	333,480
Mercury	0.3871	0.2408	0.2056	0.0543
Venus	0.7233	0.6152	0.0068	0.8137
Earth	1.0000	1.0000	0.0167	1.000
Eros (asteroid)	1.4583	1.7610	0.2230	2×10^{-9} (?)
Mars	1.5237	1.8809	0.0934	0.1071
Ceres (asteroid)	c	4.6035	0.0765	1/8000 (?)
Jupiter	5.2028	c	0.0484	318.35
Saturn	9.5388	29.458	0.0557	c
Uranus	19.182	84.013	0.0472	14.58
Neptune	30.058	164.794	0.0086	17.26
Pluto	39.518	248.430	0.2486	<0.1

[a] One astronomical unit is defined as the length of the semimajor axis of the Earth's orbit. One A.U. $\cong 1.495 \times 10^{13}$ cm $\cong 93 \times 10^{6}$ miles.

[b] The mass of the Earth is approximately 5.976×10^{27} g.

[c] See Problem 8-18.

of the pertinent data regarding the major objects of the solar system.) Comets, of course, have eccentricities that are close to unity. Therefore, if Eq. (8.42) is used, it will in general be necessary to take many terms in the series in order to achieve an accuracy comparable with that of astronomical observations (which is exceedingly high!). This is at best a tedious procedure, so we wish to find a less laborious method for calculating $\theta(t)$. The solution to this problem was sought by Kepler (although, of course, he did not know the mathematical relations stated by the equations above), who devised an ingenious geometrical method to calculate the anomaly as a function of time. We shall not give his geometrical solution here, but instead we will obtain the same result by algebraic means.

Figure 8-9 shows Kepler's construction. The motion takes place in the elliptical orbit with the force center located at the focus O which is also the origin for a rectangular coordinate system. In this system the equation of the orbit is

$$\frac{(x + a\varepsilon)^2}{a^2} + \frac{y^2}{b^2} = 1 \tag{8.43}$$

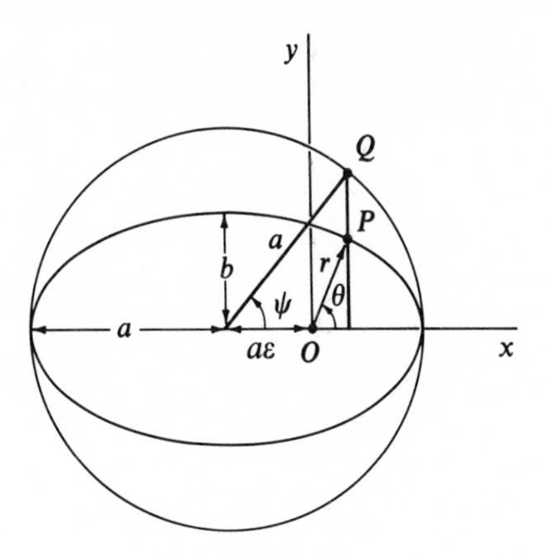

FIG. 8-9

Next, we circumscribe the ellipse with a circle of radius a and project the point P (defined by r and θ) onto the circle at point Q. The angle between the x-axis and the line connecting the center of the circle with the point Q is called the *eccentric anomaly* ψ and is defined by

$$\left.\begin{aligned} \cos\psi &= \frac{x + a\varepsilon}{a} \\ \sin\psi &= \frac{y}{b} \end{aligned}\right\} \tag{8.44}$$

From these relations we can write

$$\left.\begin{aligned} x &= a(\cos\psi - \varepsilon) \\ y &= b\sin\psi \\ &= a\sqrt{1-\varepsilon^2}\sin\psi \end{aligned}\right\} \tag{8.45}$$

Squaring these equations and adding, we find

$$\begin{aligned} r^2 &= x^2 + y^2 \\ &= a^2(1 - \varepsilon\cos\psi)^2 \end{aligned}$$

so that in terms of the eccentric anomaly, the quantity r is

$$r = a(1 - \varepsilon\cos\psi) \tag{8.46}$$

We now wish to obtain an explicit relationship between ψ and θ. First, we rewrite Eq. (8.40) as

$$\varepsilon r\cos\theta = a(1 - \varepsilon^2) - r \tag{8.47}$$

If we add εr to both sides of this equation, we have

$$\varepsilon r(1 + \cos\theta) = (1 - \varepsilon)[a(1 + \varepsilon) - r]$$

And substituting for r from Eq. (8.46) in the right-hand side of this expression, we find

$$\varepsilon r(1 + \cos\theta) = (1 - \varepsilon)[a(1 + \varepsilon) - a(1 - \varepsilon\cos\psi)]$$

or,

$$r(1 + \cos\theta) = a(1 - \varepsilon)(1 + \cos\psi) \tag{8.48a}$$

If we subtract εr from both sides of Eq. (8.47) and simplify, we obtain

$$r(1 - \cos\theta) = a(1 + \varepsilon)(1 - \cos\psi) \tag{8.48b}$$

Upon dividing Eq. (8.48b) by Eq. (8.48a), we find

$$\frac{1 - \cos\theta}{1 + \cos\theta} = \frac{1 + \varepsilon}{1 - \varepsilon} \cdot \frac{1 - \cos\psi}{1 + \cos\psi}$$

We can use the half-angle formula for the tangent to write this equation as

$$\tan\frac{\theta}{2} = \sqrt{\frac{1 + \varepsilon}{1 - \varepsilon}}\tan\frac{\psi}{2} \tag{8.49}$$

which gives ψ uniquely in terms of θ (if we confine ourselves to principal values of the tangent functions). Therefore, $\theta(t)$ can be easily obtained once $\psi(t)$ is found. In order to calculate $\psi(t)$ we can transform Eq. (8.39a) into an equation for ψ by computing the integrand in terms of ψ. Differentiating Eq. (8.49) yields

$$d\theta = \sqrt{\frac{1 + \varepsilon}{1 - \varepsilon}} \cdot \frac{\cos^2(\theta/2)}{\cos^2(\psi/2)}\,d\psi \tag{8.50}$$

From Eq. (8.48a) we can write

$$r = a(1 - \varepsilon)\frac{1 + \cos\psi}{1 + \cos\theta}$$

$$= a(1 - \varepsilon)\frac{\cos^2(\psi/2)}{\cos^2(\theta/2)} \tag{8.51}$$

where we have used the half-angle formula for the cosine functions.

In order to express $r^2\,d\theta$ in terms of ψ, we take one factor of r from Eq.

(8.46), the other factor of r from Eq. (8.51), and $d\theta$ from Eq. (8.50). Thus,

$$r^2\,d\theta = [a(1-\varepsilon\cos\psi)]\left[a(1-\varepsilon)\,\frac{\cos^2(\psi/2)}{\cos^2(\theta/2)}\right]$$

$$\cdot\left[\sqrt{\frac{1+\varepsilon}{1-\varepsilon}}\cdot\frac{\cos^2(\theta/2)}{\cos^2(\psi/2)}\,d\psi\right]$$

$$= a^2\sqrt{1-\varepsilon^2}\,(1-\varepsilon\cos\psi)\,d\psi \tag{8.52}$$

In view of this result, Eq. (8.39a) can now be written as

$$\frac{\pi ab}{\tau}\,t = \frac{a^2\sqrt{1-\varepsilon^2}}{2}\int_0^{\psi}(1-\varepsilon\cos\psi)\,d\psi$$

Integrating, and again using $b = a\sqrt{1-\varepsilon^2}$, we have the result

$$\frac{2\pi t}{\tau} = \psi - \varepsilon\sin\psi \tag{8.53}$$

The quantity $2\pi t/\tau$ is called the *mean anomaly* since it measures the angular deviation of a body moving in a circular orbit with a period τ. Following astronomical practice,* we denote the mean anomaly by M. Thus,

$$\boxed{M = \psi - \varepsilon\sin\psi} \tag{8.54}$$

This is *Kepler's equation.* In order to find $\psi(t)$, this result must be inverted by some approximation procedure. Then, since Eq. (8.49) relates ψ and θ, the time dependence of the true anomaly can be found. One of the possible approximation methods is discussed in the following section.

Kepler's equation can be used to obtain a simple expression for the velocity of a body in its orbit in terms of the magnitude of the radius vector. Referring to Fig. 8-9, we can write

$$v^2 = \dot{x}^2 + \dot{y}^2 \tag{8.55}$$

Using Eqs. (8.45) for x and y, the square of the velocity becomes

$$\begin{aligned} v^2 &= a^2\dot{\psi}^2\sin^2\psi + a^2(1-\varepsilon^2)\dot{\psi}^2\cos^2\psi \\ &= a^2\dot{\psi}^2(1-\varepsilon^2\cos^2\psi) \end{aligned} \tag{8.56}$$

If we differentiate Kepler's equation (8.53) with respect to the time, we have

$$\frac{2\pi}{\tau} = \dot{\psi}(1-\varepsilon\cos\psi) \tag{8.57}$$

* It is also customary to denote the eccentric anomaly by E and the true anomaly by v or f.

Solving this equation for $\dot{\psi}$ and substituting into Eq. (8.56), we obtain

$$\begin{aligned} v^2 &= \left(\frac{2\pi}{\tau}\right)^2 a^2 \frac{1 - \varepsilon^2 \cos^2 \psi}{(1 - \varepsilon \cos \psi)^2} \\ &= \left(\frac{2\pi}{\tau}\right)^2 a^2 \frac{1 + \varepsilon \cos \psi}{1 - \varepsilon \cos \psi} \\ &= \left(\frac{2\pi}{\tau}\right)^2 a^2 \frac{2 - (1 - \varepsilon \cos \psi)}{1 - \varepsilon \cos \psi} \end{aligned} \tag{8.58}$$

Substituting $r/a = 1 - \varepsilon \cos \psi$ from Eq. (8.46), there results

$$v^2 = \left(\frac{2\pi}{\tau}\right)^2 a^3 \left(\frac{2}{r} - \frac{1}{a}\right) \tag{8.59}$$

Finally, Kepler's Third Law [Eq. (8.38)] can be used to reduce this expression to

$$\boxed{v^2 = \frac{k}{\mu}\left(\frac{2}{r} - \frac{1}{a}\right)} \tag{8.60}$$

■ 8.9 Approximate Solution of Kepler's Equation

If we wish to calculate $\theta(t)$ for the motion of a body whose orbit has an eccentricity that is not too large ($\varepsilon \approx 0.1$, say), and if we wish to achieve an accuracy of, say 1 part in 10^6, then many terms in Eq. (8.42) will be necessary. The use of Kepler's equation in such a situation is somewhat easier. In fact, a rather simple result can be obtained which is accurate to order ε^4. A method devised by E. W. Brown* begins by setting the eccentric anomaly equal to the mean anomaly plus a correction term:

$$\psi = M + \xi \tag{8.61}$$

Then, Kepler's equation can be written as

$$\xi = \varepsilon \sin(M + \xi) \tag{8.62}$$

We can expand sin ξ according to

$$\begin{aligned} \sin \xi &= \xi - \frac{\xi^3}{3!} + \frac{\xi^5}{5!} - \cdots \\ &= \varepsilon \sin(M + \xi) - \frac{\varepsilon^3}{6} \sin^3(M + \xi) + \frac{\varepsilon^4}{120} \sin^5(M - \xi) - \cdots \end{aligned} \tag{8.63}$$

* E. W. Brown, *Monthly Notices Roy. Astron. Soc.* **92**, 104 (1931).

If we transpose the term $\varepsilon \sin(M + \xi)$ and use the expression for the sine of the sum of two angles, we find

$$\begin{aligned}(1 - \varepsilon \cos M)\sin \xi - (\varepsilon \sin M)\cos \xi \\ = -\frac{\varepsilon^3}{6}\sin^3(M + \xi) + \frac{\varepsilon^5}{120}\sin^5(M + \xi) - \cdots\end{aligned} \tag{8.64}$$

We can express the quantities ε and M in terms of two other quantities η and ξ_0 according to

$$\left.\begin{aligned}\eta \sin \xi_0 &\equiv \varepsilon \sin M \\ \eta \cos \xi_0 &\equiv 1 - \varepsilon \cos M\end{aligned}\right\} \tag{8.65}$$

Thus, ξ_0 can be calculated from

$$\xi_0 = \tan^{-1}\left(\frac{\varepsilon \sin M}{1 - \varepsilon \cos M}\right) \tag{8.66a}$$

and η is obtained from

$$\eta = \sqrt{1 - 2\varepsilon \cos M + \varepsilon^2} \tag{8.66b}$$

Substituting Eqs. (8.65) into Eq. (8.64), we find

$$\sin(\xi - \xi_0) = -\frac{\varepsilon^3}{6\eta}\sin^3(M + \xi) + \frac{\varepsilon^5}{120\eta}\sin^5(M + \xi) - \cdots \tag{8.67}$$

Now, Eq. (8.66b) indicates that η is of the order of unity, so Eq. (8.67) shows that $\xi - \xi_0$ is of the order of ε^3. Therefore, if we neglect the term proportional to ε^5 and write ξ_0 for ξ in the first term on the right-hand side of Eq. (8.67), we obtain

$$\sin(\xi - \xi_0) \cong -\frac{\varepsilon^3}{6\eta}\sin^3(M + \xi_0) \tag{8.68}$$

which is correct to the order ε^4; if $\varepsilon = 0.1$, the first neglected term is of the order of $\varepsilon^5/120 \cong 10^{-7}$. The quantity ξ can be calculated from Eq. (8.68) and then ψ is given by Eq. (8.61). Finally, $\theta(t)$ is obtained from Eq. (8.49). Although this procedure involves several steps, each is a comparatively easy calculation. Astronomical calculations of orbits are almost always based on Kepler's equation. Further details of approximation procedures for Kepler's equation can be found in various texts on celestial mechanics.*

* More than 120 methods of obtaining approximate solutions to Kepler's equation are discussed in the literature. See, for example, Moulton (Mo58, p. 164 ff) for details.

8.10 Apsidal Angles and Precession

If a particle executes bounded, noncircular motion in a central-force field, then the radial distance from the force center to the particle must always be in the range $r_{max} \geq r \geq r_{min}$; that is, r is bounded by the apsidal distances. That there are only *two* apsidal distances for bounded, noncircular motion is clear from Fig. 8-6. In executing one complete revolution in θ, however, the particle may not return to its original position (see Fig. 8-4). The angular separation between two successive values of $r = r_{max}$ will depend upon the exact nature of the force. The angle between any two consecutive apsides is called the *apsidal angle*, and since a closed orbit must be symmetrical about any apsis, it follows that all apsidal angles for such motion must be equal. The apsidal angle for elliptic motion, for example, is just π. If the orbit is not closed, the particle will reach the apsidal distances at different points in each revolution; the apsidal angle will not then be a rational fraction of 2π, as is required for a closed orbit. If the orbit is *almost* closed, the apsides will *precess* or rotate slowly in the plane of the motion. This effect is exactly anologous to the slow rotation of the elliptic motion of a two-dimensional harmonic oscillator whose natural frequencies for the x and y motions are almost equal (see Section 3.4).

Since an inverse-square-law force requires that all elliptic orbits be exactly closed, then the apsides must stay fixed in space for all time. If the apsides are found to move with time, however slowly, this indicates that the force law under which the body moves does not vary exactly as the inverse square of the distance. This important fact was realized by Newton who pointed out that any advance or regression of a planet's perihelion would require the radial dependence of the force law to be slightly different from $1/r^2$. Thus, Newton argued, the observation of the time dependence of the perihelia of the planets would be a sensitive test of the validity of the form of the universal gravitation law.

In point of fact, for planetary motion within the solar system, one expects that, due to the perturbations introduced by the existence of all of the other planets, the force experienced by any planet does not vary exactly as $1/r^2$, if r is measured from the Sun. This effect is small, however, and only slight variations of planetary perihelia have been observed. The perihelion of Mercury, for example, which shows the largest effect, advances only about 574 sec of arc per century.* Detailed calculations of the influence of the other

* This precession is in addition to the general precession of the equinox with respect to the "fixed" stars which amounts to 5025.645 ± 0.050 sec of arc per century.

planets on the motion of Mercury predict that the rate of advance of the perihelion should be approximately 531 sec of arc per century. The uncertainties in this calculation are considerably less than the difference of 43 sec between observation and calculation,*† and for a considerable time this discrepancy was the outstanding unresolved difficulty in the Newtonian theory. We now know that the modification which is introduced into the equation of motion of a planet by the general theory of relativity almost exactly accounts for the difference of 43 sec; this result is one of the major triumphs of relativity theory. We shall indicate below the way in which the advance of the perihelion can be calculated from the modified equation of motion.

In order to perform this calculation we will find it convenient to use the equation of motion in the form of Eq. (8.20). If we use the universal gravitational law for $F(r)$, we can write

$$\begin{aligned} \frac{d^2u}{d\theta^2} + u &= -\frac{m}{l^2}\frac{1}{u^2}F(u) \\ &= \frac{\gamma m^2 M}{l^2} \end{aligned} \tag{8.69}$$

where we consider the motion of a body of mass m in the gravitational field of a body of mass M. The quantity u is therefore the reciprocal of the distance between m and M.

Now, the modification of the gravitational force law required by the general theory of relativity introduces into the force a small component which varies as $1/r^4$ $(=u^4)$. Thus, we have

$$\frac{d^2u}{d\theta^2} + u = \frac{\gamma m^2 M}{l^2} + \frac{3\gamma M}{c^2}u^2 \tag{8.70}$$

where c is the velocity of propagation of the gravitational interaction and is

* In 1845, the French astronomer Urbain Jean Joseph LeVerrier (1811–1877) first called attention to the irregularity in the motion of Mercury. Similar studies by LeVerrier and by the English astronomer John Couch Adams of irregularities in the motion of Uranus led to the discovery of the planet Neptune in 1846. An interesting account of this episode is given by Turner (Tu04, Chapter 2).

† We must note, in this regard, that perturbations may be either *periodic* or *secular* (i.e., ever increasing with time). Laplace showed in 1773 (published, 1776) that any perturbation of a planet's mean motion that is caused by the attraction of another planet must be periodic in nature, although the period may be extremely long. This is the case for Mercury; the precession of 531 sec of arc per century is periodic, but the period is so long that the change from century to century is small compared to the residual effect of 43 sec.

identified with the velocity of light.* In order to simplify the notation we define

$$\frac{1}{\alpha} \equiv \frac{\gamma m^2 M}{l^2}$$
$$\delta \equiv \frac{3\gamma M}{c^2} \tag{8.71}$$

so that we can write Eq. (8.70) as

$$\frac{d^2u}{d\theta^2} + u = \frac{1}{\alpha} + \delta u^2 \tag{8.72}$$

This is a nonlinear equation and we use a successive approximation procedure in order to obtain a solution. The first trial solution is chosen to be the solution of Eq. (8.72) in the case that the term δu^2 is neglected†:

$$u_1 = \frac{1}{\alpha}(1 + \varepsilon \cos \theta) \tag{8.73}$$

which is the familiar result [cf. Eq. (8.32)] for the pure inverse-square-law-force. [Note that α is here the same as that defined in Eq. (8.31) except that μ has been replaced by m.] If we substitute this expression into the right-hand side of Eq. (8.72) we find

$$\frac{d^2u}{d\theta^2} + u = \frac{1}{\alpha} + \frac{\delta}{\alpha^2}[1 + 2\varepsilon \cos \theta + \varepsilon^2 \cos^2 \theta]$$
$$= \frac{1}{\alpha} + \frac{\delta}{\alpha^2}\left[1 + 2\varepsilon \cos \theta + \frac{\varepsilon^2}{2}(1 + \cos 2\theta)\right] \tag{8.74}$$

where $\cos^2 \theta$ has been expanded in terms of $\cos 2\theta$. Now, the first trial function u_1, when substituted into the left-hand side of Eq. (8.72), reproduces only the first term on the right-hand side, viz., $1/\alpha$. Therefore, we can construct a second trial function by adding to u_1 a term that will reproduce

* One half of the relativistic term is due to effects understandable in terms of special relativity, viz., time dilation (1/3) and the variation of mass with velocity (1/6) (the velocity is greatest at perihelion and least at aphelion). (See Chapter 10.) The other half of the term arises from general relativistic effects and may be considered to be associated with the finite propagation time of gravitational interactions. Thus, the agreement between theory and experiment constitutes a confirmation of the prediction that the gravitational propagation velocity is the same as that for light.

† We eliminate the necessity for introducing an arbitrary phase into the argument of the cosine term by choosing to measure θ from the position of perihelion; i.e., u_1 is a maximum (and, hence, r is a minimum) at $\theta = 0$.

the remainder of the right-hand side [in Eq. (8.74)]. It is easy to verify that such a particular integral is

$$u_p = \frac{\delta}{\alpha^2}\left[\left(1 + \frac{\varepsilon^2}{2}\right) + \varepsilon\theta \sin\theta - \frac{\varepsilon^2}{6}\cos 2\theta\right] \tag{8.75}$$

Therefore, the second trial function is

$$u_2 = u_1 + u_p$$

If we stop the approximation procedure at this point, we have

$$\begin{aligned} u \cong u_2 &= u_1 + u_p \\ &= \left[\frac{1}{\alpha}(1 + \varepsilon\cos\theta) + \frac{\delta\varepsilon}{\chi^2}\theta\sin\theta\right] \\ &\quad + \left[\frac{\delta}{\alpha^2}\left(1 + \frac{\varepsilon^2}{2}\right) - \frac{\delta\varepsilon^2}{6\alpha^2}\cos 2\theta\right] \end{aligned} \tag{8.76}$$

where we have regrouped the terms in u_1 and u_p.

Now, consider the terms in the second set of brackets in Eq. (8.76): the first of these is just a constant and the second is only a small and periodic disturbance of the normal Keplerian motion. Therefore, on a long time scale neither of these terms will contribute, on the average, to any change in the positions of the apsides. In the first set of brackets, however, the term proportional to θ will give rise to secular and therefore observable effects. Let us consider the first set of brackets:

$$u_{\text{secular}} = \frac{1}{\alpha}\left[1 + \varepsilon\cos\theta + \frac{\delta\varepsilon}{\alpha}\theta\sin\theta\right] \tag{8.77}$$

Next, we can expand the quantity

$$\begin{aligned} 1 + \varepsilon\cos\left(\theta - \frac{\delta}{\alpha}\theta\right) &= 1 + \varepsilon\left(\cos\theta\cos\frac{\delta}{\alpha}\theta + \sin\theta\sin\frac{\delta}{\alpha}\theta\right) \\ &\cong 1 + \varepsilon\cos\theta + \frac{\delta\varepsilon}{\alpha}\theta\sin\theta \end{aligned} \tag{8.78}$$

where we have used the fact that δ is small in order to approximate

$$\cos\frac{\delta}{\alpha}\theta \cong 1; \qquad \sin\frac{\delta}{\alpha}\theta \cong \frac{\delta}{\alpha}\theta$$

Hence, we can write u_{secular} as

$$u_{\text{secular}} \cong \frac{1}{\alpha}\left[1 + \varepsilon\cos\left(\theta - \frac{\delta}{\alpha}\theta\right)\right] \tag{8.79}$$

We have chosen to measure θ from the position of perihelion at $t = 0$. Successive appearances at perihelion will result when the argument of the cosine term in u_{secular} has increased to 2π, $4\pi, \ldots,$ etc. But an increase of the argument by 2π requires that

$$\theta - \frac{\delta}{\alpha}\theta = 2\pi$$

or,

$$\theta = \frac{2\pi}{1 - (\delta/\alpha)} \cong 2\pi\left(1 + \frac{\delta}{\alpha}\right)$$

Therefore, the effect of the relativistic term in the force law is to displace the perihelion, in each revolution, by an amount

$$\Delta \cong \frac{2\pi\delta}{\alpha} \tag{8.80}$$

That is, the apsides rotate slowly in space. If we refer to the definitions of α and δ [Eqs. (8.71)], we find

$$\Delta \cong 6\pi\left(\frac{\gamma m M}{cl}\right)^2 \tag{8.80a}$$

From Eqs. (10.31) and (10.33a) we can write $l^2 = \mu k a(1 - \varepsilon^2)$; then since $k = \gamma m M$ and $\mu \cong m$, we have

$$\boxed{\Delta \cong \frac{6\pi\gamma M}{ac^2(1 - \varepsilon^2)}} \tag{8.80b}$$

We see therefore that the effect is enhanced if the semimajor axis a is small and if the eccentricity is large. Therefore, Mercury, which is the planet nearest the Sun and which has the most eccentric orbit of any planet (except Pluto), provides the most sensitive test of the theory.* The calculated value of the precessional rate for Mercury is 43.03 ± 0.03 seconds of arc per century. The observed value (corrected for the influence of the other planets) is 43.11 ± 0.45 seconds,† so that the prediction of relativity theory is confirmed in striking fashion. The precessional rates for some of the planets are given in Table 8.2.

* Alternatively, we can say that the relativistic advance of the perihelion is a maximum for Mercury because of the fact that the orbital velocity is greatest for Mercury so that the relativistic parameter v/c is largest (see Chapter 10 and Problem 10-14).

† R. L. Duncombe, *Astron. J.* **61**, 174 (1956); see also G. M. Clemence, *Rev. Mod. Phys.* **19**, 361 (1947).

Table 8.2

PRECESSIONAL RATES FOR THE PERIHELIA OF SOME PLANETS

Planet	Precessional rate (seconds of arc/century)	
	Calculated	Observed
Mercury	43.03 ± 0.03	43.11 ± 0.45
Venus	8.63	8.4 ± 4.8
Earth	3.84	5.0 ± 1.2
Mars	1.35	—
Jupiter	0.06	—

8.11 Stability of Circular Orbits

In Section 8.6 it was pointed out that the orbit will be *circular* in the event that the total energy equals the minimum value of the effective potential energy, $E = V_{\min}$. More generally, however, it is true that a circular orbit will be allowed for *any* attractive potential since it is clear that the attractive force can *always* be made to just balance the centrifugal force by the proper choice of radial velocity. Although circular orbits are therefore always possible in a central, attractive force field, such orbits are not necessarily stable. A circular orbit at $r = \rho$ will exist if $\dot{r}|_{r=\rho} = 0$ for all t; this will be possible if $(\partial V/\partial r)|_{r=\rho} = 0$. But only in the event that the effective potential has a *true minimum* will stability result; all other equilibrium circular orbits will be unstable.

Let us consider an attractive central force which has the form

$$F(r) = -\frac{k}{r^n} \tag{8.81}$$

The potential function for such a force is therefore

$$U(r) = -\frac{k}{n-1} \cdot \frac{1}{r^{(n-1)}} \tag{8.82}$$

and the effective potential function is

$$V(r) = -\frac{k}{n-1} \cdot \frac{1}{r^{(n-1)}} + \frac{l^2}{2\mu r^2} \tag{8.83}$$

The conditions for a minimum of $V(r)$ and hence for a stable circular orbit with a radius ρ are

$$\left.\frac{\partial V}{\partial r}\right|_{r=\rho} = 0 \qquad \text{and} \qquad \left.\frac{\partial^2 V}{\partial r^2}\right|_{r=\rho} > 0 \tag{8.84}$$

Applying these criteria to the effective potential of Eq. (8.83), we have

$$\left.\frac{\partial V}{\partial r}\right|_{r=\rho} = \frac{k}{\rho^n} - \frac{l^2}{\mu\rho^3} = 0$$

or,

$$\rho^{(n-3)} = \frac{\mu k}{l^2} \tag{8.85}$$

and,

$$\left.\frac{\partial^2 V}{\partial r^2}\right|_{r=\rho} = -\frac{nk}{\rho^{(n+1)}} + \frac{3l^2}{\mu\rho^4} > 0$$

so that

$$-\frac{nk}{\rho^{(n-3)}} + \frac{3l^2}{\mu} > 0 \tag{8.86}$$

Substituting $\rho^{(n-3)}$ from Eq. (8.85) into (8.86) we have

$$(3-n)\frac{l^2}{\mu} > 0 \tag{8.87}$$

Thus, the condition that a stable circular orbit exist is that $n < 3$.

Next, we apply a more general procedure and inquire as to the frequency of oscillation about a circular orbit in a general force field. We write the force as

$$F(r) = -\mu g(r) = -\frac{\partial U}{\partial r} \tag{8.88}$$

Then, Eq. (8.18) can be written as

$$\ddot{r} - r\dot{\theta}^2 = -g(r) \tag{8.89}$$

Substituting for $\dot{\theta}$ from Eq. (8.10),

$$\ddot{r} - \frac{l^2}{\mu^2 r^3} = -g(r) \tag{8.90}$$

We now consider the particle to be initially in a circular orbit with radius ρ and apply a perturbation of the form $r \to \rho + x$ where x is a small quantity.

Since $\rho =$ const., we also have $\ddot{r} \rightarrow \ddot{x}$. Thus,

$$\ddot{x} - \frac{l^2}{\mu^2 \rho^3 [1 + (x/\rho)]^3} = -g(\rho + x) \tag{8.91}$$

But, by hypothesis $(x/\rho) \ll 1$, so that we can expand the quantity

$$[1 + (x/\rho)]^{-3} = 1 - 3(x/\rho) + \cdots \tag{8.92}$$

Now, we also assume that $g(r) = g(\rho + x)$ can be expanded in a Taylor series about the point $r = \rho$:

$$g(\rho + x) = g(\rho) + xg'(\rho) + \cdots \tag{8.93}$$

where

$$g'(\rho) \equiv \left. \frac{dg}{dr} \right|_{r=\rho}$$

If we neglect all terms in x^2 and higher powers, then the substitution of Eqs. (8.92) and (8.93) into Eq. (8.91) yields

$$\ddot{x} - \frac{l^2}{\mu^2 \rho^3} [1 - 3(x/\rho)] \cong -[g(\rho) + xg'(\rho)] \tag{8.94}$$

Recall that we assumed the particle to be initially in a circular orbit with $r = \rho$. Under such a condition there is no radial motion, i.e., $\dot{r}|_{r=\rho} = 0$. Then, also, $\ddot{r}|_{r=\rho} = 0$. Therefore, evaluating Eq. (8.90) at $r = \rho$ we have

$$g(\rho) = \frac{l^2}{\mu^2 \rho^3} \tag{8.95}$$

Substituting this relation into Eq. (8.94), we have, approximately,

$$\ddot{x} - g(\rho)[1 - 3(x/\rho)] \cong -[g(\rho) + xg'(\rho)]$$

or,

$$\ddot{x} + \left[\frac{3g(\rho)}{\rho} + g'(\rho) \right] x \cong 0 \tag{8.96}$$

If we define

$$\omega_0^2 \equiv \frac{3g(\rho)}{\rho} + g'(\rho) \tag{8.97}$$

then Eq. (8.96) becomes the familiar equation for the undamped harmonic oscillator:

$$\ddot{x} + \omega_0^2 x = 0 \tag{8.98}$$

The solution to this equation is

$$x(t) = Ae^{+i\omega_0 t} + Be^{-i\omega_0 t} \tag{8.99}$$

If $\omega_0^2 < 0$ so that ω_0 is imaginary, then the second term becomes $B\exp(|\omega_0|t)$, which clearly increases without limit as time increases. Therefore, the condition for oscillation is that $\omega_0^2 > 0$, or

$$\frac{3g(\rho)}{\rho} + g'(\rho) > 0 \tag{8.100}$$

Since $g(\rho) > 0$ [cf. Eq. (8.95)], we can divide through by $g(\rho)$ and write this inequality as

$$\frac{g'(\rho)}{g(\rho)} + \frac{3}{\rho} > 0 \tag{8.100a}$$

or, since $g(r)$ and $F(r)$ are related by a constant multiplicative factor, stability will result if

$$\boxed{\frac{F'(\rho)}{F(\rho)} + \frac{3}{\rho} > 0} \tag{8.101}$$

We now compare the condition on the force law imposed by Eq. (8.101) with that previously obtained for a power-law force:

$$F(r) = -\frac{k}{r^n} \tag{8.102}$$

Then, Eq. (8.101) becomes

$$\frac{nk\rho^{-(n+1)}}{-k\rho^{-n}} + \frac{3}{\rho} > 0$$

or,

$$(3 - n) \cdot \frac{1}{\rho} > 0 \tag{8.103}$$

and we are led to the same condition as before, viz., $n < 3$. (We must note, however, that the case $n = 3$ needs further examination; see Problem 8-23.)

It should be noted that the expansions in powers of x that were made in Eqs. (8.92) and (8.93) are another application of the method of perturbations which was first discussed in Section 5.5. We therefore have here another example of the power and elegance of this very important technique.

▶ Example 8.11(a) Stability in a Screened Potential

Let us investigate the stability of circular orbits in a force field described by the potential function

$$U(r) = -\frac{k}{r} e^{-(r/a)} \tag{1}$$

where $k > 0$ and $a > 0$. This is called the *screened Coulomb potential* (when $k = Ze^2$, where Z is the atomic number and e is the electronic charge) since it falls off with distance more rapidly than $1/r$ and hence approximates the electrostatic potential of the atomic nucleus in the vicinity of the nucleus by taking into account the partial "cancellation" or "screening" of the nuclear charge by the atomic electrons. Then,

$$F(r) = -\frac{\partial U}{\partial r}$$

$$= -k\left[\frac{1}{ar} + \frac{1}{r^2}\right] e^{-(r/a)} \tag{2}$$

and,

$$\frac{\partial F}{\partial r} = k\left[\frac{1}{a^2 r} + \frac{2}{ar^2} + \frac{2}{r^3}\right] e^{-(r/a)} \tag{3}$$

The condition for stability is [cf. Eq. (8.101)]

$$3 + \rho \frac{F'(\rho)}{F(\rho)} > 0 \tag{4}$$

Therefore,

$$3 + \frac{\rho k\left[\dfrac{1}{a^2\rho} + \dfrac{2}{a\rho^2} + \dfrac{2}{\rho^3}\right]}{-k\left[\dfrac{1}{a\rho} + \dfrac{1}{\rho^2}\right]} > 0 \tag{5}$$

which simplifies to

$$a^2 + a\rho - \rho^2 > 0 \tag{6}$$

We may write this as

$$\frac{a^2}{\rho^2} + \frac{a}{\rho} - 1 > 0 \tag{7}$$

Hence, stability will result for all $q \equiv a/\rho$ which exceed the value that satisfies the equation

$$q^2 + q - 1 = 0 \tag{8}$$

The positive (and therefore the only physically meaningful) solution is

$$q = \tfrac{1}{2}(\sqrt{5} - 1) \cong 0.62 \tag{9}$$

and therefore if the angular momentum and energy are such as to allow a circular orbit at $r = \rho$, the motion will be stable if

$$\frac{a}{\rho} \gtrsim 0.62$$

or,

$$\rho \lesssim 1.62\, a \tag{10}$$

The stability condition for orbits in a screened potential is illustrated graphically in Fig. 8.10 which shows the potential $V(r)$ for various values of ρ/a. The force constant k is the same for all of the curves, but $l^2/2\mu$ has been adjusted to maintain the minimum of the potential at the same value of the radius as a is changed. It is apparent that for $\rho/a < 1.62$, there is a true minimum for the potential indicating that the circular orbit is stable with respect to small oscillations. For $\rho/a > 1.62$ there is no minimum so that circular orbits cannot exist. For $\rho/a = 1.62$ the potential has zero slope at the position which a circular orbit would occupy. The orbit is unstable at this position since ω_0^2 is zero in Eq. (8.98) and the displacement x would increase linearly with time.

An interesting feature of this potential function is the fact that under certain conditions there can exist bound orbits for which the total energy is positive. (See, for example, curve 4 in Fig. 8-10.) In all the cases which we have discussed previously, the potential increased monotonically from its minimum value toward $V = 0$ as r increased toward infinity. Figure 8-10 indicates, however, that as the minimum in the screened potential becomes more shallow (due to a decrease in the angular momentum l), a point is reached at which the potential crosses into the positive energy region, thus allowing bound orbits with $E > 0$. In any event, we have, of course, that $V \to 0$ as $r \to \infty$.

▶ Example 8.11(b) Stable Circular Orbits on the Surface of a Cone

Let us return to Example 7.4 of the particle constrained to move on the surface of a cone. We found that the angular momentum about the z-axis was a constant of the motion:

$$l = mr^2\dot{\theta} = \text{const.} \tag{1}$$

We also found the equation of motion for the coordinate r:

$$\ddot{r} - r\dot{\theta}^2 \sin^2\alpha - g \sin\alpha \cos\alpha = 0 \tag{2}$$

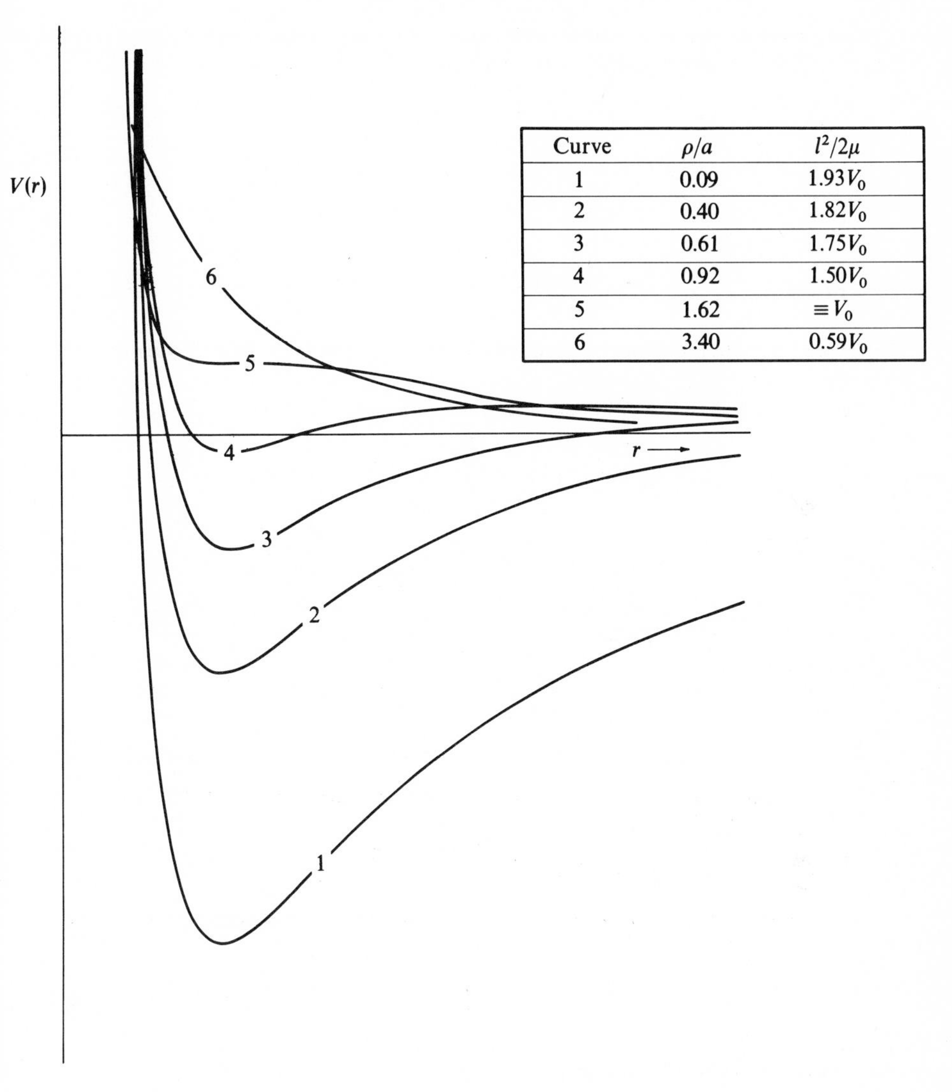

Curve	ρ/a	$l^2/2\mu$
1	0.09	$1.93V_0$
2	0.40	$1.82V_0$
3	0.61	$1.75V_0$
4	0.92	$1.50V_0$
5	1.62	$\equiv V_0$
6	3.40	$0.59V_0$

FIG. **8-10**

It is clear that if the initial conditions are appropriately selected, the particle can move in a circular orbit about the vertical axis with the plane of the orbit at a constant height z_0 above the horizontal plane passing through the apex of the cone. Although this problem does not involve a central force, certain aspects of the motion are the same as for the central-force case. Thus, we may discuss, for example, the stability of circular orbits for the particle. In order to do this, we perform a perturbation calculation.

First, we assume that a circular orbit exists for $r = \rho$. Then we apply the perturbation $r \to \rho + x$. The quantity $r\dot{\theta}^2$ which appears in Eq. (2) can be expressed as

$$\begin{aligned} r\dot{\theta}^2 &= r \cdot \frac{l^2}{m^2 r^4} = \frac{l^2}{m^2 r^3} \\ &= \frac{l^2}{m^2}(\rho + x)^{-3} = \frac{l^2}{m^2 \rho^3}\left(1 + \frac{x}{\rho}\right)^{-3} \\ &\cong \frac{l^2}{m^2 \rho^3}\left(1 - 3\frac{x}{\rho}\right) \end{aligned} \tag{3}$$

where we have retained only the first term in the expansion since x/ρ is by hypothesis a small quantity.

Then, since $\ddot{\rho} = 0$, Eq. (2) becomes, approximately,

$$\ddot{x} - \frac{l^2 \sin^2\alpha}{m^2 \rho^3}\left(1 - 3\frac{x}{\rho}\right) + g \sin\alpha \cos\alpha = 0 \tag{4}$$

or,

$$\ddot{x} + \left(\frac{3l^2 \sin^2\alpha}{m^2 \rho^4}\right)x - \frac{l^2 \sin\alpha}{m^2 \rho^3} + g \sin\alpha \cos\alpha = 0 \tag{5}$$

If we evaluate Eq. (2) at $r = \rho$, then $\ddot{r} = 0$, and we have

$$\begin{aligned} g \sin\alpha \cos\alpha &= \rho\dot{\theta}^2 \sin^2\alpha \\ &= \frac{l^2}{m^2 \rho^3} \sin^2\alpha \end{aligned} \tag{6}$$

In view of this result, the last two terms in Eq. (5) cancel, and there remains

$$\ddot{x} + \left(\frac{3l^2 \sin^2\alpha}{m^2 \rho^4}\right)x = 0 \tag{7}$$

The solution to this equation is just a harmonic oscillation with a frequency ω, where

$$\omega = \frac{\sqrt{3}\, l}{m\rho^2} \sin\alpha \tag{8}$$

Thus, the circular orbit is stable.

■ 8.12 The Problem of Three Bodies

In Sections 8.4 and 8.5 it was shown that the motion of two bodies which interact *via* central forces and which are subject to no external forces is completely soluble. If the interaction force is not of a simple power-law form, then the equations of motion may not be expressible in terms of simple functions, but nevertheless the motion can always be given in terms of an integral [cf. Eq. (8.17)] which can in principle be solved. The addition of a third body to the system, however, in general renders the problem insoluble in finite terms by means of any elementary functions. Thus, the problem of the motion of three bodies mutually interacting through gravitational forces is still unsolved after over two hundred years of study by a most distinguished list of mathematicians and physicists.* In spite of the insolubility of the general problem, certain aspects of the three-body system are still subject to analysis.† For example, if the initial conditions are such that the velocity vectors of the three bodies all lie in the plane defined by the bodies, then the motion will always be in that plane. The *restricted* problem of three gravitating bodies treats the case in which the mass of one of the bodies‡ is negligibly small compared to either of the other two masses, and in which the two larger masses move in circular orbits about their common center of mass. Thus, the small mass is assumed not to disturb the motion of the larger masses. We wish to treat here only the stability aspects of the restricted problem of three bodies; i.e., we wish to discover at what points the planetoid (of mass m) can be placed in the field of the larger bodies (of masses M_1 and M_2) so that it is in a position of equilibrium.

The masses M_1 and M_2 are assumed to be executing circular orbits about their center of mass O as in Fig. 8-11. Let the angular velocity of this motion be ω. Then if m is to be in a position of equilibrium, it must also be revolving about O with a constant angular velocity ω. The mass M_1 at A exerts a force $\mathbf{F}_1$ on m which is located at the point $P(x, y)$, and the mass M_2 at B exerts a force $\mathbf{F}_2$. Thus, in order for m to be in equilibrium with respect to M_1 and M_2, the centrifugal force $mr\omega^2$ must be balanced by the gravitational forces $\mathbf{F}_1$ and $\mathbf{F}_2$; that is,

$$mr\omega^2\mathbf{e}_r + \mathbf{F}_1 + \mathbf{F}_2 = 0 \tag{8.104}$$

If we let $M_1 + M_2 \equiv M$, and if M_1 and M_2 are separated by a distance s, then M_1 lies at a distance $(M_2/M)s$ from the center of mass O, and M_2 lies at a distance $[1 - (M_2/M)]s$ from O. Let the ratio M_2/M be p so that

$$\left.\begin{aligned} \overline{OB} &= (1 - p)s \\ \overline{OA} &= ps \end{aligned}\right\} \tag{8.105}$$

* The names of Lagrange, Laplace, Jacobi, and particularly Poincaré are associated with the efforts to solve the three-body system. Much valuable work was done by K. F. Sundman in the early 1900's.

† Perturbation calculations can always be used to numerically determine the effects introduced by a third body or even many bodies.

‡ Historically referred to as the *planetoid*, but recently more frequently identified with an *artificial satellite*.

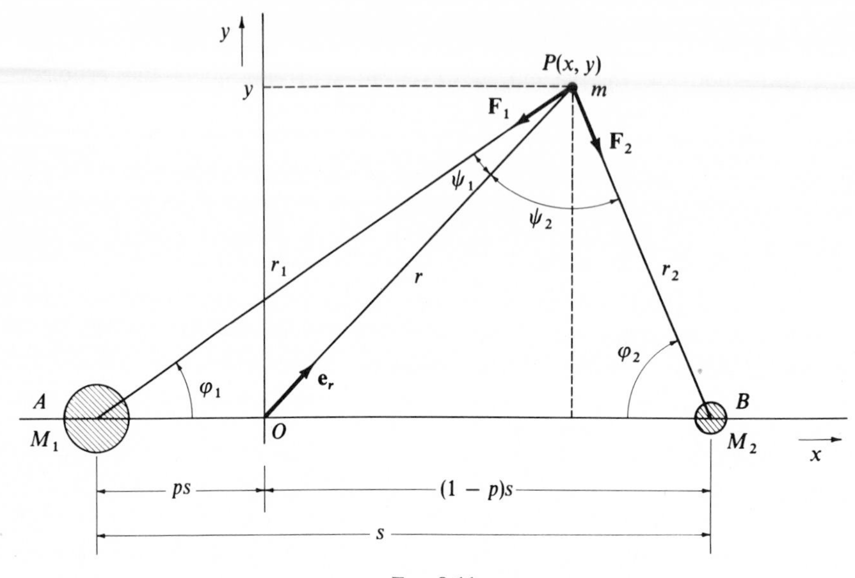

FIG. **8-11**

Now, the magnitudes of the forces $\mathbf{F}_1$ and $\mathbf{F}_2$ are

$$\left.\begin{aligned} F_1 &= \frac{\gamma m M_1}{r_1^2} = \frac{\gamma m M(1-p)}{r_1^2} \\ F_2 &= \frac{\gamma m M_2}{r_2^2} = \frac{\gamma m M p}{r_2^2} \end{aligned}\right\} \tag{8.106}$$

where γ is the gravitational constant. In order for m to be in equilibrium, the components of $\mathbf{F}_1$ and $\mathbf{F}_2$ perpendicular to $\mathbf{e}_r$ must be equal, and the sum of the components along $\mathbf{e}_r$ must be equal to $mr\omega^2$. Clearly, then, if m is to be in equilibrium, it must lie in the plane of motion of M_1 and M_2, for otherwise there would be components of $\mathbf{F}_1$ and $\mathbf{F}_2$ that could not be balanced by the centrifugal force. Let us first consider the components of the force perpendicular to $\mathbf{e}_r$:

$$\frac{F_1}{F_2} = \frac{\sin \psi_2}{\sin \psi_1} \tag{8.107}$$

Applying the law of sines to be the triangle OAP, we have

$$\frac{\sin \varphi_1}{r} = \frac{\sin \psi_1}{ps} \tag{8.108}$$

But we also have

$$\sin \varphi_1 = \frac{y}{r_1} \tag{8.109}$$

Therefore, combining Eqs. (8.108) and (8.109), we find

$$\sin \psi_1 = \frac{psy}{rr_1} \tag{8.110a}$$

And similarly,

$$\sin \psi_2 = \frac{(1-p)sy}{rr_2} \tag{8.110b}$$

If we write Eq. (8.107) by expressing F_1/F_2 from Eqs. (8.106) and $\sin \psi_2/\sin \psi_1$ from Eqs. (8.110), we obtain

$$\frac{\gamma mM(1-p)/r_1^2}{\gamma mMp/r_2^2} = \frac{(1-p)sy/rr_2}{psy/rr_1}$$

and upon simplifying, we find

$$\frac{y}{r_1^3} = \frac{y}{r_2^3} \tag{8.111}$$

We therefore have two possible (finite) solutions: (a) $r_1 = r_2 \equiv \rho$, and (b) $y = 0$.* For case (a) the three masses form an isosceles triangle, and for case (b) the three masses are co-linear.

Let us first consider solution (a) and calculate the value of the common distance ρ. In order to do this, we must equate the centrifugal force to the sum of the components of $\mathbf{F}_1$ and $\mathbf{F}_2$ along $\mathbf{e}_r$:

$$mr\omega^2 = F_1 \cos \psi_1 + F_2 \cos \psi_2 \tag{8.112}$$

First, we apply the cosine law to triangle OAP with the result

$$\cos \psi_1 = \frac{r^2 + \rho^2 - p^2 s^2}{2r\rho} \tag{8.113a}$$

and for $\cos \psi_2$ from triangle OBP we find

$$\cos \psi_2 = \frac{r^2 + \rho^2 - (1-\rho)^2 s^2}{2r\rho} \tag{8.113b}$$

Next, we need to calculate the value of ω^2 in terms of the other parameters

* Note that the possible solution $r = 0$ is included in the class of solutions $y = 0$. Note also that $r = 0$ will be a position of equilibrium only in the special case $M_1 = M_2$.

of the problem. We can do this by equating the gravitational and centrifugal forces on, say, M_1:

$$\frac{\gamma M_1 M_2}{s^2} = M_1 p s \omega^2$$

or,

$$\omega^2 = \frac{\gamma M_2}{p s^2} = \frac{\gamma M}{s^3} \tag{8.114}$$

where we have used $p = M_2/M$. Substituting Eqs. (8.106) for F_1 and F_2, Eqs. (8.113) for $\cos\psi_1$ and $\cos\psi_2$, and Eq. (8.114) for ω^2 into Eq. (8.112), we find

$$mr\left(\frac{\gamma M}{s^2}\right) = \left(\frac{\gamma m M(1-p)}{\rho^2}\right)\left(\frac{r^2+\rho^2-p^2s^2}{2r\rho}\right) + \left(\frac{\gamma m M p}{\rho^2}\right)\left(\frac{r^2+\rho^2-(1-p)^2s^2}{2r\rho}\right)$$

which simplifies to

$$\rho^3 = \frac{r^2+\rho^2-ps^2+p^2s^2}{2r^2}\cdot s^3 \tag{8.115}$$

Now, from triangle OAP, we can write

$$r^2 = \rho^2 + p^2s^2 - 2\rho p s \cos\varphi_1 \tag{8.116}$$

But we also have

$$\cos\varphi_1 = \frac{x+ps}{\rho} \tag{8.117}$$

And since r_1 and r_2 are equal, the value of x must correspond to the midpoint of the line $\overline{AB}$. Thus,

$$x = \frac{s}{2} - ps \tag{8.118}$$

so that

$$\cos\varphi_1 = \frac{s}{2\rho} \tag{8.119}$$

and then

$$r^2 = \rho^2 + p^2s^2 - ps^2 \tag{8.120}$$

Using this result in Eq. (8.115), we obtain $\rho^3 = s^3$, from which we conclude

$$\rho = s \tag{8.121}$$

Therefore, solution (a) states that equilibrium will result if the mass m is placed in the plane of motion of M_1 and M_2 so that PAB is an *equilateral* triangle. It can further be shown that such an equilibrium point is *stable* if $M_2 < 0.0385M$.* Thus, there are two positions of stable equilibrium (for M_2 sufficiently small), designated by S_1 and S_2 in Fig. 8-12.

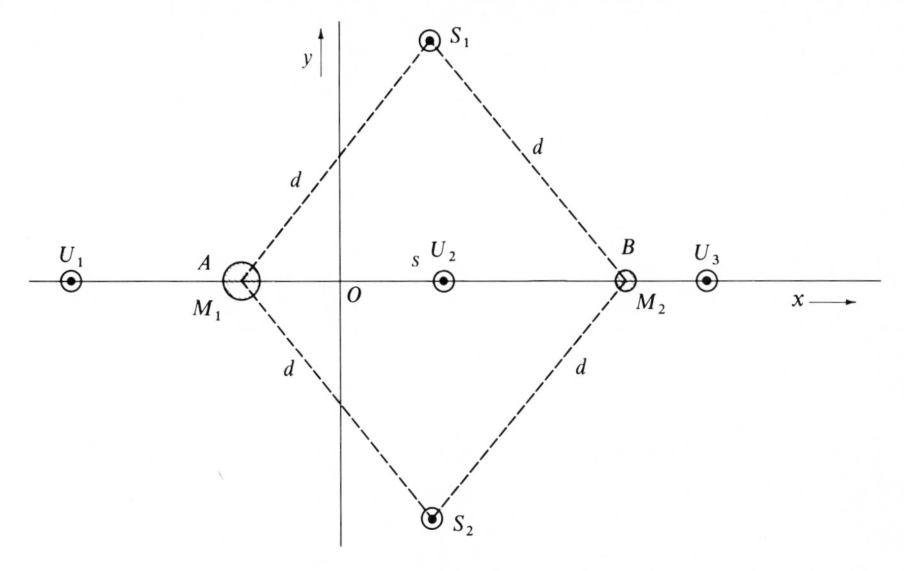

FIG. **8-12**

For solution (b) we have $y = 0$ so that all three masses must lie on the x-axis. In Fig. 8-13 we plot the radial forces on m due to M_1 and M_2, and due to the centrifugal force. In the region between the two large masses the gravitational forces combine as shown. The centrifugal force increases linearly with r and so is represented by a straight line. Note that the *negative* of the centrifugal force is plotted, so that at those points for which the gravitational and centrifugal curves intersect, the net force on the planetoid is zero and the position is one of equilibrium. There are three such positions of equilibrium, and they can be shown to be unstable; they are represented by U_1, U_2, and U_3 in Figs. 8-12 and 8-13.

* This is the case, for example, in the Earth-Moon system, or in the Jupiter-Sun system, etc. Some details regarding the question of stability are given by Symon (Sy60, pp. 500–509) and in various texts on celestial mechanics: Danby (Da62, pp. 194–198), McCuskey (Mc63, pp. 118–126).

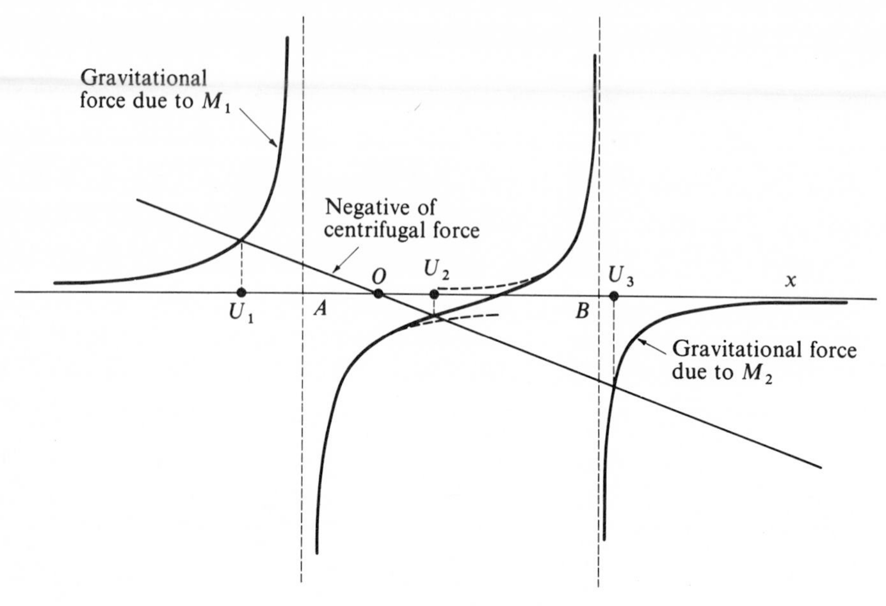

FIG. **8-13**

The calculational procedure that is necessary in order to solve for the values of r corresponding to the unstable points is straightforward but quite complicated since inversions of series expansions are required; the details of the results themselves are not very illuminating except in some special cases (see, for example, Problem 8-30).

We may summarize the situation as follows: If a planetoid of negligible mass is introduced into a system which consists of two large masses executing circular orbits about their common center of mass, then there are *five* points of equilibrium for the planetoid, all of which lie in the plane of the motion of the larger masses. Two of these points are *stable* (for M_2 sufficiently small) and are the so-called *equilateral points*. The other three points are *co-linear* with the larger masses, lying, respectively, outside of one or the other mass and between the two masses.* These latter points are positions of *unstable* equilibrium. For a real physical case, such as the Earth-Moon system in which the larger masses do not move precisely in circular orbits and in which there are perturbations due to the Sun and other planets, it is not known whether the equilateral points are in fact stable.

* These solutions to this special case of the three-body problem were discovered by Lagrange in 1772; they are known, respectively, as *Lagrange's equidistant particles* and *Lagrange's co-linear particles*.

It is interesting that there is a natural example in the solar system of the stability (or, at least, quasi-stability) of the equilateral points. It is well known that between the orbits of Mars and Jupiter there exists large numbers of asteroids and these objects tend to cluster in two positions which are the equilateral points of the Jupiter-Sun system (the "Trojan" group*). Some disturbance of these clusters results from the perturbations caused by other members of the solar system (primarily Saturn), but the general configuration persists since the equilateral points are stable positions, or at least are nearly so.

Suggested References

The central-force problem and planetary motion are treated with varying degrees of completeness in most texts. At the intermediate level, see Becker (Be54, Chapter 10), Bradbury (Br68, Chapter 9), Hauser (Ha65, Chapter 7), Lindsay (Li61, Sections 3.6–3.10), McCuskey (Mc59, Chapter 3), Slater and Frank (Sl47, Chapter 3), Symon (Sy60, Chapter 3), and Wangsness (Wa63, Chapter 8).

Many texts on advanced dynamics give only brief discussions of central-force motion; an exception is Goldstein (Go50, Chapter 3), which is quite detailed. Shorter treatments are those of Corben and Stehle (Co60, Sections 36 and 37) and of Landau and Lifshitz (La60, Sections 14 and 15). Whittaker (Wh37, Chapter 4), as usual, presents an exhaustive account.

The primary application of the theory of central-force motion is, of course, in astronomy. Texts on celestial mechanics therefore present many details not usually found in the standard physics texts. A good introductory account presented from a modern viewpoint is given by van de Kamp (Ka63). At the intermediate level, see, for example, Danby (Da62) or McCuskey (Mc63). Somewhat more advanced treatments are given by Brouwer and Clemence (Br61) and by Smart (Sm53). A standard treatise, which is highly mathematical, is Wintner (Wi41). Modern problems in astrodynamics are discussed, for example, by Baker and Makemson (Ba60).

The restricted three-body problem is discussed by Danby (Da62, Chapter 8), McCuskey (Mc63, Chapter 5), Symon (Sy60, pp. 285–290, 500–509), Whittaker (Wh37, Chapter 13), and Wintner (Wi41, Chapter 6).

The precession of the perihelion of Mercury is treated in Bergmann (Be46, pp. 212–218), Danby (Da62, pp. 66–67), and Weber (We61, pp. 64–67), as well as in most texts on relativity.

Background material in general astronomy may be found, for example, in Baker (Ba59), Blanco and McCuskey (Bl61), and McLaughlin (Mc61). An interesting account of the historical development of the theory of planetary motion is given by Feather (Fe59, Chapter 7). Sketches of some important astronomical discoveries are given by Turner (Tu04).

* It was not until 1906 that the first of these asteroids was discovered. Prior to this time Lagrange's solution to the three-body problem was considered to be only of academic interest.

Problems

8-1. In Section 8.2 it was shown that the motion of two bodies which interact only with each other via central forces can be reduced to an equivalent one-body problem. Show by explicit calculation that such a reduction is also possible for the case in which the bodies move in an external uniform gravitational field.

8-2. Perform the integration of Eq. (8.29) to obtain Eq. (8.30).

8-3. A particle moves in a circular orbit in a force field given by

$$F(r) = -k/r^2$$

If suddenly k decreases to half its original value, show that the particle's orbit becomes parabolic.

8-4. Perform an explicit calculation of the time average (i.e., the average over one complete period) of the potential energy for a particle moving in an elliptic orbit in a central, inverse-square-law force field. Express the result in terms of the force constant of the field and the semimajor axis of the ellipse. Perform a similar calculation for the kinetic energy. Compare the results and thereby verify the virial theorem for this case.

8-5. Two particles moving under the influence of their mutual gravitational force describe circular orbits about one another with a period τ. If they are suddenly stopped in their orbits and allowed to gravitate toward each other, show that they will collide after a time $\tau/4\sqrt{2}$.

8-6. Two gravitating masses m_1 and m_2 ($m_1 + m_2 = M$) are separated by a distance r_0 and released from rest. Show that when the separation is $r(<r_0)$, the velocities are

$$v_1 = m_2\sqrt{\frac{2\gamma}{M}\left(\frac{1}{r} - \frac{1}{r_0}\right)}; \qquad v_2 = m_1\sqrt{\frac{2\gamma}{M}\left(\frac{1}{r} - \frac{1}{r_0}\right)}$$

8-7. Show that the areal velocity is constant for a particle moving under the influence of an attractive force given by $F(r) = -kr$. Calculate the time averages of the kinetic and potential energies and compare with the results of the virial theorem.

8-8. Investigate the motion of a particle which is *repelled* by a force center according to the law $F(r) = kr$. Show that the orbit can only be hyperbolic.

8-9. A particle moves under the influence of a central force given by $F(r) = -k/r^n$. If the particle's orbit is circular and passes through the force center, show that $n = 5$.

8-10. Consider a comet which moves in a parabolic orbit in the plane of the Earth's orbit. If the distance of closest approach of the comet to the Sun is βr_e, where r_e is the radius of the Earth's (assumed) circular orbit and where $\beta < 1$, show that the length of time which the comet spends within the orbit of the Earth is given by

$$\sqrt{2(1-\beta)} \cdot (1+2\beta)/3\pi \times 1 \text{ year}$$

If the comet approaches the Sun to the distance of the perihelion of Mercury, how many days is it within the Earth's orbit?

8-11. Discuss the motion of a particle in a central inverse-square-law force field for the case in which there is a superimposed force whose magnitude is inversely proportional to the cube of the distance from the particle to the force center. That is,

$$F(r) = -\frac{k}{r^2} - \frac{\lambda}{r^3}, \qquad k, \lambda > 0$$

Show that the motion is described by a precessing ellipse. Consider the cases $\lambda < l^2/\mu$, $\lambda = l^2/\mu$, and $\lambda > l^2/\mu$.

8-12. Find the force law for a central-force field which allows a particle to move in a spiral orbit given by $r = k\theta^2$, where k is a constant.

8-13. Find the force law for a central-force field which allows a particle to move in a logarithmic spiral orbit given by $r = ke^{\alpha\theta}$, where k and α are constants.

8-14. A particle of unit mass moves from infinity along a straight line which, if continued, would allow it to pass a distance $b\sqrt{2}$ from a point P. If the particle is attracted toward P with a force which varies as k/r^5, and if the angular momentum about the point P is $\sqrt{k}/b$, show that the trajectory is given by

$$r = b \coth(\theta/\sqrt{2})$$

8-15. A particle executes elliptic (but almost circular) motion about a force center. At some point in the orbit a *tangential* impulse is applied to the particle, changing the velocity from v to $v + \delta v$. Show that the resulting relative change in the major and minor axes of the orbit is twice the relative change in the velocity and that the axes are *increased* if $\delta v < 0$.

8-16. A particle moves in an elliptic orbit in an inverse-square-law central-force field. If the ratio of the maximum angular velocity to the minimum angular velocity of the particle in its orbit is n, then show that the eccentricity of the orbit is

$$\varepsilon = \frac{\sqrt{n} - 1}{\sqrt{n} + 1}$$

8-17. Use the results of Kepler (i.e., his first and second laws) to show that the gravitational force must be central and that the radial dependence must be $1/r^2$. Thus, perform an inductive derivation of the gravitational force law.

8-18. Calculate the missing entries of Table 8.1.

8-19. Show that the product of the maximum and minimum (linear) velocities of a body moving in an elliptic orbit is $(2\pi a/\tau)^2$.

8-20. If η is defined as the angle between the direction of motion of a planet (in an elliptic orbit) and the direction perpendicular to the planet's radius vector, show that

$$\tan \eta = \frac{\varepsilon \sin \psi}{\sqrt{1 - \varepsilon^2}}$$

where ψ is the eccentric anomaly.

8-21. For a particle moving in an elliptic orbit with semimajor axis a and eccentricity ε, show that

$$\langle (a/r)^4 \cos \theta \rangle = \varepsilon/(1 - \varepsilon^2)^{5/2}$$

where the slanted brackets denote a time average over one complete period.

8-22. Consider the family of orbits in a central potential for which the total energy is a constant. Show that if a stable circular orbit exists, the angular momentum associated with this orbit is larger than that for any other orbit of the family.

8-23. Discuss the motion of a particle which moves in an attractive central-force field described by $F(r) = -k/r^3$.* Sketch some of the orbits for different

* This particular force law was extensively investigated by Roger Cotes (1682–1716), and the orbits are known as *Cotes' spirals*.

values of the total energy. Can a circular orbit be stable in such a force field? [Investigate the higher-order terms in Eq. (8.92).]

8-24. Consider a force law of the form

$$F(r) = -\frac{k}{r^2} - \frac{k'}{r^4}$$

Show that if $\rho^2 k > k'$, then a particle can move in a stable circular orbit at $r = \rho$.

8-25. Consider a force law of the form $F(r) = -(k/r^2)\exp(-r/a)$. Investigate the stability of circular orbits in this force field.

8-26. Consider a particle of mass m that is constrained to move on the surface of a paraboloid whose equation (in cylindrical coordinates) is $r^2 = 4az$. If the particle is subject to a gravitational force, show that the frequency of small oscillations about a circular orbit with radius $\rho = \sqrt{4az_0}$ is

$$\omega = \sqrt{\frac{2mg}{a + z_0}}$$

8-27. Consider the problem of the particle moving on the surface of a cone, as discussed in Examples 7.4 and 8.11(b). Show that the effective potential is

$$V(r) = \frac{l^2}{2mr^2} + mgr \cot \alpha$$

(Note that here r is the radial distance in cylindrical coordinates, not spherical coordinates; see Fig. 7-1.) Show that the turning points of the motion can be found from the solution of a cubic equation in r. Show further that only two of the roots are physically meaningful, so that the motion is confined to lie within two horizontal planes that cut the cone.

8-28. An orbit which is almost circular (i.e., $\varepsilon \ll 1$) can be considered to be a circular orbit to which a small perturbation has been applied. Then the frequency of the radial motion is given by Eq. (8.97). Consider a case in which the force law is $F(r) = -k/r^n$ (where n is an integer) and show that the apsidal angle is $\pi/\sqrt{3 - n}$. Thus, show that a closed orbit will result in general only for the harmonic oscillator force and the inverse-square-law force (if values of n equal to or smaller than -6 are excluded).

8-29. A particle moves in an almost-circular orbit in a force field described by $F(r) = -(k/r^2)\exp(-r/a)$. Show that the apsides will advance by an amount approximately equal to $\pi\rho/a$ in each revolution, where ρ is the radius of the circular orbit.

8-30. Refer to Fig. 8-12. Show that if the mass M_2 is much smaller than M_1 (i.e., $p \ll 1$), then the equilibrium points U_2 and U_3 are approximately symmetrical about M_2 and lie at distances from M_2 approximately equal to $s(p/3)^{\frac{1}{3}}$.

8-31. Generalize the problem of the motion of three bodies in a plane as discussed in Section 8.12 to the case in which the three masses, m_1, m_2, and m_3 are all comparable. Show that a solution exists if the particles form an equilateral triangle of side h and execute circular orbits about the common center of mass with an angular velocity ω, where $\omega^2 = \gamma M/h^3$ and $M = m_1 + m_2 + m_3$. (This solution of the three-body problem is due to Laplace.)

8-32. Define an effective potential $V(x, y)$ appropriate for the restricted problem of three bodies. Sketch the function $V(x, 0)$ and show that there are two "valleys" (corresponding to the positions of M_1 and M_2) and three "hills" (corresponding to the three points of unstable equilibrium). Also, sketch the function $V(\frac{1}{2}s - ps, y)$ and show that there are two "valleys" (corresponding to the two points of stable equilibrium). Compare the two curves and show that the "hill" in the latter curve is not a point of equilibrium. Use the above results to sketch an equipotential "contour map."

CHAPTER 9

Kinematics of Two-Particle Collisions

9.1 Introduction

When two particles interact, the motion of one particle relative to the other is governed by the force law which describes the interaction. This interaction may result from actual contact, as in the collision of two billiard balls, or the interaction may take place through the intermediary of a force field. For example, a *free* object (i.e., one not bound in a solar orbit) may "scatter" from the Sun via a gravitational interaction, or an α-particle may be scattered by the electric field of an atomic nucleus. As was demonstrated in the previous chapter, once the force law is known, the two-body problem can be completely solved. On the other hand, even if the force of interaction between two particles is not known, a great deal can still be learned about the relative motion by using only the results of the conservation of momentum and energy. Thus, if the initial state of the system is known (i.e., if the velocity vector of each of the particles is specified), the conservation laws allow us to obtain information regarding the velocity vectors in the final state.*

* The "initial state" of the system refers to the condition of the particles when they are not yet sufficiently close to interact appreciably, and the "final state" is the condition after the interaction has taken place. For a contact interaction these conditions are obvious, but if the interaction takes place via a force field, then the rate of decrease of the force with distance must be taken into account in specifying the initial and final states.

On the basis of the conservation theorems alone, it is not possible to predict, for example, the angle between the initial and final velocity vectors of one of the particles; knowledge of the force law is required for such details. We shall derive in this chapter those relationships which require only the conservation of momentum and energy, and then we shall examine the features of the collision process which demand that the force law be specified. The discussion here is limited to elastic collisions. Of course, collision processes in which kinetic energy is absorbed or rest-mass energy is released are also quite important. However, all of the essential features of two-particle kinematics are adequately demonstrated by elastic collisions, and so the more complicated processes are omitted. (The interested reader will find satisfactory treatments in the Suggested References.)

It should be noted that the results which are obtained under the assumption only of momentum and energy conservation are valid (in the nonrelativistic velocity region) even for quantum mechanical systems, since these conservation theorems are applicable to quantum as well as to classical systems.

9.2 Elastic Collisions—Center-of-Mass and Laboratory Coordinate Systems

It has been demonstrated previously on several occasions that the descriptions of many physical processes are considerably simplified if one chooses a coordinate system that is at rest with respect to the center of mass of the system. In the problem which we shall now discuss, viz., the elastic collision of two particles,*† the usual situation (and the one to which we shall confine our attention) is one in which the collision is between a moving particle and a particle at rest. Although it is indeed simpler to describe the effects of the collision in a coordinate system in which the center of mass is at rest, the actual measurements in such a case would be made in the system in which one of the particles is moving and in which the struck particle is initially at rest. This latter system is called the *laboratory coordinate system*. We shall frequently refer to these two coordinate systems simply as the *C.M.* and the *lab* systems.

* A collision is *elastic* if no change in the internal energy of the particles results; thus, the conservation of energy may be applied without regard to the internal energy. Notice that *heat* may be generated when two mechanical bodies collide inelastically. Heat is just a manifestation of the agitation of a body's constituent particles and may therefore be considered as a part of the internal energy.

† The laws governing the elastic collision of two bodies were first investigated by Rev. John Wallis (1668), Sir Christopher Wren (1668), and Christian Huygens (1669).

Now, we wish to take advantage of the simplifications that result by describing an elastic collision in the C.M. system. It is therefore necessary to derive the equations which connect the center-of-mass and laboratory coordinate systems.

We shall use the following notation:

$$\begin{matrix} m_1 = \\ m_2 = \end{matrix} \text{ mass of the } \begin{Bmatrix} \text{moving} \\ \text{struck} \end{Bmatrix} \text{ particle}$$

In general, primed quantities will refer to the C.M. system:

$$\left.\begin{matrix} \mathbf{u}_1 = \text{initial} \\ \mathbf{v}_1 = \text{final} \end{matrix}\right\} \text{velocity of } m_1 \text{ in the lab system}$$

$$\left.\begin{matrix} \mathbf{u}_1' = \text{initial} \\ \mathbf{v}_1' = \text{final} \end{matrix}\right\} \text{velocity of } m_1 \text{ in the C.M. system}$$

and similarly for $\mathbf{u}_2$, $\mathbf{v}_2$, $\mathbf{u}_2'$, and $\mathbf{v}_2'$ (but $\mathbf{u}_2 = 0$).

$$\begin{matrix} T_0 = \\ T_0' = \end{matrix} \text{ total initial kinetic energy in } \begin{Bmatrix} \text{lab} \\ \text{C.M.} \end{Bmatrix} \text{ system}$$

$$\begin{matrix} T_1 = \\ T_1' = \end{matrix} \text{ final kinetic energy of } m_1 \text{ in } \begin{Bmatrix} \text{lab} \\ \text{C.M.} \end{Bmatrix} \text{ system}$$

and similarly for T_2 and T_2'.

$\mathbf{V}$ = velocity of the center of mass in the lab system
ψ = angle through which m_1 is deflected in the lab system
ζ = angle through which m_2 is deflected in the lab system
θ = angle through which m_1 and m_2 are deflected in the C.M. system

Figure 9-1 illustrates the geometry of an elastic collision* in both the laboratory and center-of-mass coordinate systems. The final state in the lab and C.M. systems for the scattered particle m_1 may be conveniently summarized by the diagrams in Fig. 9-2. We can interpret these diagrams in the following manner. To the velocity $\mathbf{V}$ of the C.M. we can add the final C.M. velocity $\mathbf{v}_1'$ of the scattered particle. Depending upon the angle θ at which the scattering takes place, the possible vectors $\mathbf{v}_1'$ lie on the circle of radius v_1' whose center is at the terminus of the vector $\mathbf{V}$. The lab velocity $\mathbf{v}_1$ and lab scattering angle

* We assume throughout that the scattering is axially symmetric so that no azimuthal angle need be introduced. It should be noted, however, that axial symmetry is not always found in scattering problems; this is particularly true in certain quantum mechanical systems.

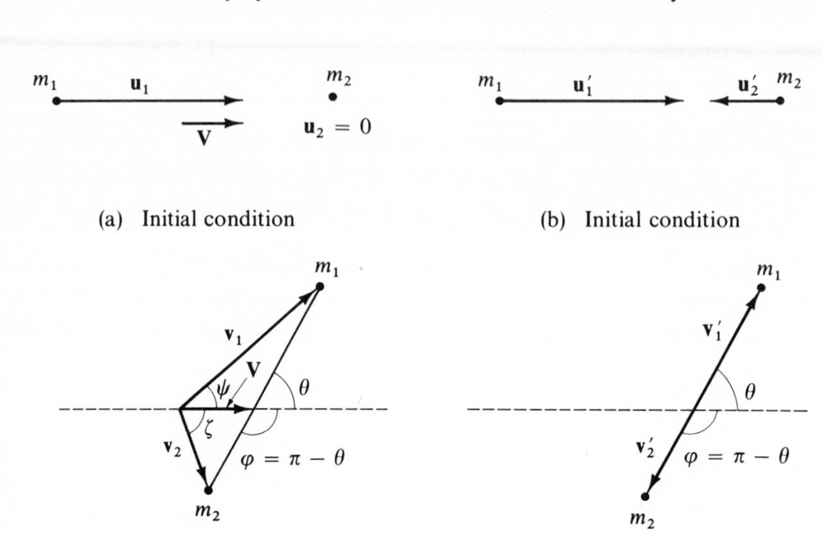

FIG. 9-1

ψ are then obtained by connecting the point or origin of $\mathbf{V}$ with the terminus of $\mathbf{v}_1'$.

If $V < v_1'$, there is only one possible relationship between $\mathbf{V}$, $\mathbf{v}_1$, $\mathbf{v}_1'$, and θ (see Fig. 9-2a). However, if $V > v_1'$, then for every set, $\mathbf{V}$, $\mathbf{v}_1'$, there exist two possible scattering angles and laboratory velocities: $\mathbf{v}_{1,b}$, θ_b, and $\mathbf{v}_{1,f}$, θ_f (see Fig. 9-2b), where the designations b and f stand for *backward* and *forward*. This situation results from the fact that if the final C.M. velocity $\mathbf{v}_1'$ is

Possible backscattering

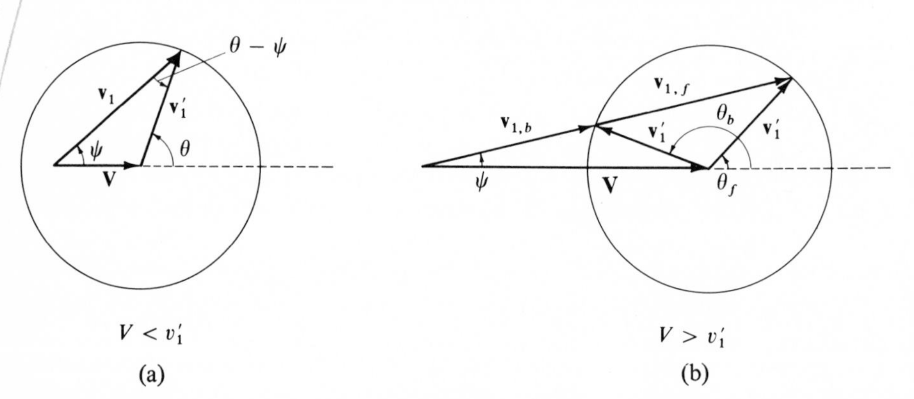

FIG. 9-2

insufficient to overcome the velocity $\mathbf{V}$ of the center of mass, then, even if m_1 is scattered into the backward direction in the C.M. system ($\theta > \pi/2$), the particle will appear at a forward angle in the lab system ($\psi < \pi/2$). Thus, for $V > v_1'$, the velocity $\mathbf{v}_1$, in the lab system, is a double-valued function of $\mathbf{v}_1'$. In an experiment, one usually measures ψ, not the velocity *vector* $\mathbf{v}_1$, so that a single value of ψ can correspond to two different values of θ. Note, however, that a specification of the vectors $\mathbf{V}$ and $\mathbf{v}_1'$ always leads to a unique combination $\mathbf{v}_1$, θ; but a specification of $\mathbf{V}$ and only the *direction* of $\mathbf{v}_1'$ (i.e., ψ) will allow the possibility of two final vectors, $\mathbf{v}_{1,b}$ and $\mathbf{v}_{1,f}$, if $V > v_1'$.

Having given a qualitative description of the scattering process, we now proceed to obtain some of the equations relating the various quantities.

According to the definition of center-of-mass coordinates [Eq. (2.25)], we have

$$m_1\mathbf{r}_1 + m_2\mathbf{r}_2 = M\mathbf{R} \tag{9.1}$$

Differentiating with respect to the time, we find

$$m_1\mathbf{u}_1 + m_2\mathbf{u}_2 = M\mathbf{V} \tag{9.1a}$$

But $\mathbf{u}_2 = 0$ and $M = m_1 + m_2$; therefore, the center of mass must be moving (in the lab system) toward m_2 with a velocity

$$V = \frac{m_1 u_1}{m_1 + m_2} \tag{9.2a}$$

Since m_2 is initially at rest, then by the same reasoning, the initial C.M. velocity of m_2 must just equal V:

$$u_2' = V = \frac{m_1 u_1}{m_1 + m_2} \tag{9.2b}$$

Note, however, that vectorially, $\mathbf{u}_2' = -\mathbf{V}$, since the motions are in opposite directions.

The great advantage of using the C.M. coordinate system lies in the fact that the total linear momentum in such a system is zero, so that before collision the particles move directly toward each other and after collision they move in exactly opposite directions. If the collision is elastic, as we have specified, then the masses do not change, and the conservation of linear momentum and kinetic energy is sufficient to provide that the C.M. velocities before and after collision are equal:

$$u_1' = v_1'; \qquad u_2' = v_2' \tag{9.3}$$

Now, u_1 is the *relative velocity* of the two particles in either the C.M. or the lab system, $u_1 = u_1' + u_2'$. Therefore, we have for the final C.M. velocities,

$$v_2' = \frac{m_1 u_1}{m_1 + m_2} \tag{9.4a}$$

$$v_1' = u_1 - u_2' = \frac{m_2 u_1}{m_1 + m_2} \tag{9.4b}$$

Referring to Fig. 9-2a, we have

$$v_1' \sin \theta = v_1 \sin \psi \tag{9.5a}$$

and

$$v_1' \cos \theta + V = v_1 \cos \psi \tag{9.5b}$$

Dividing Eq. (9.5a) by (9.5b),

$$\tan \psi = \frac{v_1' \sin \theta}{v_1' \cos \theta + V} = \frac{\sin \theta}{\cos \theta + (V/v_1')} \tag{9.6}$$

According to Eqs. (9.2a) and (9.4b), V/v_1' is given by

$$\frac{V}{v_1'} = \frac{m_1 u_1/(m_1 + m_2)}{m_2 u_1/(m_1 + m_2)} = \frac{m_1}{m_2} \tag{9.7}$$

Thus, we see that the ratio m_1/m_2 governs whether Fig. 9-2a or Fig. 9-2b describes the scattering process:

Fig. 9-2a: $\quad V < v_1', \quad m_1 < m_2$

Fig. 9-2b: $\quad V > v_1', \quad m_1 > m_2$

If we combine Eqs. (9.6) and (9.7) and write

$$\boxed{\tan \psi = \frac{\sin \theta}{\cos \theta + (m_1/m_2)}} \tag{9.8}$$

then, we see that if $m_1 \ll m_2$, the lab and C.M. scattering angles are approximately equal; that is, the particle m_2 is but little affected by the collision with m_1 and acts essentially as a fixed scattering center. Thus,

$$\boxed{\psi \cong \theta, \qquad m_1 \ll m_2} \tag{9.9a}$$

On the other hand, if $m_1 = m_2$, then

$$\tan \psi = \frac{\sin \theta}{\cos \theta + 1} = \tan \frac{\theta}{2}$$

so that

$$\boxed{\psi = \frac{\theta}{2}, \qquad m_1 = m_2} \tag{9.9b}$$

and the lab scattering angle is one-half the C.M. scattering angle. Since the maximum value of θ is 180°, Eq. (9.9b) indicates that for $m_1 = m_2$ there can be no scattering in the lab system at angles greater than 90°.

Let us now refer to Fig. 9-1c and construct a diagram for the recoil particle m_2 which is similar to Fig. 9-2a. The situation is illustrated in Fig. 9-3, from which we find

$$v_2 \sin \zeta = v_2' \sin \theta \tag{9.10a}$$

$$v_2 \cos \zeta = V - v_2' \cos \theta \tag{9.10b}$$

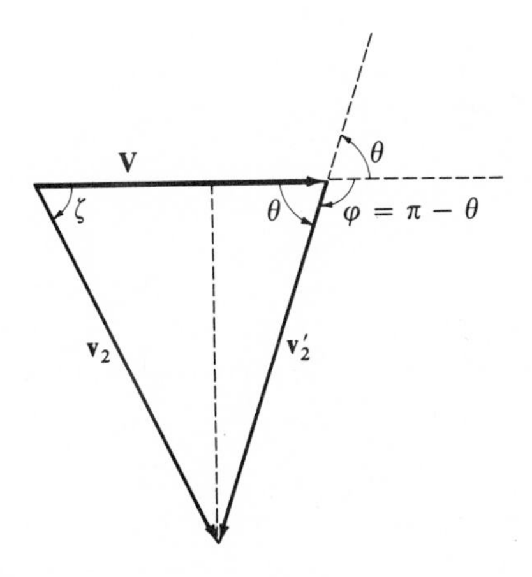

FIG. 9-3

Dividing Eq. (9.10a) by (9.10b), we have

$$\tan \zeta = \frac{v_2' \sin \theta}{V - v_2' \cos \theta} = \frac{\sin \theta}{(V/v_2') - \cos \theta}$$

But, according to Eqs. (9.2b) and (9.4a), V and v_2' are equal. Therefore,

$$\tan \zeta = \frac{\sin \theta}{1 - \cos \theta} = \cot \frac{\theta}{2} \tag{9.11}$$

which we may write as

$$\tan \zeta = \tan\left(\frac{\pi}{2} - \frac{\theta}{2}\right)$$

Thus,

$$2\zeta = \pi - \theta = \varphi \tag{9.12}$$

For the case of particles with equal mass, $m_1 = m_2$, we had $\theta = 2\psi$. Combining this result with Eq. (9.12), we have

$$\boxed{\zeta + \psi = \frac{\pi}{2}, \qquad m_1 = m_2} \tag{9.13}$$

Hence, the scattering of particles of equal mass always produces a final state in which the velocity vectors of the particles are at right angles if one of the particles is initially at rest* (see Fig. 9-4).

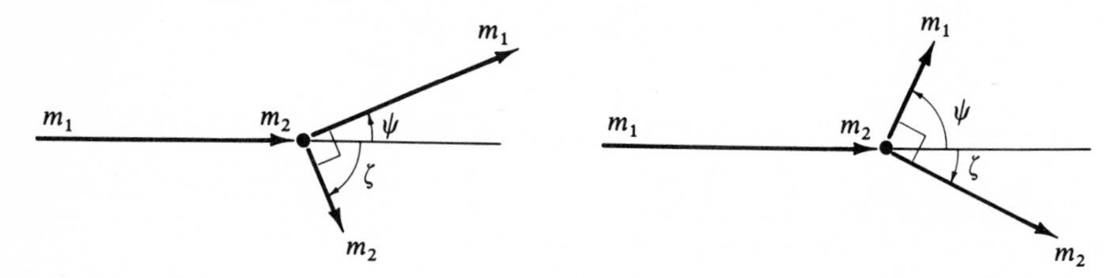

FIG. 9-4

If we refer to Fig. 9-2b and ask what is the maximum value that ψ can attain for the case in which $V > v_1'$, we have the situation of Fig. 9-5. Clearly, the angle between $\mathbf{v}_1'$ and $\mathbf{v}_1$ is 90°, so that

$$\sin\psi_{\max} = \frac{v_1'}{V} = \frac{m_2}{m_1} \tag{9.14}$$

from which

$$\boxed{\psi_{\max} = \sin^{-1}(m_2/m_1)} \tag{9.15}$$

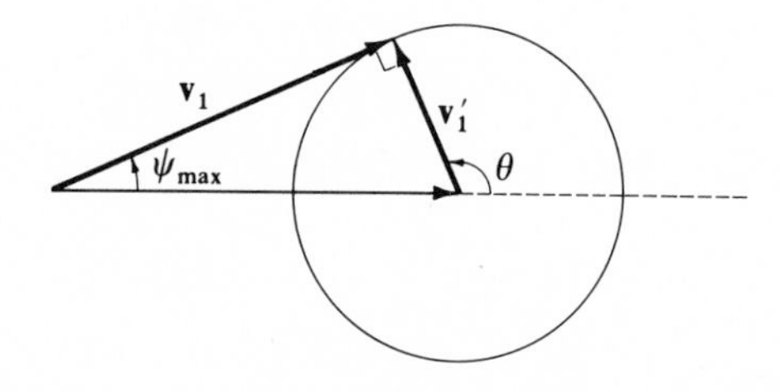

FIG. 9-5

* This result is valid only in the nonrelativistic limit; see Eq. (10.76) for the relativistic expression which governs this case.

9.3 Kinematics of Elastic Collisions

Relationships involving the energies of the particles may be obtained as follows. First, we have simply,

$$T_0 = \tfrac{1}{2}m_1u_1^2 \tag{9.16a}$$

and, in the C.M. system,

$$T_0' = \tfrac{1}{2}(m_1u_1'^2 + m_2u_2'^2)$$

which, upon using Eqs. (9.4a) and (9.4b), becomes

$$T_0' = \frac{1}{2}\frac{m_1m_2}{m_1 + m_2}u_1^2 = \frac{m_2}{m_1 + m_2}T_0 \tag{9.16b}$$

This result shows that the initial kinetic energy in the C.M. system, T_0', is always a fraction $m_2/(m_1 + m_2) < 1$ of the initial lab energy. For the final C.M. energies, we find

$$T_1' = \tfrac{1}{2}m_1v_1'^2 = \tfrac{1}{2}m_1\left(\frac{m_2}{m_1 + m_2}\right)^2 u_1^2 = \left(\frac{m_2}{m_1 + m_2}\right)^2 T_0 \tag{9.17}$$

and

$$T_2' = \tfrac{1}{2}m_2v_2'^2 = \tfrac{1}{2}m_2\left(\frac{m_1}{m_1 + m_2}\right)^2 u_1^2 = \frac{m_1m_2}{(m_1 + m_2)^2}T_0 \tag{9.18}$$

In order to obtain T_1 in terms of T_0, we write

$$\frac{T_1}{T_0} = \frac{\tfrac{1}{2}m_1v_1^2}{\tfrac{1}{2}m_1u_1^2} = \frac{v_1^2}{u_1^2} \tag{9.19}$$

Now, referring to Fig. 9-2a, and using the cosine law, we can write

$$v_1'^2 = v_1^2 + V^2 - 2v_1V\cos\psi$$

or,

$$\frac{T_1}{T_0} = \frac{v_1^2}{u_1^2} = \frac{v_1'^2}{u_1^2} - \frac{V^2}{u_1^2} + 2\frac{v_1V}{u_1^2}\cos\psi \tag{9.20}$$

From the previous definitions, we have

$$\frac{v_1'}{u_1} = \frac{m_2}{m_1 + m_2} \quad \text{and} \quad \frac{V}{u_1} = \frac{m_1}{m_1 + m_2} \tag{9.21}$$

The squares of these quantities give the desired expressions for the first two terms on the right-hand side of Eq. (9.20). In order to evaluate the third term, we write, using Eq. (9.5a),

$$2\frac{v_1 V}{u^2}\cos\psi = 2\left(v_1'\frac{\sin\theta}{\sin\psi}\right)\cdot\frac{V}{u_1^2}\cos\psi \tag{9.22}$$

The quantity $v_1' V/u_1^2$ can be obtained from the product of the equations in (9.21), and using Eq. (9.8), we have

$$\frac{\sin\theta\cos\psi}{\sin\psi} = \frac{\sin\theta}{\tan\psi} = \cos\theta + \frac{m_1}{m_2}$$

so that

$$2\frac{v_1 V}{u_1^2}\cos\psi = \frac{2m_1 m_2}{(m_1+m_2)^2}\left(\cos\theta + \frac{m_1}{m_2}\right) \tag{9.23}$$

Substituting Eqs. (9.21) and (9.23) into Eq. (9.20), we obtain

$$\frac{T_1}{T_0} = \left(\frac{m_2}{m_1+m_2}\right)^2 - \left(\frac{m_1}{m_1+m_2}\right)^2 + \frac{2m_1 m_2}{(m_1+m_2)^2}\left(\cos\theta + \frac{m_1}{m_2}\right)$$

which simplifies to

$$\frac{T_1}{T_0} = 1 - \frac{2m_1 m_2}{(m_1+m_2)^2}(1-\cos\theta) \tag{9.24a}$$

Similarly, we can also obtain the ratio T_1/T_0 in terms of the lab scattering angle ψ:

$$\frac{T_1}{T_0} = \frac{m_1^2}{(m_1+m_2)^2}\left[\cos\psi \pm \sqrt{\left(\frac{m_2}{m_1}\right)^2 - \sin^2\psi}\right]^2 \tag{9.24b}$$

where the plus (+) sign for the radical is to be taken unless $m_1 > m_2$, in which case the result is double-valued and Eq. (9.15) specifies the maximum value allowed for ψ.

The lab energy of the recoil particle m_2 can be calculated from

$$\frac{T_2}{T_0} = 1 - \frac{T_1}{T_0} = \frac{4m_1 m_2}{(m_1+m_2)^2}\cos^2\zeta, \qquad \zeta \le \pi/2 \tag{9.25}$$

If $m_1 = m_2$, we have the simple relation,

$$\boxed{\frac{T_1}{T_0} = \cos^2\psi, \qquad m_1 = m_2} \tag{9.26a}$$

with the restriction noted in the discussion following Eq. (9.9b) that $\psi \leq 90°$. Also,

$$\boxed{\frac{T_2}{T_0} = \sin^2 \psi, \qquad m_1 = m_2} \tag{9.26b}$$

Several further relationships are

$$\sin \zeta = \sqrt{\frac{m_1 T_1}{m_2 T_2}} \sin \psi \tag{9.27}$$

$$\tan \psi = \frac{\sin 2\zeta}{(m_1/m_2) - \cos 2\zeta} \tag{9.28}$$

$$\sin \varphi = \frac{m_1 + m_2}{m_2} \sin \psi \tag{9.29}$$

As an example of the application of the kinematic relations which we have derived, consider the following situation. Suppose that we have a beam of projectiles, all with mass m_1 and energy T_0. We direct this beam toward a target which consists of a group of particles whose masses m_2 may not all be the same. Some of the incident particles interact with the target particles and are scattered. The incident particles all move in the same direction in a beam of small cross-sectional area, and we assume that the target particles are localized in space so that the scattered particles emerge from a small region. If we position a detector at, say, 90° to the incident beam and with this detector measure the energies of the scattered particles, then the results can be displayed as in the lower portion of Fig. 9-6. This graph is a *histogram* in which is plotted the number of particles detected within a range of energy ΔT at the energy T. This particular histogram shows that three energy groups were observed in the scattered particles detected at $\psi = 90°$. The upper portion of the figure shows a curve which gives the scattered energy T_1 in terms of T_0 as a function of the mass ratio m_2/m_1 [Eq. (9.24b)]. The curve can be used to determine the mass m_2 of the particle from which one of the incident particles was scattered in order to fall into one of the three energy groups. Thus, the energy group with $T_1 \cong 0.8T_0$ results from the scattering by target particles with mass $m_2 = 10m_1$, and the other two groups arise from target masses $5m_1$ and $2m_1$.

The measurement of the energies of scattered particles is therefore a method of *qualitative analysis* of the target material. Indeed, this method is useful in practice when the incident beam consists of particles (protons, say) that have been given high velocities in an accelerator of some sort.

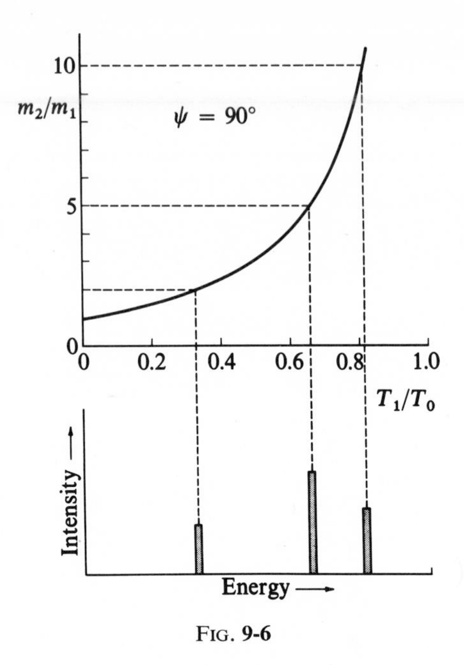

FIG. 9-6

If the detector is capable of precision energy measurements, then the method will yield accurate information regarding the composition of the target. (*Quantitative analysis* can also be made from the intensities of the groups if the cross sections are known; see the following section.)

9.4 Cross Sections

In the preceding sections various relationships were derived which connect the initial state of a moving particle with the final states of the original particle and a struck particle. Only kinematic relationships were involved; that is, no attempt was made to *predict* a scattering angle or a final velocity—only equations *connecting* these quantities were obtained. We now look more closely at the collision process and investigate the scattering in the event that the particles interact via a specified force field. Consider the situation depicted in Fig. 9-7 which illustrates such a collision in the laboratory coordinate system when a repulsive force exists between m_1 and m_2. The particle m_1 approaches the vicinity of m_2 in such a way that if there were no force acting between the particles, m_1 would pass m_2 with a distance of closest approach b. The quantity b is called the *impact parameter*. If the velocity of m_1 is u_1,

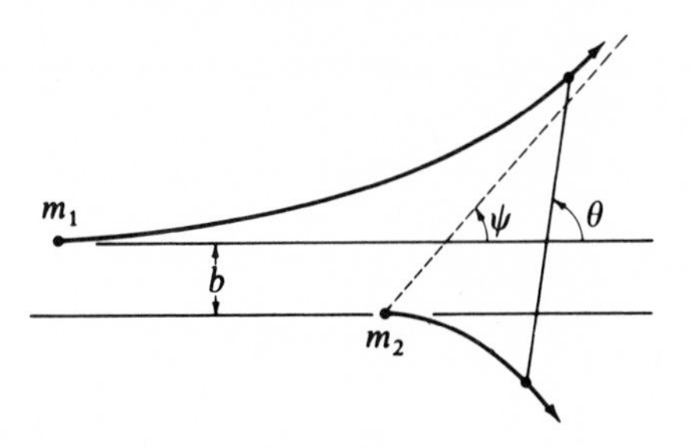

FIG. 9-7

then the impact parameter b clearly specifies the angular momentum l of particle m_1 about m_2:

$$l = m_1 u_1 b \tag{9.30}$$

We may express u_1 in terms of the incident energy T_0 by using Eq. (9.16):

$$l = b\sqrt{2m_1 T_0} \tag{9.31}$$

Evidently, for a given energy T_0, the angular momentum and hence the scattering angle θ (or ψ) will be uniquely specified* by the impact parameter b if the force law is known.

We now consider the distribution of scattering angles which would result from collisions with various impact parameters. In order to accomplish this, let us assume that we have a narrow beam of particles, each of which has mass m_1 and energy T_0. We direct this beam toward a small region of space which contains a collection of particles, each of which has mass m_2 and is at rest (in the laboratory system). The *intensity* (or *flux density*) I of the incident particles is defined as the number of particles which pass in unit time through a unit area that is normal to the direction of the beam. If we assume that the force law between m_1 and m_2 falls off with distance sufficiently rapidly, then after an encounter, the motion of a scattered particle will asymptotically approach a straight line with a well-defined angle θ between the initial and final directions of motion. We now define a *differential scattering cross section* $\sigma(\theta)$ in the C.M. system for the scattering into an element of solid angle $d\Omega'$ at a particular C.M. angle θ:

$$\sigma(\theta) = \frac{\begin{pmatrix}\text{Number of interactions per target particle that} \\ \text{lead to scattering into } d\Omega' \text{ at the angle } \theta\end{pmatrix}}{\text{Number of incident particles per unit area}} \tag{9.32}$$

* Clearly, in the scattering of atomic or nuclear particles, we can neither choose nor measure directly the impact parameter. We are therefore reduced, in such situations, to speaking in terms of the *probability* for scattering at various angles θ.

If dN is the number of particles scattered into $d\Omega'$ per unit time, then

$$\sigma(\theta)\,d\Omega' = \frac{dN}{I} \tag{9.33a}$$

We sometimes write, alternatively,

$$\sigma(\theta) = \frac{d\sigma}{d\Omega'} = \frac{1}{I}\frac{dN}{d\Omega'} \tag{9.33b}$$

(The fact that $\sigma(\theta)$ has the dimensions of *area* gives rise to the term "cross section.") If the scattering has axial symmetry (as will be the case for central forces), we can immediately perform the integration over the azimuthal angle to obtain 2π, and then the element of solid angle $d\Omega'$ is given by

$$d\Omega' = 2\pi \sin\theta\,d\theta \tag{9.34}$$

If we return, for the moment, to the equivalent one-body problem as discussed in the preceding chapter, we can consider the scattering of a particle of mass μ by a force center. Then, Fig. 9-8 shows that the number of particles

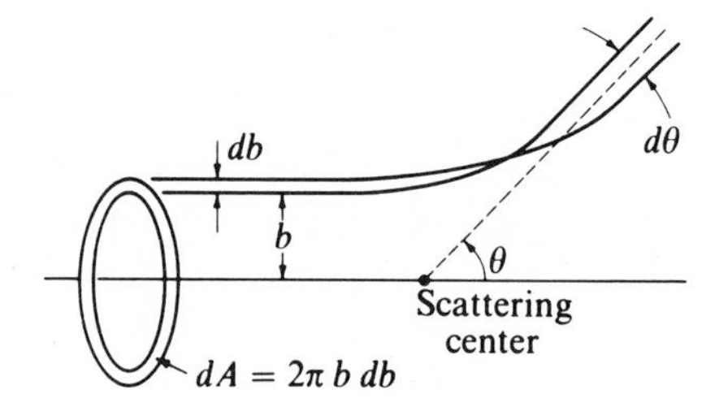

FIG. 9-8

with impact parameters within a range db at a distance b must correspond to the number of particles scattered into the angular range $d\theta$ at an angle θ. Therefore,

$$I \cdot 2\pi b\,db = -I \cdot \sigma(\theta) \cdot 2\pi \sin\theta\,d\theta \tag{9.35}$$

where $db/d\theta$ is negative since we assume that the force law is such that the amount of angular deflection decreases (monotonically) with increasing impact parameter. Hence,

$$\sigma(\theta) = \frac{b}{\sin\theta}\left|\frac{db}{d\theta}\right| \tag{9.35a}$$

We can obtain the relationship between the impact parameter b and the scattering angle θ by making use of Fig. 9-9. In the preceding chapter we

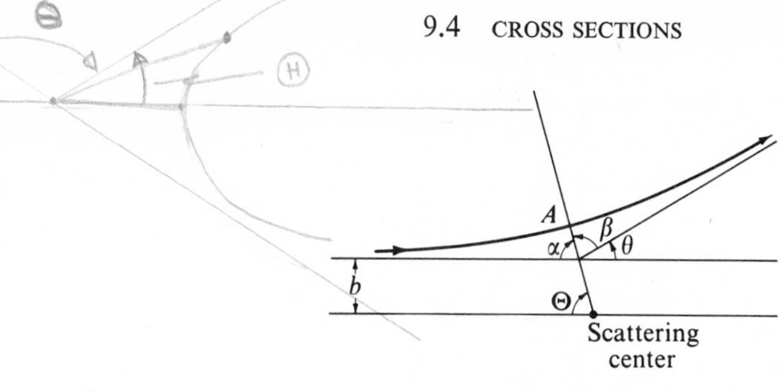

FIG. 9-9

found [Eq. (8.22)] that the change in angle for a particle of mass μ moving in a central-force field is given by

$$\Delta\Theta = \int_{r_{\min}}^{r_{\max}} \frac{(l/r^2)\,dr}{\sqrt{2\mu[E - U - (l^2/2\mu r^2)]}} \tag{9.36}$$

Now, the motion of a particle in a central-force field is symmetric about the point of closest approach to the force center (see point A in Fig. 9-9). Therefore, the angles α and β are equal and, in fact, are equal to Θ. Thus,

$$\theta = \pi - 2\Theta \tag{9.37}$$

For the case that $r_{\max} = \infty$, the angle Θ is given by

$$\Theta = \int_{r_{\min}}^{\infty} \frac{(b/r^2)\,dr}{\sqrt{1 - (b^2/r^2) - (U/T_0')}} \tag{9.38}$$

where use has been made of the one-body equivalent of Eq. (9.31), viz.,

$$l = b\sqrt{2\mu T_0'}$$

where, as in Eq. (9.16b), $T_0' = \frac{1}{2}\mu u_1^2$. We have also used $E = T_0'$ since the total energy E must equal the kinetic energy T_0' at $r = \infty$ where $U = 0$. The value of $r_{\min}$, it will be recalled, is a root of the radical in the denominator in Eq. (9.36) or (9.38), i.e., $r_{\min}$ is a turning point of the motion and corresponds to the distance of closest approach of the particle to the force center.

Thus, Eqs. (9.37) and (9.38) give the dependence of the scattering angle θ on the impact parameter b. Once $b = b(\theta)$ is known for a given potential $U(r)$ and a given value of T_0', the scattering cross section can be calculated from Eq. (9.35a). This procedure, of course, leads to the scattering cross section in the C.M. system since we have been considering m_2 as a fixed force center. If $m_2 \gg m_1$, the cross section so obtained will be very close to the lab system cross section; but if m_1 cannot be considered to be negligible compared to m_2, the proper transformation of solid angles must be made. We now obtain the general relations.

Since the total number of particles scattered into a unit solid angle must be the same in the lab system as in the C.M. system we have

$$\sigma(\theta)\, d\Omega' = \sigma(\psi)\, d\Omega$$

$$\sigma(\theta) \cdot 2\pi \sin\theta\, d\theta = \sigma(\psi) \cdot 2\pi \sin\psi\, d\psi \tag{9.39}$$

where θ and ψ represent the *same* scattering angle but measured in the C.M. or lab system, respectively, and where $d\Omega'$ and $d\Omega$ represent the *same* element of solid angle but measured in the C.M. or lab system, respectively. Therefore, $\sigma(\theta)$ and $\sigma(\psi)$ are the differential cross sections for the scattering in the C.M. and lab systems, respectively. Thus,

$$\sigma(\psi) = \sigma(\theta) \cdot \frac{\sin\theta\, d\theta}{\sin\psi\, d\psi} \tag{9.39a}$$

The derivative $d\theta/d\psi$ can be evaluated by first referring to Fig. 9-2a and writing, from the sine law,

$$\frac{\sin(\theta - \psi)}{\sin\psi} = \frac{m_1}{m_2} \equiv x \tag{9.40}$$

Differentiating this equation, we find

$$\frac{d\theta}{d\psi} = \frac{\sin(\theta - \psi)\cos\psi}{\cos(\theta - \psi)\sin\psi} + 1$$

Expanding $\sin(\theta - \psi)$ and simplifying, we have

$$\frac{d\theta}{d\psi} = \frac{\sin\theta}{\cos(\theta - \psi)\sin\psi}$$

so that

$$\sigma(\psi) = \sigma(\theta) \cdot \frac{\sin^2\theta}{\cos(\theta - \psi)\sin^2\psi} \tag{9.41}$$

Multiplying both sides of Eq. (9.40) by $\cos\psi$ and then adding $\cos(\theta - \psi)$ to both sides, we have

$$\frac{\sin(\theta - \psi)\cos\psi}{\sin\psi} + \cos(\theta - \psi) = x\cos\psi + \cos(\theta - \psi)$$

Upon expanding $\sin(\theta - \psi)$ and $\cos(\theta - \psi)$ on the left-hand side, we obtain

$$\frac{\sin\theta}{\sin\psi} = x\cos\psi + \cos(\theta - \psi)$$

Substituting this result into Eq. (9.41),

$$\sigma(\psi) = \sigma(\theta) \cdot \frac{[x \cos \psi + \cos(\theta - \psi)]^2}{\cos(\theta - \psi)}, \qquad (x < 1) \tag{9.42}$$

Also, from Eq. (9.40), we have

$$\cos(\theta - \psi) = \sqrt{1 - x^2 \sin^2 \psi}$$

Hence,

$$\boxed{\sigma(\psi) = \sigma(\theta) \cdot \frac{\left[x \cos \psi + \sqrt{1 - x^2 \sin^2 \psi}\right]^2}{\sqrt{1 - x^2 \sin^2 \psi}}} \tag{9.43}$$

Now if include inelastic collisions we must write: $x = \frac{V}{v_1'}$ instead of $x = \frac{m_1}{m_2}$!

Equation (9.40) can be used to write

$$\boxed{\theta = \sin^{-1}(x \sin \psi) + \psi} \tag{9.44}$$

Therefore, Eqs. (9.43) and (9.44) specify the cross section entirely in terms of the angle ψ.* Clearly, for the general case (i.e., for an arbitrary value of x), the evaluation of $\sigma(\psi)$ is complicated. Tables exist, however, so that the particular cases can be computed with relative ease.†

The transformation represented by Eqs. (9.43) and (9.44) assumes a simple form for two cases. For $x = m_1/m_2 = 1$, we have from Eq. (9.9b), $\theta = 2\psi$, and Eq. (9.43) becomes

$$\sigma(\psi) = \sigma(\theta)|_{\theta = 2\psi} \cdot 4 \cos \psi, \qquad m_1 = m_2 \tag{9.45a}$$

and for $m_1 \ll m_2$, $x \cong 0$, and $\theta \cong \psi$, so that

$$\sigma(\psi) \cong \sigma(\theta)|_{\theta = \psi}, \qquad m_1 \ll m_2 \tag{9.45b}$$

9.5 The Rutherford Scattering Formula‡

One of the most important problems which makes use of the formulae developed in the preceding section is the scattering of charged particles in

* It should be noted that these equations apply not only for elastic collisions, but also for inelastic collisions (in which the internal potential energy of one or both of the particles is altered as a result of the interaction) if the parameter x is written as V/v_1' instead of m_1/m_2 [cf. Eq. (9.7)]. Note that the equations above refer only to the usual case $x < 1$.

† See, for example, the tables by Marion *et al.* (Ma59).

‡ E. Rutherford, *Phil. Mag.* **21**, 669 (1911).

a Coulomb or electrostatic field. The potential for this case is

$$U(r) = \frac{k}{r} \tag{9.46}$$

where $k = q_1 q_2$, with q_1 and q_2 the amounts of charge that the two particles carry (k may be either positive or negative, depending on whether the charges are of the same or opposite sign; $k > 0$ corresponds to a repulsive force and $k < 0$ to an attractive force). Equation (9.38) then becomes

$$\Theta = \int_{r_{\min}}^{\infty} \frac{(b/r)\,dr}{\sqrt{r^2 - (k/T'_0)r - b^2}} \tag{9.47}$$

which can be integrated to obtain [cf. the integration of Eq. (8.29)]:

$$\cos\Theta = \frac{(\kappa/b)}{\sqrt{1 + (\kappa/b)^2}} \tag{9.48}$$

where

$$\kappa \equiv \frac{k}{2T'_0} \tag{9.49}$$

Equation (9.48) can be rewritten as

$$b^2 = \kappa^2 \tan^2 \Theta \tag{9.50}$$

But Eq. (9.37) states that $\Theta = \pi/2 - \theta/2$, so that

$$b = \kappa \cot(\theta/2) \tag{9.51}$$

Thus,

$$\frac{db}{d\theta} = -\frac{\kappa}{2} \frac{1}{\sin^2(\theta/2)} \tag{9.52}$$

Then, Eq. (9.35) becomes

$$\sigma(\theta) = \frac{\kappa^2}{2} \cdot \frac{\cot(\theta/2)}{\sin\theta \sin^2(\theta/2)}$$

Now,

$$\sin\theta = 2\sin(\theta/2)\cos(\theta/2)$$

Hence,

$$\sigma(\theta) = \frac{\kappa^2}{4} \cdot \frac{1}{\sin^4(\theta/2)}$$

or,

$$\boxed{\sigma(\theta) = \frac{k^2}{(4T_0')^2} \cdot \frac{1}{\sin^4(\theta/2)}} \tag{9.53}$$

which is the Rutherford scattering formula* and demonstrates the dependence of the C.M. scattering cross section on the inverse fourth power of $\sin(\theta/2)$. Note that $\sigma(\theta)$ is independent of the sign of k, so that the form of the scattering distribution is the same for an attractive force as for a repulsive one. It is also rather remarkable that the quantum mechanical treatment of Coulomb scattering leads to exactly the same result as does the classical derivation.† This is indeed a fortunate circumstance since, if it were otherwise, the disagreement at this early stage between classical theory and experiment might have seriously delayed the progress of nuclear physics.

For the case $m_1 = m_2$, Eq. (9.16b) states that $T_0' = \frac{1}{2}T_0$, so that

$$\sigma(\theta) = \frac{k^2}{4T_0^2} \cdot \frac{1}{\sin^4(\theta/2)}, \qquad m_1 = m_2 \tag{9.54}$$

Or, from Eq. (9.45a),

$$\sigma(\psi) = \frac{k^2}{T_0^2} \frac{\cos\psi}{\sin^4\psi}, \qquad m_1 = m_2 \tag{9.55}$$

9.6 The Total Cross Section

All of the above discussion applies to the calculation of *differential* scattering cross sections. If it is desired to know the probability that *any* interaction *whatsoever* will take place, then it is necessary to integrate $\sigma(\theta)$ [or $\sigma(\psi)$] over all possible scattering angles. The resulting quantity is called the *total scattering cross section* σ_t and is equal to the effective area of the target particle for producing a scattering event:

$$\begin{aligned} \sigma_t &= \int_{4\pi} \sigma(\theta)\, d\Omega' \\ &= 2\pi \int_0^{\pi} \sigma(\theta) \sin\theta \, d\theta \end{aligned} \tag{9.56}$$

* This form of the scattering law was verified for the interaction of α particles and heavy nuclei by the experiments of H. Geiger and E. Marsden, *Phil. Mag.* **25**, 605 (1913).

† N. Bohr showed that the identity of the results is a consequence of the $1/r^2$ nature of the force; it cannot be expected for any other type of force law.

where the integration over θ runs from 0 to π. The *total* cross section is, of course, the same in the lab as in the C.M. system. If we wish to express the total cross section in terms of an integration over the lab quantities,

$$\sigma_t = \int \sigma(\psi)\, d\Omega$$

then if $m_1 < m_2$, ψ also runs from 0 to π. In the event that $m_1 \geq m_2$, ψ runs only up to $\psi_{\max}$ [given by Eq. (9.15)], and we have

$$\sigma_t = 2\pi \int_0^{\psi_{\max}} \sigma(\psi) \sin\psi\, d\psi \tag{9.57}$$

If we attempt to calculate σ_t for the case of Rutherford scattering, we find that the result is infinite. This arises from the fact that the Coulomb potential which varies as $1/r$, falls off so slowly that as the impact parameter b is allowed to become indefinitely large, the decrease in scattering angle is too slow to prevent the integral from diverging. We have, however, pointed out in Example 8.11(a) that the Coulomb field of a real atomic nucleus is screened by the surrounding electrons so that the potential is effectively cut off at large distances. The inclusion of the screening term, $\exp(-r/a)$, in the potential then yields a finite value for σ_t. The evaluation of the scattering cross section for a screened Coulomb potential according to the classical theory is quite complicated and will not be discussed here; the quantum mechanical treatment is actually easier for this case.

Suggested References

The kinematics of collisions is a topic not often treated in detail in texts on classical mechanics. Two exceptions are Konopinski (Ko69, Chapter 5) and Landau and Lifshitz (La60, Chapter 4). A brief discussion is also given by Goldstein (Go50, Chapter 3).

Considerable detail is given in some books devoted to nuclear reaction processes; see, for example, Baldin, Goldanskii, and Rozental (Ba61) and Mather and Swan (Ma58).

Quantum mechanics texts frequently include short discussions of collision kinematics (which do not require a knowledge of quantum mechanics to be understood); see, for example, Schiff (Sc55, Chapter 5).

Problems

9-1. In an elastic collision of two particles, with masses m_1 and m_2, the initial velocities are $\mathbf{u}_1$ and $\mathbf{u}_2 = \alpha\mathbf{u}_1$ ($\alpha \neq 0$). If the initial kinetic energies of the two particles are equal, find the conditions on u_1/u_2 and m_1/m_2 so that m_1 will be at rest after the collision.

9-2. Show that

$$\frac{T_1}{T_0} = \frac{m_1^2}{(m_1 + m_2)^2} \cdot S^2$$

where

$$S \equiv \cos \psi + \frac{\cos(\theta - \psi)}{(m_1/m_2)}$$

9-3. Show that T_1/T_0 can be expressed in terms of $m_2/m_1 \equiv \alpha$ and $\cos \psi \equiv y$ as

$$\frac{T_1}{T_0} = (1 + \alpha)^{-2}[2y^2 + \alpha^2 - 1 + 2y\sqrt{\alpha^2 + y^2 - 1}]$$

Plot T_1/T_0 as a function of ψ for $\alpha = 1, 2, 4$, and 12. These plots correspond to the energies of protons or neutrons after scattering from hydrogen ($\alpha = 1$), deuterium ($\alpha = 2$), helium ($\alpha = 4$), and carbon ($\alpha = 12$), or of alpha particles scattered from helium ($\alpha = 1$), oxygen ($\alpha = 4$), etc.

9-4. A particle of mass m_1 with initial laboratory velocity u_1 collides with a particle of mass m_2 which is at rest in the laboratory system. The particle m_1 is scattered through a laboratory angle ψ and has a final velocity v_1, where $v_1 = v_1(\psi)$. Find the surface such that the time of travel of the scattered particle from the point of collision to the surface is independent of the scattering angle. Consider the cases (a) $m_2 = m_1$, (b) $m_2 = 2m_1$, (c) $m_2 = \infty$. Suggest an application of this result in terms of a detector for nuclear particles.

9-5. Show that the equivalent of Eq. (9.44) expressed in terms of θ, rather than ψ, is

$$\sigma(\theta) = \sigma(\psi) \cdot \frac{1 + x \cos \theta}{(1 + 2x \cos \theta + x^2)^{3/2}}$$

9-6. Calculate the differential cross section $\sigma(\theta)$ and the total cross section σ_t for the elastic scattering of a particle from an impenetrable sphere; i.e., the potential is given by

$$U(r) = \begin{cases} 0, & r > a \\ \infty, & r < a \end{cases}$$

9-7. If, in the previous problem, the energy lost by the scattered particle to the sphere is ε, show that

$$d\sigma_{\text{C.M.}}(\varepsilon) = \frac{\pi a^2}{\varepsilon_{\max}} \, d\varepsilon$$

Thus, show that in the center-of-mass system the energies of the scattered particles are distributed uniformly.

9-8. Show that the Rutherford scattering cross section (for the case $m_1 = m_2$) can be expressed in terms of the recoil angle as

$$\sigma_{\text{lab}}(\zeta) = \frac{k^2}{T_0^2} \cdot \frac{1}{\cos^3 \zeta}$$

9-9. Consider the case of Rutherford scattering in the event that $m_1 \gg m_2$ (i.e., the mass of the incident particle is much greater than that of the target). Obtain an approximate expression for the differential cross section in the *laboratory* coordinate system.

9-10. Consider the case of Rutherford scattering in the event that $m_2 \gg m_1$. Obtain an expression for the differential cross section in the center-of-mass system that is correct to first order in the quantity m_1/m_2. Compare this result with Eq. (9.53).

9-11. A fixed force center scatters a particle of mass m according to the force law $F(r) = k/r^3$. If the initial velocity of the particle is u_0, show that the differential scattering cross section is

$$\sigma(\theta) = \frac{k\pi}{2mu_0^2\,\theta^2 \sin\theta}$$

The integral of this expression gives an infinite result for the total cross section. However, if the force vanishes for $r > r_0$, show that there is some minimum scattering angle θ_0; express the result in terms of m, u_0, r_0, and k. Then, show that the total cross section is

$$\sigma_t = 2\pi \int_{\theta_0}^{\pi} \sigma(\theta) \sin\theta \, d\theta = \pi r_0^2$$

9-12. It is found experimentally that in the elastic scattering of neutrons by protons ($m_n \cong m_p$) at relatively low energies, the energy distribution of the recoiling protons in the lab system is constant up to a maximum energy which is the energy of the incident neutrons. What is the angular distribution of the scattering in the C.M. system?

9-13. Show that the energy distribution of particles recoiling from an elastic collision is always directly proportional to the differential scattering cross section in the C.M. system.*

* This result is due to H. H. Barschall and M. H. Kanner, *Phys. Rev.* **58**, 590 (1940).

CHAPTER 10

The Special Theory of Relativity

10.1 Introduction

In Section 2.12 it was pointed out that the Newtonian idea of the complete separability of space and time and the concept of the absoluteness of time break down when they are subjected to a critical analysis. The final overthrow of the Newtonian system as the ultimate description of dynamics was the result of several crucial experiments, culminating with the work of Michelson and Morley in 1881–1887. The results of these experiments indicated that the velocity of light is independent of any relative uniform motion between source and observer. This fact, coupled with the finite velocity of light, required a fundamental reorganization of the structure of dynamics. This was provided during the period 1904–1905 by H. Poincaré, H. A. Lorentz, and A. Einstein*

* Although Albert Einstein (1879–1955) is usually accorded the credit for the formulation of relativity theory (see, however, Wh53, Chapter 2), the basic *formalism* had been discovered by Poincaré and Lorentz by 1904. Einstein was unaware of some of this previous work at the time (1905) of the publication of his first paper on relativity. (Einstein's friends often remarked that "he read little, but thought much.") The important contribution of Einstein to special relativity theory was the clearing away of the many *ad hoc* assumptions made by Lorentz and others and replacing them with but two basic postulates from which all of the results could be *derived*. [The question of precedence in relativity theory is discussed by G. Holton, *Am. J. Phys.* **28**, 627 (1960); see also Am63.] In addition, Einstein later provided the fundamental contribution to the formulation of the *general* theory of relativity in 1916. (His first publication on a topic of importance in general relativity, viz., speculations on the influence of gravity on light, was in 1907.) It is interesting to note that Einstein's 1921 Nobel prize was awarded, not for contributions to relativity theory, but for his work on the photoelectric effect.

who formulated the *theory of relativity* in order to provide a consistent description of the experimental facts.

The basis of relativity theory is contained in two postulates:

(i) The laws of physical phenomena are the same in all inertial reference frames. (That is, only the relative motion of inertial frames can be measured; the concept of motion relative to "absolute rest" is meaningless.)
(ii) The velocity of light (in free space) is a universal constant, independent of any relative motion of the source and the observer.

Using these postulates as a foundation, Einstein was able to construct a beautiful theory which is a model of logical precision. A wide variety of phenomena which take place at high velocity and which cannot be interpreted in the Newtonian scheme are accurately described by relativity theory.

We shall not attempt to give the experimental background for the theory of relativity; such information can be found in essentially every textbook on "modern physics" and in many that are concerned with electrodynamics.* Rather, we shall adopt a postulational approach and simply accept as correct the above two postulates. Starting from this point, we will work out some of the consequences which apply to the area of mechanics.† The discussion here will be limited to the case of *special relativity* in which we consider only inertial references frames, i.e., frames which are in uniform motion with respect to one another. The more general treatment of accelerated reference frames is the subject of the *general theory of relativity.*

10.2 Galilean Invariance

In Newtonian mechanics, the concepts of space and time are supposed to be completely separable, and it is further assumed that time is an absolute quantity, susceptible of precise definition independent of the reference frame. This assumption leads to the invariance of the laws of mechanics under coordinate transformations of the following type. Consider two inertial reference frames K and K' which move along their x_1- and x_1'-axes with a uniform relative velocity v, as in Fig. 10-1. The transformation of the coordinates of a point from one system to the other is clearly of the form

$$\left.\begin{aligned} x_1' &= x_1 - vt \\ x_2' &= x_2 \\ x_3' &= x_3 \end{aligned}\right\} \tag{10.1a}$$

* A particularly good discussion of the experimental necessity for relativity theory may be found in Panofsky and Phillips (Pa62, Chapter 15).

† Relativistic effects in electrodynamics are discussed in Marion (Ma65b, Chapter 13).

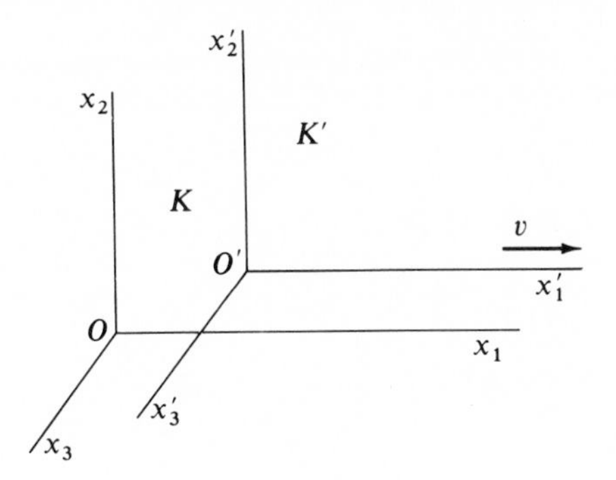

FIG. **10-1**

In addition we have

$$t' = t \tag{10.1b}$$

Equations (10.1) define a *Galilean transformation.* Furthermore, the element of length in the two systems is the same and is given by

$$\begin{aligned} ds^2 &= \sum_j dx_j^2 \\ &= \sum_j dx_j'^2 = ds'^2 \end{aligned} \tag{10.2}$$

The fact that Newton's laws are invariant with respect to Galilean transformations is termed *the principle of Newtonian relativity* or *Galilean invariance.* Newton's equations of motion in the two systems are

$$\begin{aligned} F_j &= m\ddot{x}_j \\ &= m\ddot{x}_j' = F_j' \end{aligned} \tag{10.3}$$

Thus, the form of the law of motion is *invariant* to a Galilean transformation. The individual terms are not invariant, however, but they transform according to the same scheme and are said to be *covariant.*

10.3 The Lorentz Transformations

The principle of Galilean invariance predicts that the velocity of light is different in two inertial reference frames that are in relative motion. This result is in contradiction to the second postulate of relativity. Therefore, a new transformation law must be found which will render physical laws *relativistically* covariant. Such a transformation law is the *Lorentz transformation.*

Historically, the use of the Lorentz transformation preceded the development of Einsteinian relativity theory,* but it also follows from the basic postulates of relativity, and we shall derive it on this basis.

If a light pulse is emitted from the common origin of the moving systems K and K' (see Fig. 10-1) when they are coincident, then according to the second postulate, the wavefronts observed in the two systems must be described by

$$\left.\begin{aligned} \sum_{j=1}^{3} x_j^2 - c^2t^2 = 0 \\ \sum_{j=1}^{3} x_j'^2 - c^2t'^2 = 0 \end{aligned}\right\} \tag{10.4}$$

If we define a new coordinate in each system, $x_4 \equiv ict$ and $x_4' \equiv ict'$, then we can write Eqs. (10.4) as†

$$\left.\begin{aligned} \sum_{\mu=1}^{4} x_\mu^2 = 0 \\ \sum_{\mu=1}^{4} x_\mu'^2 = 0 \end{aligned}\right\} \tag{10.5}$$

From these equations it is clear that the two sums must be proportional, and since the motion is symmetric between the systems, the proportionality constant is unity.‡ Thus,

$$\sum_\mu x_\mu^2 = \sum_\mu x_\mu'^2 \tag{10.6}$$

This relation is analogous to the three-dimensional, distance-preserving, orthogonal rotations that we have studied previously (see Section 1.4) and indicates that the transformation which we are now seeking corresponds to a rotation in a *four-dimensional* space (called *world space* or *Minkowski space*§).

* This transformation was originally postulated by Hendrik Anton Lorentz (1853–1928) in 1904 in order to explain certain electromagnetic phenomena, but the formulae had been set up as early as 1900 by J. J. Larmor. The complete generality of the transformation was not realized until Einstein *derived* the result. W. Voigt was actually the first to use the equations in a discussion of oscillatory phenomena in 1887.

† In accordance with standard convention we use Greek indices (usually μ or ν) to indicate summations that run from 1 to 4; in relativity theory Latin indices are usually reserved for summations that run from 1 to 3.

‡ A "proof" is given in Appendix G.

§ After Herman Minkowski (1864–1909), who made important contributions to the mathematical theory of relativity and who introduced *ict* as a fourth component.

Therefore, *the Lorentz transformations are orthogonal transformations in Minkowski space*. That is,

$$\boxed{x'_\mu = \sum_\nu \lambda_{\mu\nu} x_\nu} \tag{10.7}$$

where the $\lambda_{\mu\nu}$ are the elements of the *Lorentz transformation matrix*. If we consider a translation of the K' system in the x_1-direction, then $x'_2 = x_2$ and $x'_3 = x_3$ so that the transformation matrix $\boldsymbol{\lambda}$ is of the form

$$\boldsymbol{\lambda} = \begin{pmatrix} \lambda_{11} & 0 & 0 & \lambda_{14} \\ 0 & 1 & 0 & 0 \\ 0 & 0 & 1 & 0 \\ \lambda_{41} & 0 & 0 & \lambda_{44} \end{pmatrix} \tag{10.8}$$

Since the transformation is orthogonal, we have in analogy with Eqs. (1.13) and (1.15),

$$\sum_\nu \lambda_{\mu\nu} \lambda_{\gamma\nu} = \delta_{\mu\gamma} \tag{10.9a}$$

or,

$$\sum_\nu \lambda_{\nu\mu} \lambda_{\nu\gamma} = \delta_{\mu\gamma} \tag{10.9b}$$

If we apply these conditions (recall that they are actually equivalent) to the $\boldsymbol{\lambda}$ matrix in Eq. (10.8), we find

$$\lambda_{11}^2 + \lambda_{14}^2 = \lambda_{11}^2 + \lambda_{41}^2 = \lambda_{41}^2 + \lambda_{44}^2 = \lambda_{14}^2 + \lambda_{44}^2 = 1 \tag{10.10a}$$

and

$$\lambda_{11}\lambda_{14} + \lambda_{41}\lambda_{44} = \lambda_{11}\lambda_{41} + \lambda_{14}\lambda_{44} = 0 \tag{10.10b}$$

We have written down all the orthogonality relations from *both* of the equations in (10.9), even though they are not all independent. [There are only *three* independent relations, viz., those from *either* Eq. (10.9a) *or* Eq. (10.9b).]

If we apply the transformation matrix $\boldsymbol{\lambda}$ to the matrix $\mathbf{x}$, we find for x'_1,

$$\begin{aligned} x'_1 &= \lambda_{11} x_1 + \lambda_{14} x_4 \\ &= \lambda_{11}\left(x_1 + ic\frac{\lambda_{14}}{\lambda_{11}}t\right) \end{aligned} \tag{10.11}$$

Now, when $x'_1 = 0$, we must have $x_1 = vt$; i.e., the point which is the origin of the K' system moves with a uniform velocity along the x_1-axis. Therefore, Eq. (10.11) yields

$$v = -ic\frac{\lambda_{14}}{\lambda_{11}}$$

or,

$$\frac{\lambda_{14}}{\lambda_{11}} = i\beta \tag{10.12}$$

where

$$\beta \equiv \frac{v}{c} \tag{10.13}$$

From the equations in (10.10a) we can also write

$$\lambda_{11}^2 + \lambda_{14}^2 = 1$$

or, using Eq. (10.12),

$$\lambda_{11}^2 = \frac{1}{1 + (\lambda_{14}^2/\lambda_{11}^2)}$$

$$= \frac{1}{1 - \beta^2}$$

Therefore,

$$\lambda_{11} = \frac{1}{\sqrt{1 - \beta^2}} \tag{10.14}$$

where the positive square root must be chosen in order that Eq. (10.11) reduce to $x_1' = x_1$ when $v = 0$.

Next, from the relations in Eqs. (10.10a), we have

$$\lambda_{14} = \pm \lambda_{41} \tag{10.15}$$

And from Eq. (10.10b),

$$\lambda_{44} = -\frac{\lambda_{11}\lambda_{41}}{\lambda_{14}} = \pm \lambda_{11}$$

$$= \pm \frac{1}{\sqrt{1 - \beta^2}}$$

We can choose the proper sign of λ_{44} by writing the expression for x_4' from $\boldsymbol{\lambda}\mathbf{x}$:

$$x_4' = \lambda_{41} x_1 + \lambda_{44} x_4$$

or,

$$ict' = \lambda_{41} x_1 + ic\lambda_{44} t \tag{10.16}$$

Now, when $v = 0$, then $\lambda_{44} \rightarrow \pm 1$. But we must have $t = t'$ at the common origin ($x_1 = x'_1 = 0$) in such a case. Therefore, in this limit λ_{44} must reduce to $+1$, and so

$$\lambda_{44} = \frac{1}{\sqrt{1-\beta^2}} \tag{10.17}$$

Since both λ_{11} and λ_{44} are positive numbers, Eq. (10.10b) requires that λ_{14} and λ_{41} be of opposite sign. Therefore, Eq. (10.15) becomes

$$\lambda_{14} = -\lambda_{41} \tag{10.18}$$

Combining this result with Eqs. (10.12) and (10.14), we have finally,

$$\lambda_{14} = i\beta\lambda_{11} = \frac{i\beta}{\sqrt{1-\beta^2}} = -\lambda_{41} \tag{10.19}$$

Using Eqs. (10.14), (10.17), and (10.19), the Lorentz transformation matrix is

$$\boxed{\lambda = \begin{pmatrix} \gamma & 0 & 0 & i\beta\gamma \\ 0 & 1 & 0 & 0 \\ 0 & 0 & 1 & 0 \\ -i\beta\gamma & 0 & 0 & \gamma \end{pmatrix}} \tag{10.20}$$

where, in the customary notation

$$\gamma \equiv \frac{1}{\sqrt{1-\beta^2}} \tag{10.21}$$

Therefore, the space-time coordinates in the K' system are

$$\boxed{\begin{aligned} x'_1 &= \frac{x_1 - vt}{\sqrt{1-\beta^2}} = \gamma(x_1 - vt) \\ x'_2 &= x_2 \\ x'_3 &= x_3 \\ t' &= \frac{t - (v/c^2)x_1}{\sqrt{1-\beta^2}} = \gamma\left(t - \frac{\beta}{c}x_1\right) \end{aligned}} \tag{10.22}$$

As required, these equations reduce to the Galilean equations (10.1) when $v \to 0$ (or when $c \to \infty$).

In electrodynamics, the fields propagate with the velocity of light, so that Galilean transformations are never allowed. Indeed, the fact that the electrodynamic field equations (*Maxwell's equations*) are not covariant to Galilean transformations was a major factor in the realization of the necessity for a new theory. It seems rather extraordinary that Maxwell's equations, which are a complete set of equations for the electromagnetic field and are *covariant to Lorentz transformations*, were deduced from experiment long before the advent of relativity theory.

10.4 Momentum and Energy in Relativity

A quantity is called a *four-vector* if it consists of four components, each of which transforms according to the relation*

$$A'_\mu = \sum_\nu \lambda_{\mu\nu} A_\nu \tag{10.23}$$

where the $\lambda_{\mu\nu}$ define a Lorentz transformation. Such a four-vector† is

$$\mathbb{X} = (x_1, x_2, x_3, ict) \tag{10.24}$$

or,

$$\mathbb{X} = (\mathbf{x}, ict) \tag{10.24a}$$

where the notation of the last line means that the first three (space) components of $\mathbb{X}$ define the ordinary three-dimensional position vector $\mathbf{x}$ and that the fourth component is ict. Similarly, the differential of $\mathbb{X}$ is a four-vector:

$$d\mathbb{X} = (d\mathbf{x}, ic\,dt) \tag{10.25}$$

Now, in Minowski space the four-dimensional element of length is an *invariant* (i.e., the magnitude is unaffected by a Lorentz transformation):

$$ds = \sqrt{\sum_\mu dx_\mu^2} = \sqrt{\sum_j dx_j^2 - c^2\,dt^2} \tag{10.26}$$

Furthermore,

$$d\tau = \sqrt{dt^2 - \frac{1}{c^2}\sum_j dx_j^2} = \frac{i}{c}\sqrt{\sum_\mu dx_\mu^2} \tag{10.27}$$

is an invariant since it is simply i/c times the element of length ds. The quantity $d\tau$ is called the element of *proper time* in Minkowski space. The ratio of the

* We shall not distinguish here between *covariant* and *contravariant* vector components; see, for example, Bergmann (Be46, Chapter 5).

† Four-vectors are denoted exclusively by open-face capital letters.

four-vector $d\mathbb{X}$ to the invariant $d\tau$ is therefore also a four-vector, called the four-vector velocity $\mathbb{V}$:

$$\mathbb{V} = \frac{d\mathbb{X}}{d\tau} = \left(\frac{d\mathbf{x}}{d\tau}, ic\frac{dt}{d\tau}\right) \tag{10.28}$$

Now, the components of the ordinary velocity $\mathbf{v}$ are

$$v_j = \frac{dx_j}{dt} \tag{10.29}$$

so that $d\tau$ can be expressed as

$$d\tau = dt\sqrt{1 - \frac{1}{c^2}\sum_j \frac{dx_j^2}{dt^2}}$$

or,

$$d\tau = dt\sqrt{1 - \beta^2} \tag{10.30}$$

The four-vector velocity can therefore be written as

$$\mathbb{V} = \frac{1}{\sqrt{1 - \beta^2}}(\mathbf{v}, ic) \tag{10.31}$$

where $\mathbf{v}$ represents the three space components of ordinary velocity, v_1, v_2, v_3.

In Newtonian mechanics we obtained the momentum of a particle by taking the product of its mass and its velocity. We may do the same in relativistic mechanics, but in order that the "mass" of the particle be truly a characteristic of the *particle* and not of its velocity in some arbitrary reference frame, the "mass" must be that measured in the frame of reference which is at rest with respect to the particle, i.e., the particle's *rest frame* (or *proper frame*). We call this mass the *rest mass* (or *proper mass*) of the particle and denote it by m_0. The four-vector momentum is therefore

$$\mathbb{P} = \left(\frac{m_0 \mathbf{v}}{\sqrt{1 - \beta^2}}, ip_4\right) \tag{10.32}$$

where

$$p_4 \equiv \frac{m_0 c}{\sqrt{1 - \beta^2}} \tag{10.33}$$

The first three components of the four-vector momentum $\mathbb{P}$ are just the components of the ordinary momentum:

$$P_j = p_j = mv_j, \qquad j = 1, 2, 3 \tag{10.34}$$

where

$$\boxed{m \equiv \frac{m_0}{\sqrt{1-\beta^2}}} \tag{10.35}$$

Therefore, if we wish to interpret the momentum of a particle in the classical sense, then the mass is no longer an invariant but depends upon the velocity in the particular reference frame.* Thus, the covariant formulation of momentum leads immediately to the variation of mass with velocity. It should be emphasized that it is the *inertial* mass m which is a function of the velocity.

Next, by taking the time derivative of the three space components of the momentum, we obtain the equations of motion from $F_j = \dot{p}_j$. Thus,

$$\boxed{\mathbf{F} = \frac{d}{dt}\left(\frac{m_0 \mathbf{v}}{\sqrt{1-\beta^2}}\right)} \tag{10.36}$$

where $\mathbf{F}$ is the three-dimensional force vector.

The relativistic relation for energy can be derived by noting that $\mathbf{F} \cdot \mathbf{v}$ is just the work done on the particle by the force per unit time and is equal to the time rate of change of the kinetic energy T. Using Eq. (10.36),

$$\mathbf{F} \cdot \mathbf{v} = \frac{dT}{dt} = \mathbf{v} \cdot \frac{d}{dt}\left(\frac{m_0 \mathbf{v}}{\sqrt{1-\beta^2}}\right) \tag{10.37}$$

It is easily verified by direct calculation that this expression is equivalent to

$$\frac{dT}{dt} = m_0 c^2 \frac{d}{dt}\left(\frac{1}{\sqrt{1-\beta^2}}\right) \tag{10.38}$$

If we integrate this equation with respect to time, we obtain

$$\int_{t_1}^{t_2} \frac{dT}{dt}\, dt = T_2 - T_1$$

$$= \left.\frac{m_0 c^2}{\sqrt{1-\beta^2}}\right|_{t_1}^{t_2} \tag{10.38a}$$

* This result was first obtained by Lorentz in 1904, but under very special assumptions that are not necessary in Einsteinian relativity theory.

If we take t_1 to correspond to the time at which the particle was at rest, then the kinetic energy T can be written in general form as

$$\boxed{T = \frac{m_0 c^2}{\sqrt{1-\beta^2}} - m_0 c^2} \tag{10.39}$$

The first term in this expression is just c^2 times the mass m defined by Eq. (10.35) so that

$$\boxed{T = mc^2 - m_0 c^2} \tag{10.40}$$

That is, the kinetic energy T is the difference between mc^2 and the *rest energy* m_0c^2. Hence, the quantity mc^2 is to be interpreted as the *total energy* E of the particle:

$$E \text{ (total energy)} = mc^2 = T \text{ (kinetic energy)} + m_0 c^2 \text{ (rest energy)} \tag{10.40a}$$

This is the simplest example of the equivalence of mass and energy, a result which is of paramount importance in all theories and applications of nuclear physics.*

If, in Eq. (10.39), the velocity v is small so that $\beta \ll 1$, then we may expand the radical and obtain

$$\begin{aligned} T &= m_0 c^2(1 + \tfrac{1}{2}\beta^2 + \tfrac{3}{8}\beta^4 + \cdots) - m_0 c^2 \\ &= \tfrac{1}{2}m_0 v^2 + \tfrac{3}{8}m_0 \frac{v^4}{c^2} + \cdots \end{aligned} \tag{10.41}$$

Therefore, for sufficiently small velocities, only the first term is significant and the relation becomes the same as the Newtonian result $T = \frac{1}{2}m_0v^2$ to a high degree of accuracy.

According to Eq. (10.40a), the fourth component of the momentum can be expressed as

$$p_4 = \frac{m_0 c}{\sqrt{1-\beta^2}} = mc = \frac{E}{c} \tag{10.42}$$

Therefore, the four-vector momentum can be written as

$$\begin{aligned} \mathbb{P} &= m_0 \mathbb{V} \\ &= \left(\mathbf{p}, i\frac{E}{c}\right) \end{aligned} \tag{10.43}$$

where $\mathbf{p}$ stands for the three space components of momentum. Thus, in

* The mass-energy relation was first obtained by Einstein in 1905.

relativity theory momentum and energy are linked in a manner similar to that which joins the concepts of space and time. If we apply the Lorentz transformation matrix in Eq. (10.20) to the momentum $\mathbb{P}$, we find

$$\boxed{\begin{aligned} p_1' &= \frac{p_1 - (v/c^2)E}{\sqrt{1-\beta^2}} \\ p_2' &= p_2 \\ p_3' &= p_3 \\ E' &= \frac{E - vp_1}{\sqrt{1-\beta^2}} \end{aligned}} \tag{10.44}$$

The square of the four-vector velocity [Eq. (10.31)] is an invariant:

$$\mathbb{V}^2 = \sum_\mu V_\mu^2 = \frac{v^2 - c^2}{1-\beta^2} = -c^2 \tag{10.45}$$

Hence, the square of the four-vector momentum is an invariant:

$$\mathbb{P}^2 = \sum_\mu P_\mu^2 = m_0^2 \mathbb{V}^2 = -m_0^2 c^2 \tag{10.46a}$$

From Eq. (10.43) we also have, using $\mathbf{p}\cdot\mathbf{p} = p^2 \equiv p_1^2 + p_2^2 + p_3^2$,

$$\mathbb{P}^2 = p^2 - \frac{E^2}{c^2} \tag{10.46b}$$

and combining these expressions leads to the important result

$$\boxed{E^2 = p^2c^2 + m_0^2 c^4} \tag{10.47}$$

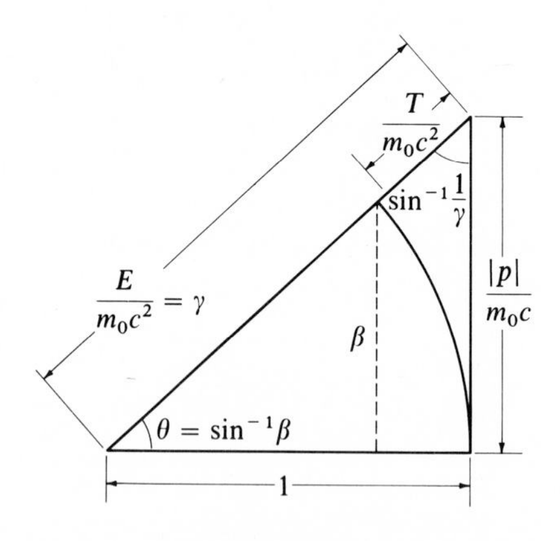

FIG. 10-2

In normalized form, we can write

$$\left(\frac{E}{m_0 c^2}\right)^2 = \left(\frac{|p|}{m_0 c}\right)^2 + 1 \tag{10.47a}$$

If we define an angle θ such that $\beta = \sin\theta$, the relativistic relations between velocity, momentum, and energy can be obtained from trigonometric relations involving the so-called "relativistic triangle," shown in Fig. 10-2.

10.5 Some Consequences of the Lorentz Transformation

The application of the Lorentz transformation leads immediately to a wide variety of interesting results. We shall not attempt to present here a catalog of these results but will only discuss three effects of particular interest.

▶ Example 10.5(a) The FitzGerald–Lorentz Contraction of Length*

Consider a rod of length l which lies along the x_1-axis of an inertial frame K. An observer in a system K' moving with a uniform velocity v along the x_1-axis (as in Fig. 10-1), measures the length of the rod in *his* coordinate system. He accomplishes this by determining *at a given instant of time* t' the difference in the coordinates of the ends of the rod, $x_1'(2) - x_1'(1)$. According to the transformation equations (10.22),

$$x_1'(2) - x_1'(1) = \frac{[x_1(2) - x_1(1)] - v[t(2) - t(1)]}{\sqrt{1-\beta^2}}$$

where $x_1(2) - x_1(1) = l$. Note that the times $t(2)$ and $t(1)$ are the times in the K system at which the observations are made and do not correspond to the instant in K' at which the observer makes his measurements. In fact, since $t'(2) = t'(1)$, Eqs. (10.22) give

$$t(2) - t(1) = [x_1(2) - x_1(1)]\frac{v}{c^2}$$

Therefore, the length l' as measured in the K' system is

$$l' = x_1'(2) - x_1'(1)$$

or,

$$\boxed{l' = l\sqrt{1-\beta^2}} \tag{10.48}$$

* The contraction of length in the direction of motion was proposed by G. F. FitzGerald (1851–1901) in 1892 as a possible explanation of the Michelson–Morley ether-drift experiment. Almost immediately this hypothesis was adopted by Lorentz who proceeded to apply it in his theory of electrodynamics.

Thus, to an observer in motion relative to an object, the dimensions of objects are contracted by a factor $\sqrt{1-\beta^2}$ in the direction of motion.

An interesting consequence of the FitzGerald–Lorentz contraction of length was discovered in 1959 by James Terrell.* Consider a cube of side l which moves with a uniform velocity v with respect to an observer who is some distance away. Figure 10-3a shows the projection of the cube on the plane

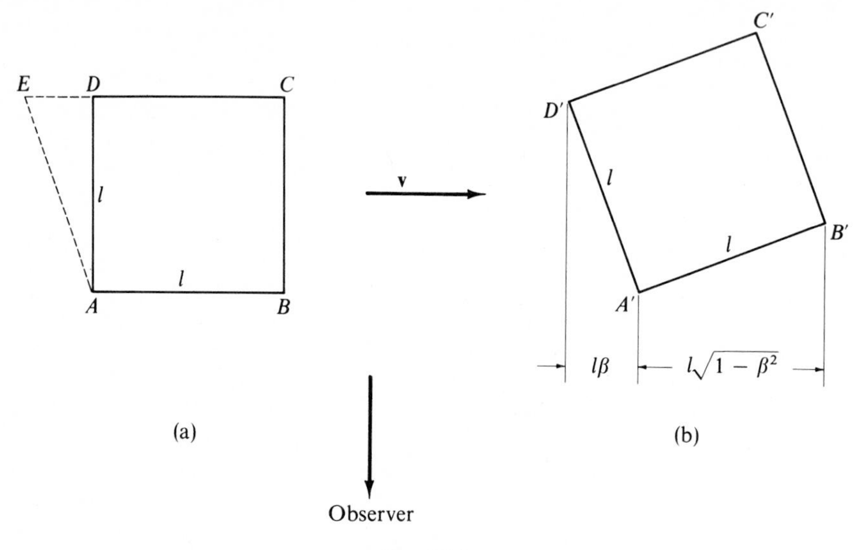

FIG. **10-3**

containing the velocity vector **v** and the observer. The cube moves with its side AB perpendicular to the observer's line of sight. We wish to determine what the observer "sees"; that is, at a given instant of time in the observer's rest frame, we wish to determine the relative orientation of the corners A, B, C, and D. The traditional view (which went unquestioned for over 50 years!) was that the only effect is a foreshortening of the sides AB and CD so that the observer sees a distorted cube of height l but of length $l\sqrt{1-\beta^2}$. Terrell pointed out that this interpretation overlooks the fact that in order for light from the corners A and D to reach the observer at the same instant, the light from D, which must travel a distance l farther than that from A, must have been emitted when the corner D was at the position E. The length DE is equal to $(l/c)v = l\beta$. Therefore, the observer sees not only the face AB which is perpendicular to his line of sight but also the face AD which is *parallel* to his line of sight. In addition, the length of the side AB is foreshortened in the normal

* J. Terrell, *Phys. Rev.* **116**, 1041 (1959).

way to $l\sqrt{1-\beta^2}$. The net result is shown in Fig. 10-3b and corresponds exactly to the view that the observer would have if the cube were rotated through an angle $\sin^{-1}\beta$. Therefore, the cube is not distorted; it undergoes an *apparent* rotation. The customary statement* that a moving sphere appears as an ellipsoid is incorrect; it appears still as a sphere.†

▶ Example 10.5(b) Time Dilation

Consider a clock at a certain position in the K system which produces signals with the interval

$$\Delta t = t(2) - t(1)$$

To an observer in the moving system K', the time interval will be

$$\Delta t' = t'(2) - t'(1) = \frac{t(2) - t(1)}{\sqrt{1-\beta^2}}$$

or,

$$\boxed{\Delta t' = \frac{\Delta t}{\sqrt{1-\beta^2}}} \tag{10.49}$$

Thus, to an observer in motion relative to the clock, the time intervals appear to be lengthened.

A striking example of time dilation is afforded by observations made on muons. These particles are known to be generated in large numbers from the decay of π mesons produced by cosmic-ray particles interacting with the tenuous gas near the top of the atmosphere, i.e., at heights of 10 to 20 km. Although muons are unstable particles and eventually decay, the fact that they are observed at the surface of the Earth implies that the lifetime must be at least 10 km$/c \cong 3 \times 10^{-5}$ sec. But it is also known that the mean lifetime of muons in their *proper frame* is approximately 2.2×10^{-6} sec. The increase by an order of magnitude of the lifetime in the Earth reference system is due to the high velocity ($\beta \approx 0.98$) with which the muons traverse the atmosphere.

▶ Example 10.5(c) The Velocity Addition Rule

Suppose that there are three inertial reference frames, K, K', and K'', which are in co-linear motion along their respective x_1-axes. Let the velocity of K' relative to K be v_1 and let the velocity of K'' relative to K' be v_2. Now, the velocity of K'' relative to K cannot be $v_1 + v_2$ since it must be possible to propagate a signal between

* See, for example, Joos and Freeman (Jo50, p. 242).

† An interesting discussion of apparent rotations at high velocity is given by V. F. Weisskopf, *Phys. Today* **13**, No. 9, 24 (1960); reprinted in Am63.

any two inertial frames, and if both v_1 and v_2 are greater than $c/2$ (but less than c), then $v_1 + v_2 > c$. Therefore, the rule for the addition of velocities in relativity must be different from that in Galilean theory. The relativistic velocity addition rule can be obtained by considerihg the Lorentz transformation matrix which connects K and K''. The individual transformation matrices are:

$$\lambda_{K' \to K} = \begin{pmatrix} \gamma_1 & 0 & 0 & i\beta_1\gamma_1 \\ 0 & 1 & 0 & 0 \\ 0 & 0 & 1 & 0 \\ -i\beta_1\gamma_1 & 0 & 0 & \gamma_1 \end{pmatrix}$$

$$\lambda_{K'' \to K'} = \begin{pmatrix} \gamma_2 & 0 & 0 & i\beta_2\gamma_2 \\ 0 & 1 & 0 & 0 \\ 0 & 0 & 1 & 0 \\ -i\beta_2\gamma_2 & 0 & 0 & \gamma_2 \end{pmatrix}$$

Therefore, the transformation from K'' to K is just the product of these two transformations:

$$\lambda_{K'' \to K} = \lambda_{K'' \to K'}\, \lambda_{K' \to K} = \begin{pmatrix} \gamma_1\gamma_2(1 + \beta_1\beta_2) & 0 & 0 & i\gamma_1\gamma_2(\beta_1 + \beta_2) \\ 0 & 1 & 0 & 0 \\ 0 & 0 & 1 & 0 \\ -i\gamma_1\gamma_2(\beta_1 + \beta_2) & 0 & 0 & \gamma_1\gamma_2(1 + \beta_1\beta_2) \end{pmatrix}$$

In order that the elements of this matrix correspond to those of the normal Lorentz matrix [Eq. (10.20)], we must identify β and γ for the $K'' \to K$ transformation as

$$\left.\begin{aligned} \gamma &= \gamma_1\gamma_2(1 + \beta_1\beta_2) \\ \beta\gamma &= \gamma_1\gamma_2(\beta_1 + \beta_2) \end{aligned}\right\} \qquad (10.50)$$

from which we obtain

$$\beta = \frac{\beta_1 + \beta_2}{1 + \beta_1\beta_2} \qquad (10.51)$$

If we multiply this last expression by c, we have the usual form of the velocity addition rule:

$$\boxed{v = \frac{v_1 + v_2}{1 + (v_1 v_2/c^2)}} \qquad (10.51a)$$

It follows that if $v_1 < c$ and $v_2 < c$, then $v < c$ also.

Even though *signal* velocities can never exceed c, there are certain other types of velocities that can be greater than c. For example, the *phase velocity* of a light wave in a medium in which the index of refraction is less than unity will be greater than c, but the phase velocity does not correspond to the signal velocity in such a medium; the signal velocity is indeed less than c. Also, consider an electron gun which emits a beam of electrons. If the gun is rotated, then the electron beam will describe a certain path on a screen placed at some appropriate distance. If the angular velocity of the gun and the distance to the screen are sufficiently large, then the velocity of the spot traveling across the screen can be *any* velocity, arbitrarily large. Thus, the *writing speed* of an oscilloscope can exceed c, but, again, the writing speed does not correspond to the signal velocity. That is, information cannot be transmitted from one point on the screen to another by means of the electron beam. In such a device, a signal can be transmitted only from the gun to the screen and this transmission takes place at the velocity of the electrons in the beam (i.e., $< c$).

■ 10.6 The Lagrangian Function in Special Relativity

The discussion of Lagrangian and Hamiltonian dynamics in Chapter 7 assumed the constancy of mass and is therefore correct only in the nonrelativistic limit. We can extend the Lagrangian formalism into the realm of special relativity in the following way. For a single (nonrelativistic) particle moving in a velocity-independent potential, the rectangular momentum components may be written as [cf. Eq. (7.50)]:

$$p_i = \frac{\partial L}{\partial v_i} \tag{10.52}$$

Now, according to Eq. (10.32), the relativistic expression for the ordinary (i.e., space) momentum components is

$$p_i = \frac{m_0 v_i}{\sqrt{1 - \beta^2}} \tag{10.53}$$

We now require that the *relativistic* Lagrangian, when differentiated with respect to v_i as in Eq. (10.52), yield the momentum components given by Eq. (10.53):

$$\frac{\partial L}{\partial v_i} = \frac{m_0 v_i}{\sqrt{1 - \beta^2}} \tag{10.54}$$

This requirement which we have imposed involves only the *velocity* of the particle, so we expect that the velocity-*independent* part of the relativistic Lagrangian is unchanged from the nonrelativistic case. The velocity-*dependent* part, however, may no longer be equal to the kinetic energy. Therefore, we write

$$L = T^* - U \tag{10.55}$$

where $U = U(x_i)$ and $T^* = T^*(v_i)$. The function T^* must satisfy the relation

$$\frac{\partial T^*}{\partial v_i} = \frac{m_0 v_i}{\sqrt{1-\beta^2}} \tag{10.56}$$

It can be easily verified that a suitable expression for T^* (apart from a possible constant of integration which can be suppressed) is

$$T^* = -m_0 c^2 \sqrt{1-\beta^2} \tag{10.57}$$

Hence, the relativistic Lagrangian can be written as

$$\boxed{L = -m_0 c^2 \sqrt{1-\beta^2} - U} \tag{10.58}$$

and the equations of motion are obtained in the standard way from Lagrange's equations.

Notice that the Lagrangian is *not* given by $T - U$, since the relativistic expression for the kinetic energy is [Eq. (10.39)]:

$$T = \frac{m_0 c^2}{\sqrt{1-\beta^2}} - m_0 c^2 \tag{10.59}$$

The Hamiltonian can be calculated from [cf. Eq. (7.53)]:

$$\begin{aligned} H &= \sum_i v_i p_i - L \\ &= \sum_i \frac{m_0 v_i^2}{\sqrt{1-\beta^2}} + m_0 c^2 \sqrt{1-\beta^2} + U \\ &= \frac{m_0 c^2}{\sqrt{1-\beta^2}} + U \\ &= T + m_0 c^2 + U = E \end{aligned} \tag{10.60}$$

so that the relativistic Hamiltonian is equal to the total energy (in this case *including* the rest energy).

■ 10.7 Relativistic Kinematics

In the event that the velocities in a collision process are not negligible with respect to the velocity of light, it becomes necessary to use *relativistic* kinematics. In the discussion in Chapter 9 we took advantage of the properties of the center-of-mass coordinate system in deriving many of the kinematic relations. Since mass is a variable quantity in relativity theory, it no longer is meaningful to speak of a "center-of-mass" system; in relativistic kinematics, one uses a "center-of-momentum" coordinate system instead. Such a system possesses the same essential property as does the previously used C.M. system, viz., that the total linear momentum in the system is zero. Therefore, if a particle of rest mass m_1 collides with a particle of rest mass m_2, then in the center-of-momentum system we have

$$p_1' = p_2' \tag{10.61}$$

Using Eq. (10.32), the space components of the momentum four-vector can be written as

$$m_1 u_1' \gamma_1' = m_2 u_2' \gamma_2' \tag{10.62}$$

where, as before, $\gamma \equiv 1/\sqrt{1-\beta^2}$ and $\beta \equiv u/c$.

In a collision problem it is convenient to associate the laboratory coordinate system with the inertial system K and the center-of-momentum system with K'. Then, a simple Lorentz transformation connects the two systems. To derive the relativistic kinematic expressions, the procedure is to obtain the center-of-momentum relations and then to perform a Lorentz transformation back to the lab system. We choose the coordinate axes so that m_1 moves along the x-axis in K with a velocity u_1. Since m_2 is initially at rest in K, $u_2=0$. In K', m_2 moves with a velocity u_2' and so K' moves with respect to K also with a velocity u_2' and in the same direction as the initial motion of m_1.

Using the fact that $\beta\gamma = \sqrt{\gamma^2 - 1}$, we have

$$p_1' = m_1 u_1' \gamma_1' = m_1 c \beta_1' \gamma_1' = m_1 c \sqrt{\gamma_1'^2 - 1} = m_2 c \sqrt{\gamma_2'^2 - 1} = p_2' \tag{10.63}$$

which expresses the equality of the momenta in the center-of-momentum system.

Now, according to Eq. (10.44), the transformation of the momentum p_1 (from K to K') is

$$p_1' = \left(p_1 - \frac{u_2'}{c^2} E_1\right)\gamma_2' \tag{10.64}$$

We also have

$$\left.\begin{aligned} p_1 &= m_1 u_1 \gamma_1 \\ E_1 &= m_1 c^2 \gamma_1 \end{aligned}\right\} \tag{10.65}$$

so that Eq. (10.63) can be used to obtain

$$\begin{aligned} m_1 c\sqrt{\gamma_1'^2 - 1} &= (m_1 c \beta_1 \gamma_1 - \beta_2' m_1 c \gamma_1)\gamma_2' \\ &= m_1 c\left(\gamma_2' \sqrt{\gamma_1^2 - 1} - \gamma_1 \sqrt{\gamma_2'^2 - 1}\right) = m_2 c \sqrt{\gamma_2'^2 - 1} \end{aligned} \tag{10.66}$$

These equations can be solved for γ_1' and γ_2' in terms of γ_1:

$$\gamma_1' = \frac{\gamma_1 + \dfrac{m_1}{m_2}}{\sqrt{1 + 2\gamma_1\left(\dfrac{m_1}{m_2}\right) + \left(\dfrac{m_1}{m_2}\right)^2}} \tag{10.67a}$$

$$\gamma_2' = \frac{\gamma_1 + \dfrac{m_2}{m_1}}{\sqrt{1 + 2\gamma_1\left(\dfrac{m_1}{m_2}\right) + \left(\dfrac{m_1}{m_2}\right)^2}} \tag{10.67b}$$

Next, we write the equations of the transformation of the momentum components from K' back to K after the scattering. We now have both x- and y-components:

$$\begin{aligned} p_{1,x} &= \left(p_{1,x}' + \frac{u_2'}{c^2} E_1'\right)\gamma_2' \\ &= (m_1 c\beta_1'\gamma_1' \cos\theta + m_1 c\beta_2'\gamma_1')\gamma_2' \\ &= m_1 c\gamma_1'\gamma_2'(\beta_1' \cos\theta + \beta_2') \end{aligned} \tag{10.68a}$$

[Note that the transformation is from K' to K so that a plus sign occurs before the second term, in contrast to Eq. (10.64).] Also,

$$p_{1,y} = m_1 c\beta_1'\gamma_1' \sin\theta \tag{10.68b}$$

Now, the tangent of the laboratory scattering angle ψ is given by $p_{1,y}/p_{1,x}$; therefore, dividing Eq. (10.68b) by (10.68a), we obtain

$$\tan\psi = \frac{1}{\gamma_2'} \frac{\sin\theta}{\cos\theta + (\beta_2'/\beta_1')}$$

Using Eq. (10.62) to express β_2'/β_1', the result is

$$\tan\psi = \frac{1}{\gamma_2'} \frac{\sin\theta}{\cos\theta + (m_1\gamma_1'/m_2\gamma_2')} \tag{10.69}$$

For the recoil particle, we have

$$\begin{aligned} p_{2,x} &= \left(p_{2,x}' + \frac{u_2'}{c^2} E_2'\right)\gamma_2' \\ &= (-m_2 c\beta_2'\gamma_2' \cos\theta + m_2 c\beta_2'\gamma_2')\gamma_2' \\ &= m_2 c\beta_2'\gamma_2'^2(1 - \cos\theta) \end{aligned} \tag{10.70a}$$

where a minus sign occurs in the first term since $p'_{2,x}$ is directed opposite to $p_{1,x}$. Also,

$$p_{2,y} = -m_2 c\beta'_2 \gamma'_2 \sin\theta \tag{10.70b}$$

As before, the tangent of the laboratory angle recoil ζ is given by $p_{2,y}/p_{2,x}$:

$$\tan\zeta = -\frac{1}{\gamma'_2}\frac{\sin\theta}{1-\cos\theta} \tag{10.71}$$

The overall minus sign indicates that if m_1 is scattered toward positive values of y, then m_2 recoils in the negative y direction.

A case of special interest is that in which $m_1 = m_2$. From Eqs. (10.67), we find

$$\gamma'_1 = \gamma'_2 = \sqrt{\frac{1+\gamma_1}{2}}, \qquad m_1 = m_2 \tag{10.72}$$

The tangents of the scattering angles become

$$\tan\psi = \sqrt{\frac{2}{1+\gamma_1}}\cdot\frac{\sin\theta}{1+\cos\theta} \tag{10.73}$$

$$\tan\zeta = -\sqrt{\frac{2}{1+\gamma_1}}\cdot\frac{\sin\theta}{1-\cos\theta} \tag{10.74}$$

Therefore, the product is

$$\tan\psi\tan\zeta = -\frac{2}{1+\gamma_1}, \qquad m_1 = m_2 \tag{10.75}$$

(The minus sign is of no essential importance; it only indicates that ψ and ζ are measured in opposite directions.)

We previously found that in the nonrelativistic limit, there was always a right angle between the final velocity vectors in the scattering of particles of equal mass. Indeed, in the limit $\gamma_1 \to 1$, Eqs. (10.73) and (10.74) become equal to Eqs. (9.8) and (9.11), respectively, so that $\psi + \zeta = \pi/2$. Equation (10.75), on the other hand, shows that in the relativistic case $\psi + \zeta < \pi/2$; thus, the included angle in the scattering is always smaller than in the non-relativistic limit. For equal scattering and recoil angles, $\psi = \zeta$, Eq. (10.75) becomes

$$\tan\psi = \left(\frac{2}{1+\gamma_1}\right)^{\frac{1}{2}}, \qquad m_1 = m_2$$

and the included angle between the directions of the scattered and recoil particles is

$$\begin{aligned}\varphi &= \psi + \zeta = 2\psi \\ &= 2 \tan^{-1}\left(\frac{2}{1+\gamma_1}\right)^{\frac{1}{2}}, \qquad m_1 = m_2\end{aligned} \tag{10.76}$$

Figure 10-4 shows φ as a function of γ_1 up to $\gamma_1 = 20$. At $\gamma_1 = 10$, the included angle is approximately 46°. This value of γ_1 corresponds to an initial velocity which is 99.5% of the velocity of light. According to Eq. (10.39), the kinetic

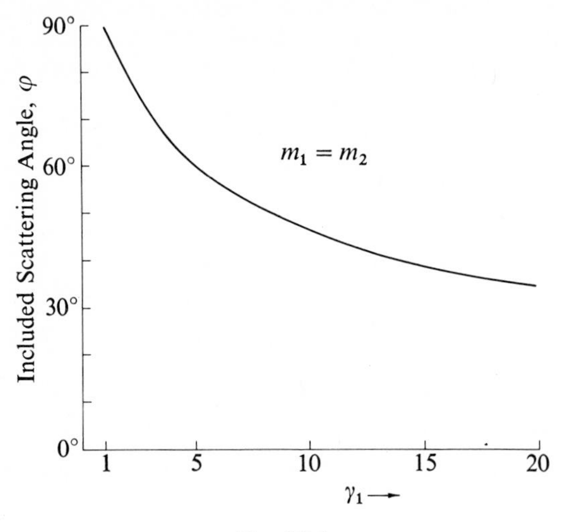

FIG. **10-4**

energy is given by $T_1 = m_1c^2(\gamma_1 - 1)$; therefore, a proton with $\gamma_1 = 10$ would have a kinetic energy of approximately 8.4 GeV, whereas an electron with the same velocity would have $T_1 \cong 4.6$ MeV.*

By using the transformation properties of the fourth component of the momentum four-vector (i.e., the total energy), it is possible to obtain the relativistic analogues of all the energy equations which we have previously derived in the nonrelativistic limit.

Suggested References

The foundations of relativity theory are discussed, for example, by Lindsay and Margenau (Li36, Chapter 7).

At the intermediate level, Joos and Freeman (Jo50, Chapter 10) give a short but comprehensive discussion. A very readable account is to be found in Tolman (To34, Chapters 1 and 2). Bergmann (Be46) gives a general treatment, but because

* These units of energy are defined in Problem 10-10: 1 GeV $= 10^3$ MeV $= 10^9$ eV $= 1.602 \times 10^{-3}$ erg.

it is designed for general relativity the notation tends to be a little strange in places. Bradbury (Br68, Chapter 13) gives a more-or-less standard treatment. Goldstein's discussion (Go50, Chapter 6) is concise and elegant. The treatment by Wangsness (Wa63a) is brief and to the point. In addition, almost every text on "modern physics" discusses relativity to some extent; see, for example, Richtmyer, Kennard, and Cooper (Ri69, Chapters 2 and 3), or Leighton (Le59, Chapter 1).

A delightful little book, well worth reading, is that by Ney (Ne62). Sherwin (Sh61, Chapter 4) also gives a very interesting discussion, complete with many clever diagrams and examples. An interesting introductory account is given by Taylor (Ta63, Chapters 11–13). The treatments given by Bondi (Bo64) and by Bohm (Bo65) are both readable and instructive. One of the best treatments available is that by Taylor and Wheeler (Ta66).

Several recent paperbacks treat special relativity at the intermediate level: Good (Go68), Kacser (Ka67), Katz (Ka64), and Resnick (Re68).

Michelson's measurement of the speed of light (for which he became the first American to win a Nobel prize in physics) and the Michelson–Morley experiment are described in the short and readable book by Jaffe (Ja60).

Translations of some of the original papers by Lorentz, Einstein, Minkowski, and Weyl are available (Lo23); some of the more recent important papers on special relativity have also been collected (Am63).

At the intermediate-to-advanced level, the book by Rosser (Ro64) is particularly complete.

Among the advanced treatments, those of Pauli (Pa58) and Møller (Mø52) are probably the most useful.

Two of the better discussions of relativity in electrodynamics are those of Jackson (Ja62, Chapters 11 and 12) and Panofsky and Phillips (Pa62, Chapters 15–18). See also Ma65b.

Relativistic kinematics is discussed by Morrison (Mo53a, pp. 3–9) and by Monahan (Mo60). An exhaustive treatment is that of Baldin, Goldanskii, and Rozental (Ba61).

Problems

10-1. Show that the transformation equations connecting the K' and K systems [Eqs. (10.22)] can be expressed as

$$x_1' = x_1 \cosh\alpha - ct \sinh\alpha$$

$$x_2' = x_2\,; \qquad x_3' = x_3$$

$$t' = t\cosh\alpha - \frac{x_1}{c}\sinh\alpha$$

where $\tanh\alpha = v/c$. Show that the Lorentz transformation corresponds to a rotation through an angle $i\alpha$ in four-dimensional space.

10-2. Show that the equation

$$\nabla^2 \Psi - \frac{1}{c^2}\frac{\partial^2 \Psi}{\partial t^2} = 0$$

is invariant under a Lorentz transformation but not under a Galilean transformation. (This is the wave equation which describes the propagation of light waves in free space.)

10-3. Show that the expression for the FitzGerald–Lorentz contraction [Eq. (10.48)] can also be obtained if the observer in the K' system measures the time necessary for the rod to pass a fixed point in his system and then multiplies the result by v.

10-4. What are the apparent dimensions of a cube of side l (in its own proper frame) which moves with a uniform velocity v directly *toward* or *away from* an observer?

10-5. Consider two events that take place at different points in the K system at the same instant t. If these two points are separated by a distance Δx, show that in the K' system the events are not simultaneous but are separated by a time interval $\Delta t' = -v\gamma\,\Delta x/c^2$.

10-6. Two clocks, located at the origins of the K and K' systems (which have a relative velocity v), are synchronized when the origins coincide. After a time t, an observer at the origin of the K system observes the K' clock by means of a telescope. What does the K' clock read?

10-7. In his 1905 paper (see the translation in Lo23), Einstein states: "We conclude that a balance-clock at the equator must go more slowly, by a very small amount, than a precisely similar clock situated at one of the poles under otherwise identical conditions." Neglect the fact that the equator clock does not undergo uniform motion and show that after a century the clocks will differ by approximately 0.0025 sec.

10-8. Consider a relativistic rocket whose velocity with respect to a certain inertial frame is v and whose exhaust gases are emitted with a constant velocity V with respect to the rocket. Show that the equation of motion is

$$m_0 \frac{dv}{dt} + V\frac{dm_0}{dt}(1 - \beta^2) = 0$$

where $m_0 = m_0(t)$ is the mass of the rocket in its rest frame and $\beta = v/c$.

10-9. Consider an inertial frame K which contains a number of particles with rest masses $m_{0,\alpha}$, ordinary momentum components $p_{\alpha,j}$, and total energies E_α. The center-of-mass system of such a group of particles is defined to be that system in which the net ordinary momentum is zero. Show that the velocity components of the center-of-mass system with respect to K are given by

$$\frac{v_j}{c} = \frac{\sum_\alpha p_{\alpha,j}\, c}{\sum_\alpha E_\alpha}$$

10-10. A common unit of energy used in atomic and nuclear physics is the *electron volt* (eV), the energy acquired by an electron in falling through a potential difference of one volt: 1 MeV = 10^6 eV = 1.602×10^{-6} erg. In these units the rest mass of an electron is $m_e c^2 = 0.511$ MeV and that of a proton is $m_p c^2 = 931$ MeV. Calculate the kinetic energy and the quantities β and γ for an electron and for a proton each of which has a momentum of 100 MeV/c. Show that the electron is "relativistic" whereas the proton is "nonrelativistic."

10-11. The energy of a light quantum (or *photon*) is expressed by $E = h\nu$, where h is Planck's constant and ν is the frequency of the photon. The momentum of the photon is $h\nu/c$. Show that if the photon scatters from a free electron (of mass m_e), the scattered photon will have an energy

$$E' = E\left[1 + \frac{E}{m_e c^2}(1 - \cos\theta)\right]^{-1}$$

where θ is the angle through which the photon scatters. Show also that the electron acquires a kinetic energy

$$T = \frac{E^2}{m_e c^2}\left[\frac{1 - \cos\theta}{1 + \dfrac{E}{m_e c^2}(1 - \cos\theta)}\right]$$

10-12. The expression for the ordinary force is [Eq. (10.36)]

$$\mathbf{F} = \frac{d}{dt}\left(\frac{m_0 \mathbf{v}}{\sqrt{1 - \beta^2}}\right)$$

Take $\mathbf{v}$ to be in the x_1-direction and compute the components of the force. Show that

$$F_1 = m_l\, \dot{v}_1; \qquad F_2 = m_t\, \dot{v}_2; \qquad F_3 = m_t\, \dot{v}_3$$

where m_l and m_t are, respectively, the *longitudinal mass* and the *transverse mass*:

$$m_l = \frac{m_0}{(1 - \beta^2)^{3/2}}; \qquad m_t = \frac{m_0}{\sqrt{1 - \beta^2}}$$

10-13. The average rate at which solar radiant energy reaches the Earth is approximately 1.4×10^6 ergs/cm^2-sec. Assume that all of this energy results from the conversion of mass to energy. Calculate the rate at which solar mass is being consumed. If this rate is maintained, calculate the remaining lifetime of the Sun. (Pertinent numerical data will be found in Table 8-1.)

10-14. Use Eq. (8.60) to show that the rate of advance of the perihelion may be expressed as

$$\Delta \cong 6\pi \left(\frac{v_{\max}}{c} \right)^2$$

if ε is small. That the value for Mercury is greater than for any other planet is attributable to the fact that the orbital velocity of Mercury is the greatest of any planet. Consequently, the relativistic parameter v/c is largest for Mercury. Compute $v_{\max}/c$ for Mercury and compare with results for Venus, Earth, and Saturn. Use the results of this calculation to verify that $\Delta \cong 40$ sec of arc per century for Mercury.

10-15. Consider a one-dimensional, relativistic harmonic oscillator for which the Lagrangian is

$$L = m_0 c^2 (1 - \sqrt{1 - \beta^2}) - \tfrac{1}{2} k x^2$$

Obtain the Lagrange equation of motion and show that it can be integrated to yield

$$E = m_0 c^2 + \tfrac{1}{2} k a^2$$

where a is the maximum excursion from equilibrium of the oscillating particle. Show that the period

$$\tau = 4 \int_{x=0}^{x=a} dt$$

can be expressed as

$$\tau = \frac{2a}{\kappa c} \int_0^{\pi/2} \frac{(1 + 2\kappa^2 \cos^2 \varphi)}{\sqrt{1 + \kappa^2 \cos^2 \varphi}} \, d\varphi$$

Expand the integrand in powers of $\kappa \equiv (a/2)\sqrt{k/m_0 c^2}$ and show that, to first order in κ,

$$\tau \cong \tau_0\left(1 + \frac{3}{16}\frac{ka^2}{m_0 c^2}\right)$$

where τ_0 is the nonrelativistic period for small oscillations, $2\pi\sqrt{m_0/k}$.

10-16. Show that the relativistic expression for the kinetic energy of a particle scattered through an angle ψ by a target particle of equal rest mass is

$$\frac{T_1}{T_0} = \frac{2\cos^2\psi}{(\gamma_1 + 1) - (\gamma_1 - 1)\cos^2\psi}$$

The expression evidently reduces to Eq. (9.25a) in the nonrelativistic limit $(\gamma_1 \to 1)$. Sketch $T_1(\psi)$ for neutron-proton scattering for incident neutron energies of 100 MeV, 1 GeV, and 10 GeV.

CHAPTER 11

Motion in a Noninertial Reference Frame

11.1 Introduction

The advantage of choosing an inertial reference frame for the description of dynamical processes were made evident in the discussions in Chapters 2 and 7. It is, of course, always possible to express the equations of motion for a system in an inertial frame; however, there are types of problems for which these equations would be extremely complex, and it becomes easier to treat the motion of the system in a noninertial frame of reference.

In order to describe, for example, the motion of a particle on or near the surface of the Earth, it is clearly tempting to do so by choosing a coordinate system fixed with respect to the Earth. We know, however, that the Earth undergoes a complicated motion, compounded of many different rotations (and, hence, accelerations) with respect to an inertial reference frame identified with the "fixed" stars. The Earth coordinate system is, therefore, a *noninertial* frame of reference; and, although the solutions to many problems can be obtained to the desired degree of accuracy by ignoring this distinction, there are many important effects that result from the noninertial nature of the Earth coordinate system.

In the analysis of the motion of rigid bodies in the following chapter, we shall also find it convenient to use noninertial reference frames and we will therefore make use of much of the development presented here.

11.2 Rotating Coordinate Systems

Let us consider two sets of coordinate axes; let one set be the "fixed" or inertial axes, and let the other be an arbitrary set which may be in motion with respect to the inertial system. We shall designate these axes as the "fixed" and "rotating" axes, respectively, and shall use x_i' as coordinates in the fixed system and x_i as coordinates in the rotating system. If we choose some point P, as in Fig. 11-1, we clearly have

$$\mathbf{r}' = \mathbf{R} + \mathbf{r} \tag{11.1}$$

where $\mathbf{r}'$ is the radius vector of P in the fixed system and where $\mathbf{r}$ is the radius vector of P in the rotating system. The vector $\mathbf{R}$ locates the origin of the rotating system in the fixed system.

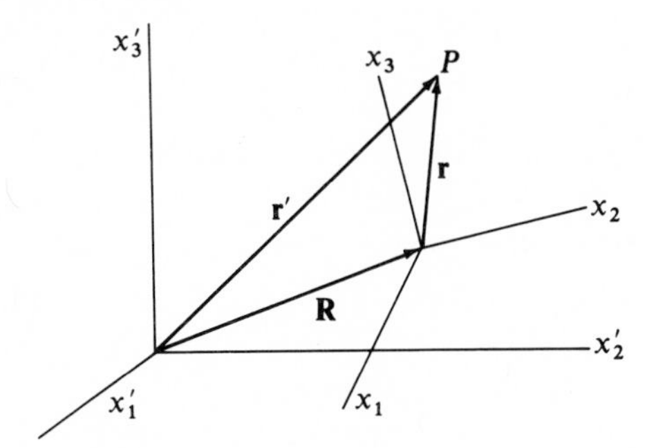

FIG. 11-1

An arbitrary infinitesimal displacement may always be represented by a pure rotation about some axis which is called the *instantaneous axis of rotation.* For example, the instantaneous motion of a disk rolling down an inclined plane can be described as a rotation about the point of contact between the disk and the plane. Therefore, if the x_i system undergoes an infinitesimal rotation $\delta\boldsymbol{\theta}$, corresponding to some arbitrary infinitesimal displacement, the motion of P (which, for the moment, we consider to be at rest in the x_i system) can be described in terms of Eq. (1.106) as

$$(d\mathbf{r})_{\text{fixed}} = d\boldsymbol{\theta} \times \mathbf{r} \tag{11.2}$$

where the designation "fixed" is explicitly included to indicate that the quantity $d\mathbf{r}$ is measured in the x_i', or *fixed* coordinate system. Dividing this equation by dt, the time interval during which the infinitesimal rotation takes place, we obtain the time of rate change of $\mathbf{r}$ as measured in the fixed coordinate system:

$$\left(\frac{d\mathbf{r}}{dt}\right)_{\text{fixed}} = \frac{d\boldsymbol{\theta}}{dt} \times \mathbf{r} \tag{11.3}$$

or, since the angular velocity of the rotation is

$$\boldsymbol{\omega} \equiv \frac{d\boldsymbol{\theta}}{dt} \tag{11.4}$$

we have

$$\left(\frac{d\mathbf{r}}{dt}\right)_{\text{fixed}} = \boldsymbol{\omega} \times \mathbf{r} \qquad \text{(for } P \text{ fixed in } x_i \text{ system)} \tag{11.5}$$

Now, if we allow the point P to have a velocity $(d\mathbf{r}/dt)_{\text{rotating}}$ with respect to the x_i system, this velocity must be added to $\boldsymbol{\omega} \times \mathbf{r}$ to obtain the time rate of change of $\mathbf{r}$ in the fixed system:

$$\left(\frac{d\mathbf{r}}{dt}\right)_{\text{fixed}} = \left(\frac{d\mathbf{r}}{dt}\right)_{\text{rotating}} + \boldsymbol{\omega} \times \mathbf{r} \tag{11.6}$$

Although we chose the displacement vector $\mathbf{r}$ for the derivation of Eq. (11.6), the validity of this expression is not limited to the vector $\mathbf{r}$. In fact, for an arbitrary vector $\mathbf{Q}$, we have

$$\boxed{\left(\frac{d\mathbf{Q}}{dt}\right)_{\text{fixed}} = \left(\frac{d\mathbf{Q}}{dt}\right)_{\text{rotating}} + \boldsymbol{\omega} \times \mathbf{Q}} \tag{11.7}$$

We note, for example, that the angular acceleration $\dot{\boldsymbol{\omega}}$ is the same in both the fixed and rotating systems:

$$\left(\frac{d\boldsymbol{\omega}}{dt}\right)_{\text{fixed}} = \left(\frac{d\boldsymbol{\omega}}{dt}\right)_{\text{rotating}} + \boldsymbol{\omega} \times \boldsymbol{\omega} \equiv \dot{\boldsymbol{\omega}} \tag{11.8}$$

since $\boldsymbol{\omega} \times \boldsymbol{\omega}$ vanishes and where $\dot{\boldsymbol{\omega}}$ designates the common value in the two systems.

Equation (11.7) may now be used to obtain the expression for the velocity of the point P as measured in the fixed coordinate system. From Eq. (11.1) we have

$$\left(\frac{d\mathbf{r}'}{dt}\right)_{\text{fixed}} = \left(\frac{d\mathbf{R}}{dt}\right)_{\text{fixed}} + \left(\frac{d\mathbf{r}}{dt}\right)_{\text{fixed}} \tag{11.9}$$

so that

$$\left(\frac{d\mathbf{r}'}{dt}\right)_{\text{fixed}} = \left(\frac{d\mathbf{R}}{dt}\right)_{\text{fixed}} + \left(\frac{d\mathbf{r}}{dt}\right)_{\text{rotating}} + \boldsymbol{\omega} \times \mathbf{r} \tag{11.10}$$

If we define

$$\mathbf{v}_f \equiv \dot{\mathbf{r}}_f \equiv \left(\frac{d\mathbf{r}'}{dt}\right)_{\text{fixed}} \tag{11.11a}$$

$$\mathbf{V} \equiv \dot{\mathbf{R}}_f \equiv \left(\frac{d\mathbf{R}}{dt}\right)_{\text{fixed}} \tag{11.11b}$$

$$\mathbf{v}_r \equiv \dot{\mathbf{r}}_r \equiv \left(\frac{d\mathbf{r}}{dt}\right)_{\text{rotating}} \tag{11.11c}$$

we may write

$$\boxed{\mathbf{v}_f = \mathbf{V} + \mathbf{v}_r + \boldsymbol{\omega} \times \mathbf{r}} \tag{11.12}$$

where

$\mathbf{v}_f$ = velocity relative to the fixed axes
$\mathbf{V}$ = linear velocity of the moving origin
$\mathbf{v}_r$ = velocity relative to the rotating axes
$\boldsymbol{\omega}$ = angular velocity of the rotating axes
$\boldsymbol{\omega} \times \mathbf{r}$ = velocity due to the rotation of the moving axes

11.3 The Coriolis Force

We have seen that Newton's equation $\mathbf{F} = m\mathbf{a}$ is valid only in an inertial frame of reference. Therefore, the expression for the force on a particle can be obtained from

$$\mathbf{F} = m\mathbf{a}_f = m\left(\frac{d\mathbf{v}_f}{dt}\right)_{\text{fixed}} \tag{11.13}$$

where the differentiation must be carried out with respect to the fixed system. Differentiating Eq. (11.12) and specializing to the case of *constant angular velocity* (so that $\dot{\boldsymbol{\omega}} = 0$, a restriction of no great limitation in the problems considered here), we have

$$\mathbf{F} = m\ddot{\mathbf{R}}_f + m\left(\frac{d\mathbf{v}_r}{dt}\right)_{\text{fixed}} + m\boldsymbol{\omega} \times \left(\frac{d\mathbf{r}}{dt}\right)_{\text{fixed}} \tag{11.14}$$

The second term can be evaluated by substituting $\mathbf{v}_r$ for $\mathbf{Q}$ in Eq. (11.7):

$$\left(\frac{d\mathbf{v}_r}{dt}\right)_{\text{fixed}} = \left(\frac{d\mathbf{v}_r}{dt}\right)_{\text{rotating}} + \boldsymbol{\omega} \times \mathbf{v}_r$$

$$= \mathbf{a}_r + \boldsymbol{\omega} \times \mathbf{v}_r \tag{11.15}$$

where $\mathbf{a}_r$ is the acceleration in the rotating coordinate system. The last term in Eq. (11.14) can be obtained directly from Eq. (11.6):

$$\boldsymbol{\omega} \times \left(\frac{d\mathbf{r}}{dt}\right)_{\text{fixed}} = \boldsymbol{\omega} \times \left(\frac{d\mathbf{r}}{dt}\right)_{\text{rotating}} + \boldsymbol{\omega} \times (\boldsymbol{\omega} \times \mathbf{r})$$

$$= \boldsymbol{\omega} \times \mathbf{v}_r + \boldsymbol{\omega} \times (\boldsymbol{\omega} \times \mathbf{r}) \tag{11.16}$$

Combining Eqs. (11.14)–(11.16), we obtain

$$\mathbf{F} = m\ddot{\mathbf{R}}_f + m\mathbf{a}_r + m\boldsymbol{\omega} \times (\boldsymbol{\omega} \times \mathbf{r}) + 2m\boldsymbol{\omega} \times \mathbf{v}_r, \qquad \dot{\boldsymbol{\omega}} = 0 \tag{11.17}$$

where $\ddot{\mathbf{R}}_f$ is the acceleration of the origin of the moving coordinate system relative to the fixed system.

For most applications, we shall be concerned with x_i systems the origins of which do not have significant accelerations with respect to the inertial coordinate system. That is, the x_i systems will either have only rotation or, at most, rotation plus a uniform velocity with respect to the fixed system. Under these conditions $\ddot{\mathbf{R}}_f = 0$, and we have

$$\mathbf{F} = m\mathbf{a}_f = m\mathbf{a}_r + 2m\boldsymbol{\omega} \times \mathbf{v}_r + m\boldsymbol{\omega} \times (\boldsymbol{\omega} \times \mathbf{r}) \tag{11.18}$$

To an observer in the rotating coordinate system, the effective force on a particle is given by*

$$\boxed{\mathbf{F}_{\text{eff}} \equiv m\mathbf{a}_r = m\mathbf{a}_f - m\boldsymbol{\omega} \times (\boldsymbol{\omega} \times \mathbf{r}) - 2m\boldsymbol{\omega} \times \mathbf{v}_r} \tag{11.19}$$

The first quantity in the expression is the usual term from Newton's equation. The quantity $-m\boldsymbol{\omega} \times (\boldsymbol{\omega} \times \mathbf{r})$ is the usual *centrigugal force* term and reduces to $-m\omega^2 r$ for the case in which $\boldsymbol{\omega}$ is normal to the radius vector. Note that the minus sign implies that the centrifugal force is directed *outwards* from the center of rotation (see Fig. 11-2).

The last term in Eq. (11.19) is a totally new quantity which arises from the motion of the particle in the rotation coordinate system. This term is called the *Coriolis force*. Note that the Coriolis force does indeed arise from the *motion* of the particle since the force is proportional to v_r and hence vanishes if there is no motion.

Since we have used (on several occasions) the term "centrifugal force" and have now introduced the "Coriolis force," it is necessary to inquire

* This result was published by G. G. Coriolis in 1835. The theory of the composition of accelerations was an outgrowth of Coriolis' study of water wheels.

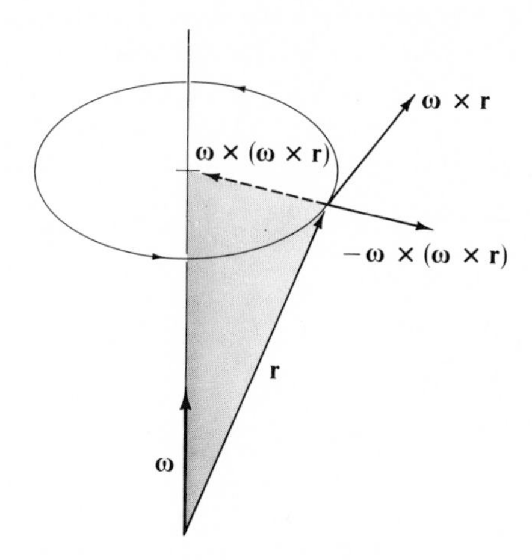

FIG. **11-2**

as to the physical meaning of these quantities. It is important to realize that the centrifugal and Coriolis forces are not "forces" in the usual sense of the word; they have been introduced in an artifical manner as a result of our arbitrary requirement that we be able to write an equation which resembles Newton's equation and which at the same time is valid in a noninertial reference frame. That is, the equation

$$\mathbf{F} = m\mathbf{a}_f$$

is valid only in an inertial frame. If, in a rotating reference frame, we wish to write

$$\mathbf{F}_{\text{eff}} = m\mathbf{a}_r$$

then we can express such an equation in terms of the real force $m\mathbf{a}_f$ as

$$\mathbf{F}_{\text{eff}} = m\mathbf{a}_f + (\text{noninertial terms})$$

where the "noninertial terms" are identified as the centrifugal and Coriolis "forces." Thus, for example, if a body rotates about a fixed force center, the only real force on the body is the force of attraction toward the force center (and gives rise to the *centripetal* acceleration). An observer moving with the rotating body, however, measures this central force and, in addition, notes that the body does not fall toward the force center. In order to reconcile this result with the requirement that the net force on the body vanish, the observer must postulate an additional force—the centrifugal force. But the "requirement" is an artificial one; it arises solely from an attempt to extend the form of Newton's equation to a noninertial system, and this can be done only by introducing a fictitious "correction force." The same comments apply for the Coriolis force; this "force" arises when an attempt is made to describe motion relative to the rotating body.

In spite of their artificiality, the usefulness of the concepts of centrifugal and Coriolis forces is obvious. To describe the motion of a particle relative to a body that is rotating with respect to an inertial reference frame is clearly a complicated matter. On the other hand, the problem can be made relatively easy by the simple expedient of introducing the "noninertial forces" which then allows the use of an equation of motion that resembles Newton's equation.

11.4 Motion Relative to the Earth

The motion of the Earth with respect to an inertial reference frame is dominated by the Earth's rotation about its axis, the effects of the other motions (revolution about the Sun, motion of the solar system with respect to the local galaxy, etc.) being small by comparison. Therefore, to a good approximation (see Problem 11-1) we can consider a coordinate system fixed relative to the Earth to be in pure rotation with respect to an inertial frame of reference, and we can therefore apply Eq. (11.19) to the problems of motion on or near the surface of the Earth.

The angular velocity vector $\boldsymbol{\omega}$ which represents the Earth's rotation about its axis is directed in a northerly direction. Therefore, in the Northern Hemisphere, $\boldsymbol{\omega}$ has a component ω_z which is directed *outward* along the local vertical. If a particle is projected in a horizontal plane (in the local coordinate system at the surface of the Earth) with a velocity $\mathbf{v}_r$, then the Coriolis force $-2m\boldsymbol{\omega} \times \mathbf{v}_r$ will have a component in the plane of magnitude $2m\omega_z v_r$ which will be directed toward the *right* of the particle's motion (see Fig. 11-3), and a deflection from the original direction of motion will result.*

Since the magnitude of the horizontal component of the Coriolis force is proportional to the vertical component of $\boldsymbol{\omega}$, the portion of the Coriolis force which is effective in producing deflections depends upon the latitude, being a maximum at the North Pole and zero at the Equator. In the Southern Hemisphere, the component ω_z is directed *inward* along the local vertical, and hence all deflections are in the opposite sense from those in the Northern Hemisphere.†

* Poisson discussed the deviation of projectile motion in 1837.

† During the naval engagement near the Falkland Islands which occurred early in World War I, the British gunners were surprised to see their accurately aimed salvos falling 100 yards to the left of the German ships. The designers of the sighting mechanisms were well aware of the Coriolis deflection and had carefully taken this into account, but they apparently were under the impression that all sea battles took place near 50° N latitude and never near 50° S latitude. The British shots, therefore, fell at a distance from the targets equal to *twice* the Coriolis deflection.

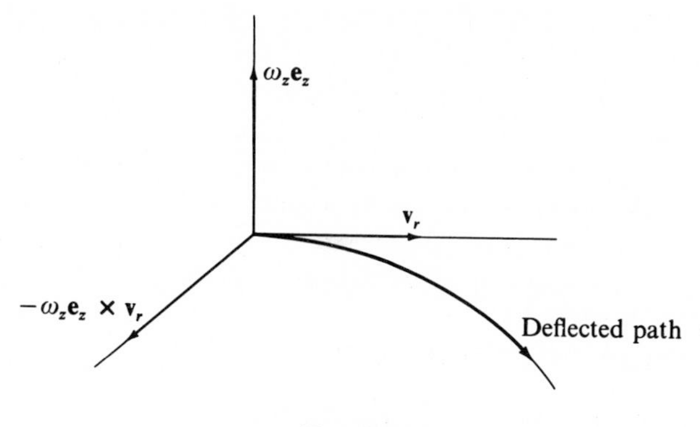

FIG. **11-3**

It is also interesting to note that the radial flow of air masses from high pressure regions into low pressure regions, since they are always deflected to the right by the Coriolis force (in the Northern Hemisphere), produce cyclonic motion, as indicated in Fig. 11-4. The actual motion of air masses is of course, much more complicated, but the qualitative features of cyclonic motion are correctly given by considering the effects of the Coriolis force. The motion of water in whirlpools is (at least, in principle) a similar situation, but in actuality other factors (various perturbations and residual angular momentum) dominate the Coriolis force, and whirlpools are found with both directions of flow. (Even under laboratory conditions it is extremely difficult to isolate the Coriolis effect.)

We note that if we are considering motion in the Earth's gravitational field, then the quantity that we call the acceleration due to gravity (i.e., g or the

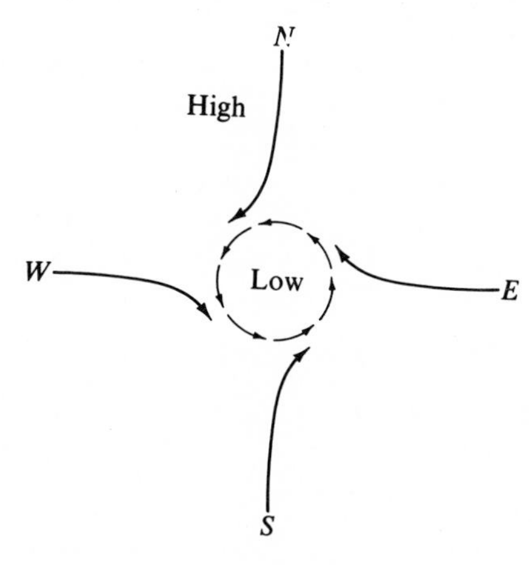

FIG. **11-4**

vector **g**) is actually a combination of the gravitational acceleration proper (as defined from the universal gravitation law) and the apparent outward acceleration (i.e., the centrifugal acceleration) which results from the fact that our coordinate system is fixed with respect to the rotating Earth. That is, **g** is defined only in terms of measurements that we make: the magnitude is determined by the period of a pendulum and the direction by the direction that a plumb-bob assumes at equilibrium. Thus, **g** already includes the term $-\boldsymbol{\omega} \times (\boldsymbol{\omega} \times \mathbf{r})$. Because of this fact, the direction of **g** at a given point is in general slightly different from the true vertical (defined as the direction of the line connecting the point with the center of the Earth). (See Problem 11-7.)

▶ Example 11.4(a) Deflection of a Falling Particle

Let us consider the horizontal deflection due to the Coriolis force of a particle falling freely in the Earth's gravitational field. The value of ω that occurs in the force equation (11.19) will be that of the Earth's rotation:

$$\omega = \frac{2\pi \text{ rad/day}}{86{,}400 \text{ sec/day}} \cong 7.29 \times 10^{-5} \text{ rad/sec}$$

The acceleration of the particle is given by

$$\mathbf{a}_r = \mathbf{g} - 2\boldsymbol{\omega} \times \mathbf{v}_r \tag{1}$$

where **g** is the acceleration due to gravity. We choose a z-axis directed vertically outward from the surface of the Earth. With this definition of $\mathbf{e}_z$, we complete the construction of a right-handed coordinate system by specifying that $\mathbf{e}_x$ be in a southerly and $\mathbf{e}_y$ in an easterly direction, as in Fig. 11-5. We make the approximation that the distance of fall is sufficiently small so that g remains constant during the process.

Since we have chosen the origin O of the rotating coordinate system to lie in the Northern Hemisphere,* we have

$$\left.\begin{aligned} \omega_x &= -\omega \cos \lambda \\ \omega_y &= 0 \\ \omega_z &= \omega \sin \lambda \end{aligned}\right\} \tag{2}$$

Although the Coriolis force will produce small velocity components in the $\mathbf{e}_y$ and $\mathbf{e}_x$ directions, we can certainly neglect $\dot{x}$ and $\dot{y}$ compared to $\dot{z}$, the vertical velocity. Then, approximately,

$$\left.\begin{aligned} \dot{x} &\cong 0 \\ \dot{y} &\cong 0 \\ \dot{z} &\cong -gt \end{aligned}\right\} \tag{3}$$

* Since the point O does not move uniformly with respect to an inertial reference frame, we are not really justified in using Eq. (11.19) which was obtained under the assumption that $\ddot{\mathbf{R}}_f = 0$. For motion near the surface of the Earth, however, the term proportional to $\ddot{\mathbf{R}}_f$ gives only a small contribution and can be neglected.

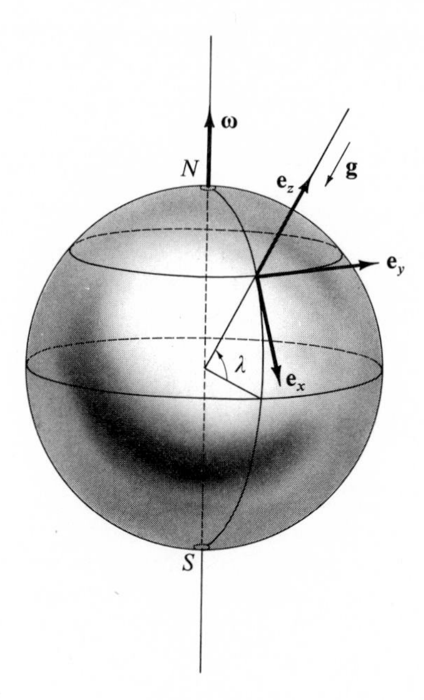

FIG. **11-5**

where we obtain $\dot{z}$ by considering a fall from rest. Therefore, we have

$$\boldsymbol{\omega} \times \mathbf{v}_r \cong \begin{vmatrix} \mathbf{e}_x & \mathbf{e}_y & \mathbf{e}_z \\ -\omega \cos \lambda & 0 & \omega \sin \lambda \\ 0 & 0 & -gt \end{vmatrix}$$

$$\cong -(\omega gt \cos \lambda)\mathbf{e}_z \tag{4}$$

The components of **g** are (neglecting terms in ω^2; see Problem 11-7):

$$\left.\begin{aligned} g_x &= 0 \\ g_y &= 0 \\ g_z &= -g \end{aligned}\right\} \tag{5}$$

so that the equations for the components of $\mathbf{a}_r$ become

$$\left.\begin{aligned} (\mathbf{a}_r)_x &= \ddot{x} \cong 0 \\ (\mathbf{a}_r)_y &= \ddot{y} \cong 2\omega gt \cos \lambda \\ (\mathbf{a}_r)_z &= \ddot{z} \cong -g \end{aligned}\right\} \tag{6}$$

Thus, the effect of the Coriolis force is to produce an acceleration in the $\mathbf{e}_y$, or easterly, direction. Integrating $\ddot{y}$ twice, we have

$$y(t) \cong \tfrac{1}{3}\omega gt^3 \cos \lambda \tag{7}$$

where $y = 0$ and $\dot{y} = 0$ at $t = 0$. The integration of $\dot{z}$ yields the familiar result for the distance of fall,

$$z(t) \cong z(0) - \tfrac{1}{2}gt^2 \tag{8}$$

so that the time of fall from a height $h = z(0)$ is given by

$$t \cong \sqrt{2h/g} \tag{9}$$

Hence, the result for the eastward deflection d of a particle dropped from rest at a height h and at a northern latitude λ is*

$$d \cong \tfrac{1}{3}\omega \cos \lambda \sqrt{8h^3/g} \tag{10}$$

Therefore, an object dropped from a height of 100 m at latitude 45° will be deflected approximately 1.55 cm (neglecting the effects of air resistance).

▶ Example 11.4(b) An Alternative Method of Calculation

In order to demonstrate the power of the Coriolis method for obtaining the equations of motion in a noninertial reference frame, let us rework the last example but use only the formalism previously developed, viz., the theory of central-force motion.

If we release a particle of small mass from a height h above the Earth's surface, the path that the particle will describe is a conic section—an ellipse with $\varepsilon \cong 1$ and with one focus very close to the Earth's center. If r_0 is the Earth's radius and if λ is the (northern) latitude, then at the moment of release, the particle has a horizontal velocity in the eastward direction:

$$v_{\text{hor}} = r\omega \cos \lambda = (r_0 + h)\omega \cos \lambda \tag{1}$$

and the angular momentum about the polar axis is

$$l = mrv_{\text{hor}} = m(r_0 + h)^2\omega \cos \lambda \tag{2}$$

The equation of the path is†

$$\frac{\alpha}{r} = 1 - \varepsilon \cos \theta \tag{3}$$

if we measure θ from the initial position of the particle (see Fig. 11-6). At $t = 0$ we have

$$\frac{\alpha}{r_0 + h} = 1 - \varepsilon \tag{4}$$

* The eastward deflection was predicted by Newton (1679), and several experiments (notably, those of Robert Hooke) appeared to confirm the results. The most careful measurements were probably those of F. Reich (1831, published 1833) who dropped pellets down a mine shaft 188 m deep. He observed a mean deflection of 28 mm. [This is smaller than the value calculated from Eq. (10), the decrease being due to air resistance effects. In all the experiments a small southerly component of the deflection was observed, which remained unaccounted for until Coriolis' theorem was appreciated (see Problem 11-7).

† Notice that there is a change of sign between Eq. (3) and Eq. (8.32) due to the different origins for θ in the two cases.

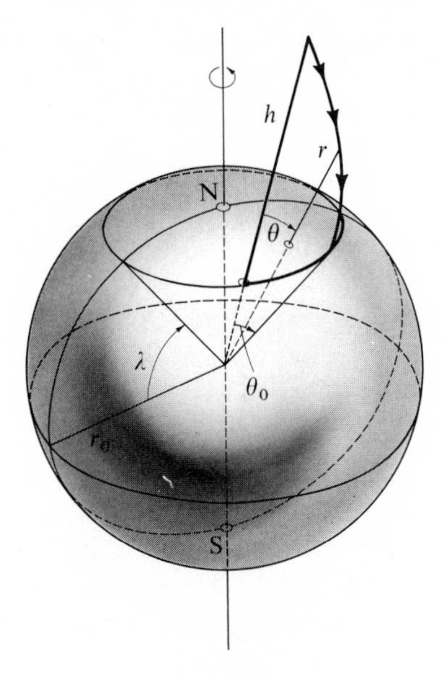

FIG. **11-6**

so that Eq. (3) can be written as

$$r = \frac{(1-\varepsilon)(r_0 + h)}{1 - \varepsilon \cos \theta} \tag{5}$$

Now, from Eq. (8.12) for the areal velocity we can write

$$\tfrac{1}{2} r^2 \frac{d\theta}{dt} = \frac{l}{2m} \tag{6}$$

Thus, the time t required to describe an angle θ is

$$t = \frac{m}{l} \int_0^\theta r^2 \, d\theta \tag{7}$$

Substituting into this expression the value of l from Eq. (2) and r from Eq. (5) we find

$$t = \frac{1}{\omega \cos \lambda} \int_0^\theta \left(\frac{1-\varepsilon}{1 - \varepsilon \cos \theta} \right)^2 d\theta \tag{8}$$

If we let $\theta = \theta_0$ when the particle has reached the Earth's surface ($r = r_0$), then Eq. (5) becomes

$$\frac{r_0}{r_0 + h} = \frac{1 - \varepsilon}{1 - \varepsilon \cos \theta_0} \tag{9}$$

or, upon inverting,

$$1+\frac{h}{r_0}=\frac{1-\varepsilon\cos\theta_0}{1-\varepsilon}$$

$$=\frac{1-\varepsilon[1-2\sin^2(\theta_0/2)]}{1-\varepsilon}$$

$$=1+\frac{2\varepsilon}{1-\varepsilon}\sin^2\frac{\theta_0}{2} \tag{10}$$

from which we have

$$\frac{h}{r_0}=\frac{2\varepsilon}{1-\varepsilon}\sin^2\frac{\theta_0}{2} \tag{11}$$

Now, since the path described by the particle is almost vertical, there is little change in the angle θ between the position of release and the point at which the particle reaches the surface of the Earth; therefore, θ_0 is small and $\sin(\theta_0/2)$ can be approximated by its argument:

$$\frac{h}{r_0}\cong\frac{\varepsilon\theta_0^2}{2(1-\varepsilon)} \tag{12}$$

If we expand the integrand in Eq. (8) by the same method that was used to obtain Eq. (10), we find

$$t=\frac{1}{\omega\cos\lambda}\int_0^\theta\frac{d\theta}{\{1+[2\varepsilon/(1-\varepsilon)]\sin^2(\theta/2)\}^2} \tag{13}$$

and since θ is small, we have

$$t\cong\frac{1}{\omega\cos\lambda}\int_0^\theta\frac{d\theta}{[1+\varepsilon\theta^2/2(1-\varepsilon)]^2} \tag{14}$$

Substituting for $\varepsilon/2(1-\varepsilon)$ from Eq. (12), and writing $t(\theta=\theta_0)=T$ for the total time of fall, we obtain

$$T\cong\frac{1}{\omega\cos\lambda}\int_0^{\theta_0}\frac{d\theta}{[1+(h\theta^2/r_0\,\theta_0^2)]^2}$$

$$\cong\frac{1}{\omega\cos\lambda}\int_0^{\theta_0}\left(1-\frac{2h}{r_0\,\theta_0^2}\theta^2\right)d\theta$$

$$=\frac{1}{\omega\cos\lambda}\left(1-\frac{2h}{3r_0}\right)\theta_0 \tag{15}$$

Solving for θ_0, we find

$$\theta_0\cong\frac{\omega T\cos\lambda}{1-2h/3r_0}\cong\omega T\cos\lambda\left(1+\frac{2h}{3r_0}\right) \tag{16}$$

During the time of fall T, the Earth turns through an angle ωT, so that the point on the Earth directly beneath the initial position of the particle moves toward the east by an amount $r_0\,\omega T\cos\lambda$. During the same time, the particle is deflected toward the East by an amount $r_0\theta_0$. Thus, the net easterly deviation d is

$$\begin{aligned} d &= r_0\theta_0 - r_0\,\omega T\cos\lambda \\ &= \tfrac{2}{3}h\omega T\cos\lambda \end{aligned} \tag{17}$$

and using $T \cong \sqrt{2h/g}$ as in the preceding example, we have, finally,

$$d \cong \tfrac{1}{3}\omega\cos\lambda\sqrt{\frac{8h^3}{g}} \tag{18}$$

which is identical with the result obtained previously.

▶ Example 11.4(c) The Foucault Pendulum*

The effect of the Coriolis force on the motion of a pendulum is to produce a *precession*, or rotation with time of the plane of oscillation. In order to describe this effect, let us select a set of coordinate axes with origin at the equilibrium point of the pendulum and with a z-axis along the local vertical. We are interested only in the rotation of the plane of oscillation, i.e., we wish to consider the motion of the pendulum bob in the x-y plane (the horizontal plane). Therefore, we limit the motion to oscillations of small amplitude, with the horizontal excursions small compared to the length of the pendulum. Under this condition, $\dot{z}$ is small compared to $\dot{x}$ and $\dot{y}$ and can be neglected.

The equation of motion is

$$\mathbf{a}_r = \mathbf{g} + \frac{\mathbf{T}}{m} - 2\boldsymbol{\omega}\times\mathbf{v}_r \tag{1}$$

where $\mathbf{T}/m$ is the acceleration produced by the force of tension $\mathbf{T}$ in the pendulum suspension, as shown in Fig. 11-7. We therefore have, approximately,

$$\left.\begin{aligned} T_x &= -T\cdot\frac{x}{l} \\ T_y &= -T\cdot\frac{y}{l} \\ T_z &\cong T \end{aligned}\right\} \tag{2}$$

As before,

$$\left.\begin{aligned} g_x &= 0 \\ g_y &= 0 \\ g_z &= -g \end{aligned}\right\} \tag{3}$$

* Devised in 1851 by the French physicist, Jean Lèon Foucault (1819–1868).

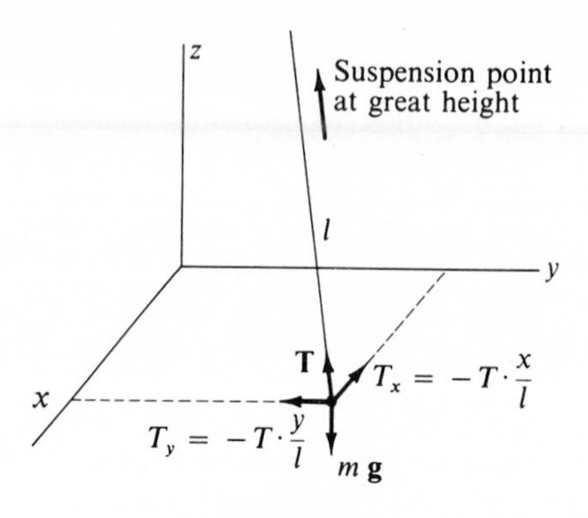

FIG. **11-7**

and,

$$\left.\begin{aligned} \omega_x &= -\omega \cos \lambda \\ \omega_y &= 0 \\ \omega_z &= \omega \sin \lambda \end{aligned}\right\} \tag{4}$$

with

$$\left.\begin{aligned} (\mathbf{v}_r)_x &= \dot{x} \\ (\mathbf{v}_r)_y &= \dot{y} \\ (\mathbf{v}_r)_z &= \dot{z} \cong 0 \end{aligned}\right\} \tag{5}$$

Therefore,

$$\boldsymbol{\omega} \times \mathbf{v}_r \cong \begin{vmatrix} \mathbf{e}_x & \mathbf{e}_y & \mathbf{e}_z \\ -\omega \cos \lambda & 0 & \omega \sin \lambda \\ \dot{x} & \dot{y} & 0 \end{vmatrix} \tag{6}$$

so that

$$\left.\begin{aligned} (\boldsymbol{\omega} \times \mathbf{v}_r)_x &\cong -\dot{y}\omega \sin \lambda \\ (\boldsymbol{\omega} \times \mathbf{v}_r)_y &\cong \dot{x}\omega \sin \lambda \\ (\boldsymbol{\omega} \times \mathbf{v}_r)_z &\cong -\dot{y}\omega \cos \lambda \end{aligned}\right\} \tag{7}$$

Thus, the equations of interest are

$$\left.\begin{aligned} (\mathbf{a}_r)_x &= \ddot{x} \cong -\frac{T}{m} \cdot \frac{x}{l} + 2\dot{y}\omega \sin \lambda \\ (\mathbf{a}_r)_y &\cong \ddot{y} \cong -\frac{T}{m} \cdot \frac{y}{l} - 2\dot{x}\omega \sin \lambda \end{aligned}\right\} \tag{8}$$

For small displacements, $T \cong mg$. Defining $\alpha^2 \equiv T/ml \cong g/l$, and writing $\omega_z = \omega \sin \lambda$ we have

$$\left.\begin{aligned} \ddot{x} + \alpha^2 x &\cong \quad 2\omega_z \dot{y} \\ \ddot{x} + \alpha^2 y &\cong -2\omega_z \dot{x} \end{aligned}\right\} \tag{9}$$

We note that the equation for $\ddot{x}$ contains a term in $\dot{y}$, and that the equation for $\ddot{y}$ contains a term in $\dot{x}$. Such equations are called *coupled equations*. A solution for this pair of coupled equations can be effected by adding the first of the above equations to i times the second:

$$\begin{aligned} (\ddot{x} + i\ddot{y}) + \alpha^2(x + iy) &\cong -2\omega_z(i\dot{x} - \dot{y}) \\ &= -2i\omega_z(\dot{x} + i\dot{y}) \end{aligned} \tag{10}$$

If we write

$$q \equiv x + iy \tag{11}$$

we then have

$$q + 2i\omega_z \dot{q} + \alpha^2 q \cong 0 \tag{12}$$

This equation is identical with the equation (3.38) that describes damped oscillations, except that here the term which corresponds to the damping factor is purely imaginary. The solution is [cf. Eq. (3.40)]:

$$q(t) \cong \exp[-i\omega_z t][A \exp(\sqrt{-\omega_z^2 - \alpha^2}\, t) + B \exp(-\sqrt{-\omega_z^2 - \alpha^2}\, t)] \tag{13}$$

Now, if the Earth were not rotating, so that $\omega_z = 0$, then the equation for q would become

$$\ddot{q}' + \alpha^2 q' \cong 0, \qquad \omega_z = 0 \tag{14}$$

from which it is seen that α corresponds to the oscillation frequency of the pendulum. This frequency is clearly much greater than the angular frequency of the Earth's rotation. Therefore $\alpha \gg \omega_z$, and the equation for $q(t)$ becomes

$$q(t) \cong e^{-i\omega_z t}\,(Ae^{i\alpha t} + Be^{-i\alpha t}) \tag{15}$$

We can interpret this equation more easily if we note that the equation for q' above has the solution

$$q'(t) = x'(t) + iy'(t) = Ae^{i\alpha t} + Be^{-i\alpha t} \tag{16}$$

Thus,

$$q(t) = q'(t) \cdot e^{-i\omega_z t} \tag{17}$$

or,

$$\begin{aligned} x(t) + iy(t) &= [x'(t) + iy'(t)] \cdot e^{-i\omega_z t} \\ &= [x' + iy'][\cos \omega_z t - i \sin \omega_z t] \\ &= [x' \cos \omega_z t + y' \sin \omega_z t] \\ &\quad + i[-x' \sin \omega_z t + y' \cos \omega_z t] \end{aligned} \tag{18}$$

Equating real and imaginary parts,

$$\left.\begin{aligned} x(t) &= x' \cos \omega_z t + y' \sin \omega_z t \\ y(t) &= -x' \sin \omega_z t + y' \cos \omega_z t \end{aligned}\right\} \tag{19}$$

We can write these equations in matrix form as

$$\begin{pmatrix} x(t) \\ y(t) \end{pmatrix} = \begin{pmatrix} \cos \omega_z t & \sin \omega_z t \\ -\sin \omega_z t & \cos \omega_z t \end{pmatrix} \begin{pmatrix} x'(t) \\ y'(t) \end{pmatrix} \tag{20}$$

from which it is evident that (x, y) may be obtained from (x', y') by the application of a rotation matrix of the familiar form

$$\lambda = \begin{pmatrix} \cos \theta & \sin \theta \\ -\sin \theta & \cos \theta \end{pmatrix} \tag{21}$$

Thus, the angle of rotation is $\theta = \omega_z t$, and the plane of oscillation of the pendulum therefore rotates with a frequency $\omega_z = \omega \sin \lambda$. The observation of this rotation gives a clear demonstration of the rotation of the Earth.*

Suggested References

Motion with respect to accelerated reference frames is discussed in most books on dynamics. The accounts which are the most detailed are those of Becker (Be54, Chapter 11), Fowles (Fo62, Chapter 5), Halfman (Ha62a, Chapter 5), and Symon (Sy60, Chapter 7).

Briefer treatments are given by Bradbury (Br68, Chapter 10), Constant (Co54, Section 5–8), Goldstein (Go50, Sections 4–8 and 4–9), McCuskey (Mc59, Sections 1–14 and 1–15), and Page (Pa52, Sections 31–33).

Problems

11-1. Calculate the centrifugal acceleration, due to the Earth's rotation, on a particle on the surface of the Earth at the Equator. Compare this result with the gravitational acceleration. Compute also the centrifugal acceleration due to the motion of the Earth about the Sun and justify the remark made in the text that this acceleration may be neglected compared to the acceleration due to axial rotation.

* Vincenzo Viviani (1622–1703), a pupil of Galileo, had noticed as early as about 1650 that a pendulum undergoes a slow rotation, but there is no evidence that he correctly interpreted the phenomenon. Foucault's invention of the gyroscope in the year following the demonstration of his pendulum provided the means for an even more striking visual proof of the Earth's rotation.

11-2. Obtain an expression for the angular deviation of a particle projected from the North Pole in a path which lies close to the Earth. Is the deviation significant for a missile which makes a 3000-mile flight in 10 minutes? What is the "miss distance" if the missile is aimed directly at the target? Is the "miss distance" greater for a 12,000-mile flight at the same velocity?

11-3. If a particle is projected vertically upward from a point on the Earth's surface at a northern latitude λ, show that it strikes the ground at a point $\frac{4}{3}\omega \cos \lambda \sqrt{8h^3/g}$ to the West. (Neglect air resistance and consider only small vertical heights.)

11-4. If a projectile is fired due East from a point on the surface of the Earth at a northern latitude λ, with a velocity of magnitude V_0 and at an angle of inclination to the horizontal of α, show that the lateral deflection when the projectile strikes the Earth is

$$d = \frac{4V_0^3}{g^2} \cdot \omega \sin \lambda \cdot \sin^2 \alpha \cos \alpha$$

where ω is the rotation frequency of the Earth.

11-5. In the preceding problem, if the range of the projectile is R for the case $\omega = 0$, show that the change of range due to the rotation of the Earth is

$$\Delta R = \sqrt{\frac{2R^3}{g}} \cdot \omega \cos \lambda [\cot^{\frac{1}{2}} \alpha - \tfrac{1}{3} \tan^{\frac{3}{2}} \alpha]$$

11-6. Show that the angular deviation ε of a plumb line from the true vertical at a point on the Earth's suface at a latitude λ is

$$\varepsilon = \frac{r_0 \omega^2 \sin \lambda \cos \lambda}{g - r_0 \omega^2 \cos^2 \lambda}$$

where r_0 is the radius of the Earth. What is the value (in seconds of arc) of the maximum deviation?

11-7. Refer to Example 11.4(a) concerning the deflection of a particle falling in the Earth's gravitational field. Perform a calculation in second approximation (i.e., retain terms in ω^2) and show that there is a *southerly* deflection

$$d_S \cong \frac{3}{2} \frac{h^2 \omega^2}{g} \sin \lambda \cos \lambda$$

11-8. Consider the description of the motion of a particle in a coordinate system that is in uniform rotation with respect to an inertial reference frame. Obtain the Lagrangian for the particle. Next, calculate the Hamiltonian and attempt to identify this quantity with the total energy. (Are all of the requirements for this identification met?) The expression for the total energy thus obtained is the standard formula $\frac{1}{2}mv^2 + U$ plus an additional term. Show that the extra term is the *centrifugal potential energy*. Finally, show that it is possible to define an *effective potential* for the problem which is exactly that used in the central-force problem [see Eq. (8.25)].

CHAPTER 12

Dynamics of Rigid Bodies

12.1 Introduction

A rigid body is defined as a collection of particles whose relative distances are constrained to remain absolutely fixed. Such bodies do not, of course, exist in Nature, since the ultimate component particles which comprise every body (the atoms) are always undergoing some relative motion. This motion, however, is of a microscopic nature, and therefore usually may be ignored for the purposes of describing the macroscopic motion of the body. On the other hand, macroscopic displacement within the body (such as elastic deformations) can also take place. For many bodies of interest we can safely neglect the changes in size and shape due to such deformations and obtain equations of motion which are valid to a high degree of accuracy.

It is also clear that there is a relativistic limitation to the concept of an absolutely rigid body. Consider, for example, a long bar of some material. If we strike a blow at one end of the bar, and if the bar were absolutely rigid, the effect would be felt instantaneously at the opposite end. But this corresponds to the transmission of a signal with an infinite velocity, a situation which from relativity theory we know is impossible. (Actually, the velocity of transmission of such a signal in a metal bar is rather low compared with the velocity of light—being $\sim 10^5$ cm/sec—and depends on the elastic properties of the material.)

We shall use interchangeably the idealized concept of a rigid body as a collection of discrete particles or as a continuous distribution of matter. The only change is the replacement of summations over particles by integrations over mass density distributions. The equations of motion are equally valid for either viewpoint.

For the description of the motion of a rigid body, we shall use two coordinate systems—an inertial frame and a coordinate system fixed with respect to the body. Six quantities must be specified in order to denote the position of the body. These can be taken to be the coordinates of the center of mass (which can often conveniently be made to coincide with the origin of the body coordinate system) and three independent angles which give the orientation of the body coordinate system with respect to the fixed (or inertial) system.* The three independent angles can conveniently be taken to be the *Eulerian angles*, which are described in Section 12.7.

It is intuitively obvious that any arbitrary finite motion of a rigid body can be considered to be the sum of two independent motions—a linear translation of some point of the body plus a rotation about that point.† If the point is chosen to be the center of mass of the body, then such a separation of the motion into two parts allows the use of the development in Chapter 2, which indicates that the angular momentum [see Eq. (2.38)] and the kinetic energy [see Eq. (2.54)] can be separated into portions relating to the motion *of* the center of mass and to the motion *around* the center of mass.

If the potential energy can also be separated (as will always be the case, for example, for the potential energy in a uniform force field), then the Lagrangian separates, and the entire problem conveniently divides into two parts, one involving only translation and the other only rotation. Each portion of the problem can then be solved independently of the other.‡ This type of separation is essential for a relatively uncomplicated description of rigid-body motion.

12.2 The Inertia Tensor

We now direct our attention to a rigid body which is composed of n particles of masses m_α, $\alpha = 1, 2, 3, \ldots, n$. If the body rotates with an instantaneous angular velocity $\boldsymbol{\omega}$ about some point which is fixed with respect to

* We shall use in this chapter the designation *body system* in place of the term *rotating system* used in the preceding chapter. The term *fixed system* will be retained.

† *Chasles' theorem*, which is even more general than this statement (it says that the line of translation and the axis of rotation can be made to coincide), was proven by the French mathematician Michel Chasles (1793–1880) in 1830. The proof is given, e.g., by Whittaker (Wh37, p. 4).

‡ This important point was first realized by Euler, 1749.

the body coordinate system, and if this point moves with an instantaneous linear velocity $\mathbf{V}$ with respect to the fixed coordinate system, then the instantaneous velocity of the αth particle in the fixed system can be obtained by using Eq. (11.12). But we are now considering a rigid body so that

$$\mathbf{v}_r = \left(\frac{d\mathbf{r}}{dt}\right)_{\text{rotating}} \equiv 0$$

Therefore,

$$\boxed{\mathbf{v}_\alpha = \mathbf{V} + \boldsymbol{\omega} \times \mathbf{r}_\alpha} \tag{12.1}$$

where the subscript f, denoting the fixed coordinate system, has been deleted from the velocity $\mathbf{v}_\alpha$, it now being understood that all velocities are measured in the fixed system; all velocities with respect to the rotating or body system now vanish because the body is *rigid.*

Since the kinetic energy of the αth particle is given by

$$T_\alpha = \tfrac{1}{2} m_\alpha v_\alpha^2 \tag{12.2}$$

we have, for the total kinetic energy,

$$T = \tfrac{1}{2} \sum_\alpha m_\alpha (\mathbf{V} + \boldsymbol{\omega} \times \mathbf{r}_\alpha)^2 \tag{12.3}$$

Expanding the squared term, we find

$$T = \tfrac{1}{2} \sum_\alpha m_\alpha V^2 + \sum_\alpha m_\alpha \mathbf{V} \cdot \boldsymbol{\omega} \times \mathbf{r}_\alpha + \tfrac{1}{2} \sum_\alpha m_\alpha (\boldsymbol{\omega} \times \mathbf{r}_\alpha)^2 \tag{12.4}$$

This is a general expression for the kinetic energy and is valid for any choice of the origin from which the vectors $\mathbf{r}_\alpha$ are measured. However, if we elect to make the origin of the body coordinate system coincide with the center of mass of the object, then a considerable simplification results. First, we note that in the second term on the right-hand side of this equation neither $\mathbf{V}$ nor $\boldsymbol{\omega}$ is characteristic of the αth particle and therefore these quantities may be taken outside the summation:

$$\sum_\alpha m_\alpha \mathbf{V} \cdot \boldsymbol{\omega} \times \mathbf{r}_\alpha = \mathbf{V} \cdot \boldsymbol{\omega} \times \left(\sum_\alpha m_\alpha \mathbf{r}_\alpha\right) \tag{12.5}$$

But now the term

$$\sum m_\alpha \mathbf{r}_\alpha = M\mathbf{R}$$

is the center-of-mass vector [cf. Eq. (2.25)] which vanishes in the body system since the vectors $\mathbf{r}_\alpha$ are measured from the center of mass. The kinetic energy can then be written as

$$T = T_{\text{trans}} + T_{\text{rot}}$$

where

$$T_{\text{trans}} = \tfrac{1}{2} \sum_{\alpha} m_\alpha V^2 = \tfrac{1}{2} M V^2 \tag{12.6a}$$

$$T_{\text{rot}} = \tfrac{1}{2} \sum_{\alpha} m_\alpha (\boldsymbol{\omega} \times \mathbf{r}_\alpha)^2 \tag{12.6b}$$

T_{trans} and T_{rot} designate the translational and rotational kinetic energies, respectively. Thus, the kinetic energy separates into two independent parts as mentioned in the first section of this chapter.

The rotational kinetic energy term can be evaluated by noting that

$$\begin{aligned}(\mathbf{A} \times \mathbf{B})^2 &= (\mathbf{A} \times \mathbf{B}) \cdot (\mathbf{A} \times \mathbf{B}) \\ &= A^2 B^2 - (\mathbf{A} \cdot \mathbf{B})^2\end{aligned}$$

Therefore,

$$T_{\text{rot}} = \tfrac{1}{2} \sum_{\alpha} m_\alpha [\omega^2 r_\alpha^2 - (\boldsymbol{\omega} \cdot \mathbf{r}_\alpha)^2] \tag{12.7}$$

We now express T_{rot} by making use of the components ω_i and $r_{\alpha,i}$ of the vectors $\boldsymbol{\omega}$ and $\mathbf{r}_\alpha$. We also note that $\mathbf{r}_\alpha = (x_{\alpha,1}, x_{\alpha,2}, x_{\alpha,3})$ in the body system so that we can write $r_{\alpha,i} = x_{\alpha,i}$. Thus,

$$T_{\text{rot}} = \tfrac{1}{2} \sum_{\alpha} m_\alpha \left[\left(\sum_i \omega_i^2 \right) \left(\sum_k x_{\alpha,k}^2 \right) - \left(\sum_i \omega_i x_{\alpha,i} \right) \left(\sum_j \omega_j x_{\alpha,j} \right) \right] \tag{12.8}$$

Now, clearly, we can write $\omega_i = \sum_j \omega_j \delta_{ij}$, so that

$$\begin{aligned}T_{\text{rot}} &= \tfrac{1}{2} \sum_{\alpha} \sum_{i,j} m_\alpha \left[\omega_i \omega_j \delta_{ij} \left(\sum_k x_{\alpha,k}^2 \right) - \omega_i \omega_j x_{\alpha,i} x_{\alpha,j} \right] \\ &= \tfrac{1}{2} \sum_{i,j} \omega_i \omega_j \sum_{\alpha} m_\alpha \left[\delta_{ij} \sum_k x_{\alpha,k}^2 - x_{\alpha,i} x_{\alpha,j} \right]\end{aligned} \tag{12.9}$$

If we define the ijth element of the sum over α to be I_{ij},

$$I_{ij} \equiv \sum_{\alpha} m_\alpha \left[\delta_{ij} \sum_k x_{\alpha,k}^2 - x_{\alpha,i} x_{\alpha,j} \right] \tag{12.10}$$

then we have

$$T_{\text{rot}} = \tfrac{1}{2} \sum_{i,j} I_{ij} \omega_i \omega_j \tag{12.11}$$

This equation in its most restricted form becomes

$$T_{\text{rot}} = \tfrac{1}{2} I \omega^2 \tag{12.12}$$

where I is the (scalar) moment of inertia about the axis of rotation. This equation will be recognized as the familiar expression for the rotational kinetic energy given in elementary treatments.

The nine terms I_{ij} constitute the elements of a quantity which we designate by $\{\mathsf{I}\}$. In form, $\{\mathsf{I}\}$ is similar to a 3×3 matrix. Now, $\{\mathsf{I}\}$ is the proportionality factor between the rotational kinetic energy and the angular velocity and has the dimensions (mass) $\times$ (length)^2. Since $\{\mathsf{I}\}$ relates two quite different physical quantities, it is to be expected that $\{\mathsf{I}\}$ is a member of a somewhat higher class of functions than has heretofore been encountered. Indeed, $\{\mathsf{I}\}$ is a *tensor* and is known as the *inertia tensor*.* Note, however, that T_{rot} can be calculated, without regard to any of the special properties of tensors, by using Eq. (12.9) which completely specifies the necessary operations.

The elements of $\{\mathsf{I}\}$ can be obtained directly from Eq. (12.10). We write the elements in a 3×3 array for clarity:

$$\{\mathsf{I}\} = \begin{Bmatrix} \sum_\alpha m_\alpha(x_{\alpha,2}^2 + x_{\alpha,3}^2) & -\sum_\alpha m_\alpha x_{\alpha,1}x_{\alpha,2} & -\sum_\alpha m_\alpha x_{\alpha,1}x_{\alpha,3} \\ -\sum_\alpha m_\alpha x_{\alpha,2}x_{\alpha,1} & \sum_\alpha m_\alpha(x_{\alpha,1}^2 + x_{\alpha,3}^2) & -\sum_\alpha m_\alpha x_{\alpha,2}x_{\alpha,3} \\ -\sum_\alpha m_\alpha x_{\alpha,3}x_{\alpha,1} & -\sum_\alpha m_\alpha x_{\alpha,3}x_{\alpha,2} & \sum_\alpha m_\alpha(x_{\alpha,1}^2 + x_{\alpha,2}^2) \end{Bmatrix} \tag{12.13}$$

The diagonal elements, I_{11}, I_{22}, and I_{33}, are called the *moments of inertia* about the x_1-, x_2-, and x_3-axes, respectively, and the negatives of the off-diagonal elements I_{12}, I_{13}, etc., are termed the *products of inertia*.† Clearly, the inertia tensor is symmetric; that is,

$$I_{ij} = I_{ji} \tag{12.14}$$

and, therefore, there are only six independent elements in $\{\mathsf{I}\}$. Furthermore, the inertia tensor is composed of additive elements; the inertia tensor for a body can be considered to be the sum of the tensors for the various portions of the body. Therefore, if we consider a body as a continuous distribution of matter with mass density $\rho = \rho(\mathbf{r})$, then

$$\boxed{I_{ij} = \int_V \rho(\mathbf{r})\left[\delta_{ij}\sum_k x_k^2 - x_i x_j\right] dv} \tag{12.15}$$

where $dv = dx_1\, dx_2\, dx_3$ is the element of volume at the position defined by the vector $\mathbf{r}$, and where V is the volume of the body.

* The true test of a tensor is in its behavior under a coordinate transformation (see Section 12.6).

† Introduced by Huygens in 1673; Euler coined the name.

▶ Example 12.2 The Inertia Tensor for a Cube

As an example of the calculation of the elements of the inertia tensor, let us consider a homogeneous cube of density ρ, mass M, and side of length b. Let one corner be at the origin and let the three adjacent edges lie along the coordinate axes (see Fig. 12-1). (For this choice of the coordinate axes, obviously the origin does not lie at the center of mass; we shall return to this point later.)

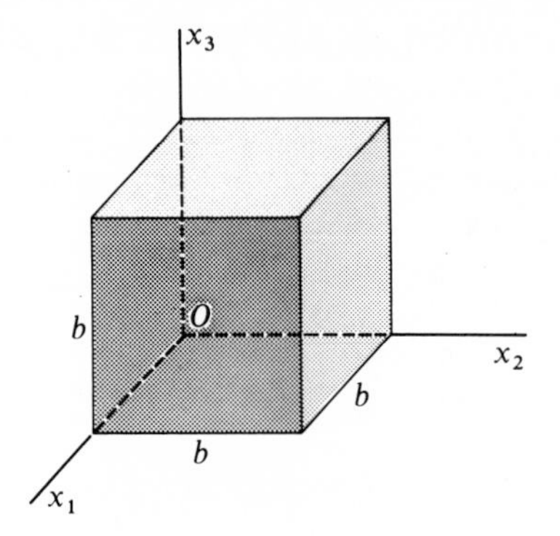

FIG. **12-1**

According to Eq. (12.15), we have

$$I_{11} = \rho \int_0^b dx_3 \int_0^b dx_2 (x_2^2 + x_3^2) \int_0^b dx_1$$

$$= \tfrac{2}{3}\rho b^5 = \tfrac{2}{3}Mb^2 \tag{1}$$

$$I_{12} = -\rho \int_0^b x_1\, dx_1 \int_0^b x_2\, dx_2 \int_0^b dx_3$$

$$= -\tfrac{1}{4}\rho b^5 = -\tfrac{1}{4}Mb^2 \tag{2}$$

It is easy to see that all of the diagonal elements are equal and, furthermore, that all the off-diagonal elements are equal. If we define $\beta \equiv Mb^2$, we have

$$\left.\begin{aligned} I_{11} = I_{22} = I_{33} = \tfrac{2}{3}\beta \\ I_{12} = I_{13} = I_{23} = -\tfrac{1}{4}\beta \end{aligned}\right\} \tag{3}$$

The moment of inertia tensor then becomes

$$\{\mathrm{I}\} = \begin{Bmatrix} \tfrac{2}{3}\beta & -\tfrac{1}{4}\beta & -\tfrac{1}{4}\beta \\ -\tfrac{1}{4}\beta & \tfrac{2}{3}\beta & -\tfrac{1}{4}\beta \\ -\tfrac{1}{4}\beta & -\tfrac{1}{4}\beta & \tfrac{2}{3}\beta \end{Bmatrix} \tag{4}$$

We shall continue the investigation of the moment of inertia tensor for the cube in later sections.

12.3 Angular Momentum

With respect to some point O that is fixed in the body coordinate system, the angular momentum of the body is

$$\mathbf{L} = \sum_{\alpha} \mathbf{r}_{\alpha} \times \mathbf{p}_{\alpha} \tag{12.16}$$

The most convenient choice for the position of the point O depends on the particular problem. There are only two choices of importance: (a) If one or more points of the body are fixed (in the fixed coordinate system), O is chosen to coincide with one such point (as in the case of the rotating top, Section 12.10); (b) if no point of the body is fixed, O is chosen to be the center of mass.

Relative to the body coordinate system, the linear momentum $\mathbf{p}_{\alpha}$ is

$$\mathbf{p}_{\alpha} = m_{\alpha}\mathbf{v}_{\alpha} = m_{\alpha}\boldsymbol{\omega} \times \mathbf{r}_{\alpha}$$

Hence, the angular momentum of the body is

$$\mathbf{L} = \sum_{\alpha} m_{\alpha}\mathbf{r}_{\alpha} \times (\boldsymbol{\omega} \times \mathbf{r}_{\alpha}) \tag{12.17}$$

The vector identity

$$\mathbf{A} \times (\mathbf{B} \times \mathbf{A}) = A^2\mathbf{B} - \mathbf{A}(\mathbf{A} \cdot \mathbf{B})$$

can be used to express $\mathbf{L}$ as

$$\boxed{\mathbf{L} = \sum_{\alpha} m_{\alpha}[r_{\alpha}^2 \boldsymbol{\omega} - \mathbf{r}_{\alpha}(\mathbf{r}_{\alpha} \cdot \boldsymbol{\omega})]} \tag{12.18}$$

The same technique that was used to write T_{rot} in tensor form can now be applied here. But the angular momentum is a vector, so that for the ith component we write

$$\begin{aligned} L_i &= \sum_{\alpha} m_{\alpha}\left[\omega_i \sum_k x_{\alpha,k}^2 - x_{\alpha,i} \sum_j x_{\alpha,j}\omega_j\right] \\ &= \sum_{\alpha} m_{\alpha} \sum_j \left[\omega_j \delta_{ij} \sum_k x_{\alpha,k}^2 - \omega_j x_{\alpha,i} x_{\alpha,j}\right] \\ &= \sum_j \omega_j \sum_{\alpha} m_{\alpha}\left[\delta_{ij} \sum_k x_{\alpha,k}^2 - x_{\alpha,i} x_{\alpha,j}\right] \end{aligned} \tag{12.19}$$

The summation over α will be recognized [cf. Eq. (12.10)] as the ijth element of the inertia tensor. Therefore,

$$\boxed{L_i = \sum_j I_{ij}\omega_j} \tag{12.20}$$

or, in tensor notation

$$\mathbf{L} = \{\mathbf{I}\} \cdot \boldsymbol{\omega} \tag{12.20a}$$

Thus, the inertia tensor relates a *sum* over the components of the angular velocity vector to the *i*th component of the angular momentum vector. This may at first seem a somewhat unexpected result; for, if we consider a rigid body for which the inertia tensor has nonvanishing off-diagonal elements, then even if $\boldsymbol{\omega}$ is directed along, say, the x_1-direction, $\boldsymbol{\omega} = (\omega_1, 0, 0)$, the angular momentum vector will be general have nonvanishing components in all three directions: $\mathbf{L} = (L_1, L_2, L_3)$. That is, the angular momentum vector does not in general have the same direction as the angular velocity vector. (It should be emphasized that this statement depends upon $I_{ij} \neq 0$ for $i \neq j$; we shall return to this point in the next section.)

As an example of a situation in which $\boldsymbol{\omega}$ and $\mathbf{L}$ are not co-linear, consider the rotating dumbbell in Fig. 12-2. (The shaft connecting m_1 and m_2 is

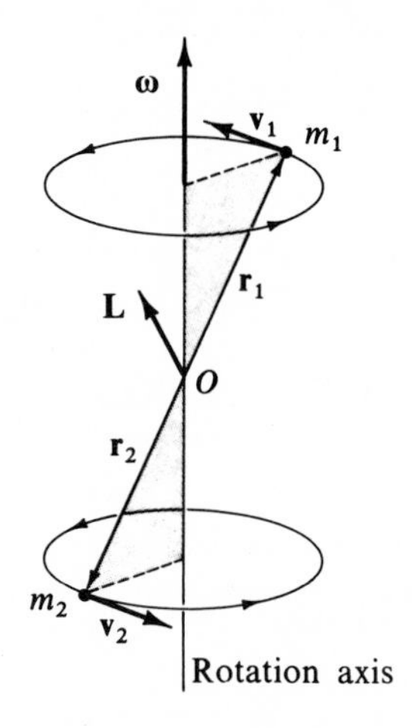

FIG. 12-2

considered to be weightless and extensionless.) The relation connecting $\mathbf{r}_\alpha$, $\mathbf{v}_\alpha$, and $\boldsymbol{\omega}$ is

$$\mathbf{v}_\alpha = \boldsymbol{\omega} \times \mathbf{r}_\alpha$$

and the relation connecting $\mathbf{r}_\alpha$, $\mathbf{v}_\alpha$, and $\mathbf{L}$ is

$$\mathbf{L} = \sum_\alpha m_\alpha \mathbf{r}_\alpha \times \mathbf{v}_\alpha$$

Then, clearly, $\boldsymbol{\omega}$ is directed along the axis of rotation, while $\mathbf{L}$ is perpendicular to the line connecting m_1 and m_2.

We note, for this example, that the angular momentum vector **L** does not remain constant in time, but rotates with an angular velocity ω in such a way that it traces out a cone whose axis is the axis of rotation. Therefore, $\dot{\mathbf{L}} \neq 0$; but Eq. (2.11) states that

$$\dot{\mathbf{L}} = \mathbf{N}$$

where **N** is the torque applied to the body. Thus, in order to keep the dumbbell rotating as in Fig. 12-2, a torque must be constantly applied.

We can obtain another result from Eq. (12.20) by multiplying L_i by $\frac{1}{2}\omega_i$ and summing over i:

$$\tfrac{1}{2}\sum_i \omega_i L_i = \tfrac{1}{2}\sum_{i,j} I_{ij}\,\omega_i\,\omega_j = T_{\text{rot}} \tag{12.21}$$

where the second equality is just Eq. (12.11). Thus,

$$\boxed{T_{\text{rot}} = \tfrac{1}{2}\boldsymbol{\omega} \cdot \mathbf{L}} \tag{12.21a}$$

Equations (12.20a) and (12.21a) illustrate two important properties of tensors. The product of a tensor and a vector yields a vector, as in

$$\mathbf{L} = \{\mathbf{I}\} \cdot \boldsymbol{\omega}$$

and the product of a tensor and two vectors yields a scalar, as in

$$T_{\text{rot}} = \tfrac{1}{2}\boldsymbol{\omega} \cdot \mathbf{L} = \tfrac{1}{2}\boldsymbol{\omega} \cdot \{\mathbf{I}\} \cdot \boldsymbol{\omega}$$

We shall not, however, have occasion to use here tensor equations written in the above form, but will always use the summation (or integral) expressions as in Eqs. (12.11), (12.15), (12.20), etc.

12.4 Principal Axes of Inertia*

It is clear that a considerable simplification in the expressions for T and **L** would result if the inertia tensor consisted only of diagonal elements. If we could write

$$I_{ij} = I_i\,\delta_{ij} \tag{12.22}$$

then the inertia tensor would be

$$\{\mathbf{I}\} = \begin{Bmatrix} I_1 & 0 & 0 \\ 0 & I_2 & 0 \\ 0 & 0 & I_3 \end{Bmatrix} \tag{12.23}$$

* Discovered by Euler, 1750.

Therefore, we would have

$$L_i = \sum_j I_i \delta_{ij} \omega_j = I_i \omega_i \tag{12.24a}$$

and

$$T_{\text{rot}} = \tfrac{1}{2} \sum_{i,j} I_i \delta_{ij} \omega_i \omega_j = \tfrac{1}{2} \sum_i I_i \omega_i^2 \tag{12.24b}$$

Thus, the condition that $\{\mathsf{I}\}$ have only diagonal elements provides quite simple expressions for the angular momentum and the rotational kinetic energy. We now determine the conditions under which Eq. (12.22) becomes the description of the inertia tensor. This involves finding a set of body axes for which the products of inertia (i.e., the off-diagonal elements of $\{\mathsf{I}\}$) vanish. Such axes are called the *principal axes of inertia*.

If a body rotates around a principal axis, both the angular velocity and the angular momentum are, according to Eq. (12.24a), directed along this axis. Then, if I is the moment of inertia about this axis, we can write

$$\mathbf{L} = I\boldsymbol{\omega} \tag{12.25}$$

Equating the components of $\mathbf{L}$ in Eqs. (12.20) and (12.25), we have

$$\left.\begin{aligned} L_1 &= I\omega_1 = I_{11}\omega_1 + I_{12}\,\omega_2 + I_{13}\,\omega_3 \\ L_2 &= I\omega_2 = I_{21}\omega_1 + I_{22}\,\omega_2 + I_{23}\,\omega_3 \\ L_3 &= I\omega_3 = I_{31}\omega_1 + I_{32}\,\omega_2 + I_{33}\,\omega_3 \end{aligned}\right\} \tag{12.26}$$

Or, upon collecting terms, we obtain

$$\left.\begin{aligned} (I_{11} - I)\omega_1 + I_{12}\,\omega_2 + I_{13}\,\omega_3 &= 0 \\ I_{21}\omega_1 + (I_{22} - I)\omega_2 + I_{23}\,\omega_3 &= 0 \\ I_{31}\omega_1 + I_{32}\,\omega_2 + (I_{33} - I)\omega_3 &= 0 \end{aligned}\right\} \tag{12.27}$$

The condition that these equations have a nontrivial solution is that the determinant of the coefficients vanish:

$$\begin{vmatrix} (I_{11} - I) & I_{12} & I_{13} \\ I_{21} & (I_{22} - I) & I_{23} \\ I_{31} & I_{32} & (I_{33} - I) \end{vmatrix} = 0 \tag{12.28}$$

The expansion of this determinant leads to the *secular equation** for I, which is a cubic. Each of the three roots corresponds to a moment of inertia about one of the principal axes. These values, I_1, I_2, and I_3, are called the *principal moments of inertia*. If the body rotates about the axis corresponding to the principal moment I_1, then Eq. (12.25) becomes $\mathbf{L} = I_1\boldsymbol{\omega}$; i.e., both $\boldsymbol{\omega}$ and $\mathbf{L}$ are directed along this axis. The direction of $\boldsymbol{\omega}$ with respect to the body

* So called because a similar equation is used to describe secular perturbations in celestial mechanics. The mathematical terminology is the *characteristic polynomial*.

coordinate system will then be the same as the direction of the principal axis corresponding to I_1. Therefore, we can determine the direction of this principal axis by substituting I_1 for I in Eq. (12.27), and determining the ratios of the components of the angular velocity vector: $\omega_1 : \omega_2 : \omega_3$. We thereby determine the direction cosines of the axis about which the moment of inertia is I_1. The directions corresponding to I_2 and I_3 can be found in a similar fashion. That the principal axes determined in this manner are indeed *real* and *orthogonal* will be proved in Section 12.6; these results also follow from the more general considerations given in Section 13.6.

The fact that the diagonalization procedure just described yields only the *ratios* of the components of $\boldsymbol{\omega}$ is no handicap, since the ratios completely determine the direction of each of the principal axes, and it is only the directions of these axes that is required. Indeed, we would not expect the *magnitudes* of the ω_i to be determined since the actual rate of the angular motion of the body cannot be specified by the geometry alone; we are free to impress on the body any magnitude of the angular velocity that we wish.

For most of the problems that are encountered in rigid-body dynamics, the bodies are of some regular shape so that the principal axes can be determined merely by examining the symmetry of the body. For example, any body which is a solid of revolution (e.g., a cylindrical rod) has one principal axis which lies along the symmetry axis (e.g., the center-line of the cylindrical rod), and the other two axes are in a plane perpendicular to the symmetry axis. Obviously, since the body is symmetric, the choice of the angular placement of these other two axes is arbitrary. If the moment of inertia along the symmetry is I_1, then $I_2 = I_3$ for a solid of revolution; i.e., the secular equation has a double root.

If a body has $I_1 = I_2 = I_3$, it is termed a *spherical top*; if $I_1 = I_2 \neq I_3$, it is termed a *symmetrical top*; if the principal moments of inertia are all distinct, it is termed an *asymmetrical top*. If a body has $I_1 = 0$, $I_2 = I_3$, it is called a *rotor*; for example, two point masses connected by a weightless shaft, or a diatomic molecule.

▶ Example 12.4 Principal Axes for a Cube

In Example 12.2(b) we found that the moment of inertia tensor for a cube (with origin at one corner) had nonzero off-diagonal elements. Evidently, the coordinate axes chosen for that calculation were not principal axes. If, for example, the cube were rotating about the x_3-axis, then $\boldsymbol{\omega} = \omega_3 \mathbf{e}_3$, and the angular momentum vector $\mathbf{L}$ [see Eq. (12.26)] has the components

$$\left.\begin{aligned} L_1 &= -\tfrac{1}{4}\beta\omega_3 \\ L_2 &= -\tfrac{1}{4}\beta\omega_3 \\ L_3 &= \tfrac{2}{3}\beta\omega_3 \end{aligned}\right\} \tag{1}$$

Thus,

$$\mathbf{L} = Mb^2\omega_3(-\tfrac{1}{4}\mathbf{e}_1 - \tfrac{1}{4}\mathbf{e}_2 + \tfrac{2}{3}\mathbf{e}_3) \tag{2}$$

which is obviously not in the same direction as $\boldsymbol{\omega}$.

In order to find the principal moments of inertia, we must solve the secular equation:

$$\begin{vmatrix} \tfrac{2}{3}\beta - I & -\tfrac{1}{4}\beta & -\tfrac{1}{4}\beta \\ -\tfrac{1}{4}\beta & \tfrac{2}{3}\beta - I & -\tfrac{1}{4}\beta \\ -\tfrac{1}{4}\beta & -\tfrac{1}{4}\beta & \tfrac{2}{3}\beta - I \end{vmatrix} = 0 \tag{3}$$

The value of a determinant is not affected by adding (or subtracting) any row (or column) from any other row (or column). Equation (3) can be solved more easily if we subtract the first row from the second:

$$\begin{vmatrix} \tfrac{2}{3}\beta - I & -\tfrac{1}{4}\beta & -\tfrac{1}{4}\beta \\ -\tfrac{11}{12}\beta + I & \tfrac{11}{12}\beta - I & 0 \\ -\tfrac{1}{4}\beta & -\tfrac{1}{4}\beta & \tfrac{2}{3}\beta - I \end{vmatrix} = 0 \tag{4}$$

We can factor $(\tfrac{11}{12}\beta - I)$ from the second row:

$$(\tfrac{11}{12}\beta - I)\begin{vmatrix} \tfrac{2}{3}\beta - I & -\tfrac{1}{4}\beta & -\tfrac{1}{4}\beta \\ -1 & 1 & 0 \\ -\tfrac{1}{4}\beta & -\tfrac{1}{4}\beta & \tfrac{2}{3}\beta - I \end{vmatrix} = 0 \tag{5}$$

Expanding, we have

$$(\tfrac{11}{12}\beta - I)[(\tfrac{2}{3}\beta - I)^2 - \tfrac{1}{8}\beta^2 - \tfrac{1}{4}\beta(\tfrac{2}{3}\beta - I)] = 0 \tag{6}$$

which can be factored to obtain

$$(\tfrac{1}{6}\beta - I)(\tfrac{11}{12}\beta - I)(\tfrac{11}{12}\beta - I) = 0 \tag{7}$$

Thus, we have the following roots, which give the principal moments of inertia:

$$I_1 = \tfrac{1}{6}\beta; \quad I_2 = \tfrac{11}{12}\beta; \qquad I_3 = \tfrac{11}{12}\beta \tag{8}$$

And the diagonalized moment of inertia tensor becomes

$$\{\mathrm{I}\} = \begin{Bmatrix} \tfrac{1}{6}\beta & 0 & 0 \\ 0 & \tfrac{11}{12}\beta & 0 \\ 0 & 0 & \tfrac{11}{12}\beta \end{Bmatrix} \tag{9}$$

Since two of the roots are identical, $I_2 = I_3$, the principal axis associated with I_1 must be an axis of symmetry.

In order to find the direction of the principal axis associated with I_1, we substitute for I in Eq. (12.27) the value $I = I_1 = \tfrac{1}{6}\beta$:

$$\left.\begin{aligned} (\tfrac{2}{3}\beta - \tfrac{1}{6}\beta)\omega_{11} - \tfrac{1}{4}\beta\omega_{21} - \tfrac{1}{4}\beta\omega_{31} &= 0 \\ -\tfrac{1}{4}\beta\omega_{11} + (\tfrac{2}{3}\beta - \tfrac{1}{6}\beta)\omega_{21} - \tfrac{1}{4}\beta\omega_{31} &= 0 \\ -\tfrac{1}{4}\beta\omega_{11} - \tfrac{1}{4}\beta\omega_{21} + (\tfrac{2}{3}\beta - \tfrac{1}{6}\beta)\omega_{31} &= 0 \end{aligned}\right\} \tag{10}$$

where the second subscript 1 on the ω_i signifies that we are considering the principal axis associated with I_1.

Dividing the first two of these equations by $\beta/4$, we have

$$\left.\begin{aligned} 2\omega_{11} - \omega_{21} - \omega_{31} &= 0 \\ -\omega_{11} + 2\omega_{21} - \omega_{31} &= 0 \end{aligned}\right\} \tag{11}$$

Subtracting the second of these equations from the first, we find $\omega_{11} = \omega_{21}$. Using this result in either of the Eqs. (11), we obtain $\omega_{11} = \omega_{21} = \omega_{31}$, and the desired ratios are

$$\omega_{11} : \omega_{21} : \omega_{31} = 1 : 1 : 1 \tag{12}$$

Therefore, when the cube is rotating about an axis which has associated with it the moment of inertia $I_1 = \frac{1}{6}\beta = \frac{1}{6}Mb^2$, the projections of $\boldsymbol{\omega}$ on the three coordinate axes are all equal. Hence, this principal axis corresponds to the diagonal of the cube.

Since the moments I_2 and I_3 are equal, the orientation of the principal axes associated with these moments is arbitrary; they need only lie in a plane normal to the diagonal of the cube.

12.5 Moments of Inertia for Different Body Coordinate Systems

In order for the kinetic energy to be separable into translational and rotational portions [see Eq. (12.5)], it is, in general, necessary to choose a body coordinate system whose origin is the center of mass of the body. For certain geometrical shapes, it may not always be convenient to compute the elements of the inertia tensor using such a coordinate system. Therefore, consider some other set of coordinate axes X_i, also fixed with respect to the body and which have the same orientation as the x_i-axes, but whose origin Q does not correspond with the origin O (which is located at the center of mass of the body coordinate system). Q may be located either within or outside of the body under consideration.

The elements of the inertia tensor relative to the X_i-axes can be written as

$$J_{ij} = \sum_\alpha m_\alpha \left[\delta_{ij} \sum_k X_{\alpha,k}^2 - X_{\alpha,i} X_{\alpha,j} \right] \tag{12.29}$$

Now, if the vector connecting Q with O is $\mathbf{a}$, then the general vector $\mathbf{R}$ can be written as (see Fig. 12-3)

$$\mathbf{R} = \mathbf{a} + \mathbf{r} \tag{12.30}$$

with components

$$X_i = a_i + x_i \tag{12.30a}$$

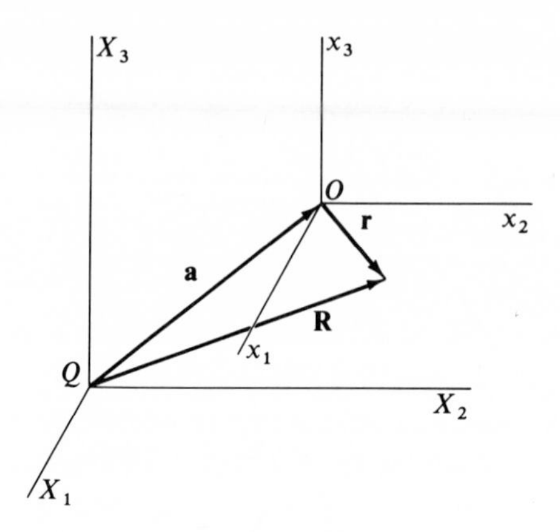

FIG. **12-3**

Using Eq. (12.30a), the tensor element J_{ij} becomes

$$J_{ij} = \sum_{\alpha} m_{\alpha} \left[\delta_{ij} \sum_{k} (x_{\alpha,k} + a_k)^2 - (x_{\alpha,i} + a_i)(x_{\alpha,j} + a_j) \right]$$

$$= \sum_{\alpha} m_{\alpha} \left[\delta_{ij} \sum_{k} x_{\alpha,k}^2 - x_{\alpha,i} x_{\alpha,j} \right]$$

$$+ \sum_{\alpha} m_{\alpha} \left[\delta_{ij} \sum_{k} (2x_{\alpha,k} a_k + a_k^2) \right.$$

$$\left. - (a_i x_{\alpha,j} + a_j x_{\alpha,i} + a_i a_j) \right] \tag{12.31}$$

Identifying the first summation as I_{ij}, we have, upon regrouping,

$$J_{ij} = I_{ij} + \sum_{\alpha} m_{\alpha} \left[\delta_{ij} \sum_{k} a_k^2 - a_i a_j \right]$$

$$+ \sum_{\alpha} m_{\alpha} \left[2\delta_{ij} \sum_{k} x_{\alpha,k} a_k - a_i x_{\alpha,j} - a_j x_{\alpha,i} \right] \tag{12.32}$$

But each term in the last summation involves a sum of the form

$$\sum_{\alpha} m_{\alpha} x_{\alpha,k}$$

We know, however, that since O is located at the center of mass,

$$\sum_{\alpha} m_{\alpha} \mathbf{r}_{\alpha} = 0$$

or, for the kth component,

$$\sum_{\alpha} m_{\alpha} x_{\alpha,k} = 0$$

Therefore, all such terms in Eq. (12.32) vanish and we have

$$J_{ij} = I_{ij} + \sum_{\alpha} m_{\alpha}\left[\delta_{ij} \sum_{k} a_k^2 - a_i a_j\right] \tag{12.33}$$

But,

$$\sum_{\alpha} m_{\alpha} = M \qquad \text{and} \qquad \sum_{k} a_k^2 \equiv a^2$$

Solving for I_{ij}, we have the result

$$\boxed{I_{ij} = J_{ij} - M[a^2 \delta_{ij} - a_i a_j]} \tag{12.34}$$

which allows the calculation of the elements I_{ij} of the desired inertia tensor (with origin at the center of mass) once those with respect to the X_i-axes are known. The second term on the right-hand side of Eq. (12.34) is the inertia tensor referred to the origin Q for a point mass M.

Equation (12.34) is the general form of Steiner's *parallel-axis theorem*,* the simplified form of which is given in elementary treatments. Consider, for example, Fig. 12-4. The element I_{11} is

$$\begin{aligned} I_{11} &= J_{11} - M[(a_1^2 + a_2^2 + a_3^2)\,\delta_{11} - a_1^2] \\ &= J_{11} - M(a_2^2 + a_3^2) \end{aligned}$$

which states that the difference between the elements is equal to the mass of the body multiplied by the square of the distance between the parallel axes (in this case, between the x_1- and X_1-axes).

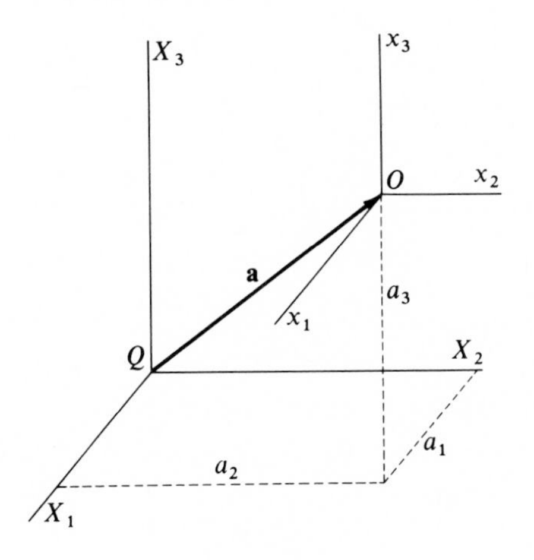

FIG. 12-4

* Jacob Steiner (1796–1863).

▶ Example 12.5 Another Look at the Cube

Let us again return to the case of the homogeneous cube for which we found the inertia tensor (for the origin at one corner) to be [Eq. (4) of Example 12.2(b)]:

$$\{\mathrm{J}\} = \begin{Bmatrix} \frac{2}{3}Mb^2 & -\frac{1}{4}Mb^2 & -\frac{1}{4}Mb^2 \\ -\frac{1}{4}Mb^2 & \frac{2}{3}Mb^2 & -\frac{1}{4}Mb^2 \\ -\frac{1}{4}Mb^2 & -\frac{1}{4}Mb^2 & \frac{2}{3}Mb^2 \end{Bmatrix} \tag{1}$$

We may now use Eq. (12.34) to obtain the inertia tensor {I} referred to a coordinate system with origin at the center of mass. In keeping with the notation of this section, we call the new axes x_i with origin O, and call the previous axes X_i with origin Q at one corner of the cube (see Fig. 12.5).

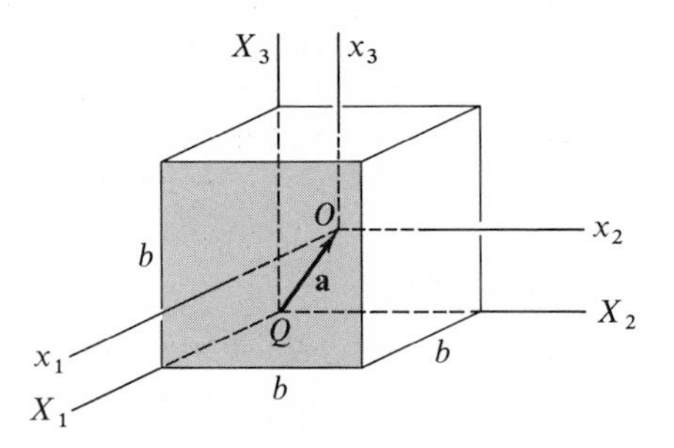

FIG. **12-5**

Now, the center of mass of the cube is at the point $(b/2, b/2, b/2)$ in the X_i coordinate system, and the components of the vector **a** therefore are

$$a_1 = a_2 = a_3 = b/2 \tag{2}$$

From Eq. (1), we have

$$\left.\begin{aligned} J_{11} &= J_{22} = J_{33} = \tfrac{2}{3}Mb^2 \\ J_{12} &= J_{13} = J_{23} = -\tfrac{1}{4}Mb^2 \end{aligned}\right\} \tag{3}$$

And applying Eq. (12.34), we find

$$\begin{aligned} I_{11} &= J_{11} - M[a^2 - a_1^2] \\ &= J_{11} - M[a_2^2 + a_3^2] \\ &= \tfrac{2}{3}Mb^2 - \tfrac{1}{2}Mb^2 = \tfrac{1}{6}Mb^2 \end{aligned} \tag{4}$$

Also,

$$\begin{aligned} I_{12} &= J_{12} - M[-a_1a_2] \\ &= -\tfrac{1}{4}Mb^2 + \tfrac{1}{4}Mb^2 = 0 \end{aligned} \tag{5}$$

So that we have altogether

$$\left.\begin{aligned} I_{12} = I_{22} = I_{33} &= \tfrac{1}{6}Mb^2 \\ I_{12} = I_{13} = I_{23} &= 0 \end{aligned}\right\} \tag{6}$$

The inertia tensor is therefore diagonal:

$$\{\mathsf{I}\} = \begin{Bmatrix} \tfrac{1}{6}Mb^2 & 0 & 0 \\ 0 & \tfrac{1}{6}Mb^2 & 0 \\ 0 & 0 & \tfrac{1}{6}Mb^2 \end{Bmatrix} \tag{7}$$

If we factor out the common term $\frac{1}{6}Mb^2$ from this expression, we can write

$$\{\mathsf{I}\} = \tfrac{1}{6}Mb^2\{\mathbf{1}\} \tag{8}$$

where $\{\mathbf{1}\}$ is the *unit tensor*:

$$\{\mathbf{1}\} \equiv \begin{Bmatrix} 1 & 0 & 0 \\ 0 & 1 & 0 \\ 0 & 0 & 1 \end{Bmatrix} \tag{9}$$

Thus, we find that for the choice of the origin at the center of mass of the cube, the principal axes are perpendicular to the faces of the cube. Since from a physical standpoint there is nothing to distinguish any one of these axes from another, the principal moments of inertia are all equal for this case. We note further, that as long as we maintain the origin at the center of mass, then the inertia tensor is the same for *any* orientation of the coordinate axes and these axes are equally valid principal axes.*

12.6 Further Properties of the Inertia Tensor

Before beginning an attack on the problems of rigid-body dynamics by obtaining the general equations of motion, we should consider the fundamental importance of some of the operations that we have been discussing. Let us begin by examining the properties of the inertia tensor under coordinate transformations.†

Now, we have already obtained the fundamental relation connecting the

* In this regard, the cube is similar to a sphere as far as the inertia tensor is concerned; i.e., for an origin at the center of mass, the structure of the inertia tensor elements is not sufficiently detailed to discriminate between a cube and a sphere.

† We will confine our attention to rectangular coordinate systems so that we may ignore some of the more complicated properties of tensors that manifest themselves in general curvilinear coordinates.

inertia tensor and the angular momentum and angular velocity vectors [Eq. (12.20)] which we can write as

$$L_k = \sum_l I_{kl}\,\omega_l \tag{12.35}$$

Since this is a vector equation, in a coordinate system rotated with respect to the system for which Eq. (12.35) applies, we must have an entirely analogous relation:

$$L'_i = \sum_j I'_{ij}\,\omega'_j \tag{12.35a}$$

where the primed quantities all refer to the rotated system. Now, both **L** and **ω** obey the standard transformation equation for vectors [Eq. (1.8)]:

$$x_i = \sum_j \lambda^t_{ij}\,x'_j = \sum_j \lambda_{ji}\,x'_j$$

Therefore, we can write

$$L_k = \sum_m \lambda_{mk}\,L'_m \tag{12.36a}$$

and

$$\omega_l = \sum_j \lambda_{jl}\,\omega'_j \tag{12.36b}$$

If we substitute Eqs. (12.36) into Eq. (12.35), we obtain

$$\sum_m \lambda_{mk}\,L'_m = \sum_l I_{kl} \sum_j \lambda_{jl}\,\omega'_j \tag{12.37}$$

Next, we multiply both sides of this equation by λ_{ik} and sum over k:

$$\sum_m \left(\sum_k \lambda_{ik}\,\lambda_{mk}\right) L'_m = \sum_j \left(\sum_{k,\,l} \lambda_{ik}\,\lambda_{jl}\,I_{kl}\right)\omega'_j \tag{12.38}$$

The term in parentheses on the left-hand side is just δ_{im}, so that upon performing the summation over m we obtain

$$L'_i = \sum_j \left(\sum_{k,\,l} \lambda_{ik}\,\lambda_{jl}\,I_{kl}\right)\omega'_j \tag{12.39}$$

In order that this equation be identical with Eq. (12.35a), we see that we must have

$$I'_{ij} = \sum_{k,\,l} \lambda_{ik}\,\lambda_{jl}\,I_{kl} \tag{12.40}$$

This is therefore the rule that the inertia tensor must obey under a coordinate transformation. Equation (12.40) is, in fact, the *general* rule that specifies the manner in which any second-rank tensor must transform.

For a tensor $\{\mathsf{T}\}$ of arbitrary rank the statement is*†

$$T'_{abcd\ldots} = \sum_{i,j,k,l,\ldots} \lambda_{ai}\lambda_{bj}\lambda_{ck}\lambda_{dl}\cdots T_{ijkl\ldots} \tag{12.41}$$

Note that we can write Eq. (12.40) as

$$\boxed{I'_{ij} = \sum_{k,l} \lambda_{ik} I_{kl} \lambda^t_{lj}} \tag{12.42}$$

Although matrices and tensors are distinct types of mathematical objects, the manipulation of tensors is in many respects the same as for matrices. Thus, Eq. (12.42) can be expressed as a matrix equation:

$$\mathsf{I}' = \boldsymbol{\lambda}\mathsf{I}\boldsymbol{\lambda}^t \tag{12.43}$$

where we understand I to be the matrix which consists of the elements of the tensor $\{\mathsf{I}\}$. Now, since we are considering only orthogonal transformation matrices, the transpose of $\boldsymbol{\lambda}$ is equal to its inverse, so that we can express Eq. (12.43) as

$$\boxed{\mathsf{I}' = \boldsymbol{\lambda}\mathsf{I}\boldsymbol{\lambda}^{-1}} \tag{12.44}$$

A transformation of this general type is called a *similarity transformation* (I' is *similar* to I).

▶ Example 12.6(a) A Similarity Transformation

In Example 12.5 the assertion was made that the inertia tensor for a cube (with origin at the center of mass) was independent of the orientation of the axes. We can easily prove this as follows. The change in the inertia tensor under a rotation of the coordinate axes can be computed by making a similarity transformation. Thus, if the rotation is described by the matrix $\boldsymbol{\lambda}$, we have

$$\mathsf{I}' = \boldsymbol{\lambda}\mathsf{I}\boldsymbol{\lambda}^{-1} \tag{1}$$

But the matrix I, which is derived from the elements of the tensor $\{\mathsf{I}\}$ [Eq. (8) of Example 12.5], is just the identity matrix $\mathbf{1}$ multiplied by a constant:

$$\mathsf{I} = \tfrac{1}{6}Mb^2\begin{pmatrix} 1 & 0 & 0 \\ 0 & 1 & 0 \\ 0 & 0 & 1 \end{pmatrix} = \tfrac{1}{6}Mb^2\mathbf{1} \tag{2}$$

* Note that a tensor of the *first rank* transforms as

$$T'_a = \sum_i \lambda_{ai} T_i$$

so that such a tensor is in fact a *vector*. A tensor of zero *rank* implies that $T' = T$, or that such a tensor is a *scalar*.

† The properties of quantities which transform in this manner were first discussed by C. Niven in 1874. The application of the term *tensor* to such quantities is due to J. Willard Gibbs.

Therefore the operations specified in Eq. (1) are trivial:

$$\mathbf{I}' = \tfrac{1}{6}Mb^2\boldsymbol{\lambda}\mathbf{1}\boldsymbol{\lambda}^{-1} = \tfrac{1}{6}Mb^2\boldsymbol{\lambda}\boldsymbol{\lambda}^{-1} = \tfrac{1}{6}Mb^2\mathbf{1} = \mathbf{I} \tag{3}$$

Thus, the transformed inertia tensor is identical to the original tensor, independent of the details of the rotation.

Let us next determine what condition must be satisfied if we take an arbitrary inertia tensor and perform a coordinate rotation in such a way that the transformed inertia tensor is diagonal. Such an operation implies that the quantity I'_{ij} in Eq. (12.40) must satisfy the relation [cf. Eq. (12.22)]

$$I'_{ij} = I_i\,\delta_{ij} \tag{12.45}$$

Thus,

$$I_i\,\delta_{ij} = \sum_{k,l} \lambda_{ik}\,\lambda_{jl}\,I_{kl} \tag{12.46}$$

If we multiply both sides of this equation by λ_{im} and sum over i, we obtain

$$\sum_i I_i\,\lambda_{im}\,\delta_{ij} = \sum_{k,\,l}\left(\sum_i \lambda_{im}\,\lambda_{ik}\right)\lambda_{jl}\,I_{kl} \tag{12.47}$$

The term in parentheses is just δ_{mk}, so that the summation over i on the left-hand side of the equation, and the summation over k on the right-hand side yield

$$I_j\,\lambda_{jm} = \sum_l \lambda_{jl}\,I_{ml} \tag{12.48}$$

Now, the left-hand side of this equation can be written as

$$I_j\,\lambda_{jm} = \sum_l I_j\,\lambda_{jl}\,\delta_{ml} \tag{12.49}$$

so that Eq. (12.48) becomes

$$\sum_l I_j\,\lambda_{jl}\,\delta_{ml} = \sum_l \lambda_{jl}\,I_{ml} \tag{12.50}$$

or,

$$\sum_l (I_{ml} - I_j\,\delta_{ml})\lambda_{jl} = 0 \tag{12.50a}$$

This is a set of simultaneous, linear algebraic equations; for each value of j there are three such equations, one for each of the three possible values of m. In order for a nontrivial solution to exist, the determinant of the coefficients must vanish, so that the principal moments of inertia, I_1, I_2, and I_3, are obtained as roots of the secular determinant for I:

$$\boxed{|I_{ml} - I\delta_{ml}| = 0} \tag{12.51}$$

This equation is just Eq. (12.28); it is a cubic equation which yields the principal moments of inertia.

We see, therefore, that for any inertia tensor, the elements of which are computed for a given origin, it is possible to perform a rotation of the coordinate axes about that origin in such a way that the inertia tensor becomes diagonal; the new coordinate axes are then the principal axes of the body and the new moments are the principal moments of inertia. Thus, for *any* body and for *any* choice of origin, there *always* exists a set of principal axes.

▶ Example 12.6(b) Rotation of the Axes for a Cube

Let us take a final look at the problem of the cube and diagonalize the inertia tensor by rotating the coordinate axes. We choose the origin to lie at one corner and perform the rotation in such a manner that the x_1-axis will coincide with the diagonal of the cube. Such a rotation can conveniently be made in two steps: first, we rotate through an angle of 45° about the x_3-axis; and, secondly, we rotate through an angle of $\cos^{-1}(\sqrt{\frac{2}{3}})$ about the x_2-axis. The first rotation matrix is

$$\lambda_1 = \begin{pmatrix} \frac{1}{\sqrt{2}} & \frac{1}{\sqrt{2}} & 0 \\ -\frac{1}{\sqrt{2}} & \frac{1}{\sqrt{2}} & 0 \\ 0 & 0 & 1 \end{pmatrix} \tag{1}$$

and the second rotation matrix is

$$\lambda_2 = \begin{pmatrix} \sqrt{\frac{2}{3}} & 0 & \frac{1}{\sqrt{3}} \\ 0 & 1 & 0 \\ -\frac{1}{\sqrt{3}} & 0 & \sqrt{\frac{2}{3}} \end{pmatrix} \tag{2}$$

Therefore, the complete rotation matrix is

$$\lambda = \lambda_2\lambda_1 = \begin{pmatrix} \frac{1}{\sqrt{3}} & \frac{1}{\sqrt{3}} & \frac{1}{\sqrt{3}} \\ -\frac{1}{\sqrt{2}} & \frac{1}{\sqrt{2}} & 0 \\ -\frac{1}{\sqrt{6}} & -\frac{1}{\sqrt{6}} & \sqrt{\frac{2}{3}} \end{pmatrix} = \frac{1}{\sqrt{3}} \begin{pmatrix} 1 & 1 & 1 \\ -\sqrt{\frac{3}{2}} & \sqrt{\frac{3}{2}} & 0 \\ -\frac{1}{\sqrt{2}} & -\frac{1}{\sqrt{2}} & \sqrt{2} \end{pmatrix} \tag{3}$$

The transformed inertia tensor is [see Eq. (12.43)]:

$$\mathsf{I}' = \lambda \mathsf{I} \lambda^t \tag{4}$$

or, upon factoring β out of I, we have

$$\mathsf{I}' = \frac{\beta}{3}\begin{pmatrix} 1 & 1 & 1 \\ -\sqrt{\frac{3}{2}} & \sqrt{\frac{3}{2}} & 0 \\ -\frac{1}{\sqrt{2}} & -\frac{1}{\sqrt{2}} & \sqrt{2} \end{pmatrix}\begin{pmatrix} \frac{2}{3} & -\frac{1}{4} & -\frac{1}{4} \\ -\frac{1}{4} & \frac{2}{3} & -\frac{1}{4} \\ -\frac{1}{4} & -\frac{1}{4} & \frac{2}{3} \end{pmatrix}\begin{pmatrix} 1 & -\sqrt{\frac{3}{2}} & -\frac{1}{\sqrt{2}} \\ 1 & \sqrt{\frac{3}{2}} & -\frac{1}{\sqrt{2}} \\ 1 & 0 & \sqrt{2} \end{pmatrix}$$

$$= \frac{\beta}{3}\begin{pmatrix} 1 & 1 & 1 \\ -\sqrt{\frac{3}{2}} & \sqrt{\frac{3}{2}} & 0 \\ -\frac{1}{\sqrt{2}} & -\frac{1}{\sqrt{2}} & \sqrt{2} \end{pmatrix}\begin{pmatrix} \frac{1}{6} & -\frac{11}{12}\sqrt{\frac{3}{2}} & -\frac{11}{12}\frac{\sqrt{2}}{2} \\ \frac{1}{6} & \frac{11}{12}\sqrt{\frac{3}{2}} & -\frac{11}{12}\frac{\sqrt{2}}{2} \\ \frac{1}{6} & 0 & \frac{11\sqrt{2}}{12} \end{pmatrix}$$

$$= \begin{pmatrix} \frac{1}{6}\beta & 0 & 0 \\ 0 & \frac{11}{12}\beta & 0 \\ 0 & 0 & \frac{11}{12}\beta \end{pmatrix} \tag{5}$$

Equation (5) is just the matrix form of the inertia tensor that was found by the diagonalization procedure using the secular determinant [Eq. (7) of Example 12.4].

We have therefore demonstrated two general procedures that may be used to diagonalize the inertia tensor. As was previously pointed out, these methods are not limited to the inertia tensor but are generally valid. Either procedure can, of course, be very complicated. For example, if we wish to use the rotation procedure in the most general case, we must first construct a matrix which describes an arbitrary rotation; this will entail three separate rotations, one about each of the coordinate axes. This rotation matrix must then be applied to the tensor in a similarity transformation. The off-diagonal elements of the resulting matrix* must then be examined and values of the rotation

* A *large* sheet of paper should be used.

angles determined so that these off-diagonal elements vanish. The actual use of such a procedure can tax the limits of human patience: however, in some simple situations this method of diagonalization can be used with profit. This is particularly true if the geometry of the problem indicates that only a simple rotation about one of the coordinate axes is necessary; the rotation angle can then be evaluated without difficulty (see, for example, Problems 12-14, 12-16, and 12-17).

The example of the cube illustrates the important point that the elements of the inertia tensor, the values of the principal moments of inertia, and the orientation of the principal axes for a rigid body all depend upon the choice of origin for the system. We recall, however, that in order for the kinetic energy to be separable into translational and rotational portions, the origin of the body coordinate system must, in general, be taken to coincide with the center of mass of the body. On the other hand, for *any* choice of the origin for *any* body there always exists an orientation of the axes which diagonalizes the inertia tensor and hence these axes become principal axes for that particular origin.

Next, we seek to prove that the principal axes actually form an orthogonal set. Let us assume that we have solved the secular equation and have determined the principal moments of inertia, all of which are distinct. Now, we know that for each principal moment there exists a corresponding principal axis which has the property that if the angular velocity vector $\boldsymbol{\omega}$ lies along this axis, then the angular momentum vector $\mathbf{L}$ is similarly oriented. That is, to each I_j there corresponds an angular velocity vector $\boldsymbol{\omega}_j$ with components $\omega_{1j}, \omega_{2j}, \omega_{3j}$. (We use the subscript on the vector $\boldsymbol{\omega}$ and the second subscript on the components of $\boldsymbol{\omega}$ to designate the principal moment with which we are concerned.) Thus, for the mth principal moment we have

$$L_{im} = I_m \omega_{im} \tag{12.52}$$

In terms of the elements of the moment of inertia tensor, we also have

$$L_{im} = \sum_k I_{ik} \omega_{km} \tag{12.53}$$

so that, combining these two relations, we have

$$\sum_k I_{ik} \omega_{km} = I_m \omega_{im} \tag{12.54a}$$

Similarly, we can write for the nth principal moment:

$$\sum_i I_{ki} \omega_{in} = I_n \omega_{kn} \tag{12.54b}$$

If we multiply Eq. (12.54a) by ω_{in} and sum over i, and multiply Eq. (12.54b) by ω_{km} and sum over k, we have

$$\left.\begin{aligned} \sum_{i,k} I_{ik}\,\omega_{km}\,\omega_{in} &= \sum_{i} I_m\,\omega_{im}\,\omega_{in} \\ \sum_{i,k} I_{ki}\,\omega_{in}\,\omega_{km} &= \sum_{k} I_n\,\omega_{kn}\,\omega_{km} \end{aligned}\right\} \tag{12.55}$$

Now, the left-hand sides of these equations are identical since the inertia tensor is symmetric ($I_{ik} = I_{ki}$). Therefore, upon subtracting the second equation from the first, we have,

$$I_m \sum_{i} \omega_{im}\,\omega_{in} - I_n \sum_{k} \omega_{km}\,\omega_{kn} = 0 \tag{12.56}$$

Since i and k are both dummy indices, we can replace them by l, say, and obtain

$$(I_m - I_n) \sum_{l} \omega_{lm}\,\omega_{ln} = 0 \tag{12.57}$$

By hypothesis, the principal moments are distinct, so that $I_m \neq I_n$. Therefore Eq. (12.57) can be satisfied only if

$$\sum_{l} \omega_{lm}\,\omega_{ln} = 0 \tag{12.58}$$

But this summation is just the definition of the scalar product of the vectors $\boldsymbol{\omega}_m$ and $\boldsymbol{\omega}_n$. Hence,

$$\boldsymbol{\omega}_m \cdot \boldsymbol{\omega}_n = 0 \tag{12.58a}$$

Since the principal moments I_m and I_n were picked arbitrarily from the set of three moments, we conclude that each pair of principal axes is perpendicular, so that the three principal axes constitute an orthogonal set.

In the event that there is a double root of the secular equation so that the principal moments are I_1, $I_2 = I_3$, then the analysis above shows that the angular velocity vectors satisfy the relations

$$\boldsymbol{\omega}_1 \perp \boldsymbol{\omega}_2\,; \qquad \boldsymbol{\omega}_1 \perp \boldsymbol{\omega}_3$$

but that nothing may be said regarding the angle between $\boldsymbol{\omega}_2$ and $\boldsymbol{\omega}_3$. However, as we have previously remarked, the fact that $I_2 = I_3$ implies that the body possesses an axis of symmetry. Therefore, $\boldsymbol{\omega}_1$ lies along the symmetry axis and $\boldsymbol{\omega}_2$ and $\boldsymbol{\omega}_3$ are required only to lie in the plane perpendicular to $\boldsymbol{\omega}_1$. Consequently, there is no loss of generality if we also choose $\boldsymbol{\omega}_2 \perp \boldsymbol{\omega}_3$. Thus, the principal axes for a rigid body with an axis of symmetry can also be chosen to be an orthogonal set.

We have previously shown that the principal moments of inertia are obtained as the roots of the secular equation—a cubic equation. Mathematically, at least one of the roots of a cubic equation must be real, but there may be two imaginary roots. If the diagonalization procedures for the inertia tensor are to be physically meaningful, we must of course always obtain only real values for the principal moments. We can show in the following way that this is a general result. First, we assume the roots to be complex and use a procedure similar to that employed in the preceding proof. But now we must also allow the quantities ω_{km} to become complex; there is no mathematical reason why we cannot do this, and we shall not be concerned with any physical interpretation of these quantities. Therefore, we write Eq. (12.54a) as before, but we take the complex conjugate of Eq. (12.54b):

$$\left.\begin{aligned} \sum_k I_{ik}\,\omega_{km} &= I_m\,\omega_{im} \\ \sum_i I_{ki}^*\,\omega_{in}^* &= I_n^*\omega_{kn}^* \end{aligned}\right\} \tag{12.59}$$

Next, we multiply the first of these equations by ω_{in}^* and sum over i and multiply the second by ω_{km} and sum over k. The inertia tensor is symmetric and its elements are all real so that $I_{ik} = I_{ki}^*$. Therefore, upon subtracting the second of these equations from the first, we find

$$(I_m - I_n^*)\sum_l \omega_{lm}\,\omega_{ln}^* = 0 \tag{12.60a}$$

For the case $m = n$, we have

$$(I_m - I_m^*)\sum_l \omega_{lm}\,\omega_{lm}^* = 0 \tag{12.60b}$$

The sum is just the definition of the scalar product of $\boldsymbol{\omega}_m$ and $\boldsymbol{\omega}_m^*$:

$$\boldsymbol{\omega}_m \cdot \boldsymbol{\omega}_m^* = |\boldsymbol{\omega}_m|^2 \geq 0 \tag{12.61}$$

Therefore, since the squared magnitude of $\boldsymbol{\omega}_m$ is in general positive, it must be true that $I_m = I_m^*$ in order for Eq. (12.60b) to be satisfied. If a quantity and its complex conjugate are equal, then the imaginary parts must vanish identically. Thus, the principal moments of inertia are all real. Since $\{\mathbf{I}\}$ is real, the vectors $\boldsymbol{\omega}_m$ must also be real.

If $m = n$ in Eq. (12.60b), and if $I_m = I_n$, then the equation can be satisfied only if $\boldsymbol{\omega}_m \cdot \boldsymbol{\omega}_n = 0$; that is, these vectors are orthogonal, as before.

In all of the proofs that have been carried out in this section, reference has been made to the inertia tensor. An examination of these proofs will reveal, however, that the only properties of the inertia tensor which have actually been used are the facts that the tensor is symmetric and that the elements are real. Therefore, we may conclude that *any* real, symmetrical tensor† has the following properties:

† To be more precise, we require only that the elements of the tensor obey the relation $I_{ik} = I_{ki}^*$; thus we allow the possibility of complex quantities. Tensors (and matrices) which have this property are said to be *Hermitean*.

(a) Diagonalization may be accomplished by an appropriate rotation of axes, i.e., a similarity transformation.

(b) The eigenvalues* are obtained as roots of the secular determinant and are real.

(c) The eigenvectors* are real and orthogonal.

12.7 The Eulerian Angles

The transformation from one coordinate system to another can be represented by a matrix equation of the form

$$\mathsf{x} = \lambda \mathsf{x}'$$

If we identify the fixed system with x' and the body system with x, then the rotation matrix λ completely describes the relative orientation of the two systems. Now, the rotation matrix λ contains three independent angles. There are many possible choices for these angles; for our purposes we will find it convenient to use the so called *Eulerian angles*,† φ, θ, and ψ.

The Eulerian angles are generated in the following series of rotations which take the x_i' system into the x_i system‡:

(1) The first rotation is counterclockwise through an angle φ about the x_3'-axis, as shown in Fig. 12-6a to transform the x_i' into the x_i''. Since the rotation takes place in the x_1'-x_2' plane, the transformation matrix is

$$\lambda_\varphi = \begin{pmatrix} \cos\varphi & \sin\varphi & 0 \\ -\sin\varphi & \cos\varphi & 0 \\ 0 & 0 & 1 \end{pmatrix} \tag{12.62}$$

(2) The second rotation is counterclockwise through an angle θ about the x_1''-axis, as shown in Fig. 12-6b, to transform the x_i'' into the x_i'''. Since the rotation is now in the x_2''-x_3'' plane, the transformation matrix is

$$\lambda_\theta = \begin{pmatrix} 1 & 0 & 0 \\ 0 & \cos\theta & \sin\theta \\ 0 & -\sin\theta & \cos\theta \end{pmatrix} \tag{12.63}$$

(3) The third rotation is counterclockwise through an angle ψ about the x_3'''-axis, as shown in Fig. 12-6c, to transform the x_i''' into the x_i. The

* The terms "eigenvalues" and "eigenvectors" are the generic names of the quantities which, in the case of the inertia tensor, are the principal moments and the principal axes, respectively. We shall encounter these terms again in the discussion of small oscillations in Chapter 13.

† The rotation scheme of Euler was first published in 1776.

‡ It should be noted that the designations of the Euler angles and even the manner in which they are generated are not universally agreed upon. Therefore, some care must be taken in any comparison of results from different sources. The notation used here is that most commonly found in modern texts.

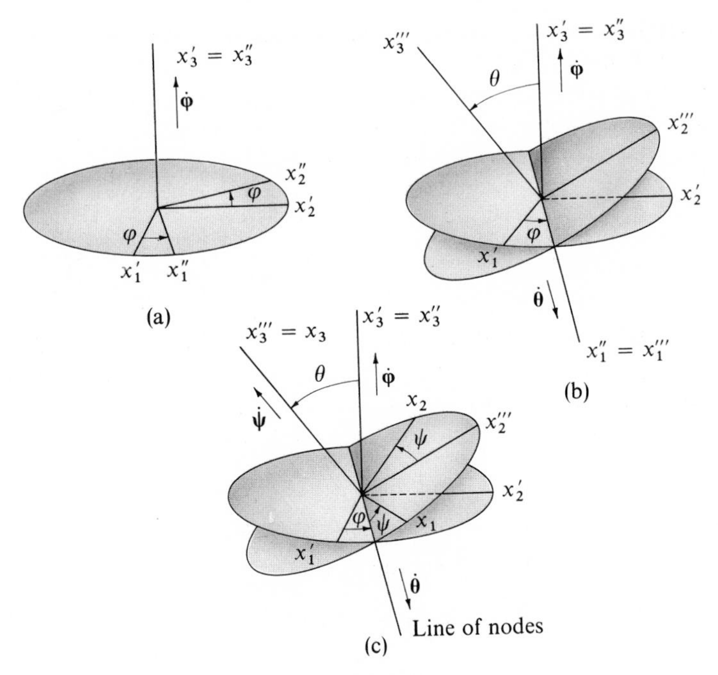

FIG. **12-6**

transformation matrix is

$$\lambda_\psi = \begin{pmatrix} \cos\psi & \sin\psi & 0 \\ -\sin\psi & \cos\psi & 0 \\ 0 & 0 & 1 \end{pmatrix} \tag{12.64}$$

The line which is common to the planes containing the x_1'- and x_2-axes and the x_1'- and x_2'-axes is called the *line of nodes.*

The complete transformation from the x_i' system to the x_i system is given by the rotation matrix λ:

$$\lambda = \lambda_\psi \lambda_\theta \lambda_\varphi \tag{12.65}$$

The components of this matrix are:

$$\left.\begin{aligned}
\lambda_{11} &= \cos\psi \cos\varphi - \cos\theta \sin\varphi \sin\psi \\
\lambda_{21} &= -\sin\psi \cos\varphi - \cos\theta \sin\varphi \cos\psi \\
\lambda_{31} &= \sin\theta \sin\varphi \\
\lambda_{12} &= \cos\psi \sin\varphi + \cos\theta \cos\varphi \sin\psi \\
\lambda_{22} &= -\sin\psi \sin\varphi + \cos\theta \cos\varphi \cos\psi \\
\lambda_{32} &= -\sin\theta \cos\varphi \\
\lambda_{13} &= \sin\psi \sin\theta \\
\lambda_{23} &= \cos\psi \sin\theta \\
\lambda_{33} &= \cos\theta
\end{aligned}\right\} \tag{12.66}$$

(The components λ_{ij} are off-set above to assist in the visualization of the complete $\boldsymbol{\lambda}$-matrix.)

Since it is possible to associate a vector with an infinitesimal rotation, we can associate the time derivatives of these rotation angles with the components of the angular velocity vector $\boldsymbol{\omega}$. Thus,

$$\left.\begin{aligned} \omega_\varphi &= \dot{\varphi} \\ \omega_\theta &= \dot{\theta} \\ \omega_\psi &= \dot{\psi} \end{aligned}\right\} \tag{12.67}$$

The rigid-body equations of motion are most conveniently expressed in the body coordinate system (i.e., the x_i system), and therefore we must express the components of $\boldsymbol{\omega}$ is this system. We note that, in Fig. 12-6, the angular velocities $\dot{\boldsymbol{\varphi}}$, $\dot{\boldsymbol{\theta}}$, and $\dot{\boldsymbol{\psi}}$ are directed along the following axes:

$\dot{\boldsymbol{\varphi}}$ along the x_3' (fixed) axis

$\dot{\boldsymbol{\theta}}$ along the line of nodes

$\dot{\boldsymbol{\psi}}$ along the x_3 (body) axis

The components of these angular velocities along the body coordinate axes are:

$$\left.\begin{aligned} \dot{\varphi}_1 &= \dot{\varphi} \sin\theta \sin\psi \\ \dot{\varphi}_2 &= \dot{\varphi} \sin\theta \cos\psi \\ \dot{\varphi}_3 &= \dot{\varphi} \cos\theta \end{aligned}\right\} \tag{12.68a}$$

$$\left.\begin{aligned} \dot{\theta}_1 &= \dot{\theta} \cos\psi \\ \dot{\theta}_2 &= -\dot{\theta} \sin\psi \\ \dot{\theta}_3 &= 0 \end{aligned}\right\} \tag{12.68b}$$

$$\left.\begin{aligned} \dot{\psi}_1 &= 0 \\ \dot{\psi}_2 &= 0 \\ \dot{\psi}_3 &= \dot{\psi} \end{aligned}\right\} \tag{12.68c}$$

Collecting the individual components of $\boldsymbol{\omega}$, we have, finally,

$$\boxed{\begin{aligned} \omega_1 &= \dot{\varphi}_1 + \dot{\theta}_1 + \dot{\psi}_1 = \dot{\varphi} \sin\theta \sin\psi + \dot{\theta} \cos\psi \\ \omega_2 &= \dot{\varphi}_2 + \dot{\theta}_2 + \dot{\psi}_2 = \dot{\varphi} \sin\theta \cos\psi - \dot{\theta} \sin\psi \\ \omega_3 &= \dot{\varphi}_3 + \dot{\theta}_3 + \dot{\psi}_3 = \dot{\varphi} \cos\theta + \dot{\psi} \end{aligned}} \tag{12.69}$$

These relations will be of use later in expressing the components of the angular momentum in the body coordinate system.

12.8 Euler's Equations for a Rigid Body

Let us first consider the force-free motion of a rigid body. In such a case, the potential energy U vanishes and the Lagrangian L becomes identical with the rotational kinetic energy T.* If we choose the x_i-axes to correspond to the principal axes of the body, then from Eq. (12.24b) we have

$$T = \frac{1}{2} \sum_i I_i \omega_i^2 \tag{12.70}$$

If we choose the Eulerian angles as the generalized coordinates, then Lagrange's equation for the coordinate ψ is

$$\frac{\partial T}{\partial \psi} - \frac{d}{dt} \frac{\partial T}{\partial \dot{\psi}} = 0 \tag{12.71}$$

which can be expressed as

$$\sum_i \frac{\partial T}{\partial \omega_i} \frac{\partial \omega_i}{\partial \psi} - \frac{d}{dt} \sum_i \frac{\partial T}{\partial \omega_i} \frac{\partial \omega_i}{\partial \dot{\psi}} = 0 \tag{12.72}$$

If we differentiate the components of $\boldsymbol{\omega}$ [Eqs. (12.69)] with respect to ψ and $\dot{\psi}$ we have

$$\left.\begin{aligned} \frac{\partial \omega_1}{\partial \psi} &= \dot{\varphi} \sin \theta \cos \psi - \dot{\theta} \sin \psi = \omega_2 \\ \frac{\partial \omega_2}{\partial \psi} &= -\dot{\varphi} \sin \theta \sin \psi - \dot{\theta} \cos \psi = -\omega_1 \\ \frac{d\omega_3}{\partial \psi} &= 0 \end{aligned}\right\} \tag{12.73}$$

and

$$\left.\begin{aligned} \frac{\partial \omega_1}{\partial \dot{\psi}} &= \frac{\partial \omega_2}{\partial \dot{\psi}} = 0 \\ \frac{\partial \omega_3}{\partial \dot{\psi}} &= 1 \end{aligned}\right\} \tag{12.74}$$

From Eq. (12.70) we also have

$$\frac{\partial T}{\partial \omega_i} = I_i \omega_i \tag{12.75}$$

* Since the motion is force-free, the translational kinetic energy is unimportant for our purposes here. (We can always transform to a coordinate system in which the center of mass of the body is at rest.)

Therefore, Eq. (12.72) becomes

$$I_1\omega_1\omega_2 + I_2\omega_2(-\omega_1) - \frac{d}{dt} I_3\omega_3 = 0$$

or,

$$(I_1 - I_2)\omega_1\omega_2 - I_3\dot{\omega}_3 = 0 \tag{12.76}$$

Since the designation of any particular principal axis as the x_3-axis is entirely arbitrary, Eq. (12.76) can be permuted to obtain relations for $\dot{\omega}_1$ and $\dot{\omega}_2$. By making use of the permutation symbol, we can write, in general,

$$\boxed{(I_i - I_j)\omega_i\omega_j - \sum_k I_k\dot{\omega}_k\varepsilon_{ijk} = 0} \tag{12.77}$$

The three equations represented by Eq. (12.77) are called *Euler's equations* for the case of force-free motion.* It must be noted that although Eq. (12.76) for $\dot{\omega}_3$ is indeed the Lagrange equation for the coordinate ψ, the Euler equations for $\dot{\omega}_1$ and $\dot{\omega}_2$, which can be obtained from Eq. (12.77), are *not* the Lagrange equations for θ and φ.

In order to obtain Euler's equations for the case of motion in a force field, we may start with the fundamental relation for the torque $\mathbf{N}$ [cf. Eq. (2.11)]:

$$\left(\frac{d\mathbf{L}}{dt}\right)_{\text{fixed}} = \mathbf{N} \tag{12.78}$$

where the designation "fixed" has been explicitly appended to $\dot{\mathbf{L}}$ since this relation is derived from Newton's equation and is therefore valid only in an inertial frame of reference. From Eq. (11.7) we have

$$\left(\frac{d\mathbf{L}}{dt}\right)_{\text{fixed}} = \left(\frac{d\mathbf{L}}{dt}\right)_{\text{body}} + \boldsymbol{\omega} \times \mathbf{L} \tag{12.79}$$

or,

$$\left(\frac{d\mathbf{L}}{dt}\right)_{\text{body}} + \boldsymbol{\omega} \times \mathbf{L} = \mathbf{N} \tag{12.80}$$

The component of this equation along the x_3-axis (note that this is a *body* axis) is

$$\dot{L}_3 + \omega_1 L_2 - \omega_2 L_1 = N_3 \tag{12.81}$$

* Leonard Euler, 1758.

But since we have chosen the x_i-axes to coincide with the principal axes of the body, we have from Eq. (12.24a),

$$L_i = I_i \omega_i$$

so that

$$I_3 \dot{\omega}_3 - (I_1 - I_2)\omega_1 \omega_2 = N_3 \tag{12.82}$$

or, in general,

$$\boxed{(I_i - I_j)\omega_i \omega_j - \sum_k (I_k \dot{\omega}_k - N_k)\varepsilon_{ijk} = 0} \tag{12.83}$$

which are the desired Euler equations for the motion of a rigid body in a force field.

It will be noted that the motion of a rigid body depends on the structure of the body only through the three numbers, I_1, I_2, and I_3, i.e., the principal moments of inertia. Thus, any two bodies which have the same principal moments will move in exactly the same manner, regardless of the fact that they may have quite different shapes. (Of course, effects such as frictional retardation may depend upon the shape of a body.) The simplest geometrical shape that a body having three given principal moments may possess is a homogeneous ellipsoid. Therefore, the motion of any rigid body can be represented by the motion of the *equivalent ellipsoid*.* The treatment of rigid-body dynamics from this point of view was originated by Poinsot in 1834. The *Poinsot construction* is sometimes useful for depicting the motion of a rigid body in a geometrical manner.†

12.9 Force-Free Motion of a Symmetrical Top

If we consider a symmetrical top, i.e., a rigid body with $I_1 = I_2 \neq I_3$, then the force-free Euler equations [Eqs. (12.77)] become

$$\left.\begin{aligned} (I_{12} - I_3)\omega_2 \omega_3 - I_{12}\dot{\omega}_1 &= 0 \\ (I_3 - I_{12})\omega_3 \omega_1 - I_{12}\dot{\omega}_2 &= 0 \\ I_3 \dot{\omega}_3 &= 0 \end{aligned}\right\} \tag{12.84}$$

where I_{12} has been substituted for both I_1 and I_2. Since for force-free motion, the center of mass of the body is either at rest or in uniform motion with

* The momental ellipsoid was introduced by the French mathematician Baron Augustin Louis Cauchy in 1827.

† See, for example, Goldstein (Go50, p. 159).

respect to the fixed or inertial frame of reference, we can, without loss of generality, specify that the center of mass of the body is at rest and located at the origin of the fixed coordinate system. We consider, of course, the case in which the angular velocity vector $\boldsymbol{\omega}$ does not lie along a principal axis of the body, since, otherwise, the motion is trivial.

The first result for the motion follows from the third of Eqs. (12.84), viz., $\dot{\omega}_3 = 0$, or

$$\omega_3(t) = \text{const.} \tag{12.85}$$

The first two of Eqs. (12.84) can be written as

$$\left.\begin{aligned}\dot{\omega}_1 &= -\left[\frac{I_3 - I_{12}}{I_{12}}\,\omega_3\right]\omega_2\\ \dot{\omega}_2 &= \left[\frac{I_3 - I_{12}}{I_{12}}\,\omega_3\right]\omega_1\end{aligned}\right\} \tag{12.86}$$

Since the terms in the brackets are identical and composed of constants, we may define

$$\Omega \equiv \frac{I_3 - I_{12}}{I_{12}}\,\omega_3 \tag{12.87}$$

so that

$$\left.\begin{aligned}\dot{\omega}_1 + \Omega\omega_2 &= 0\\ \dot{\omega}_3 - \Omega\omega_1 &= 0\end{aligned}\right\} \tag{12.88}$$

These are coupled equations of familiar form, and we can effect a solution by multiplying the second equation by i and adding to the first:

$$(\dot{\omega}_1 + i\dot{\omega}_2) - i\Omega(\omega_1 + i\omega_2) = 0 \tag{12.89}$$

If we define

$$\eta \equiv \omega_1 + i\omega_2 \tag{12.90}$$

then

$$\dot{\eta} - i\Omega\eta = 0 \tag{12.91}$$

with solution*

$$\eta(t) = Ae^{i\Omega t} \tag{12.92}$$

Thus,

$$\omega_1 + i\omega_2 = A\cos\Omega t + iA\sin\Omega t \tag{12.93}$$

* In general, the constant coefficient is complex, so that we should properly write $A\exp(i\delta)$. We will, however, set the phase δ equal to zero for simplicity; this can, of course, always be done by choosing an appropriate instant that we call $t = 0$.

and therefore,

$$\left.\begin{aligned}\omega_1(t) &= A \cos \Omega t\\ \omega_2(t) &= A \sin \Omega t\end{aligned}\right\} \tag{12.94}$$

Since $\omega_3 = \text{const.}$, we note that the magnitude of $\boldsymbol{\omega}$ is also constant:

$$|\boldsymbol{\omega}| = \omega = \sqrt{\omega_1^2 + \omega_2^2 + \omega_3^2} = \sqrt{A^2 + \omega_3^2} = \text{const.} \tag{12.95}$$

Equations (12.94) are the parametric equations of a circle, so that the projection of the vector $\boldsymbol{\omega}$ (which is of constant magnitude) onto the x_1-x_2 plane describes a circle with time, as shown in Fig. 12-7.

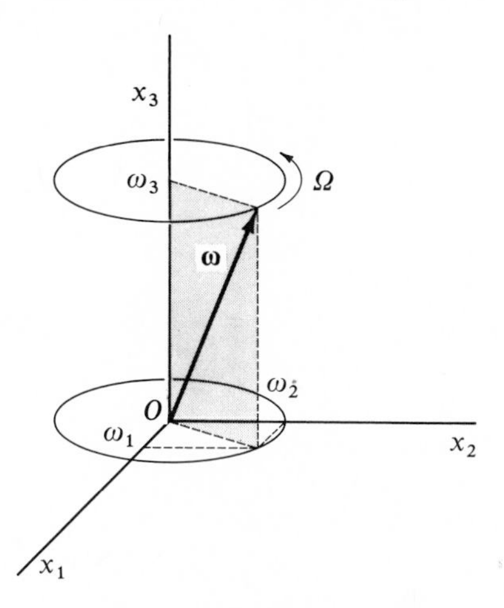

FIG. 12-7

Now, the x_3-axis is the symmetry axis of the body, so we find that the angular velocity vector $\boldsymbol{\omega}$ revolves or *precesses* about the body x_3-axis with a constant angular frequency Ω. Thus, to an observer in the body coordinate system, $\boldsymbol{\omega}$ traces out a cone around the body symmetry axis.

Since we are considering force-free motion, the angular momentum vector $\mathbf{L}$ is stationary in the fixed coordinate system and is constant in time. An additional constant of the motion for the force-free case is the kinetic energy, or, in particular, since the center of mass of the body is fixed, the *rotational* kinetic energy is constant:

$$T_{\text{rot}} = \tfrac{1}{2}\boldsymbol{\omega} \cdot \mathbf{L} = \text{const.} \tag{12.96}$$

But we have $\mathbf{L} = \text{const.}$, so that $\boldsymbol{\omega}$ must move in such a manner that its projection on the stationary angular momentum vector is constant. Thus, $\boldsymbol{\omega}$ precesses around and makes a constant angle with the vector $\mathbf{L}$. Now, $\mathbf{L}$, $\boldsymbol{\omega}$, and the x_3 (body) axis (i.e., the unit vector $\mathbf{e}_3$) all lie in a *plane* (see Problem 12-25). Therefore, if we designate the x_3'-axis in the fixed coordinate system to coincide with $\mathbf{L}$, then to an observer in the fixed system, $\boldsymbol{\omega}$ traces out a cone around the fixed x_3'-axis. The situation is then described, as in Fig. 12-8, by one cone rolling on another, such that $\boldsymbol{\omega}$ precesses around the x_3-axis in the body system and around the x_3'-axis (or $\mathbf{L}$) in the fixed system.

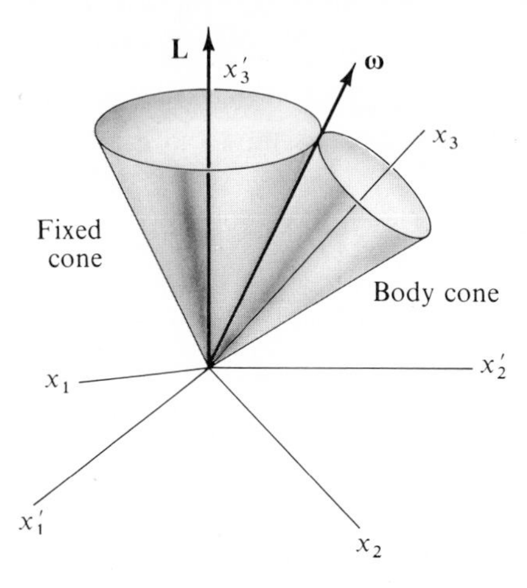

FIG. **12-8**

The rate at which $\boldsymbol{\omega}$ precesses around the body symmetry axis is given by Eq. (12.87):

$$\boldsymbol{\Omega} = \frac{I_3 - I_{12}}{I_{12}}\,\omega_3$$

If $I_{12} \cong I_3$, then Ω becomes very small compared with ω_3. The Earth is known to be slightly flattened near the poles* so that its shape can be approximated by an oblate spheriod with $I_{12} \cong I_3$, but with $I_3 > I_{12}$. If the Earth is considered to be a rigid body, then the moments I_{12} and I_3 are such that $\Omega \cong \omega_3/300$. Since the period of the Earth's rotation is $1/\omega = 1$ day, and since $\omega_3 \cong \omega$, the period predicted for the precession of the axis of rotation is $1/\Omega \cong 300$ days. The observed precession has an irregular period about 50 per cent greater than that predicted on the basis of this simple theory, the deviation being ascribed to the fact that the Earth is not a *rigid* body and to the fact that the shape is not exactly that of an oblate spheroid, but has a higher-order deformation and actually resembles a flattened pear.

* The flattening at the poles was shown by Newton to be due to the rotation of the Earth; the resulting precessional motion was first calculated by Euler.

The equatorial "bulge" of the Earth together with the fact that the Earth's rotational axis is inclined at an angle of approximately 23.5° to the plane of the Earth's orbit around the Sun (the *plane of the ecliptic*) gives rise to a gravitational torque (due to both the Sun and the Moon) which produces a slow precession of the Earth's axis. The period of this precessional motion is approximately 26,000 years. Thus, in different epochs, different stars become the "Pole Star."*

12.10 The Motion of a Symmetrical Top with One Point Fixed†

Consider a rotating symmetrical top whose tip is held fixed and which moves in a gravitational field. In our previous development we have been able to separate the kinetic energy into translational and rotational parts by taking the center of mass of the body to be the origin of the rotating or body coordinate system. Alternatively, if it is possible to choose the origins of the fixed and the body coordinate systems to coincide, then the translational kinetic energy will vanish, since $\mathbf{V} = \dot{\mathbf{R}} = 0$. Such a choice is quite convenient for the discussion of the top since the stationary tip of the top may then be taken as the origin for both coordinate systems. The Euler angles for this situation are shown in Fig. 12-9. The x_3'- (fixed) axis corresponds to the

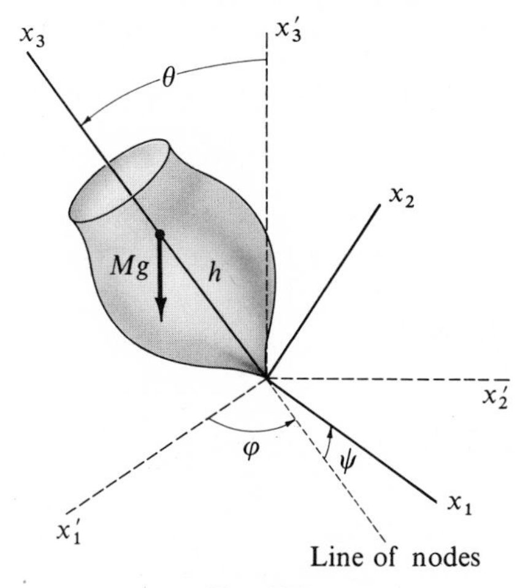

FIG. 12-9

* This *precession of the Equinoxes* was apparently discovered by the Babylonian astronomer, Cidenas, in about 343 B.C.

† This problem was first solved in detail by Lagrange in *Mechanique Analytique*.

vertical and the x_3- (body) axis is chosen to be the symmetry axis of the top. The distance from the fixed tip to the center of mass is h, and the mass of the top is M.

Since we have a symmetrical top, the principal moments of inertia about the x_1- and x_2-axes are equal: $I_1 = I_2 \equiv I_{12}$. We assume $I_3 \neq I_{12}$. The kinetic energy is then given by

$$T = \tfrac{1}{2}\sum_i I_i \omega_i^2 = \tfrac{1}{2}I_{12}(\omega_1^2 + \omega_2^2) + \tfrac{1}{2}I_3\omega_3^2 \tag{12.97}$$

According to Eqs. (12.69), we have

$$\begin{aligned}
\omega_1^2 &= (\dot{\varphi}\sin\theta\sin\psi + \dot{\theta}\cos\psi)^2 \\
&= \dot{\varphi}^2\sin^2\theta\sin^2\psi + 2\dot{\varphi}\dot{\theta}\sin\theta\sin\psi\cos\psi + \dot{\theta}^2\cos^2\psi \\
\omega_2^2 &= (\dot{\varphi}\sin\theta\cos\psi - \dot{\theta}\sin\psi)^2 \\
&= \dot{\varphi}^2\sin^2\theta\cos^2\psi - 2\dot{\varphi}\dot{\theta}\sin\theta\sin\psi\cos\psi + \dot{\theta}^2\sin^2\psi
\end{aligned}$$

so that

$$\omega_1^2 + \omega_2^2 = \dot{\varphi}^2\sin^2\theta + \dot{\theta}^2 \tag{12.98a}$$

Also,

$$\omega_3^2 = (\dot{\varphi}\cos\theta + \dot{\psi})^2 \tag{12.98b}$$

Therefore,

$$T = \tfrac{1}{2}I_{12}(\dot{\varphi}^2\sin^2\theta + \dot{\theta}^2) + \tfrac{1}{2}I_3(\dot{\varphi}\cos\theta + \dot{\psi})^2 \tag{12.99}$$

Since the potential energy is $Mgh\cos\theta$, the Lagrangian becomes

$$L = \tfrac{1}{2}I_{12}(\dot{\varphi}^2\sin^2\theta + \dot{\theta}^2) + \tfrac{1}{2}I_3(\dot{\varphi}\cos\theta + \dot{\psi})^2 - Mgh\cos\theta \tag{12.100}$$

The Lagrangian is cyclic in both the φ- and ψ-coordinates. The momenta conjugate to these coordinates are therefore constants of the motion:

$$p_\varphi = \frac{\partial L}{\partial\dot{\varphi}} = (I_{12}\sin^2\theta + I_3\cos^2\theta)\dot{\varphi} + I_3\dot{\psi}\cos\theta = \text{const.} \tag{12.101}$$

$$p_\psi = \frac{\partial L}{\partial\dot{\psi}} = I_3(\dot{\psi} + \dot{\varphi}\cos\theta) = \text{const.} \tag{12.102}$$

Since the cyclic coordinates are *angles*, the conjugate momenta are *angular momenta*, and are, of course, the angular momenta along the axes for which φ and ψ are the rotation angles, viz., the x_3'- (cr vertical) axis and the x_3- (or body symmetry) axis, respectively. We note that this result is insured by the construction shown in Fig. 12-9 since the gravitational torque is directed along the line of nodes and, hence, the torque can have no component along either the x_3'- or the x_3-axis, both of which are perpendicular to the line of nodes; thus, the angular momenta along these axes are constants of the motion.

Equations (12.101) and (12.102) can be solved for $\dot{\varphi}$ and $\dot{\psi}$ in terms of θ. From Eq. (12.102), we can write

$$\dot{\psi} = \frac{p_\psi - I_3 \dot{\varphi} \cos\theta}{I_3} \tag{12.103}$$

and substituting this result into Eq. (12.101), we find

$$(I_{12} \sin^2\theta + I_3 \cos^2\theta)\dot{\varphi} + (p_\psi - I_3 \dot{\varphi} \cos\theta) \cos\theta = p_\varphi$$

or,

$$(I_{12} \sin^2\theta)\dot{\varphi} + p_\psi \cos\theta = p_\varphi$$

so that

$$\dot{\varphi} = \frac{p_\varphi - p_\psi \cos\theta}{I_{12} \sin^2\theta} \tag{12.104}$$

Using this expression for $\dot{\varphi}$ in Eq. (12.103), we have

$$\dot{\psi} = \frac{p_\psi}{I_3} - \frac{(p_\varphi - p_\psi \cos\theta)\cos\theta}{I_{12} \sin^2\theta} \tag{12.105}$$

Now, by hypothesis, the system we are considering is conservative; therefore, we have the further property that the total energy is a constant of the motion:

$$E = \tfrac{1}{2}I_{12}(\dot{\varphi}^2 \sin^2\theta + \dot{\theta}^2) + \tfrac{1}{2}I_3 \omega_3^2 + Mgh \cos\theta = \text{const.} \tag{12.106}$$

Using the expression for ω_3 [see, e.g., Eq. (12.69)], we note that Eq. (12.102) can be written as

$$p_\psi = I_3 \omega_3 = \text{const.} \tag{12.107}$$

or,

$$I_3 \omega_3^2 = \frac{p_\psi^2}{I_3} = \text{const.} \tag{12.107a}$$

Therefore, not only is E a constant of the motion, but so is $E - \frac{1}{2}I_3 \omega_3^2$; we let this quantity be E':

$$E' \equiv E - \tfrac{1}{2}I_3 \omega_3^2 = \tfrac{1}{2}I_{12}(\dot{\varphi}^2 \sin^2\theta + \dot{\theta}^2) + Mgh \cos\theta = \text{const.} \tag{12.108}$$

Substituting into this equation the expression for $\dot{\varphi}$ [Eq. (12.104)], we have

$$E' = \tfrac{1}{2}I_{12}\dot{\theta}^2 + \frac{(p_\varphi - p_\psi \cos\theta)^2}{2I_{12} \sin^2\theta} + Mgh \cos\theta \tag{12.109}$$

which we can write as

$$E' = \tfrac{1}{2} I_{12} \dot{\theta}^2 + V(\theta) \tag{12.110}$$

where $V(\theta)$ is an "effective potential" given by

$$V(\theta) \equiv \frac{(p_\varphi - p_\psi \cos\theta)^2}{2I_{12}\sin^2\theta} + Mgh\cos\theta \tag{12.111}$$

Equation (12.110) can be solved to yield $t(\theta)$:

$$t(\theta) = \int \frac{d\theta}{\sqrt{(2/I_{12})[E' - V(\theta)]}} \tag{12.112}$$

This integral can (formally, at least) be inverted to obtain $\theta(t)$, which, in turn can be substituted into Eqs. (12.104) and (12.105) to yield $\varphi(t)$ and $\psi(t)$. Since the Euler angles θ, φ, ψ completely specify the orientation of the top, the results for $\theta(t)$, $\varphi(t)$, and $\psi(t)$ constitute a complete solution for the problem. Clearly, such a procedure is complicated and not very illuminating. We can, however, obtain some qualitative features of the motion by examining the above equations in a manner analogous to that used for the treatment of the motion of a particle in a central-force field (see Section 8-6).

Figure 12-10 shows the form of the effective potential $V(\theta)$ in the range

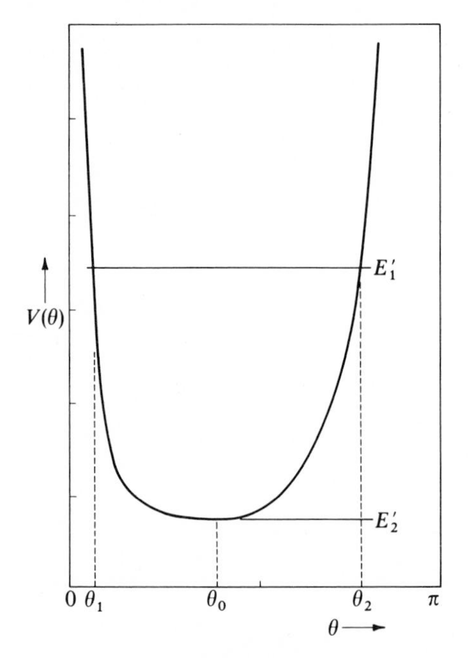

FIG. 12-10

$0 \le \theta \le \pi$, which clearly is the physically limited region for θ. This energy diagram indicates that for any general values of E' (e.g., the value represented by E'_1) the motion is limited by two extreme values of θ, viz., θ_1 and θ_2, which correspond to the turning points of the central-force problem and which are roots of the denominator in Eq. (12.112). Thus, we find that the inclination of the rotating top is, in general, confined to the region $\theta_1 \le \theta \le \theta_2$. For the case that $E' = E'_2 = V_{\min}$, θ is limited to the single value θ_0, and the motion is a steady precession at a fixed angle of inclination. Such motion is similar to the occurrence of circular orbits in the central-force problem.

The value of θ_0 can be obtained by setting the derivative of $V(\theta)$ equal to zero. Thus,

$$\left.\frac{\partial V}{\partial \theta}\right|_{\theta=\theta_0} = \frac{-\cos\theta_0(p_\varphi - p_\psi \cos\theta_0)^2 + p_\psi \sin^2\theta_0(p_\varphi - p_\psi\cos\theta_0)}{I_{12}\sin^3\theta_0} - Mgh\sin\theta_0 = 0 \tag{12.113}$$

If we define

$$\beta \equiv p_\varphi - p_\psi \cos\theta_0 \tag{12.114}$$

then Eq. (12.113) becomes

$$(\cos\theta_0)\beta^2 - (p_\psi \sin^2\theta_0)\beta + (MghI_{12}\sin^4\theta_0) = 0 \tag{12.115}$$

This is a quadratic in β and can be solved with the result

$$\beta = \frac{p_\psi \sin^2\theta_0}{2\cos\theta_0}\left[1 \pm \sqrt{1 - \frac{4MghI_{12}\cos\theta_0}{p_\psi^2}}\right] \tag{12.116}$$

Now, β must be a real quantity, so the radicand in Eq. (12.116) must be positive. If $\theta_0 < \pi/2$, we have

$$p_\psi^2 \ge 4MghI_{12}\cos\theta_0 \tag{12.117}$$

But from Eq. (12.107), $p_\psi = I_3\omega_3$; thus,

$$\omega_3 \ge \frac{2}{I_3}\sqrt{MghI_{12}\cos\theta_0} \tag{12.118}$$

We conclude therefore that there can be a steady precession at the fixed angle of inclination θ_0 only if the angular velocity of spin is larger than the limiting value given by Eq. (12.118).

From Eq. (12.104) we note that we can write (for $\theta = \theta_0$)

$$\dot{\varphi}_0 = \frac{\beta}{I_{12}\sin^2\theta_0} \tag{12.119}$$

Therefore, we have two possible values of the precessional angular velocity

$\dot{\varphi}_0$, one for each of the values of β given by Eq. (12.116):

$$\dot{\varphi}_{0(+)} \rightarrow \text{Fast precession}$$

and

$$\dot{\varphi}_{0(-)} \rightarrow \text{Slow precession}$$

If ω_3 (or p_ψ) is large (i.e., we have a "fast" top), then the second term in the radicand of Eq. (12.116) is small and we may expand the radical. Retaining only the first nonvanishing term in each case, we find

$$\left.\begin{aligned} \dot{\varphi}_{0(+)} &\cong \frac{I_3 \omega_3}{I_{12} \cos \theta_0} \\ \dot{\varphi}_{0(-)} &\cong \frac{Mgh}{I_3 \omega_3} \end{aligned}\right\} \tag{12.120}$$

It is the slower of the two possible precessional angular velocities, $\dot{\varphi}_{0(-)}$, that is usually observed.

The above results apply if $\theta_0 < \pi/2$; but in the event that* $\theta_0 > \pi/2$, the radicand in Eq. (12.116) is always positive so that there is no limiting condition on ω_3. Since the radical is greater than unity in such a case, the values of $\dot{\varphi}_0$ for fast and slow precession have opposite signs. That is, for $\theta_0 > \pi/2$, the fast precession is in the same direction as that for $\theta_0 < \pi/2$, but the slow precession takes place in the opposite sense.

For the general case, in which $\theta_1 < \theta < \theta_2$, Eq. (12.104) indicates that $\dot{\varphi}$ may or may not change sign as θ varies between its limits, depending on the values of p_φ and p_ψ. If $\dot{\varphi}$ does not change sign, the top precesses monotonically around the x_3'-axis (see Fig. 12-9) while the x_3- (or symmetry) axis oscillates between $\theta = \theta_1$ and $\theta = \theta_2$. This phenomenon is called *nutation*, and the path described by the projection of the body symmetry axis on a unit sphere in the fixed system is shown in Fig. 12-11a.

If $\dot{\varphi}$ does change sign between the limiting values of θ, the precessional angular velocity must have opposite signs at $\theta = \theta_1$ and $\theta = \theta_2$. Thus, the nutational-precessional motion gives rise to the looping motion of the symmetry axis depicted in Fig. 12-11b.

Finally, if the values of p_φ and p_ψ are such that

$$(p_\varphi - p_\psi \cos \theta)|_{\theta=\theta_1} = 0 \tag{12.121}$$

then,

$$\dot{\varphi}|_{\theta=\theta_1} = 0; \qquad \dot{\theta}|_{\theta=\theta_1} = 0 \tag{12.122}$$

* If $\theta_0 > \pi/2$, the fixed tip of the top is at a position *above* the center of mass. Such motion is possible, for example, with a gyroscopic top whose tip is actually a ball and rests in a cup that is fixed atop a pedestal.

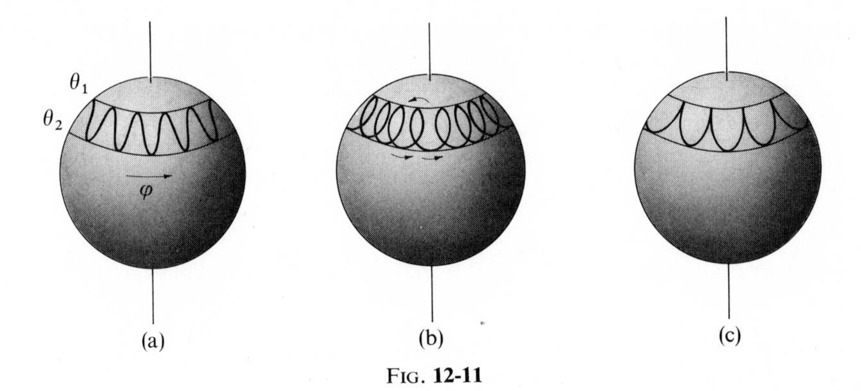

FIG. **12-11**

and Fig. 12-11c shows the resulting cusplike motion. It is just this case that corresponds to the usual method of starting a top. First, the top is set to spinning around its axis, then it is given a certain initial tilt and released. Thus, the initial conditions are $\theta = \theta_1$, and $\dot{\theta} = 0 = \dot{\varphi}$. Since the first motion of the top is to begin to fall in the gravitational field, the conditions are exactly those of Fig. 12-11c, and the cusplike motion ensues. Figures 12-11a and b correspond to the motion in the event that there is an initial angular velocity $\dot{\varphi}$ either in the direction of or opposite to the direction of precession.

12.11 The Stability of Rigid-Body Rotations*

We now consider a rigid body which is undergoing force-free rotation around one of its principal axes and inquire whether such motion is stable, "Stability" here means, as before (see Section 8.11), that if a small perturbation is applied to the system, the motion will either return to its former mode or will perform small oscillations about it.

We choose for our discussion a general rigid body for which all of the principal moments of inertia are distinct, and we label them in such a way that $I_3 > I_2 > I_1$. We let the body axes coincide with the principal axes, and we start with the body rotating around the x_1-axis, i.e., around the principal axis associated with the moment of inertia I_1. Then,

$$\boldsymbol{\omega} = \omega_1 \mathbf{e}_1 \tag{12.123}$$

If we apply a small perturbation, the angular velocity vector will assume the form

$$\boldsymbol{\omega} = \omega_1 \mathbf{e}_1 + \lambda \mathbf{e}_2 + \mu \mathbf{e}_3 \tag{12.124}$$

* This problem was first treated by Euler in 1749.

where λ and μ are small quantities and correspond to the parameters which have been used previously in other perturbation expansions. (λ and μ are sufficiently small so that their product can be neglected compared to all other quantities of interest to the discussion.)

The Euler equations become [see Eq. (12.77)]:

$$\left.\begin{aligned}(I_2 - I_3)\lambda\mu - I_1\dot{\omega}_1 &= 0\\(I_3 - I_1)\mu\omega_1 - I_2\dot{\lambda} &= 0\\(I_1 - I_2)\lambda\omega_1 - I_3\dot{\mu} &= 0\end{aligned}\right\} \tag{12.125}$$

Since $\lambda\mu \approx 0$, the first of these equations requires $\dot{\omega}_1 = 0$, or $\omega_1 = \text{const.}$ Solving the other two equations for $\dot{\lambda}$ and $\dot{\mu}$, we find

$$\dot{\lambda} = \left(\frac{I_3 - I_1}{I_2}\omega_1\right)\mu \tag{12.126}$$

$$\dot{\mu} = \left(\frac{I_1 - I_2}{I_3}\omega_1\right)\lambda \tag{12.127}$$

where the terms in parentheses are both constants. These are coupled equations, but they cannot be solved by the method employed in Section 12.9 since the constants in the two equations are different. The solution can be obtained by first differentiating the equation for $\dot{\lambda}$:

$$\ddot{\lambda} = \left(\frac{I_3 - I_1}{I_2}\omega_1\right)\dot{\mu} \tag{12.128}$$

The expression for $\dot{\mu}$ can now be substituted in this equation:

$$\ddot{\lambda} + \left(\frac{(I_1 - I_3)(I_1 - I_2)}{I_2 I_3}\omega_1^2\right)\lambda = 0 \tag{12.129}$$

The solution to this equation is

$$\lambda(t) = Ae^{i\Omega_{1\lambda}t} + Be^{-i\Omega_{1\lambda}t} \tag{12.130}$$

where

$$\Omega_{1\lambda} \equiv \omega_1\sqrt{\frac{(I_1 - I_3)(I_1 - I_2)}{I_2 I_3}} \tag{12.131}$$

and where the subscripts 1 and λ designate that we are considering the solution for λ when the rotation is around the x_1-axis.

By hypothesis, $I_1 < I_3$ and $I_1 < I_2$, so that $\Omega_{1\lambda}$ is real. Therefore, the solution for $\lambda(t)$ represents oscillatory motion with a frequency $\Omega_{1\lambda}$. We can similarly investigate $\mu(t)$ with the result that $\Omega_{1\mu} = \Omega_{1\lambda} \equiv \Omega_1$. Thus, the small perturbations introduced by forcing small x_2- and x_3-components on $\boldsymbol{\omega}$ do not increase with time, but oscillate around the equilibrium values $\lambda = 0$ and $\mu = 0$. Consequently, the rotation around the x_1-axis is stable.

If we consider rotations around the x_2- and x_3-axes, expressions for Ω_2 and Ω_3 can be obtained from Eq. (12.131) by permutation:

$$\Omega_1 = \omega_1 \sqrt{\frac{(I_1 - I_3)(I_1 - I_2)}{I_2 I_3}} \tag{12.132a}$$

$$\Omega_2 = \omega_2 \sqrt{\frac{(I_2 - I_1)(I_2 - I_3)}{I_1 I_3}} \tag{12.132b}$$

$$\Omega_3 = \omega_3 \sqrt{\frac{(I_3 - I_2)(I_3 - I_1)}{I_1 I_2}} \tag{12.132c}$$

But since $I_1 < I_2 < I_3$, we have

$$\Omega_1, \quad \Omega_3 \quad \text{real}; \qquad \Omega_2 \quad \text{imaginary}$$

Thus, when the rotation takes place around either the x_1- or x_3-axes, the perturbation produces oscillatory motion and the rotation is stable. When the rotation takes place around x_2, however, the fact that Ω_2 is imaginary results in the perturbation increasing with time without limit; such motion is unstable.

Since we have assumed a completely arbitrary rigid body for this discussion, we conclude that rotation around the principal axis corresponding to either the greatest or smallest moment of inertia is stable, while rotation around the principal axis corresponding to the intermediate moment is unstable. This effect can be easily demonstrated with, say, a book (which is kept closed by tape or a rubber band). If the book is tossed into the air with an angular velocity around one of the principal axes, the motion will be unstable for rotation around the intermediate axis and stable for the other two axes.

In the event that two of the moments of inertia are equal ($I_1 = I_2$, say), then the coefficient of λ in Eq. (12.127) vanishes, and we have $\dot{\mu} = 0$ or $\mu(t) = \text{const}$. Therefore, Eq. (12.126) for λ can be integrated to yield

$$\lambda(t) = C + Dt \tag{12.133}$$

and the perturbation increases linearly with the time; the motion around the x_1-axis is therefore unstable. We find a similar result for motion around the x_2-axis. There is stability only for the x_3-axis, independent of whether I_3 is greater or less than $I_1 = I_2$.

Suggested References

Rigid-body dynamics is a topic discussed in almost every mechanics text. Introductory accounts are given, for example, by Fowles (Fo62, Chapter 9) and by Lindsay (Li61, Chapter 8). At a slightly more advanced level are the treatments of Becker (Be54, Chapter 12), Constant (Co54, Chapter 9), Slater and Frank (Sl47, Chapter 6), and Sommerfeld (So50, Chapter 4).

For discussions at the intermediate-to-advanced level, see Bradbury (Br68, Chapter 8), Goldstein (Go50, Chapter 5), Hauser (Ha65, Chapter 8), Konopinski (Ko69, Chapters 9 and 10), Landau and Lifshitz (La60, Chapter 6), McCuskey (Mc59, Chapter 4), Symon (Sy60, Chapter 11), and Wangsness (Wa63, Chapter 9). An extensive set of problems is worked out by Whittaker (Wh37, Chapter 6).

Halfman (Ha62a, Chapters 5 and 6) gives a comprehensive discussion with a slant toward engineering physics.

The topic to which rigid-body dynamics has been applied in the greatest detail is the theory of tops and gyroscopic motion. A short bibliography (with notes) of some of the more important works is given by Goldstein (Go50, pp. 178–180).

Problems

12-1. Calculate the moments of inertia I_1, I_2, and I_3 for a homogeneous sphere of radius R and mass M. (Choose the origin at the center of the sphere.)

12-2. Calculate the moments of inertia I_1, I_2, and I_3 for a homogeneous cone of mass M whose height is h and whose base has a radius R. Choose the x_3-axis along the axis of symmetry of the cone. Choose the origin at the apex of the cone and calculate the elements of the inertia tensor. Then make a transformation such that the center of mass of the cone becomes the origin and find the principal moments of inertia.

12-3. Calculate the moments of inertia I_1, I_2, and I_3, for a homogeneous ellipsoid of mass M, the lengths of the axes being $2a > 2b > 2c$.

12-4. Consider a thin rod of length l and mass m which is pivoted about one end. Calculate the moment of inertia. Find the point at which, if all of the mass were concentrated, the moment of inertia about the pivot axis would be the same as the real moment of inertia. The distance of this point from the pivot is called the *radius of gyration.*

12-5. (a) Find the height at which a billiard ball should be struck so that it will roll with no initial slipping. (b) Calculate the optimum height of the rail of a billiard table. On what basis is the calculation predicated?

12-6. Given two spheres of the same diameter and same mass, but one of which is solid and the other is a hollow shell, describe in detail a non-destructive experiment to determine which is solid and which is hollow.

12-7. A homogeneous disk of radius R and mass M rolls without slipping on a horizontal surface and is attracted to a point which lies at a distance d below the plane. If the force of attraction is proportional to the distance from the center of mass of the disk to the force center, find the frequency of small oscillations around the position of equilibrium.

12-8. A door is constructed of a thin homogeneous slab of material; it has a width of 1 meter. If the door is opened through 90° it is found that upon release it closes itself in 2 sec. Assume that the hinges are frictionless and show that the line of hinges must make an angle of approximately 3° with the vertical.

12-9. A homogeneous slab of thickness a is placed atop a fixed cylinder of radius R whose axis is horizontal. Show that the condition for stable equilibrium of the slab under the assumption that there is no slipping is $R > a/2$. What is the frequency of small oscillations? Sketch the potential energy U as a function of the angular displacement θ. Show that there is a minimum at $\theta = 0$ for $R > a/2$ but not for $R < a/2$.

12-10. A solid sphere of mass M and radius R rotates freely in space with an angular velocity ω about a fixed diameter. A particle of mass m, initially at one pole, moves with a constant velocity v along a great circle of the sphere. Show that, when the particle has reached the other pole, the rotation of the sphere will have been retarded by an angle

$$\alpha = \omega T \left(1 - \sqrt{\frac{2M}{2M + 5m}}\right)$$

where T is the total time required for the particle to move from one pole to the other.

12-11. A homogeneous cube, each edge of which has a length l, is initially in a position of unstable equilibrium with one edge in contact with a horizontal plane. The cube is then given a small displacement and allowed to fall. Show that the angular velocity of the cube when one face strikes the plane is given by

$$\omega^2 = A \frac{g}{l} (\sqrt{2} - 1)$$

where $A = 3/2$ if the edge cannot slide on the plane and where $A = 12/5$ if sliding can occur without friction.

12-12. Show that none of the principal moments of inertia can exceed the sum of the other two.

12-13. If a physical pendulum has the same period of oscillation when pivoted about either of two points of unequal distances from the center of mass, show that the length of the simple pendulum which has the same period is equal to the separation of the pivot points. Such a physical pendulum is called *Kater's reversible pendulum* and at one time provided the most accurate way (to about 1 part in 10^5) by which measurements of the acceleration of gravity could be made.* Discuss the advantages of Kater's pendulum over a simple pendulum for such a purpose.

12-14. Consider the following inertia tensor:

$$\{\mathsf{I}\} = \begin{Bmatrix} \frac{1}{2}(A+B) & \frac{1}{2}(A-B) & 0 \\ \frac{1}{2}(A-B) & \frac{1}{2}(A+B) & 0 \\ 0 & 0 & C \end{Bmatrix}$$

Perform a rotation of the coordinate system by an angle θ about the x_3-axis. Evaluate the transformed tensor elements and show that the choice $\theta = \pi/4$ renders the inertia tensor diagonal with elements A, B, and C.

12-15. Consider a thin homogeneous plate which lies in the x_1-x_2 plane. Show that the inertia tensor takes the form

$$\{\mathsf{I}\} = \begin{Bmatrix} A & -C & 0 \\ -C & B & 0 \\ 0 & 0 & A+B \end{Bmatrix}$$

12-16. If, in the previous problem, the coordinate axes are rotated through an angle θ about the x_3-axis, show that the new inertia tensor is

$$\{\mathsf{I}\} = \begin{Bmatrix} A' & -C' & 0 \\ -C' & B' & 0 \\ 0 & 0 & A'+B' \end{Bmatrix}$$

where

$$A' = A\cos^2\theta - C\sin 2\theta + B\sin^2\theta$$

$$B' = A\sin^2\theta + C\sin 2\theta + B\cos^2\theta$$

$$C' = C\cos 2\theta - \tfrac{1}{2}(B-A)\sin 2\theta$$

* First used in 1818 by Captain Henry Kater (1777–1835), but the method was apparently suggested somewhat earlier by Bohnenberger. The theory of Kater's pendulum was treated in detail by Friedrich Wilhelm Bessel (1784–1846) in 1826.

and hence, show that the x_1- and x_2- axes become principal axes if the angle of rotation is

$$\theta = \frac{1}{2} \tan^{-1}\left(\frac{2C}{B - A}\right)$$

12-17. Consider a plane homogeneous plate of density ρ which is bounded by the logarithmic spiral $r = ke^{\alpha\theta}$ and the radii $\theta = 0$ and $\theta = \pi$. Obtain the inertia tensor for the origin at $r = 0$ if the plate lies in the x_1-x_2 plane. Perform a rotation of the coordinate axes to obtain the principal moments of inertia and use the results of the previous problem to show that they are

$$I_1' = \rho k^4 P(Q + R); \qquad I_2' = \rho k^4 P(Q - R); \qquad I_3' = I_1' + I_2'$$

where

$$P = \frac{e^{4\pi\alpha} - 1}{16(1 + 4\alpha^2)}; \qquad Q = \frac{1 + 4\alpha^2}{2\alpha}; \qquad R = \sqrt{1 + 4\alpha^2}$$

12-18. The proof represented by Eqs. (12.35)–(12.42) was expressed entirely in the summation convention. Rewrite this proof in matrix notation.

12-19. The *trace* of a tensor is defined as the sum of the diagonal elements:

$$\text{tr}\{\mathsf{I}\} \equiv \sum_k I_{kk}$$

Show, by performing a similarity transformation, that the trace is an invariant quantity; i.e., that

$$\text{tr}\{\mathsf{I}\} = \text{tr}\{\mathsf{I}'\}$$

where $\{\mathsf{I}\}$ is the tensor in one coordinate system and $\{\mathsf{I}'\}$ is the tensor in a coordinate system rotated with respect to the first system. Verify this result for the different forms of the inertia tensor for a cube that are given in several examples in the text.

12-20. Show by the same method used for the previous problem, that the *determinant* of the elements of a tensor is an invariant quantity. Verify this result also for the case of the cube.

12-21. Find the frequency of small oscillations for a thin homogeneous plate if the motion takes place in the plane of the plate and if the plate has the shape of an equilateral triangle and is suspended (a) from the mid-point of one side, and (b) from one apex.

12-22. Consider a thin disk composed of two homogeneous halves connnected along a diameter of the disk. If one half has density ρ and the other has density 2ρ, find the expression for the Lagrangian when the disk rolls without slipping along a horizontal surface. (The rotation takes place in the plane of the disk.)

12-23. Obtain the components of the angular velocity vector $\boldsymbol{\omega}$ ([see Eq. (12.69)] directly from the transformation matrix $\boldsymbol{\lambda}$ [Eq. (12.66)].

12-24. Show from Fig. 12-6c that the components of $\boldsymbol{\omega}$ along the fixed (x_i') axes are

$$\omega_1' = \dot{\theta} \cos \varphi + \dot{\psi} \sin \theta \sin \varphi$$

$$\omega_2' = \dot{\theta} \sin \varphi - \dot{\psi} \sin \theta \cos \varphi$$

$$\omega_3' = \dot{\psi} \cos \theta + \dot{\varphi}$$

12-25. Show for the case of the force-free motion of a symmetrical top that $\mathbf{L}$, $\boldsymbol{\omega}$, and $\mathbf{e}_3$ are coplanar. (See Fig. 12-8.)

12-26. Show that the situation depicted in Fig. 12-8 actually refers to the force-free motion of a symmetrical top which is a prolate spheroid, i.e., $I_3 < I_{12}$, whereas for the case of an oblate spheroid, i.e., $I_{12} < I_3$, the x_3-axis would lie *between* $\mathbf{L}$ and $\boldsymbol{\omega}$ (i.e., the body cone would revolve *inside* the fixed cone).

12-27. Refer to the discussion of the symmetrical top in Section 12.10. Investigate the equation for the turning points of the nutational motion by setting $\dot{\theta} = 0$ in Eq. (12.110). Show that the resulting equation is a cubic in $\cos \theta$ and has two real roots and one imaginary root for θ.

12-28. Investigate the motion of the symmetrical top discussed in Section 12.10 for the case in which the axis of rotation is vertical (i.e., the x_3'- and x_3-axes coincide). Show that the motion is either stable or unstable depending upon whether the quantity $4I_{12}\,Mhg/I_3^2\,\omega_3^2$ is less than or greater than unity. Sketch the effective potential $V(\theta)$ for the two cases and point out the features of these curves that determine whether or not the motion is stable. If the top is set to spinning in the stable configuration, what will be the effect as friction gradually reduces the value of ω_3? (This is the case of the "sleeping top.")

12-29. Consider a thin homogeneous plate with principal moments of inertia

$$\begin{aligned} I_1 &\quad \text{along the principal axis } x_1 \\ I_2 > I_1 &\quad \text{along the principal axis } x_2 \\ I_3 = I_1 + I_2 &\quad \text{along the principal axis } x_3 \end{aligned}$$

Let the origins of the x_i and x_i' systems coincide and be located at the center of mass O of the plate. At time $t = 0$, the plate is set rotating in a force-free manner with an angular velocity Ω about an axis inclined at an angle α from the plane of the plate and perpendicular to the x_2-axis. If $I_1/I_2 \equiv \cos 2\alpha$, show that at time t the angular velocity about the x_2-axis is

$$\omega_2(t) = \Omega \cos \alpha \tanh(\Omega t \sin \alpha)$$

CHAPTER 13

Coupled Oscillations

13.1 Introduction

In Chapter 4 we examined the motion of an oscillator subjected to an external driving force. The discussion was limited to the case in which the driving force was periodic; that is, the driver was itself a harmonic oscillator. We considered the action of the driver on the oscillator, but we did not include the *feed-back* effect of the oscillator on the driver. In many instances this neglect is unimportant, but if two (or many) oscillators are connected in such a way that energy can be transferred back and forth between (or among) them, the situation is then the more complicated case of *coupled oscillations*.* Motion of this type can be quite complex (the motion may not even be periodic), but, as we will see, it is always possible to describe the motion of any oscillatory system in terms of *normal coordinates* which have the property that each normal coordinate oscillates with a single, well-defined frequency. That is, the normal coordinates are constructed in such a way that there is

* The general theory of the oscillatory motion of a system of particles with a finite number of degrees of freedom was formulated by Lagrange during the period 1762 to 1765, but the pioneering work had been done in 1753 by Daniel Bernoulli (1700–1782).

no coupling among them, even though there is coupling among the ordinary (rectangular) coordinates that describe the positions of the particles. Initial conditions can always be prescribed for the system so that in the subsequent motion only one normal coordinate varies with time. In this circumstance we say that one of the *normal modes* of the system has been excited. If the system has n degrees of freedom (e.g., n coupled one-dimensional oscillators or $n/3$ coupled three-dimensional oscillators), there will in general be n normal modes, some of which may be identical. The general motion of the system will be a complicated superposition of all the normal modes of oscillation, but initial conditions can always be found such that any given one of the normal modes is independently excited. The identification of each of the normal modes of a system allows the construction of a revealing picture of the motion, even though the *general* motion of the system will be a complicated combination of all the normal modes.

In the following chapter we will continue the development begun here and discuss the motion of vibrating strings. This example by no means exhausts the usefulness of the normal mode approach to the description of oscillatory systems; indeed, applications can be found in many areas of mathematical physics, such as the microscopic motion in crystalline solids and the oscillations of the electromagnetic field.

13.2 Two Coupled Harmonic Oscillators

We begin by considering the simplest example of coupled motion—two identical harmonic oscillators connected by a spring. We let each of the oscillator springs have a force constant* κ; the force constant of the coupling spring is κ_{12}. The situation is shown schematically in Fig. 13-1. The motion is restricted to the line connecting the masses so that the system has only two degrees of freedom, represented by the coordinates x_1 and x_2. Each coordinate is measured from the position of equilibrium.

If m_1 and m_2 are displaced from their equilibrium positions by amounts x_1 and x_2, respectively, the force on m_1 will be $-\kappa x_1 - \kappa_{12}(x_1 - x_2)$ and the force on m_2 will be $-\kappa x_2 - \kappa_{12}(x_2 - x_1)$. Therefore, the equations of motion are

$$\left.\begin{aligned} M\ddot{x}_1 + (\kappa + \kappa_{12})x_1 - \kappa_{12}x_2 = 0 \\ M\ddot{x}_2 + (\kappa + \kappa_{12})x_2 - \kappa_{12}x_2 = 0 \end{aligned}\right\} \tag{13.1}$$

* Henceforth, we shall denote force constants by κ, rather than by k as heretofore. The symbol k is reserved for use (beginning in Chapter 15) in an entirely different context.

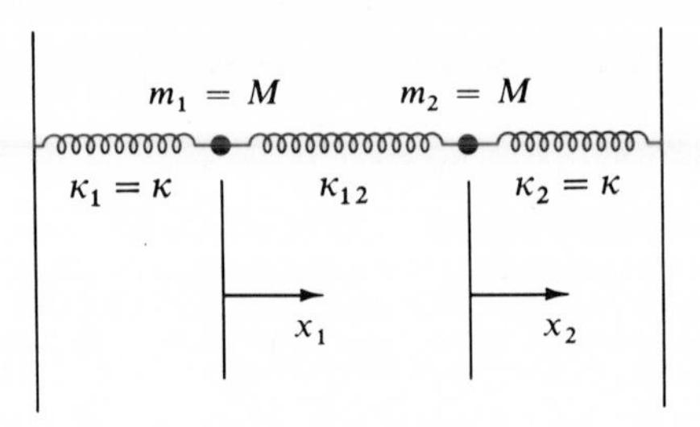

FIG. **13-1**

Since we expect the motion to be oscillatory, we attempt a solution of the form

$$\left.\begin{aligned} x_1(t) &= B_1 e^{i\omega t} \\ x_2(t) &= B_2 e^{i\omega t} \end{aligned}\right\} \tag{13.2}$$

where the frequency ω is to be determined and where the amplitudes, B_1 and B_2, can be complex.* Of course, only the *real* parts of $x_1(t)$ and $x_2(t)$ are physically meaningful. Substituting these expressions for the displacements into the equations of motion, we find

$$\left.\begin{aligned} -M\omega^2 B_1 e^{i\omega t} + (\kappa + \kappa_{12})B_1 e^{i\omega t} - \kappa_{12} B_2 e^{i\omega t} = 0 \\ -M\omega^2 B_2 e^{i\omega t} + (\kappa + \kappa_{12})B_2 e^{i\omega t} - \kappa_{12} B_1 e^{i\omega t} = 0 \end{aligned}\right\} \tag{13.3}$$

Upon collecting terms and canceling the common exponential factor, we obtain

$$\left.\begin{aligned} (\kappa + \kappa_{12} - M\omega^2)B_1 - \kappa_{12} B_2 = 0 \\ -\kappa_{12} B_1 + (\kappa + \kappa_{12} - M\omega^2)B_2 = 0 \end{aligned}\right\} \tag{13.4}$$

In order that a nontrivial solution exist for this pair of simultaneous equations, the determinant of the coefficients of B_1 and B_2 must vanish:

$$\begin{vmatrix} \kappa + \kappa_{12} - M\omega^2 & -\kappa_{12} \\ -\kappa_{12} & \kappa + \kappa_{12} - M\omega^2 \end{vmatrix} = 0 \tag{13.5}$$

The expansion of this secular determinant yields

$$(\kappa + \kappa_{12} - M\omega^2)^2 - \kappa_{12}^2 = 0 \tag{13.6}$$

* Because a complex amplitude has a *magnitude* and a *phase*, we have the two arbitrary constants that are necessary in the solution of a second-order differential equation. That is, we could equally well write $x(t) = |B| \exp[i(\omega t - \delta)]$ or $x(t) = |B| \cos(\omega t - \delta)$, as in Eq. (3.6b). Later [see Eqs. (13.9)], we will find it more convenient to employ two distinct *real* amplitudes and the time-varying factors, $\exp(i\omega t)$ and $\exp(-i\omega t)$. These various forms of solution are all entirely equivalent.

Hence,

$$\kappa + \kappa_{12} - M\omega^2 = \pm\kappa_{12}$$

Solving for ω, we obtain

$$\omega = \sqrt{\frac{\kappa + \kappa_{12} \pm \kappa_{12}}{M}} \tag{13.7}$$

Therefore, we have two *characteristic frequencies* (or *eigenfrequencies*) for the system:

$$\omega_1 = \sqrt{\frac{\kappa + 2\kappa_{12}}{M}}; \qquad \omega_2 = \sqrt{\frac{\kappa}{M}} \tag{13.8}$$

Thus, the general solution to the problem is

$$\left.\begin{aligned} x_1(t) &= B_{11}^+ e^{i\omega_1 t} + B_{11}^- e^{-i\omega_1 t} + B_{12}^+ e^{i\omega_2 t} + B_{12}^- e^{-i\omega_2 t} \\ x_2(t) &= B_{21}^+ e^{i\omega_1 t} + B_{21}^- e^{-i\omega_1 t} + B_{22}^+ e^{i\omega_2 t} + B_{22}^- e^{-i\omega_2 t} \end{aligned}\right\} \tag{13.9}$$

where we have explicitly written both positive and negative frequencies because the radicals in Eqs. (13.7) and (13.8) can carry either sign.

In Eqs. (13.9) the amplitudes are not all independent, as we may readily verify by substituting ω_1 and ω_2 into Eqs. (13.4). We find

$$\text{For} \qquad \omega = \omega_1: \quad B_{11} = -B_{21}$$

$$\text{For} \qquad \omega = \omega_2: \quad B_{12} = B_{22}$$

The only subscripts on the B's that are now necessary are those that indicate the particular eigenfrequency (i.e., the *second* subscripts). We can therefore write the general solution as

$$\left.\begin{aligned} x_1(t) &= \quad B_1^+ e^{i\omega_1 t} + B_1^- e^{-i\omega_1 t} + B_2^+ e^{i\omega_2 t} + B_2^- e^{-i\omega_2 t} \\ x_2(t) &= -B_1^+ e^{i\omega_1 t} - B_1^- e^{-i\omega_1 t} + B_2^+ e^{i\omega_2 t} + B_2^- e^{-i\omega_2 t} \end{aligned}\right\} \tag{13.10}$$

Thus, we have *four* arbitrary constants in the general solution, just as we expect, since we have *two* equations of motion which are of *second* order.

As we have mentioned earlier, it is always possible to define a set of coordinates that have a simple time dependence and correspond to the excitation of the various oscillation modes of the system. Let us examine the pair of coordinates defined by

$$\left.\begin{aligned} \eta_1 &\equiv x_1 - x_2 \\ \eta_2 &\equiv x_1 + x_2 \end{aligned}\right\} \tag{13.11}$$

or,

$$\left.\begin{aligned} x_1 &= \tfrac{1}{2}(\eta_2 + \eta_1) \\ x_2 &= \tfrac{1}{2}(\eta_2 - \eta_1) \end{aligned}\right\} \tag{13.12}$$

Substituting these expressions for x_1 and x_2 into Eqs. (13.1), we find

$$\left.\begin{aligned} M(\ddot{\eta}_1 + \ddot{\eta}_2) + (\kappa + 2\kappa_{12})\eta_1 + \kappa\eta_2 = 0 \\ M(\ddot{\eta}_1 - \ddot{\eta}_2) + (\kappa + 2\kappa_{12})\eta_1 - \kappa\eta_2 = 0 \end{aligned}\right\} \tag{13.13}$$

which can be solved (by adding and subtracting) to yield

$$\left.\begin{aligned} M\ddot{\eta}_1 + (\kappa + 2\kappa_{12})\eta_1 = 0 \\ M\ddot{\eta}_2 + \kappa\eta_2 = 0 \end{aligned}\right\} \tag{13.14}$$

The coordinates η_1 and η_2 are now *uncoupled* and are therefore *independent*. The solutions are

$$\left.\begin{aligned} \eta_1(t) = C_1^+ e^{i\omega_1 t} + C_1^- e^{-i\omega_1 t} \\ \eta_2(t) = C_2^+ e^{i\omega_2 t} + C_2^- e^{-i\omega_2 t} \end{aligned}\right\} \tag{13.15}$$

where the frequencies ω_1 and ω_2 are given by Eqs. (13.8). Thus, η_1 and η_2 are the *normal coordinates* of the problem. In a later section we will establish a general method for obtaining the normal coordinates.

If we impose the special initial conditions, $x_1(0) = -x_2(0)$ and $\dot{x}_1(0) = -\dot{x}_2(0)$, we find $\eta_2(0) = 0$ and $\dot{\eta}_2(0) = 0$ which leads to $C_2^+ = C_2^- = 0$. That is, $\eta_2(t) \equiv 0$ for all values of t. Thus, the particles oscillate always *out of phase* and with frequency ω_1; this is the *antisymmetrical* mode of oscillation. On the other hand, if we begin with $x_1(0) = x_2(0)$ and $\dot{x}_1(0) = \dot{x}_2(0)$, we find $\eta_1(t) \equiv 0$, and the particles oscillate *in phase* and with frequency ω_2; this is the *symmetrical mode* of oscillation. These results are illustrated schematically in Fig. 13-2. The general motion of the system is, of course, a linear combination of the symmetrical and antisymmetrical modes.

The fact that the antisymmetrical mode has the higher frequency and the symmetrical mode has the lower frequency is actually a general result. In a

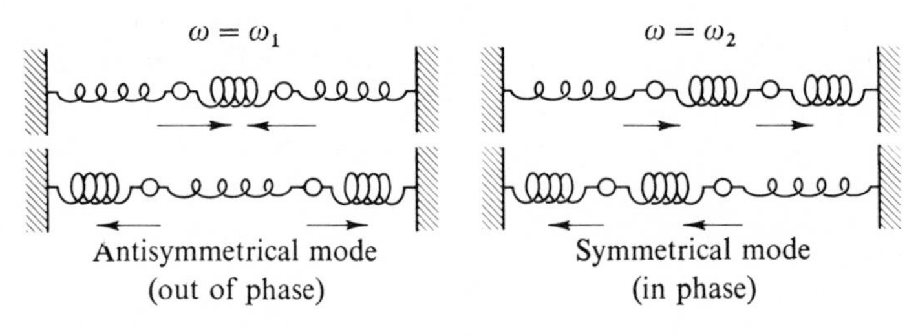

FIG. **13-2**

complex system of linearly coupled oscillators, the mode which possesses the highest degree of symmetry will have the lowest frequency. If the symmetry is destroyed, then the springs must "work harder" in the antisymmetrical modes, and the frequency is raised.

Notice that if we were to hold m_2 fixed and allow m_1 to oscillate, the frequency would be $\sqrt{(\kappa + \kappa_{12})/M}$. The same result would be obtained for the frequency of oscillation of m_2 if m_1 were held fixed. The oscillators are identical and in the absence of coupling obviously have the same oscillation frequency. The effect of coupling is to separate the common frequency, with one characteristic frequency becoming larger and one becoming smaller than the frequency for uncoupled motion. If we denote by ω_0 the frequency for uncoupled motion, then $\omega_1 > \omega_0 > \omega_2$, and we may schematically indicate the effect of the coupling as in Fig. 13-3a. The solution for the characteristic

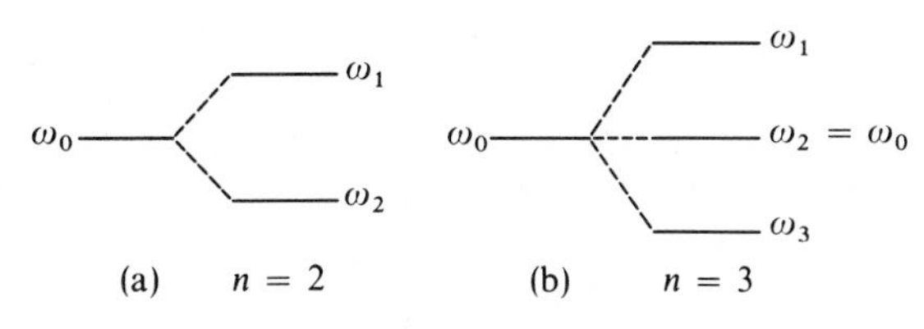

Fig. 13-3

frequencies in the problem of three coupled identical masses is illustrated in Fig. 13-3b. Again we have a splitting of the characteristic frequencies, with one greater and one smaller than ω_0. This is a general result: For an even number n of identical coupled oscillators, there will be $n/2$ characteristic frequencies greater than ω_0 and $n/2$ characteristic frequencies smaller that ω_0; if n is odd, one characteristic frequency will be equal to ω_0 and the remaining $n - 1$ characteristic frequencies will be symmetrically distributed above and below ω_3. The reader familiar with the phenomenon of the Zeeman effect in atomic spectra will appreciate the similarity with this result: in each case there is a symmetrical splitting of the frequency which is caused by the introduction of an interaction (in one case by the application of a magnetic field and in the other by the coupling of particles through the intermediary of the springs).

13.3 Weak Coupling

Some of the more interesting cases of coupled oscillations occur when the coupling is *weak*, i.e., when the force constant of the coupling spring is small compared to that of the oscillator springs: $\kappa_{12} \ll \kappa$. According to Eqs. (13.8), the frequencies ω_1 and ω_2 are

$$\omega_1 = \sqrt{\frac{\kappa + 2\kappa_{12}}{M}}; \qquad \omega_2 = \sqrt{\frac{\kappa}{M}} \tag{13.16}$$

If the coupling is weak, we may expand the expression for ω_1;

$$\begin{aligned}\omega_1 &= \sqrt{\frac{\kappa}{M}}\sqrt{1 + \frac{2\kappa_{12}}{\kappa}} \\ &\cong \sqrt{\frac{\kappa}{M}}(1 + 2\varepsilon)\end{aligned} \tag{13.17}$$

where

$$\varepsilon \equiv \frac{\kappa_{12}}{2\kappa} \ll 1 \tag{13.18}$$

Now, the natural frequency of either oscillator, when the other is held fixed, is

$$\begin{aligned}\omega_0 &= \sqrt{\frac{\kappa + \kappa_{12}}{M}} \\ &\cong \sqrt{\frac{\kappa}{M}}(1 + \varepsilon)\end{aligned} \tag{13.19}$$

or,

$$\sqrt{\frac{\kappa}{M}} \cong \omega_0(1 - \varepsilon) \tag{13.20}$$

Therefore, the two characteristic frequencies are given approximately by

$$\left.\begin{aligned}\omega_1 &\cong \sqrt{\frac{\kappa}{M}}(1 + 2\varepsilon); & \omega_2 &= \sqrt{\frac{\kappa}{M}} \\ &\cong \omega_0(1 - \varepsilon)(1 + 2\varepsilon) & &\cong \omega_0(1 - \varepsilon) \\ &\cong \omega_0(1 + \varepsilon) & &\end{aligned}\right\} \tag{13.21}$$

We can now examine the way in which a weakly-coupled system behaves. If we displace oscillator 1 a distance D and release it from rest, the initial conditions for the system are:

$$x_1(0) = D, \qquad x_2(0) = 0, \qquad \dot{x}_1(0) = 0, \qquad \dot{x}_2(0) = 0 \tag{13.22}$$

If we substitute these initial conditions into Eqs. (13.10) for $x_1(t)$ and $x_2(t)$, we find the amplitudes to be

$$B_1^+ = B_1^- = B_2^+ = B_2^- = \frac{D}{4} \tag{13.23}$$

Then, $x_1(t)$ becomes

$$\begin{aligned} x_1(t) &= \frac{D}{4}\,[(e^{i\omega_1 t} + e^{-i\omega_1 t}) + (e^{i\omega_2 t} + e^{-i\omega_2 t})] \\ &= \frac{D}{2}\,[\cos\,\omega_1 t + \cos\,\omega_2\,t] \\ &= D\,\cos\left(\frac{\omega_1 + \omega_2}{2}\,t\right)\cos\left(\frac{\omega_1 - \omega_2}{2}\,t\right) \end{aligned} \tag{13.24}$$

But, according to Eqs. (13.21),

$$\frac{\omega_1 + \omega_2}{2} = \omega_0\,; \qquad \frac{\omega_1 - \omega_2}{2} = \varepsilon\omega_0 \tag{13.25}$$

Therefore,

$$x_1(t) = (D\,\cos\,\varepsilon\omega_0\,t)\,\cos\,\omega_0\,t \tag{13.26a}$$

Similarly,

$$x_2(t) = (D\,\sin\,\varepsilon\omega_0\,t)\,\sin\,\omega_0\,t \tag{13.26b}$$

Because ε is a small quantity, the quantities $D\cos\varepsilon\omega_0 t$ and $D\sin\varepsilon\omega_0 t$ vary slowly with time. Therefore, $x_1(t)$ and $x_2(t)$ are essentially sinusoidal functions with slowly varying amplitudes. Although only x_1 is initially different from zero, as time increases the amplitude of x_1 decreases slowly with time, while the amplitude of x_2 increases slowly from zero. Hence, energy is transferred from the first oscillator to the second. When $t = \pi/2\varepsilon\omega_0$, then $D\cos\varepsilon\omega_0 t = 0$, and all of the energy has been transferred. As time increases further, energy is transferred back to the first oscillator. This is the familiar phenomenon of *beats* and is illustrated in Fig. 13-4. (In the case illustrated, $\varepsilon = 0.08$.)

13.4 Forced Vibrations of Coupled Oscillators

The phenomenon of resonance occurs in the forced motion of coupled systems just as it does in the case of a simple oscillator (Section 4.2). The essential difference in the forced vibrations of coupled oscillators is that *multiple* resonances will occur since, if there are n distinct characteristic frequencies, each of these will give rise to a resonance; $n = 1$, of course, for the simple harmonic oscillator. The discussion of resonance has physical meaning only if the system is *damped*; otherwise, infinite (and unphysical)

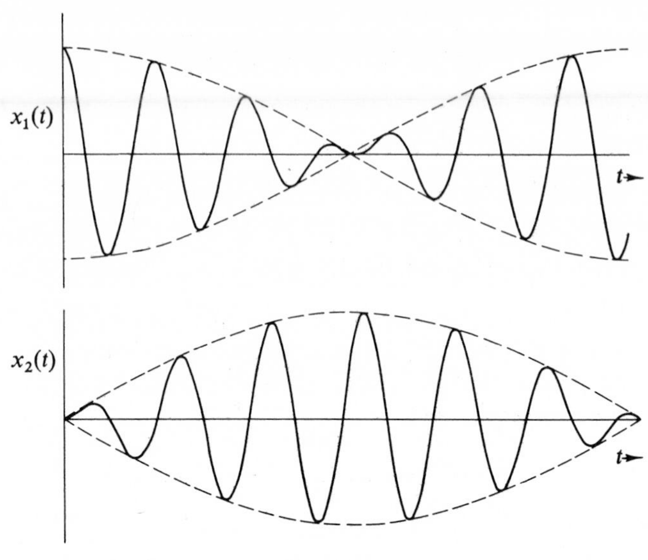

FIG. 13-4

amplitudes will result at the resonant frequencies. In the case of coupled systems, however, the explicit inclusion of damping terms greatly complicates the situation.* Rather than treat the complete problem we will confine our attention to the simpler undamped case from which we will still be able to obtain considerable qualitative information.

We again return to the pair of identical coupled oscillators pictured in Fig. 13-1 and now allow m_1 to be driven by a force, $F_0 \cos \omega t$. The equations of motion are

$$\left.\begin{aligned} M\ddot{x}_1 + (\kappa + \kappa_{12})x_1 - \kappa_{12}x_2 &= F_0 \cos \omega t \\ M\ddot{x}_2 + (\kappa + \kappa_{12})x_2 - \kappa_{12}x_1 &= 0 \end{aligned}\right\} \tag{13.27}$$

which we can write in standard notation as

$$\left.\begin{aligned} \ddot{x}_1 + \omega_0^2 x_1 - \frac{\kappa_{12}}{M} x_2 &= A \cos \omega t \\ \ddot{x}_2 + \omega_0^2 x_2 - \frac{\kappa_{12}}{M} x_1 &= 0 \end{aligned}\right\} \tag{13.27a}$$

We already know the solutions for the complementary equations [Eqs. (13.10)] so it is necessary to consider now only the particular solutions. In the event that damping is included, only the particular solutions will persist after the

* A solution in terms of the *normal coordinates* of the problem is, however, not difficult to obtain (see Problem 13-10).

transient effects have subsided; see the discussion following Eq. (4.12). Since damping has been neglected we may try particular solutions which do not include the complication of phase factors; we know that the phase changes by π at resonance (see Fig. 4-1):

$$\left.\begin{aligned} x_{1,p}(t) &= D_1 \cos \omega t \\ x_{2,p}(t) &= D_2 \cos \omega t \end{aligned}\right\} \tag{13.28}$$

Substituting these expressions into Eqs. (13.27a) and canceling the common factor, $\cos \omega t$, we obtain

$$\left.\begin{aligned} (\omega_0^2 - \omega^2)D_1 - \frac{\kappa_{12}}{M} D_2 &= A \\ -\frac{\kappa_{12}}{M} D_1 + (\omega_0^2 - \omega^2)D_2 &= 0 \end{aligned}\right\} \tag{13.29}$$

The solutions for D_1 and D_2 are

$$\left.\begin{aligned} D_1 &= \frac{A(\omega_0^2 - \omega^2)}{(\omega_0^2 - \omega^2)^2 - (\kappa_{12}/M)^2} \\ D_2 &= \frac{A(\kappa_{12}/M)}{(\omega_0^2 - \omega^2)^2 - (\kappa_{12}/M)^2} \end{aligned}\right\} \tag{13.30}$$

Using the expressions for ω_0, ω_1, and ω_2 [Eqs. (13.16) and (13.19)], it is easy to verify that D_1 and D_2 can be rewritten in the following forms:

$$\left.\begin{aligned} D_1 &= \frac{A(\omega_0^2 - \omega^2)}{(\omega_1^2 - \omega^2)(\omega_2^2 - \omega^2)} \\ D_2 &= \frac{A(\omega_0^2 - \omega_2^2)}{(\omega_1^2 - \omega^2)(\omega_2^2 - \omega^2)} \end{aligned}\right\} \tag{13.31}$$

These equations demonstrate the expected resonance effect: D_1 and D_2 become infinite at the two characteristic frequencies, ω_1 and ω_2. (The amplitudes remain finite, of course, if damping is included.) In the limit of zero coupling ($\kappa_{12} \to 0$), the three frequencies, ω_0, ω_1, and ω_2, become equal and D_1 takes on the form of the amplitude of a simple oscillator (with $Q = \infty$) while D_2 vanishes (since there is no energy transfer to m_2 in the absence of coupling):

$$\left.\begin{aligned} D_1(\kappa_{12} \to 0) &= \frac{A}{\omega_0^2 - \omega^2} \\ D_2(\kappa_{12} \to 0) &= 0 \end{aligned}\right\} \tag{13.32}$$

Equations (13.31) are illustrated in Fig. 13-5. Negative values of the amplitude factors occur because the trial solutions [Eqs. (13.28)] did not contain

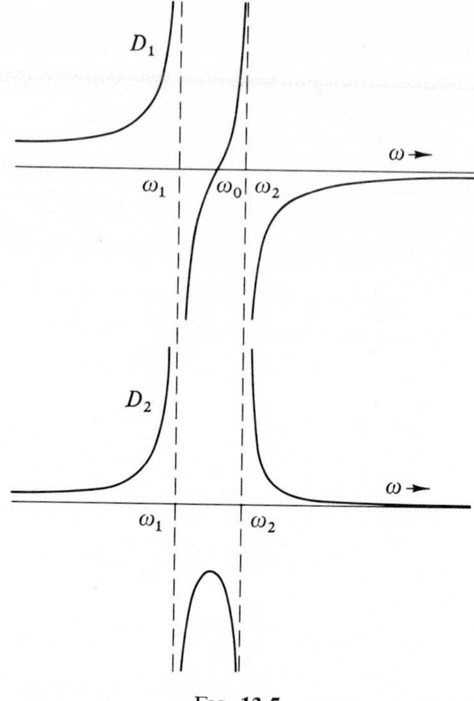

FIG. 13-5

phases; these negative values imply that the driving force and the resultant motion are *out of phase*. For $\omega < \omega_0$, the two masses have the same relative phase while for $\omega > \omega_0$, they have opposite phase.

13.5 Coupled Electrical Circuits

Figure 3-16 illustrates the equivalence of the simple, undamped oscillator and an LC electrical circuit. If such a circuit is brought into proximity with another similar circuit, oscillations will be induced in the latter by virtue of the coupling between the two coils (i.e., the *mutual inductance*). The situation is shown in Fig. 13-6; the capacitors C and the inductors L are identical, and the mutual inductance is M. Writing down the Kirchhoff circuit equations, we have

$$\left.\begin{aligned} L\dot{I}_1 + \frac{q_1}{C} + M\dot{I}_2 = 0 \\ L\dot{I}_2 + \frac{q_2}{C} + M\dot{I}_1 = 0 \end{aligned}\right\} \tag{13.33}$$

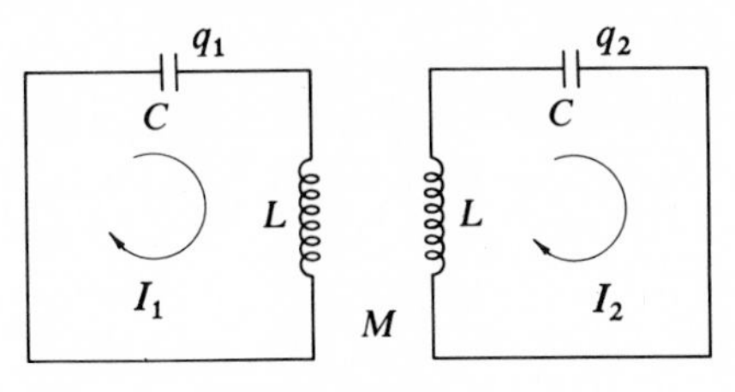

FIG. 13-6

Differentiating these equations with respect to time in order to express all quantities in terms of *currents*, we obtain (using $\dot{q} = I$)

$$\left.\begin{aligned} L\ddot{I}_1 + \frac{1}{C} I_1 + M\ddot{I}_2 = 0 \\ L\ddot{I}_2 + \frac{1}{C} I_2 + M\ddot{I}_1 = 0 \end{aligned}\right\} \tag{13.34}$$

As usual, we seek solutions of the form

$$\left.\begin{aligned} I_1 = B_1 e^{i\omega t} \\ I_2 = B_2 e^{i\omega t} \end{aligned}\right\} \tag{13.35}$$

which lead to the simultaneous equations,

$$\left.\begin{aligned} \left(\omega^2 L - \frac{1}{C}\right) B_1 + M\omega^2 B_2 = 0 \\ M\omega^2 B_1 + \left(\omega^2 L - \frac{1}{C}\right) B_2 = 0 \end{aligned}\right\} \tag{13.36}$$

Setting equal to zero the determinant of the coefficients of B_1 and B_2 yields

$$\left(\omega^2 L - \frac{1}{C}\right)^2 - (M\omega^2)^2 = 0 \tag{13.37}$$

Solving for ω, we find

$$\omega = \frac{1}{\sqrt{LC \pm MC}} \tag{13.38}$$

so that the characteristic frequencies are

$$\omega_1 = \frac{1}{\sqrt{LC - MC}}; \qquad \omega_2 = \frac{1}{\sqrt{LC + MC}} \tag{13.39}$$

Again, the characteristic frequencies are found to lie above and below the frequency for uncoupled oscillations, $1/\sqrt{LC}$.

In addition to coupling through mutual inductance, electrical circuits may be coupled in many other ways. Two possibilities, inductive and capacitive coupling, are illustrated in Fig. 13-7. Resistive coupling is also possible, for example, by replacing L_{12} in Fig. 13-7a by a resistor. (See Problem 13-14.)

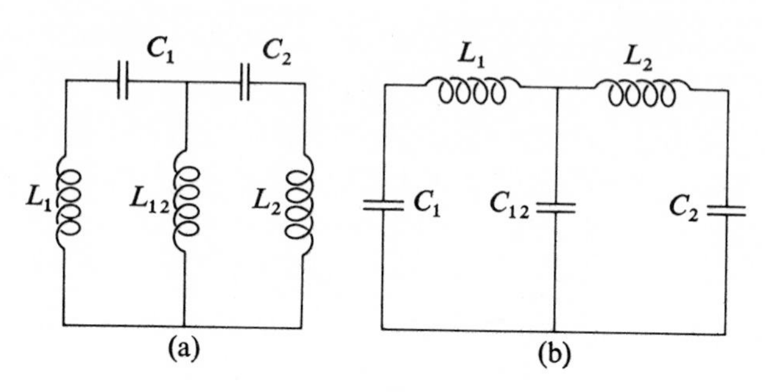

FIG. 13-7

13.6 The General Problem of Coupled Oscillations

In the preceding sections we found that the effect of coupling in a simple system with two degrees of freedom was to produce two characteristic frequencies and two modes of oscillation. We now turn our attention to the general problem of coupled oscillations. Let us consider a conservative system which can be described in terms of a set of generalized coordinates q_k and the time t. If the system has n degrees of freedom, then $k = 1, 2, \ldots, n$. We specify that a configuration of stable equilibrium exists for the system and that at equilibrium the generalized coordinates have values q_{k0}. In such a configuration, Lagrange's equations will be satisfied by

$$q_k = q_{k0}; \qquad \dot{q}_k = 0; \qquad \ddot{q}_k = 0; \qquad k = 1, 2, \ldots, n$$

Now, every nonzero term of the form $(d/dt)(\partial L/\partial \dot{q}_k)$ must contain at least either $\dot{q}_k$ or $\ddot{q}_k$, so that all such terms vanish at equilibrium. From Lagrange's equation we therefore have

$$\left.\frac{\partial L}{\partial q_k}\right|_0 = \left.\frac{\partial T}{\partial q_k}\right|_0 - \left.\frac{\partial U}{\partial q_k}\right|_0 = 0 \tag{13.40}$$

where the subscripts 0 designate that the quantity is evaluated at equilibrium.

We shall assume that the equations which connect the generalized coordinates and the rectangular coordinates do not explicitly contain the time. That is, we have

$$x_{\alpha,i} = x_{\alpha,i}(q_j) \qquad \text{or} \qquad q_j = q_j(x_{\alpha,i})$$

Then, the kinetic energy is a homogeneous quadratic function of the generalized velocities [cf. Eq. 7.32)]:

$$T = \tfrac{1}{2} \sum_{j,k} m_{jk} \dot{q}_j \dot{q}_k \tag{13.41}$$

Therefore, in general,

$$\frac{\partial T}{\partial q_k} = 0, \qquad k = 1, 2, \ldots, n \tag{13.42}$$

and, hence, from Eq. (13.40) we have

$$\left.\frac{\partial U}{\partial q_k}\right|_0 = 0, \qquad k = 1, 2, \ldots, n \tag{13.43}$$

We may further specify that the generalized coordinates q_k be measured from the equilibrium positions; i.e., we choose $q_{k0} = 0$. (If we originally had chosen a set of coordinates q_k' such that $q_{k0}' \neq 0$, we could always effect a simple linear transformation of the form $q_k = q_k' + \alpha_k$ such that $q_{k0} = 0$.)

The expansion of the potential energy in a Taylor series about the equilibrium configuration yields

$$U(q_1, q_2, \ldots, q_n) = U_0 + \sum_k \left.\frac{\partial U}{\partial q_k}\right|_0 q_k + \frac{1}{2} \sum_{j,k} \left.\frac{\partial^2 U}{\partial q_j \, \partial q_k}\right|_0 q_j q_k + \cdots \tag{13.44}$$

Now, the second term in the expansion vanishes in view of Eq. (13.43), and, without loss of generality, we may choose to measure U in such a way that $U_0 \equiv 0$. Then, if we restrict the motion of the generalized coordinates to be small, we may neglect all terms in the expansion which contain products of the q_k of degree higher than second. This is equivalent to restricting our attention to simple harmonic oscillations, in which case only terms quadratic in the coordinates appear. Thus,

$$U = \tfrac{1}{2} \sum_{j,k} A_{jk} q_j q_k \tag{13.45}$$

where we define

$$A_{jk} \equiv \left.\frac{\partial^2 U}{\partial q_j \, \partial q_k}\right|_0 \tag{13.46}$$

Since the order of differentiation is immaterial (if U has continuous second partial derivatives), the quantity A_{jk} is symmetric; that is, $A_{jk} = A_{kj}$.

We have specified that the motion of the system is to take place in the vicinity of the equilibrium configuration, and we have shown [Eq. (13.43)] that U must have a minimum value when the system is in this configuration.

Since we have chosen $U = 0$ at equilibrium, we must have in general, $U \geq 0$. It is clear that we must also have $T \geq 0$.*

Equations (13.41) and (13.45) are of a similar form:

$$\boxed{\begin{aligned} T &= \frac{1}{2}\sum_{j,k} m_{jk}\dot{q}_j\dot{q}_k \\ U &= \frac{1}{2}\sum_{j,k} A_{jk}q_j q_k \end{aligned}} \tag{13.47}$$

The quantities A_{jk} are just numbers [see Eq. (13.46)]; the m_{jk}, however, may be functions of the coordinates [see Eq. (7.30)]:

$$m_{jk} = \sum_\alpha m_\alpha \sum_i \frac{\partial x_{\alpha,i}}{\partial q_j}\frac{\partial x_{\alpha,i}}{\partial q_k}$$

We can expand the m_{jk} about the equilibrium position with the result

$$m_{jk}(q_1, q_2, \ldots, q_n) = m_{jk}(q_{l0}) + \sum_l \left.\frac{\partial m_{jk}}{\partial q_l}\right|_0 q_l + \cdots \tag{13.48}$$

We wish to retain only the first nonvanishing term in this expansion; but, in contrast to the expansion of the potential energy [Eq. (13.44)], we cannot choose the constant term $m_{jk}(q_{l0})$ to be zero, so that this leading term becomes the constant value of m_{jk} in this approximation. This is the same order of approximation as that used for U since the next higher order term in T would involve the cubic quantity $q_j\dot{q}_k\dot{q}_l$ and the next higher order term in U would contain $q_j q_k q_l$. Thus, in Eqs. (13.47), the m_{jk} and the A_{jk} are $n \times n$ arrays of *numbers* which specify the way in which the motions of the various coordinates are coupled. For example, if $m_{rs} \neq 0$ for $r \neq s$, then the kinetic energy will contain a term proportional to $\dot{q}_r\dot{q}_s$, and a coupling exists between the rth and sth coordinate. If, on the other hand, m_{jk} is *diagonal* so that† $m_{jk} \neq 0$ for $j = k$, but vanishes otherwise, then the kinetic energy is of the form

$$T = \frac{1}{2}\sum_r m_r \dot{q}_r^2$$

where m_{rr} has been abbreviated to m_r. Thus, the kinetic energy is a simple sum of the kinetic energies associated with the various coordinates. As we shall see, if, in addition, A_{jk} is diagonal so that U is also a simple sum of individual potential energies, then each coordinate will behave in an uncomplicated manner, undergoing oscillations with a single, well-defined

* That is, both U and T are *positive definite* quantities, in that they are always positive unless the coordinates (in the case of U) or the velocities (in the case of T) are zero, in which case they vanish.

† If a diagonal element of m_{jk} (say, m_{rr}) vanishes, then the problem can be reduced to one of $n - 1$ degrees of freedom.

frequency. The problem is therefore to find a coordinate transformation which simultaneously diagonalizes both m_{jk} and A_{jk} and thereby renders the system describable in the simplest possible terms. Such coordinates are the *normal coordinates*.

The equations of motion of the system which has kinetic and potential energies given by Eqs. (13.47) are obtained from Lagrange's equation

$$\frac{\partial L}{\partial q_k} - \frac{d}{dt}\frac{\partial L}{\partial \dot{q}_k} = 0$$

But, since T is a function only of the generalized velocities and U is a function only of the generalized coordinates, Lagrange's equation for the kth coordinate becomes

$$\frac{\partial U}{\partial q_k} + \frac{d}{dt}\frac{\partial T}{\partial \dot{q}_k} = 0 \tag{13.49}$$

From Eqs. (13.47) we evaluate the derivatives:

$$\left.\begin{aligned} \frac{\partial U}{\partial q_k} &= \sum_j A_{jk} q_j \\ \frac{\partial T}{\partial \dot{q}_k} &= \sum_j m_{jk} \dot{q}_j \end{aligned}\right\} \tag{13.50}$$

The equations of motion then become

$$\boxed{\sum_j (A_{jk} q_j + m_{jk} \ddot{q}_j) = 0} \tag{13.51}$$

This is a set of n second-order linear homogeneous differential equations with constant coefficients. Since we are dealing with an oscillatory system, we expect a solution of the form

$$q_j(t) = a_j e^{i(\omega t - \delta)} \tag{13.52}$$

Where the a_j are real amplitudes and where the phase δ has been included to give the two arbitrary constants (a_j and δ) required by the second-order nature of each of the differential equations.* We understand, of course, that only the real part of the right-hand side is to be considered. The frequency ω and the phase δ are to be determined by the equations of motion. If ω is a real quantity, then the solution [Eq. (13.52)] represents oscillatory motion.

* This is entirely equivalent to our previous procedure of writing $x(t) = B \exp(i\omega t)$ [see Eqs. (13.2)] with B allowed to be complex. In Eqs. (13.9) we elected to exhibit the requisite arbitrary constants as real amplitudes by using $\exp(i\omega t)$ and $\exp(-i\omega t)$ rather than by incorporating a phase factor as in Eq. (13.52).

That ω is indeed real may be seen by the following physical argument. Suppose that ω contains an imaginary part. This will give rise to terms of the form $e^{\omega t}$ and $e^{-\omega t}$ in the expression of q_j. Thus, when the total energy of the system is computed, $T + U$ will contain factors which increase or decrease monotonically with the time. But this violates the assumption that we are dealing with a conservative system; therefore, the frequency ω must be a real quantity.

With a solution of the form given by Eq. (13.52), the equations of motion become

$$\boxed{\sum_j (A_{jk} - \omega^2 m_{jk}) a_j = 0} \tag{13.53}$$

where the common factor $\exp[i(\omega t - \delta)]$ has been canceled. This is a set of n linear homogeneous *algebraic* equations which the a_j must satisfy. For a nontrivial solution to exist, the determinant of the coefficients must vanish:

$$|A_{jk} - \omega^2 m_{jk}| = 0 \tag{13.54}$$

To be more explicit, this is an $n \times n$ determinant of the form

$$\begin{vmatrix} A_{11} - \omega^2 m_{11} & A_{12} - \omega^2 m_{12} & A_{13} - \omega^2 m_{13} & \cdots \\ A_{12} - \omega^2 m_{12} & A_{22} - \omega^2 m_{22} & A_{23} - \omega^2 m_{23} & \cdots \\ A_{13} - \omega^2 m_{13} & A_{23} - \omega^2 m_{23} & A_{33} - \omega^2 m_{33} & \cdots \\ \vdots & \vdots & \vdots & \end{vmatrix} = 0$$

where the symmetry of the A_{jk} and m_{jk} has been explicitly included.

The equation which is represented by this determinant is called the *characteristic equation* or *secular equation* of the system and is an equation of degree n in ω^2. There are, in general, n roots which we may label ω_r^2. The ω_r are called the *characteristic frequencies* or *eigenfrequencies* of the system. (In some situations two or more of the ω_r can be equal; this is the phenomenon of *degeneracy* and will be discussed later.) Just as in the procedure for determining the directions of the principal axes for a rigid body, each of the roots of the characteristic equation may be substituted into Eqs. (13.54) to determine the ratios $a_1 : a_2 : a_3 : \cdots : a_n$ for each value of ω_r. Since there are n values of ω_r, we can construct n sets of ratios of the a_j. Each of these sets can be considered to define the components of an n-dimensional vector $\mathbf{a}_r$, called an *eigenvector* of the system. Thus, $\mathbf{a}_r$ is the eigenvector associated with the eigenfrequency ω_r. We designate by a_{jr} the jth component of the rth eigenvector.

Since the principle of superposition applies for the differential equation [Eq. (13.51)], we must write the general solution for q_j as a linear combination of the solutions for each of the n values of r:

$$q_j(t) = \sum_r a_{jr} e^{i(\omega_r t - \delta_r)} \tag{13.55}$$

But, of course, it is only the *real* part of $q_j(t)$ that is physically meaningful, so that we actually have

$$q_j(t) = \text{Re} \sum_r a_{jr} e^{i(\omega_r t - \delta_r)} = \sum_r a_{jr} \cos(\omega_r t - \delta_r) \tag{13.55a}$$

The motion of the coordinate q_j is therefore compounded of motions with each of the n values of the frequencies ω_r. The q_j evidently are not the normal coordinates which simplify the problem. We shall continue the search for normal coordinates in Section 13.8.

▶ Example 13.6 Eigenfrequencies for Two Coupled Oscillators

We return now to the case of the two coupled oscillators and obtain the eigenfrequencies by means of the general formalism which we have just developed. The situation is that shown in Fig. 13-1. The potential energy of the system is

$$\begin{aligned} U &= \tfrac{1}{2}\kappa x_1^2 + \tfrac{1}{2}\kappa_{12}(x_2 - x_1)^2 + \tfrac{1}{2}\kappa x_2^2 \\ &= \tfrac{1}{2}(\kappa + \kappa_{12})x_1^2 + \tfrac{1}{2}(\kappa + \kappa_{12})x_2^2 - \kappa_{12} x_1 x_2 \end{aligned} \tag{1}$$

The term proportional to $x_1 x_2$ is the factor that expresses the coupling in the system. Calculating the A_{jk}, we find

$$\left.\begin{aligned} A_{11} &= \left.\frac{\partial^2 U}{\partial x_1^2}\right|_0 = \kappa + \kappa_{12} \\ A_{12} &= \left.\frac{\partial^2 U}{\partial x_1 \partial x_2}\right|_0 = -\kappa_{12} = A_{21} \\ A_{22} &= \left.\frac{\partial^2 U}{\partial x_2^2}\right|_0 = \kappa + \kappa_{12} \end{aligned}\right\} \tag{2}$$

The kinetic energy of the system is

$$T = M\dot{x}_1^2 + \tfrac{1}{2}M\dot{x}_2^2 \tag{3}$$

According to Eq. (13.41),

$$T = \tfrac{1}{2}\sum_{j,k} m_{jk}\dot{x}_j\dot{x}_k \tag{4}$$

Identifying terms between these two expressions for T, we find

$$\left.\begin{aligned} m_{11} &= m_{22} = M \\ m_{12} &= m_{21} = 0 \end{aligned}\right\} \tag{5}$$

Thus, the secular determinant [Eq. (13.54a)] becomes

$$\begin{vmatrix} \kappa + \kappa_{12} - M\omega^2 & -\kappa_{12} \\ -\kappa_{12} & \kappa + \kappa_{12} - M\omega^2 \end{vmatrix} = 0 \tag{6}$$

This is exactly Eq. (13.5), so the solutions are the same as before:

$$\omega = \sqrt{\frac{\kappa + \kappa_{12} \pm \kappa_{12}}{M}} \tag{7}$$

The eigenfrequencies are

$$\omega_1 = \sqrt{\frac{\kappa + 2\kappa_{12}}{M}}; \qquad \omega_2 = \sqrt{\frac{\kappa}{M}} \tag{8}$$

so that the results of the two procedures are identical.

13.7 The Orthogonality of the Eigenvectors

We now wish to show that the eigenvectors $\mathbf{a}_r$ form an orthonormal set. Rewriting Eq. (13.53) for the sth root of the secular equation we have

$$\omega_s^2 \sum_k m_{jk} a_{ks} = \sum_k A_{jk} a_{ks} \tag{13.56}$$

Next, we write a comparable equation for the rth root by substituting r for s and interchanging j and k:

$$\omega_r^2 \sum_j m_{jk} a_{jr} = \sum_j A_{jk} a_{jr} \tag{13.57}$$

where use has been made of the symmetry of the m_{jk} and A_{jk}. We now multiply Eq. (13.56) by a_{jr} and sum over j, and also multiply Eq. (13.57) by a_{ks} and sum over k:

$$\left.\begin{aligned} \omega_s^2 \sum_{j,k} m_{jk} a_{jr} a_{ks} &= \sum_{j,k} A_{jk} a_{jr} a_{ks} \\ \omega_r^2 \sum_{j,k} m_{jk} a_{jr} a_{ks} &= \sum_{j,k} A_{jk} a_{jr} a_{ks} \end{aligned}\right\} \tag{13.58}$$

The right-hand sides of Eqs. (13.58) are now equal, so that subtracting the first of these equations from the second, we have

$$(\omega_r^2 - \omega_s^2) \sum_{j,k} m_{jk} a_{jr} a_{ks} = 0 \tag{13.59}$$

We now examine the two possibilities, $r = s$ and $r \neq s$. For $r \neq s$, the term $(\omega_r^2 - \omega_s^2)$ is, in general, different from zero. (The case of degeneracy, or multiple roots, will be discussed later.) Therefore, the sum must vanish identically:

$$\sum_{j,k} m_{jk} a_{jr} a_{ks} = 0, \qquad r \neq s \tag{13.60}$$

For the case $r = s$, the term $(\omega_r^2 - \omega_s^2)$ vanishes and the sum is indeterminate.

The sum, however, cannot vanish identically. To show this, we write the kinetic energy for the system and substitute the expressions for $\dot{q}_j$ and $\dot{q}_k$ from Eq. (13.55):

$$\begin{aligned} T &= \tfrac{1}{2}\sum_{j,k} m_{jk}\dot{q}_j\dot{q}_k \\ &= \tfrac{1}{2}\sum_{j,k} m_{jk}\left[\sum_r \omega_r a_{jr}\sin(\omega_r t - \delta_r)\right]\left[\sum_s \omega_s a_{ks}\sin(\omega_s t - \delta_s)\right] \\ &= \tfrac{1}{2}\sum_{r,s} \omega_r\omega_s \sin(\omega_r t - \delta_r)\sin(\omega_s t - \delta_s)\sum_{j,j} m_{jk}a_{jr}a_{ks} \end{aligned}$$

Thus, for $r = s$, the kinetic energy becomes

$$T = \tfrac{1}{2}\sum_r \omega_r^2 \sin^2(\omega_r t - \delta_r)\sum_{j,k} m_{jk}a_{jr}a_{kr} \tag{13.61}$$

We note first that

$$\omega_r^2 \sin^2(\omega_r t - \delta_r) \geq 0$$

We also know that T is positive and can become zero only if all of the velocities vanish identically. Therefore,

$$\sum_{j,k} m_{jk}a_{jr}a_{kr} \geq 0$$

Thus, the sum is, in general, positive and can vanish only in the trivial instance that the system is not in motion, i.e., that the velocities vanish identically and $T \equiv 0$.

We have previously remarked that only the ratios of the a_{jr} are determined when the ω_r are substituted into Eq. (13.53). We now remove this indeterminacy by imposing an additional condition on the a_{jr}. We require that

$$\sum_{j,k} m_{jk}a_{jr}a_{kr} = 1 \tag{13.62}$$

The a_{jr} are then said to be *normalized.* Combining Eqs. (13.60) and (13.62), we may write

$$\boxed{\sum_{j,k} m_{kj}a_{jr}a_{ks} = \delta_{rs}} \tag{13.63}$$

The vectors $\mathbf{a}_r$ defined in this way constitute an *orthonormal* set; i.e., they are *orthogonal* according to the result given by Eq. (13.60), and they have been *normalized* by setting the sum in Eq. (13.62) equal to unity.

All the above discussion bears a striking resemblance to the procedure given in the preceding chapter for determining the principal moments of inertia and the principal axes for a rigid body. Indeed, the problems are mathematically identical, except that we are now dealing with a system which has n degrees of freedom. It should be noted that the quantities m_{jk} and A_{jk} are actually tensor elements since m and A are two-dimensional arrays which relate different physical quantities,* and as such we shall write them as $\{\mathsf{m}\}$ and $\{\mathsf{A}\}$. The secular equation for the determination of the eigenfrequencies is the same as that for obtaining the principal moments of inertia, and the eigenvectors $\mathbf{a}_r$ correspond to the principal axes. Indeed, the proof of the orthogonality of the eigenvectors is merely a generalization of the proof given in Section 12.6 of the orthogonality of the principal axes. Although we have made a physical argument regarding the reality of the eigenfrequencies, we could carry out a mathematical proof using the same procedure that was employed to show that the principal moments of inertia are real.

13.8 Normal Coordinates

As we have seen [Eq. (13.55)], the general solution for the motion of the coordinate q_j must be a sum over terms each of which depends upon an individual eigenfrequency. We note, however, that we have, as a matter of convenience, normalized the a_{jr} according to Eq. (13.62). That is, we have removed all ambiguity in the solution for the q_j, so that it is no longer possible to specify an arbitrary displacement for a particle. Since such a restriction is not physically meaningful, we must introduce a constant scale factor α (which will depend on the initial conditions of the problem) in order to account for the loss of generality that has been introduced by the arbitrary normalization. Thus,

$$q_j(t) = \sum_r \alpha a_{jr} e^{i(\omega_r t - \delta_r)} \tag{13.64}$$

In order to simplify the notation, we write

$$q_j(t) = \sum_r \beta_r a_{jr} e^{i\omega_r t} \tag{13.65}$$

where the quantities β_r are new scale factors† (now complex) which incorporate the phases δ_r.

We now define a quantity η_r,

$$\boxed{\eta_r(t) \equiv \beta_r e^{i\omega_r t}} \tag{13.66}$$

* See the discussion in Section 12.6 concerning the mathematical definition of a tensor.

† There is a certain advantage in normalizing the a_{jr} to unity and introducing the scale factors α and β_r rather than leaving the normalization unspecified, in that the a_{jr} are then independent of the initial conditions so that a simple orthonormality equation results.

so that

$$\boxed{q_j(t) = \sum_r a_{jr}\eta_r(t)} \tag{13.67}$$

The η_r, by definition, are quantities which undergo oscillation at only one frequency. They may be considered as new coordinates, called *normal coordinates*, for the system. The η_r satisfy equations of the form

$$\ddot{\eta}_r + \omega_r^2 \eta_r = 0 \tag{13.68}$$

There are n independent such equations, so that the equations of motion expressed in normal coordinates become completely separable. To show this from Lagrange's equations, we note that

$$\dot{q}_j = \sum_r a_{jr}\dot{\eta}_r$$

and from Eqs. (13.47) we have for the kinetic energy,

$$\begin{aligned} T &= \tfrac{1}{2}\sum_{j,k} m_{jk}\dot{q}_j\dot{q}_k \\ &= \tfrac{1}{2}\sum_{j,k} m_{jk}\left(\sum_r a_{jr}\dot{\eta}_r\right)\left(\sum_s a_{ks}\dot{\eta}_s\right) \\ &= \tfrac{1}{2}\sum_{r,s}\left(\sum_{j,k} m_{jk}a_{jr}a_{ks}\right)\dot{\eta}_r\dot{\eta}_s \end{aligned}$$

The sum in the parentheses is just δ_{rs} according to the orthonormality condition [Eq. (13.63)]. Therefore,

$$T = \tfrac{1}{2}\sum_{r,s}\dot{\eta}_r\dot{\eta}_s\delta_{rs} = \tfrac{1}{2}\sum_r \dot{\eta}_r^2 \tag{13.69}$$

Similarly, from Eqs. (13.47) we have for the potential energy,

$$\begin{aligned} U &= \tfrac{1}{2}\sum_{j,k} A_{jk}q_j q_k \\ &= \tfrac{1}{2}\sum_{r,s}\left(\sum_{j,k} A_{jk}a_{jr}a_{ks}\right)\eta_r\eta_s \end{aligned}$$

Now, the first equation in (13.58) is

$$\begin{aligned} \sum_{j,k} A_{jk}a_{jr}a_{ks} &= \omega_s^2 \sum_{j,k} m_{jk}a_{jr}a_{ks} \\ &= \omega_s^2\,\delta_{rs} \end{aligned}$$

so that the potential energy becomes

$$U = \tfrac{1}{2}\sum_{r,s}\omega_s^2\eta_r\eta_s\delta_{rs} = \tfrac{1}{2}\sum_r \omega_r^2\eta_r^2 \tag{13.70}$$

Using Eqs. (13.69) and (13.70), the Lagrangian is

$$L = \frac{1}{2} \sum_r (\dot{\eta}_r^2 - \omega_r^2 \eta_r^2) \tag{13.71}$$

and Lagrange's equations are

$$\frac{\partial L}{\partial \eta_r} - \frac{d}{dt} \frac{\partial L}{\partial \dot{\eta}_r} = 0$$

or,

$$\ddot{\eta}_r + \omega_r^2 \eta_r = 0$$

as before.

Thus, when the configuration of a system is expressed in normal coordinates, both the potential and kinetic energies become simultaneously diagonal. Since it is the off-diagonal elements of $\{\mathsf{m}\}$ and $\{\mathsf{A}\}$ which give rise to the coupling of the particles' motions, it is evident that a choice of coordinates that renders these tensors diagonal will uncouple the coordinates and make the problem completely separable into the independent motions of the normal coordinates, each with its particular normal frequency.*

In order to completely specify the transformation to the normal coordinates [Eqs. (13.66) and (13.67)], we must evaluate the quantities β_r. We write β_r as the sum of its real and imaginary parts,

$$\beta_r \equiv \mu_r + i\nu_r \tag{13.72}$$

so that

$$q_j(t) = \sum_r a_{jr}(\mu_r + i\nu_r)e^{i\omega_r t} \tag{13.73}$$

We also have

$$\dot{q}_j(t) = \sum_r i\omega_r a_{jr}(\mu_r + i\nu_r)e^{i\omega_r t} \tag{13.74}$$

The initial value of $q_j(t)$ can be obtained from the real part of Eq. (13.73) evaluated at $t = 0$:

$$q_j(0) = \sum_r \mu_r a_{jr}$$

If we multiply this equation by $m_{jk} a_{ks}$ and sum over j and k, we have

$$\sum_{j,k} m_{jk} a_{ks} q_j(0) = \sum_r \mu_r \left(\sum_{j,k} m_{jk} a_{jr} a_{ks} \right)$$

* The German mathematician Karl Weierstrass (1815–1897) showed in 1858 that the motion of a dynamical system can always be expressed in terms of normal coordinates.

According to the orthonormality condition, Eq. (13.63), the term in parentheses is just δ_{rs}. Therefore, the sum over r leaves only the term μ_s:

$$\boxed{\mu_s = \sum_{j,k} m_{jk} a_{ks} q_j(0)} \tag{13.75}$$

Similarly, for the evaluation of ν_r, we have for the real part of Eq. (13.74), at $t = 0$,

$$\dot{q}_j(0) = -\sum_r \omega_r \nu_r a_{jr}$$

Using the same procedure as above, we find

$$\boxed{\nu_s = -\frac{1}{\omega_s} \sum_{j,k} m_{jk} a_{ks} \dot{q}_j(0)} \tag{13.76}$$

Thus, the normal coordinates can be expressed as the real part of the sum

$$\boxed{\eta_r = \sum_{j,k} m_{jk} a_{kr} e^{i\omega_r t}\left[q_j(0) - \frac{i}{\omega_r}\dot{q}_j(0)\right]} \tag{13.77}$$

We therefore see that for arbitrary initial conditions, $q_j(0)$ and $\dot{q}_j(0)$, it is possible to find a set of coordinates η_r which exhibit the property that each varies harmonically with a single frequency ω_r. Since the sum in Eq. (13.77) runs over all of the n coordinate suffixes twice, it is clear that in general the expressions for the n_r are complicated. One important case is that in which the coordinates are displaced from their equilibrium positions and then, at time $t = 0$, are released. Under these conditions we have $q_j(0) \neq 0$, $\dot{q}_j(0) = 0$, so that Eq. (13.77) becomes

$$\eta_r = e^{i\omega_r t} \sum_{j,k} m_{jk} a_{kr} q_j(0), \qquad \dot{q}_j(0) = 0 \tag{13.78}$$

or, since only the real part is physically significant,

$$\eta_r = \cos \omega_r t \sum_{j,k} m_{jk} a_{kr} q_j(0), \qquad \dot{q}_j(0) = 0 \tag{13.79}$$

▶ **Example 13.8 Eigenvectors and Normal Coordinates for Two Coupled Oscillators**

Once again we return to the case of the two coupled oscillators and now calculate the eigenvectors and normal coordinates. We use the equations of motion

$$\sum_j (A_{jk} - \omega_r^2 m_{jk}) a_{jr} = 0 \tag{1}$$

to determine the eigenvector components a_{jr}. We have two equations for each value of r, but, since we can determine only the ratios a_{1r}/a_{2r}, one equation for each r is sufficient. For $r = 1$, $k = 1$ we have

$$(A_{11} - \omega_1^2 m_{11})a_{11} + (A_{21} - \omega_1^2 m_{21})a_{21} = 0 \tag{2}$$

or, inserting the values for A_{11}, ω_1^2, and m_{11}, and with the simplification that $\kappa_{12} = \kappa$,

$$\left(2\kappa - \frac{3\kappa}{M} \cdot M\right)a_{11} - \kappa a_{21} = 0 \tag{3}$$

with the result

$$a_{11} = -a_{21} \tag{4a}$$

For $r = 2$, $k = 1$ we have

$$\left(2\kappa - \frac{\kappa}{M} \cdot M\right)a_{12} - \kappa a_{22} = 0$$

with the result

$$a_{12} = a_{22} \tag{4b}$$

The orthonormality condition is

$$\sum_{j,k} m_{jk} a_{jr} a_{ks} = \delta_{rs}$$

But $\{\mathsf{m}\}$ is diagonal with each element equal to M; hence,

$$\sum_{j,k} M\delta_{jk} a_{jr} a_{ks} = \delta_{rs} \tag{5}$$

or, summing over k,

$$M \sum_j a_{jr} a_{js} = \delta_{rs} \tag{6}$$

Thus, for $r = s = 1$, we have

$$a_{11}^2 + a_{21}^2 = \frac{1}{M} \tag{7}$$

But, $a_{11} = -a_{21}$, so that

$$a_{11} = -a_{21} = \frac{1}{\sqrt{2M}}; \qquad \mathbf{a}_1 = \frac{1}{\sqrt{2M}}(1, -1) \tag{8a}$$

Similarly,

$$a_{12} = a_{22} = \frac{1}{\sqrt{2M}}; \qquad \mathbf{a}_2 = \frac{1}{\sqrt{2M}}(1, 1) \tag{8b}$$

Note that after the equations of motion have been used to establish the ratio of a_{11} to a_{21}, viz., $a_{11}/a_{21} = -1$, then the relationship between a_{12} and a_{22} does not have to be found by using the equations of motion again, but results from the orthonormality condition

$$M \sum_j a_{jr} a_{js} = 0, \qquad r \neq s$$

or,

$$a_{11}a_{12} + a_{21}a_{22} = 0 \tag{9}$$

and substituting $a_{11} = -a_{21}$, we have

$$a_{12} = a_{22} \tag{10}$$

The expression for the normal coordinates for the case $\dot{q}_j(0) = 0$ is [Eq. (13.79)]:

$$\eta_r = \cos \omega_r t \sum_{j,k} m_{jk} a_{kr} q_j(0) \tag{11}$$

Substituting $m_{jk} = M\delta_{jk}$ and summing over k, we have

$$\eta_r = M \cos \omega_r t \sum_j a_{jr} q_j(0) \tag{12}$$

or, expanding, we find the two equations

$$\left. \begin{aligned} \eta_1 &= M(a_{11}x_{10} + a_{21}x_{20}) \cos \omega_1 t \\ \eta_2 &= M(a_{12}x_{10} + a_{22}x_{20}) \cos \omega_2 t \end{aligned} \right\} \tag{13}$$

where x_{10} and x_{20} are written for $q_1(0)$ and $q_2(0)$. If we now substitute the values for the a_{jr}, we have

$$\left. \begin{aligned} \eta_1 &= \sqrt{\frac{M}{2}}(x_{10} - x_{20}) \cos \omega_1 t \\ \eta_2 &= \sqrt{\frac{M}{2}}(x_{10} - x_{20}) \cos \omega_2 t \end{aligned} \right\} \tag{14}$$

Therefore, we see that if we choose the initial conditions $x_{10} = -x_{20} \equiv x_0$ and $\dot{x}_{10} = \dot{x}_{20} \equiv 0$, then

$$\left. \begin{aligned} \eta_1 &= x_0\sqrt{2M} \cos \omega_1 t \\ \eta_2 &= 0 \end{aligned} \right\} \quad \text{Mode 1} \tag{15}$$

whereas if we choose $x_{10} = x_{20} \equiv x_0$ and $\dot{x}_{10} = \dot{x}_{20} \equiv 0$, then

$$\left. \begin{aligned} \eta_1 &= 0 \\ \eta_2 &= x_0\sqrt{2M} \cos \omega_2 t \end{aligned} \right\} \quad \text{Mode 2} \tag{16}$$

Thus, the normal coordinates, η_1 and η_2, may be identified with the two distinct modes of oscillation that are shown in Fig. 13-2. In *Mode* 1, the particles oscillate *out of phase* (the antisymmetrical mode), and in *Mode* 2, they oscillate *in phase* (the symmetrical mode). In *Mode* 2, the distance between the particles is always the same, and they oscillate as if the spring connecting them were a rigid, weightless rod; the velocities in this mode are identical. In *Mode* 1 the velocities are equal in magnitude but are of opposite sign.

In order to obtain expressions for the coordinates $x_k(t)$, we use Eq. (13.67):

$$x_k(t) = \sum_r a_{kr} \eta_r$$

For *Mode* 1, we have

$$x_1(t) = a_{11}\eta_1 + a_{12}\eta_2$$

$$= \frac{1}{\sqrt{2M}} \cdot x_0\sqrt{2M}\cos\omega_1 t + 0$$

$$= x_0 \cos\omega_1 t = -x_2(t) \tag{17a}$$

and the motions are out of phase. For *Mode* 2, we find

$$x_1(t) = x_0 \cos\omega_2 t = x_2(t) \tag{17b}$$

and the motions are in phase. This analysis confirms our previous results (Section 13.2).

13.9 Three Linearly Coupled Plane Pendula—An Example of Degeneracy

▶ Example 13.9

Consider three identical plane pendula which are suspended from a slightly yielding support. Since the support is not rigid, there is a coupling between the pendula, and energy can be transferred from one pendulum to the other. Figure 13-8 shows the geometry of the problem.

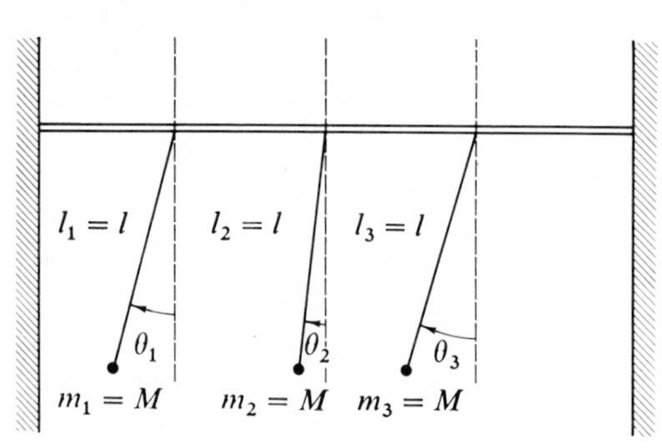

FIG. **13-8**

In order to simplify the notation, we adopt a system of units (sometimes called *natural units*) in which all lengths are measured in units of the length of the pendula l, all masses in units of the pendula masses M, and accelerations in units of g. Therefore, in our equations the values of the quantities M, l, and g are numerically equal to unity. If the coupling between each pair of the pendula is the same, we have

$$\left.\begin{aligned} T &= \tfrac{1}{2}(\dot\theta_1^2 + \dot\theta_2^2 + \dot\theta_3^2) \\ U &= \tfrac{1}{2}(\theta_1^2 + \theta_2^2 + \theta_3^2 - 2\varepsilon\theta_1\theta_2 - 2\varepsilon\theta_1\theta_3 - 2\varepsilon\theta_2\theta_3) \end{aligned}\right\} \tag{1}$$

Thus, the tensor $\{\mathsf{m}\}$ is diagonal,

$$\{\mathsf{m}\} = \begin{Bmatrix} 1 & 0 & 0 \\ 0 & 1 & 0 \\ 0 & 0 & 1 \end{Bmatrix} \tag{2}$$

but $\{\mathsf{A}\}$ has the form

$$\{\mathsf{A}\} = \begin{Bmatrix} 1 & -\varepsilon & -\varepsilon \\ -\varepsilon & 1 & -\varepsilon \\ -\varepsilon & -\varepsilon & 1 \end{Bmatrix} \tag{3}$$

The secular determinant is

$$\begin{vmatrix} 1-\omega^2 & -\varepsilon & -\varepsilon \\ -\varepsilon & 1-\omega^2 & -\varepsilon \\ -\varepsilon & -\varepsilon & 1-\omega^2 \end{vmatrix} = 0 \tag{4}$$

Upon expanding, we have

$$(1-\omega^2)^3 - 2\varepsilon^3 - 3\varepsilon^2(1-\omega^2) = 0$$

which can be factored to

$$(\omega^2 - 1 - \varepsilon)^2(\omega^2 - 1 + 2\varepsilon) = 0$$

and hence the roots are

$$\left.\begin{aligned} \omega_1 &= \sqrt{1+\varepsilon} \\ \omega_2 &= \sqrt{1+\varepsilon} \\ \omega_3 &= \sqrt{1-2\varepsilon} \end{aligned}\right\} \tag{5}$$

Notice that we have a *double root*: $\omega_1 = \omega_2 = \sqrt{1+\varepsilon}$. The normal modes corresponding to these frequencies are therefore *degenerate*; i.e., these two modes are indistinguishable.

We now evaluate the quantities a_{jr}, beginning with a_{j3}. Again we note that since the equations of motion determine only the ratios, we need consider only two of the three available equations; the third equation will automatically be satisfied. Using the equation

$$\sum_j (A_{jk} - \omega_3^2 m_{jk})a_{j3} = 0$$

we find

$$\left.\begin{aligned} 2\varepsilon a_{13} - \varepsilon a_{23} - \varepsilon a_{33} &= 0 \\ -\varepsilon a_{13} + 2\varepsilon a_{23} - \varepsilon a_{33} &= 0 \end{aligned}\right\} \tag{6}$$

Equations (6) yield

$$a_{13} = a_{23} = a_{33} \tag{7}$$

and from the normalization condition we have

$$a_{13}^2 + a_{23}^2 + a_{33}^2 = 1$$

or,

$$a_{13} = a_{23} = a_{33} = \frac{1}{\sqrt{3}} \tag{8}$$

Thus, we find that for $r = 3$ there is no problem in the evaluation of the components of the eigenvector $\mathbf{a}_3$. (This is a general rule: there is no indefiniteness in evaluating the eigenvector components for a nondegenerate mode.) Since all the components of $\mathbf{a}_3$ are equal, this corresponds to the mode in which all three pendula oscillate in phase.

Let us now attempt to evaluate the a_{j1} and a_{j2}. From the six possible equations of motion (3 values of j and 2 values of r), we obtain only two different relations:

$$\varepsilon(a_{11} + a_{21} + a_{31}) = 0 \qquad (*) \tag{9}$$

$$\varepsilon(a_{12} + a_{22} + a_{32}) = 0 \qquad (*) \tag{10}$$

Now, the orthogonality equation is

$$\sum_{j,\,k} m_{jk}\, a_{jr}\, a_{ks} = 0, \qquad r \neq s$$

but since $m_{jk} = \delta_{jk}$, this becomes

$$\sum_{j} a_{jr}\, a_{js} = 0, \qquad r \neq s \tag{11}$$

which leads to only one new equation:

$$a_{11}a_{12} + a_{21}a_{22} + a_{31}a_{32} = 0 \qquad (*) \tag{12}$$

(The other two possible equations are identical with Eqs. (9) and (10) above.) Finally, the normalization conditions yield

$$a_{11}^2 + a_{21}^2 + a_{31}^2 = 1 \qquad (*) \tag{13}$$

$$a_{12}^2 + a_{22}^2 + a_{32}^2 = 1 \qquad (*) \tag{14}$$

Thus, we have a total of only *five* (starred, *) equations for the six unknowns a_{j1} and a_{j2}. This indeterminacy in the eigenvectors corresponding to a double root is exactly the same as that encountered in constructing the principal axes for a rigid body with an axis of symmetry: the two equivalent principal axes may be placed in any direction, as long as the set of three axes is orthogonal. Therefore, we are at liberty to arbitrarily specify the eigenvectors $\mathbf{a}_1$ and $\mathbf{a}_2$, as long as the orthogonality and normalizing relations are satisfied. For a simple system such as we are discussing it is not difficult to construct these vectors, and we shall not give any general rules here.

If we arbitrarily choose $a_{31} = 0$, the indeterminancy is removed. We then find

$$\mathbf{a}_1 = \frac{1}{\sqrt{2}}(1,\ -1,\ 0); \qquad \mathbf{a}_2 = \frac{1}{\sqrt{6}}(1,\ 1,\ -2) \tag{15}$$

from which it is easily verified that the starred relations above are all satisfied.

It will be recalled that the nondegenerate mode corresponded to the in-phase oscillation of all three pendula:

$$\mathbf{a}_3 = \frac{1}{\sqrt{3}}(1, 1, 1) \tag{16}$$

We now see that the degenerate modes each correspond to out-of-phase oscillation. For example, $\mathbf{a}_2$ in Eq. (15) represents two pendula oscillating together with a certain amplitude while the third is out of phase and has twice the amplitude. Similarly, $\mathbf{a}_1$ in Eq. (15) represents one pendulum stationary and the other two in out-of-phase oscillation. It should be noted that the eigenvectors $\mathbf{a}_1$ and $\mathbf{a}_2$ given above are only one set of any infinity of sets which satisfy the conditions of the problem. But all such eigenvectors represent some sort of out-of-phase oscillation. (Further details of this example are examined in Problems 13-16 and 13-17.)

13.10 The Loaded String*

We now consider a more complex system which consists of an elastic string (or a spring) on which are placed a number of identical particles at regular intervals. The ends of the string are constrained to remain stationary. Let the mass of each of the n particles be m, and let the spacing between particles at equilibrium be d. Thus, the length of the string is $L = (n + 1)d$. The equilibrium situation is shown in Fig. 13-9.

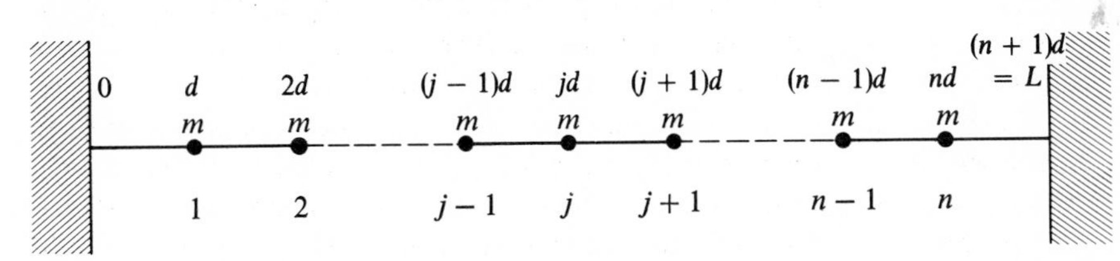

FIG. 13-9

We wish to treat the case of small transverse oscillations of the particles about their equilibrium positions. First, consider the vertical displacements of the masses numbered $j - 1$, j, and $j + 1$, as in Fig. 13-10. If the vertical displacements, q_{j-1}, q_j, and q_{j+1} are small, then the tension τ in the string is approximately constant and equal to its value at equilibrium. For small

* The first attack on the problem of the loaded string (or *one-dimensional lattice*) was by Newton (*Principia*, 1687). The work was continued by Johann Bernoulli, and his son Daniel, starting in 1727 and culminating in the latter's formulation of the principle of superposition in 1753. It is from this point that the theoretical treatment of the physics of *systems* (as distinct from *particles*) begins.

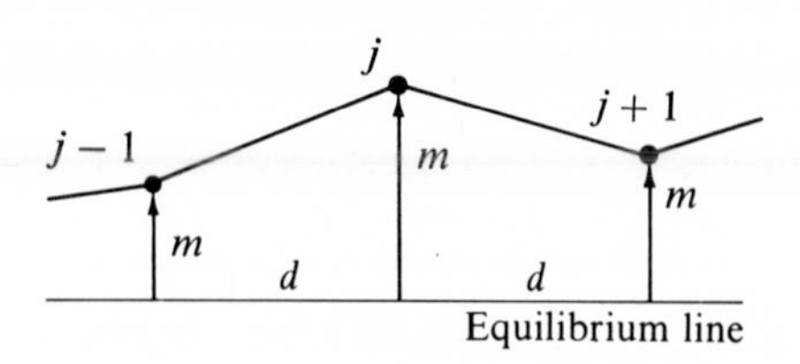

FIG. **13-10**

displacements the string section between any pair of particles will make only small angles with the equilibrium line. Approximating the sines of these angles by the tangents, the expression for the force that tends to restore the jth particle to its equilibrium position is

$$F_j = -\frac{\tau}{d}(q_j - q_{j-1}) - \frac{\tau}{d}(q_j - q_{j+1}) \tag{13.80}$$

The force F_j is, according to Newton's law, equal to $m\ddot{q}_j$; therefore, Eq. (13.80) can be written as

$$\ddot{q}_j = \frac{\tau}{md}(q_{j-1} - 2q_j + q_{j+1}) \tag{13.81}$$

which is the equation of motion for the jth particle. The system is clearly coupled, since the force on the jth particle depends on the positions of the $(j-1)$th and $(j+1)$th particles; this is therefore an example of *nearest neighbor* interaction, in which the coupling is only to the adjacent particles. It is, of course, not necessary that the interaction be confined to nearest neighbors. If the force between pairs of particles were electrostatic, for example, then each particle would be coupled to *all* of the other particles. The problem can then become quite complicated. However, even if the force is electrostatic, the $1/r^2$ dependence on distance frequently permits the neglect of interactions at distances greater than one interparticle spacing so that the simple expression for the force given in Eq. (13.80) is approximately correct.

We have considered above only the motion perpendicular to the line of the string; the oscillations are therefore *transverse*. It is easy to show that exactly the same form for the equations of motion will result if we consider longitudinal vibrations, i.e., motions along the line of the string. In this case, the factor τ/d is replaced by κ, the force constant of the string. (See Problem 13-20.)

Although Newton's equation was used to obtain the equations of motion (13.81), we may equally well use the Lagrangian method. The potential energy arises from the work done to stretch the $n+1$ string segments*:

$$U = \frac{1}{2}\frac{\tau}{d}\sum_{j=1}^{n+1}(q_{j-1} - q_j)^2 \tag{13.82}$$

where q_0 and q_{n+1} are identically zero since these positions correspond to the fixed ends of the string. We note that Eq. (13.82) yields an expression for the force on the jth particle which is the same as the previous result [Eq. (13.81)]:

$$\begin{aligned} F_j &= -\frac{\partial U}{\partial q_j} = -\frac{1}{2}\frac{\tau}{d}\frac{\partial}{\partial q_j}[(q_{j-1} - q_j)^2 + (q_j - q_{j+1})^2] \\ &= \frac{\tau}{d}(q_{j-1} - 2q_j + q_{j+1}) \end{aligned} \tag{13.83}$$

The kinetic energy is given by the sum of the kinetic energies of the n individual particles:

$$T = \tfrac{1}{2}m\sum_{j=1}^{n}\dot{q}_j^2 \tag{13.84}$$

Since $\dot{q}_{n+1} \equiv 0$, we may extend the sum in Eq. (13.84) to $j = n+1$ so that the range of j is the same as that in the expression for the potential energy. Therefore, the Lagrangian becomes

$$L = \frac{1}{2}\sum_{j=1}^{n+1}\left[m\dot{q}_j^2 - \frac{\tau}{d}(q_{j-1} - q_j)^2\right] \tag{13.85}$$

Now, the equation of motion for the jth particle obviously must arise from only those terms in the Lagrangian which contain q_j or $\dot{q}_j$. If we expand the sum in L, we find

$$L = \cdots + \frac{1}{2}m\dot{q}_j^2 - \frac{1}{2}\frac{\tau}{d}(q_{j-1} - q_j)^2 - \frac{1}{2}\frac{\tau}{d}(q_j - q_{j+1})^2 - \cdots \tag{13.86}$$

where we have written only those terms which contain either q_j or $\dot{q}_j$. Applying Lagrange's equation for the coordinate q_j, we have

$$m\ddot{q}_j - \frac{\tau}{d}(q_{j-1} - 2q_j + q_{j+1}) = 0 \tag{13.87}$$

Thus, the result is the same as that obtained by using the Newtonian method.

In order to obtain a solution to the equations of motion, we substitute, as usual,

$$q_j(t) = a_j e^{i\omega t} \tag{13.88}$$

* We consider the potential energy to be only the elastic energy in the string; i.e., we do not consider the individual masses to have any gravitational (or any other) potential energy.

where a_j can be *complex*. Substituting this expression for $q_j(t)$ into Eq. (13.87) we find

$$-\frac{\tau}{d}a_{j-1} + \left(2\frac{\tau}{d} - m\omega^2\right)a_j - \frac{\tau}{d}a_{j+1} = 0 \tag{13.89}$$

where $j = 1, 2, \ldots, n$, but since the ends of the string are fixed, we must have $a_0 = a_{n+1} = 0$.

Equation (13.89) represents a *linear difference equation* which can be solved for the eigenfrequencies ω_r by setting the determinant of the coefficients equal to zero. We therefore have the following secular determinant:

$$\begin{vmatrix} \lambda & -\frac{\tau}{d} & 0 & 0 & 0 & \cdots \\ -\frac{\tau}{d} & \lambda & -\frac{\tau}{d} & 0 & 0 & \cdots \\ 0 & -\frac{\tau}{d} & \lambda & -\frac{\tau}{d} & 0 & \cdots \\ 0 & 0 & -\frac{\tau}{d} & \lambda & -\frac{\tau}{d} & \cdots \\ 0 & 0 & 0 & \cdot & \cdot & \\ \vdots & \vdots & \vdots & \vdots & \vdots & \end{vmatrix} = 0 \tag{13.90}$$

where we have used

$$\lambda \equiv 2\frac{\tau}{d} - m\omega^2 \tag{13.91}$$

This secular determinant is a special case of the general determinant [Eq. (13.54a)] which results in the event that the tensor $\{\mathsf{m}\}$ is diagonal and the tensor $\{\mathsf{A}\}$ involves a coupling only between adjacent particles. Thus, Eq. (13.90) consists only of diagonal elements plus elements once-removed from the diagonal.

For the case $n = 1$ (i.e., a single mass suspended between two identical springs), we have $\lambda = 0$, or

$$\omega = \sqrt{\frac{2\tau}{md}}$$

We may adapt this result to the case of longitudinal motion by replacing τ/d by κ; we then obtain the familiar expression,

$$\omega = \sqrt{\frac{2\kappa}{m}}$$

For the case $n = 2$, and with τ/d replaced by κ, we have $\lambda^2 = \kappa^2$, or

$$\omega = \sqrt{\frac{2\kappa \pm \kappa}{m}}$$

which are the same frequencies as those found in Section 13.2 for two coupled masses [Eq. (13.8)].

The secular equation is relatively easy to solve directly for small values of n, but obviously the solution becomes quite complicated for large n. In such cases it is simpler to use the following method. We try a solution of the form

$$a_j = ae^{i(j\gamma - \delta)} \tag{13.92}$$

where a is *real*. The use of this device will be justified if it is possible to find a quantity γ and a phase δ such that the conditions of the problem are all satisfied. Substituting a_j in this form into Eq. (13.89) and canceling the phase factor, we find

$$-\frac{\tau}{d}e^{-i\gamma} + \left(2\frac{\tau}{d} - m\omega^2\right) - \frac{\tau}{d}e^{i\gamma} = 0$$

Solving for ω^2, we obtain

$$\left.\begin{aligned} \omega^2 &= \frac{2\tau}{md} - \frac{\tau}{md}(e^{i\gamma} + e^{-i\gamma}) \\ &= \frac{2\tau}{md}(1 - \cos\gamma) \\ &= \frac{4\tau}{md}\sin^2\frac{\gamma}{2} \end{aligned}\right\} \tag{13.93}$$

Since we know that the secular determinant is of order n and will therefore yield exactly n values for ω^2, we can write

$$\omega_r = 2\sqrt{\frac{\tau}{md}}\sin\frac{\gamma_r}{2}, \qquad r = 1, 2, \ldots, n \tag{13.94}$$

We now evaluate the quantity γ_r and the phase δ_r by applying the boundary condition that the ends of the string remain fixed. Thus, we have

$$a_{jr} = a_r e^{i(j\gamma_r - \delta_r)} \tag{13.95}$$

or, since it is only the real part that is physically meaningful,

$$a_{jr} = a_r \cos(j\gamma_r - \delta_r) \tag{13.95a}$$

The boundary condition is

$$a_{0r} = a_{(n+1)r} \equiv 0 \tag{13.96}$$

In order that Eq. (13.95a) yield $a_{jr} = 0$ for $j = 0$, it is clear that δ_r must be $\pi/2$ (or some odd integer multiple thereof). Hence,

$$\begin{aligned} a_{jr} &= a_r \cos\left(j\gamma_r - \frac{\pi}{2}\right) \\ &= a_r \sin j\gamma_r \end{aligned} \tag{13.97}$$

For $j = n + 1$, we have

$$a_{(n+1)r} = 0 = a_r \sin(n + 1)\gamma_r$$

Therefore,

$$(n + 1)\gamma_r = s\pi, \qquad s = 1, 2, \ldots$$

or,

$$\gamma_r = \frac{s\pi}{n + 1}, \qquad s = 1, 2, \ldots$$

But there are just n distinct values of γ_r since Eq. (13.94) requires n distinct values of ω_r. Therefore, the index s runs from 1 to n. Because there is a one-to-one correspondence between the values of s and the values of r, we can simply replace s in this last expression by the index r:

$$\gamma_r = \frac{r\pi}{n + 1}, \qquad r = 1, 2, \ldots, n \tag{13.98}$$

The a_{jr} then become

$$\boxed{a_{jr} = a_r \sin\left(j \frac{r\pi}{n + 1}\right)} \tag{13.99}$$

The general solution for q_j is [see Eq. (13.65)]:

$$\begin{aligned} q_j &= \sum_r \beta_r' a_{jr} e^{i\omega_r t} \\ &= \sum_r \beta_r' a_r \sin\left(j \frac{r\pi}{n + 1}\right) e^{i\omega_r t} \\ &= \sum_r \beta_r \sin\left(j \frac{r\pi}{n + 1}\right) e^{i\omega_r t} \end{aligned} \tag{13.100}$$

where we have written $\beta_r \equiv \beta_r' a_r$. Furthermore, for the frequency we have

$$\boxed{\omega_r = 2\sqrt{\frac{\tau}{md}} \sin\left(\frac{r\pi}{2(n + 1)}\right)} \tag{13.101}$$

We note that this expression yields the same results previously found for the case of two coupled oscillators [Eqs. (13.8)] when we insert $n = 2$, $r = 1, 2$, and replace τ/d by κ $(= \kappa_{12})$.

Notice also that if either $r = 0$ or $r = n + 1$ is substituted into Eq. (13.99), then all of the amplitude factors a_{jr} vanish identically. These values of r therefore refer to *null modes*. Moreover, if r takes on the values $n + 2$, $n + 3, \ldots, 2n + 1$, then the a_{jr} are the same (except for a trivial sign change and in reverse order) as for $r = 1, 2, \ldots, n$; also, $r = 2n + 2$ yields the next null mode. We conclude, therefore, that there are indeed only n distinct modes and that increasing r beyond n merely duplicates the modes for smaller n. (A similar argument applies for $r < 0$.) These conclusions are illustrated in Fig. 13-11 for the case $n = 3$. The distinct modes are specified by $r = 1, 2, 3$; $r = 4$ is a null mode. The displacement patterns are duplicated for $r =$ 7, 6, 5, 8, but with a change in sign. In Fig. 13.11 the dashed curves merely represent the sinusoidal behavior of the amplitude factors a_{rj} for various values of r; the only physically meaningful features of these curves are the values at the positions occupied by the particles ($j = 1, 2, 3$). Therefore, the "high frequency" of the sine curves for $r = 5, 6, 7, 8$ is not at all related to the frequency of the particles' motions; these latter frequencies are the same as for $r = 1, 2, 3, 4$.

The normal coordinates of the system are defined as

$$\eta_r(t) \equiv \beta_r e^{i\omega_r t} \tag{13.102}$$

so that

$$q_j(t) = \sum_r \eta_r \sin\left(j \frac{r\pi}{n+1}\right) \tag{13.103}$$

This equation for q_j is similar to the previous expression [Eq. (13.67)] except that the quantities a_{jr} are now replaced by $\sin[j(r\pi)/(n + 1)]$. Now, it was possible in the development in Section 13.5 to evaluate the coefficients β_r only because there existed an orthogonality relation among the a_{jr}. A similar relationship in the form of a trigonometric identity is available for the sine terms:

$$\sum_{j=1}^{n} \sin\left(j \frac{r\pi}{n+1}\right) \sin\left(j \frac{s\pi}{n+1}\right) = \frac{n+1}{2} \delta_{rs}; \qquad r, s = 1, 2, \ldots, n \tag{13.104}$$

In analogy with Eq. (13.73), we write for the real part of q_j,

$$q_j(t) = \sum_r \sin\left(j \frac{r\pi}{n+1}\right)(\mu_r \cos \omega_r t - \nu_r \sin \omega_r t) \tag{13.105}$$

where

$$\beta_r = \mu_r + i\nu_r \tag{13.106}$$

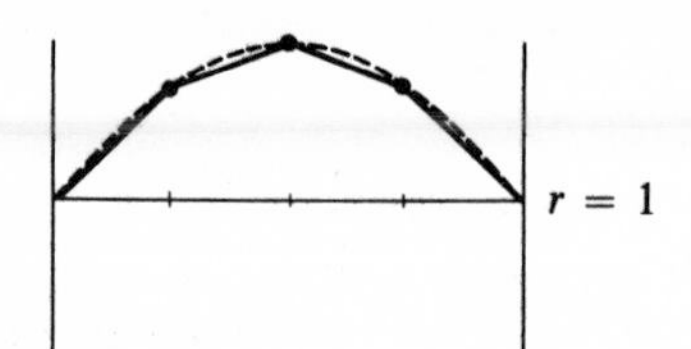

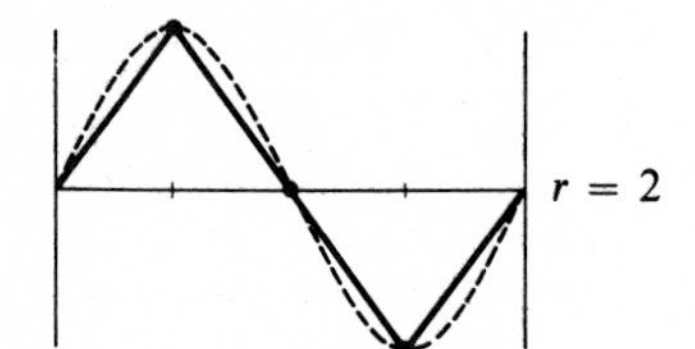

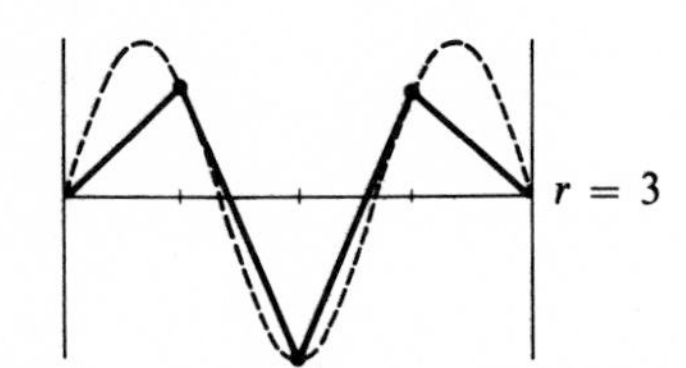

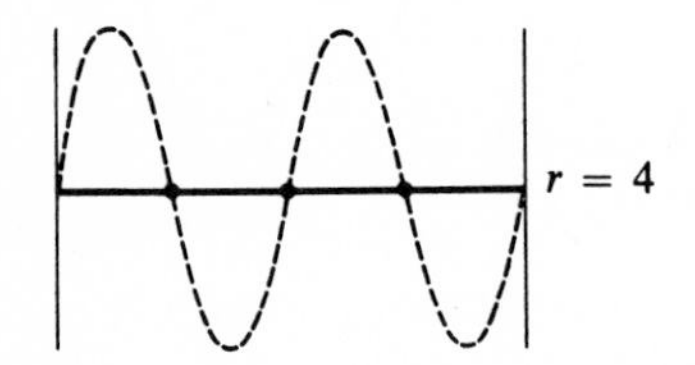

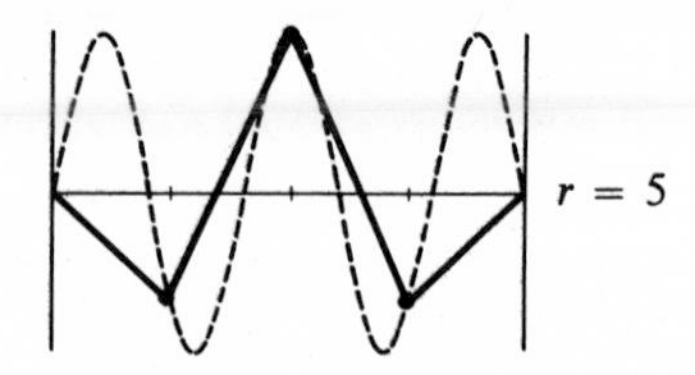

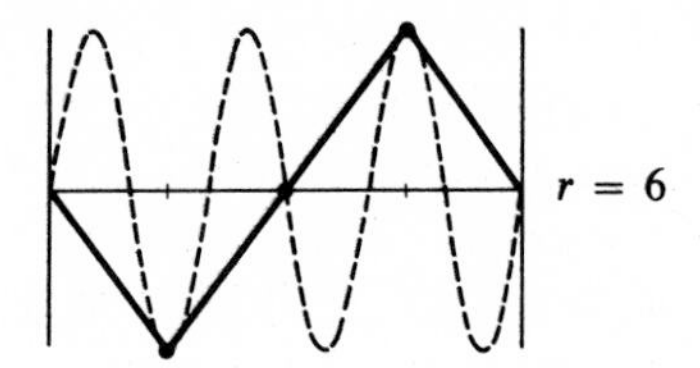

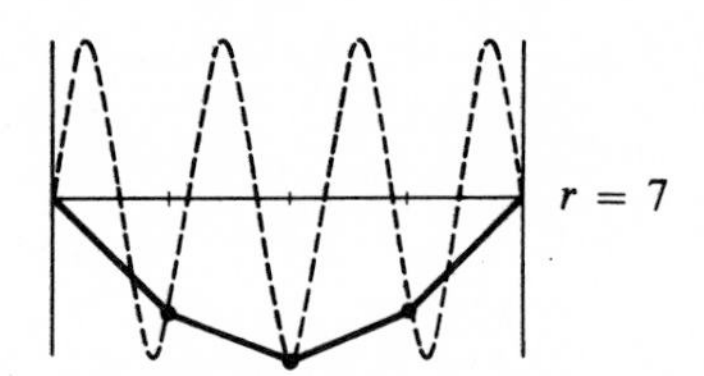

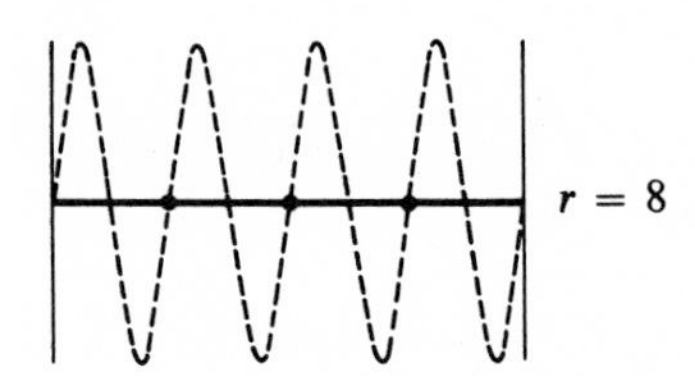

FIG. **13-11**

As before, we have

$$q_j(0) = \sum_r \mu_r \sin\left(j \frac{r\pi}{n+1}\right) \tag{13.107a}$$

$$\dot{q}_j(0) = -\sum_r \omega_r \nu_r \sin\left(j \frac{r\pi}{n+1}\right) \tag{13.107b}$$

If we multiply Eq. (13.107a) by $\sin[j(s\pi)/n + 1)]$ and sum over j, we find

$$\sum_j q_j(0) \sin\left(j \frac{s\pi}{n+1}\right) = \sum_{j,r} \mu_r \sin\left(j \frac{r\pi}{n+1}\right) \sin\left(j \frac{s\pi}{n+1}\right)$$

$$= \sum_r \mu_r \cdot \frac{n+1}{2} \delta_{rs}$$

$$= \frac{n+1}{2} \mu_s$$

or,

$$\mu_s = \frac{2}{n+1} \sum_j q_j(0) \sin\left(j \frac{s\pi}{n+1}\right) \tag{13.108a}$$

A similar procedure for ν_s yields

$$\nu_s = -\frac{2}{\omega_s(n+1)} \sum_j \dot{q}_j(0) \sin\left(j \frac{s\pi}{n+1}\right) \tag{13.108b}$$

Thus, we have evaluated all of the necessary quantities, and the description of the vibrations of a loaded string is therefore complete.

We should note the following point regarding the normalization procedures that have been used. First, in Eq. (13.62) we arbitrarily normalized the a_{jr} to unity. Thus, the a_{jr} are *required* to be independent of the initial conditions imposed upon the system. The scale factors α and β_r then allowed the magnitude of the oscillations to be varied by the selection of the initial conditions. Next, in the problem of the loaded string we found that instead of the quantities a_{jr}, there arose the sine functions, $\sin[j(r\pi)/(n+1)]$, and these functions possess a normalization property [Eq. (13.104)] that is specified by trigonometric identities. Therefore, in this case it is not possible to arbitrarily impose a normalization condition; we are automatically presented with the condition. But this is no restriction; it only means that the scale factors β_r for this case have a slightly different form. Thus, there are certain constants that occur in the two problems which, for convenience, are separated in different ways in the two cases.

▶ Example 13.10 An Initial Condition Problem

Consider a loaded string consisting of three particles regularly spaced on the string. At $t = 0$ the center particle (only) is displaced a distance a and released from rest. In order to calculate the subsequent motion we proceed as follows. First, the initial conditions are

$$\left.\begin{array}{l} q_2(0) = a, \qquad q_1(0) = q_3(0) = 0 \\ \dot{q}_1(0) = \dot{q}_2(0) = \dot{q}_3(0) = 0 \end{array}\right\} \tag{1}$$

Since the initial velocites are zero, the ν_r vanish. The μ_r are given by [Eq. (13.108a)]

$$\begin{aligned} \mu_r &= \frac{2}{n+1}\sum_j q_j(0)\sin\left(j\frac{r\pi}{n+1}\right) \\ &= \frac{1}{2}a\sin\left(\frac{r\pi}{2}\right) \end{aligned} \tag{2}$$

because only the term $j = 2$ contributes to the sum. Thus,

$$\mu_1 = \tfrac{1}{2}a; \qquad \mu_2 = 0; \qquad \mu_3 = -\tfrac{1}{2}a \tag{3}$$

The quantities $\sin[j(r\pi)/(n+1)]$ that appear in the expression for $q_j(t)$ [Eq. (13.105)] are

$j =$ \ $r =$	1	2	3
1	$\frac{\sqrt{2}}{2}$	1	$\frac{\sqrt{2}}{2}$
2	1	0	-1
3	$\frac{\sqrt{2}}{2}$	-1	$-\frac{\sqrt{2}}{2}$

(4)

The displacements of the three particles therefore are

$$\left.\begin{aligned} q_1(t) &= \frac{\sqrt{2}}{4}a(\cos\omega_1 t - \cos\omega_3 t) \\ q_2(t) &= \tfrac{1}{2}a(\cos\omega_1 t - \cos\omega_3 t) \\ q_3(t) &= \frac{\sqrt{2}}{4}a(\cos\omega_1 t - \cos\omega_3 t) = q_1(t) \end{aligned}\right\} \tag{5}$$

where the characteristic frequencies are given by Eq. (13.100):

$$\omega_r = 2\sqrt{\frac{\tau}{md}}\sin\left(\frac{r\pi}{8}\right), \qquad r = 1, 2, 3 \tag{6}$$

Notice that because the *middle* particle was initially displaced, no vibration mode occurs in which this particle is at rest; that is, Mode 2 with frequency ω_2 (see Fig. 13-11) is absent.

Suggested References

At the intermediate level, discussions of coupled oscillations and normal modes are given by Becker (Be54, Chapter 14), Houston (Ho48, Chapter 7), Magnus (Ma59c, Chapter 6), Slater and Frank (Sl47, Chapter 7), and Wangsness (Wa63, Chapter 12); Morse's book (Mo48) is thorough and excellent. An introductory account is given by Feather (Fe61, Chapter 2).

For more advanced treatments of the general theory of vibrating systems, see Goldstein (Go50, Chapter 10), Hauser (Ha65, Chapter 11), Konopiński (Ko69, Chapter 11), and Symon (Sy60, Chapter 8).

The classic treatise by Lord Rayleigh (Ra94) describes the theory of vibrations in great detail; see particularly Chapters 4 and 5.

Problems

13-1. Reconsider the problem of two coupled oscillators discussed in Section 13.2 in the event that the three springs all have different force constants. Find the two characteristic frequencies and compare the magnitudes with the natural frequencies of the two oscillators in the absence of coupling.

13-2. Continue the above problem and investigate the case of weak coupling: $\kappa_{12} \ll \kappa_1, \kappa_2$. Show that the phenomenon of beats occurs, but that the energy transfer process is incomplete.

13-3. Two identical harmonic oscillators (with masses M and natural frequencies ω_0) are coupled by adding to the system a mass m that is common to both oscillators. The equations of motion are then

$$\ddot{x}_1 + (m/M)\ddot{x}_2 + \omega_0^2 x_1 = 0$$

$$\ddot{x}_2 + (m/M)\ddot{x}_1 + \omega_0^2 x_2 = 0$$

Solve this pair of coupled equations and obtain the frequencies of the normal modes of the system.

13-4. Refer to the problem of the two coupled oscillators discussed in Section 13.2. Show that the total energy of the system is constant. (Calculate the kinetic energy of each of the particles and the potential energy stored in each of the three springs and sum the results.) Notice that the kinetic and potential energy terms that have κ_{12} as a coefficient depend on C_1 and ω_1 but not on C_2 or ω_2. Why is such a result to be expected?

13-5. Find the normal coordinates for the problem discussed in Section 13.2 and in Example 13.8 in the event that the two masses are different, $m_1 \neq m_2$.

13-6. Two identical harmonic oscillators are placed so that the two masses slide against one another, as in the figure. The frictional force provides a

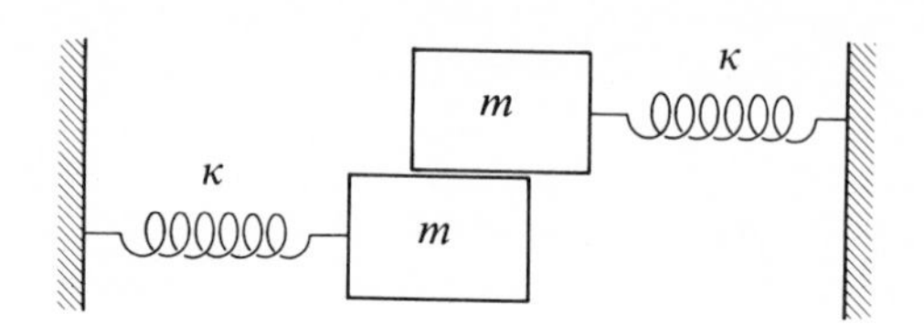

coupling of the motions that is proportional to the instantaneous relative velocity. Discuss the coupled oscillations of the system.

13-7. A particle of mass m is attached to a rigid support by means of a spring with force constant κ. At equilibrium the spring hangs vertically downward. To this mass-spring combination is attached an identical oscillator, the spring of the latter being connected to the mass of the former. Calculate the characteristic frequencies for one-dimensional, vertical oscillations and compare with the frequencies when one or the other of the particles is held fixed while the other oscillates. Describe the normal modes of motion for the system.

13-8. A simple pendulum consists of a bob of mass m suspended by an inextensible (and massless) string of length l. From the bob of this pendulum is suspended a second, identical pendulum. Consider the case of small oscillations (so that $\sin \theta \cong \theta$) and calculate the characteristic frequencies. Also, describe the normal modes of the system. (Refer to Problem 7-7.)

13-9. The motion of a pair of coupled oscillators may be described by using a method similar to that employed in constructing a phase diagram for a single oscillator (Section 3.3). For coupled oscillators, the two positions, $x_1(t)$ and $x_2(t)$, may be represented by a point (the *system point*) in the two-dimensional *configuration space* x_1-x_2. As t increases, the locus of all such points defines a certain curve. The projections of the system point onto the x_1- and x_2-axes represent the motions of m_1 and m_2, respectively. In the general case, $x_1(t)$ and $x_2(t)$ are complicated functions and so the curve is also complicated. However, it is always possible to rotate the x_1-x_2 axes to a new set, x_1'-x_2', in such a way that the projection of the system point onto each of the new axes is *simple harmonic*. The projected motions along the new axes take place with the characteristic frequencies and correspond to the normal modes of the system. The new axes are called *normal axes*. Find the normal axes for the problem discussed in Section 13.2 and verify the above statements regarding the motion relative to this coordinate system.

13-10. Consider two identical, coupled oscillators (as in Fig. 13-1). Let each of the oscillators be damped and let each have the same damping parameter β. A force $F_0 \cos \omega t$ is applied to m_1. Write down the pair of coupled differential equations that describe the motion. Obtain the solution by expressing the differential equations in terms of the normal coordinates given by Eqs. (13.12) and by comparing these equations with Eqs. (4.3). Show that the normal coordinates η_1 and η_2 exhibit resonance peaks at the characteristic frequencies ω_1 and ω_2, respectively.

13-11. Show that Eqs. (13.34) can be put into the same form as Eqs. (13.1) by solving the second of Eqs. (13.34) for $\ddot{I}_2$ and substituting the result into the first equation. Similarly, substitute for $\ddot{I}_1$ in the second equation. The characteristic frequencies may then be written down immediately in analogy with Eqs. (13.8).

13-12. Find the characteristic frequencies of the coupled circuits of Fig. 13-7a.

13-13. Discuss the normal modes of the system shown in Fig. 13-7b.

13-14. In Fig. 13-7a, replace L_{12} by a resistor and analyze the oscillations.

13-15. A thin hoop of radius R and mass M is allowed to oscillate in its own plane with one point of the hoop fixed. Attached to the hoop is a small mass M which is constrained to move (in a frictionless manner) along the hoop. Consider only small oscillations and show that the eigenfrequencies are

$$\omega_1 = \sqrt{2}\sqrt{\frac{g}{R}}, \qquad \omega_2 = \frac{\sqrt{2}}{2}\sqrt{\frac{g}{R}}$$

Find the two sets of initial conditions which allow the system to oscillate in its normal modes. Describe the physical situation for each mode.

13-16. In the problem of the three coupled pendula, consider the case in which the three coupling constants are distinct, so that the potential energy may be written as

$$U = \tfrac{1}{2}(\theta_1^2 + \theta_2^2 + \theta_3^2 - 2\varepsilon_{12}\theta_1\theta_2 - 2\varepsilon_{13}\theta_1\theta_3 - 2\varepsilon_{23}\theta_2\theta_3)$$

with ε_{12}, ε_{12}, ε_{22} all different. Show that there is no degeneracy in such a system. Show also that degeneracy can occur *only* if $\varepsilon_{12} = \varepsilon_{13} = \varepsilon_{23}$.

13-17. Construct two more of the possible eigenvectors for the degenerate modes in the case of the three coupled pendula by requiring the following conditions: (a) $a_{11} = 2a_{21}$; (b) $a_{31} = -3a_{21}$. Interpret these situations physically.

13-18. Three oscillators of equal mass m are coupled in such a manner that the potential energy of the system is given by

$$U = \tfrac{1}{2}[\kappa_1(x_1^2 + x_3^2) + \kappa_2 x_2^2 + \kappa_3(x_1 x_2 + x_2 x_3)]$$

where $\kappa_2 = \sqrt{2\kappa_1\kappa_2}$. Find the eigenfrequencies by solving the secular equation. What is the physical interpretation of the null mode?

13-19. Consider a thin homogeneous plate of mass M which lies in the x_1-x_2 plane with its center at the origin. Let the length of the plate be $2A$ (in the x_2-direction) and let the width be $2B$ (in the x_1-direction). The plate is suspended from a fixed support by four springs of equal force constant κ located at the four corners of the plate. The plate is free to oscillate, but with the constraint that its center must remain on the x_3-axis. Thus, there are there degrees of freedom: (1) vertical motion, with the center of the plate moving along the x_3-axis; (2) a tipping motion lengthwise, with the x_1-axis serving as an axis of rotation (choose an angle θ to describe this motion); and (3) a tipping motion sidewise, with the x_2-axis serving as an axis of rotation (choose an angle φ to describe this motion). Assume only small oscillations and show that the secular equation has a double root and, hence, that the system is degenerate. Discuss the normal modes of the system. (In evaluating the a_{jk} for the degenerate modes, arbitrarily set one of the a_{jk} equal to zero in order to remove the indeterminacy.) Show that the degeneracy can be removed by the addition to the plate of a thin bar of mass m and length $2A$ which is situated (at equilibrium) along the x_2-axis. Find the new eigenfrequencies for the system.

13-20. Evaluate the total energy associated with a normal mode and show that it is constant in time. Show this explicitly for the case of Example 13.8.

13-21. Show that the equations of motion for *longitudinal* vibrations of a loaded string are of exactly the same form as the equations for transverse motion [Eq. (13.81)], except that the factor τ/d must be replaced by κ, the force constant of the string.

13-22. Rework the problem in Example 13.10 in the event that all three particles are displaced a distance a and released from rest.

CHAPTER 14

Vibrating Strings

14.1 Introduction

In this chapter we will extend the discussion of the vibrations of a loaded string that was presented in Chapter 13 by examining the consequences of allowing the number of particles on the string to become infinite (while maintaining a constant linear mass density). In this way we pass to the case of a *continuous string*. We will find that all of the results of interest for such a string can be obtained by this limiting process—including the derivation of the important *wave equation*.

In these discussions we reach the various conclusions by explicitly solving an equation of motion. One of the most useful of these results, viz., the frequency of the fundamental oscillation mode, can be obtained with surprisingly good accuracy from energy considerations alone, without the necessity of solving the equation of motion. This technique, known as *Rayleigh's method*, is of considerable use in situations where the equation of motion is difficult (or impossible) to solve exactly. The method is illustrated with several examples concerning vibrating strings where the accuracy can be checked against the exact solution.

We conclude this chapter by considering the case of the nonuniform string. The equation which describes the vibrations of such a string (the *Sturm–Liouville* equation) is solved by a perturbation technique similar to that encountered previously.

14.2 The Continuous String as a Limiting Case of the Loaded String

In the preceding chapter we considered a set of equally spaced point masses suspended by a string. We now wish to allow the number of masses to become infinite so that we have a continuous string. In order to do this, we must require that as $n \to \infty$ we simultaneously let the mass of each particle and the distance between each particle approach zero ($m \to 0$, $d \to 0$) in such a manner that the ratio m/d remains constant. We note that $m/d \equiv \rho$ is just the linear mass density of the string. Thus, we have

$$\left.\begin{array}{lll} n \to \infty, \quad d \to 0 & \text{such that} & (n+1)\,d = L \\ m \to 0, \quad d \to 0 & \text{such that} & \dfrac{m}{d} = \rho = \text{const.} \end{array}\right\} \tag{14.1}$$

From Eq. (13.103) we have

$$q_j(t) = \sum_r \eta_r(t) \sin\left(j\,\frac{r\pi}{n+1}\right) \tag{14.2}$$

We can now write

$$j\,\frac{r\pi}{n+1} = r\pi\,\frac{jd}{(n+1)d} = r\pi\,\frac{x}{L} \tag{14.3}$$

where $jd = x$ now specifies the distance along the continuous string. Thus, $q_j(t)$ becomes a continuous function of the variables x and t:

$$q(x, t) = \sum_r \eta_r(t) \sin\frac{r\pi x}{L} \tag{14.4}$$

or,

$$q(x, t) = \sum_r \beta_r e^{i\omega_r t} \sin\frac{r\pi x}{L} \tag{14.5}$$

In the case of the loaded string containing n particles there are n degrees of freedom of motion and therefore n normal modes and n characteristic frequencies. Thus, in Eq. (13.103) [or Eq. (14.2)] the sum is over the range $r = 1$ to $r = n$. But now the number of particles is infinite so that there is an infinite set of normal modes and the sum in Eqs. (14.4) and (14.5) runs from $r = 1$ to $r = \infty$. That is, there are infinitely many constants (the real and imaginary parts of the β_r) that must be evaluated in order to completely specify the motion of the continuous string. This is exactly the situation encountered when it is required to represent some function as a Fourier series—the infinitely many constants are specified by certain integrals involving the original function [see Eqs. (4.62)]. We may view the situation in another way: There are infinitely many arbitrary constants in the solution of the equation of motion but there are also infinitely many initial conditions available for their evaluation, viz., the continuous functions, $q(x, 0)$ and $\dot{q}(x, 0)$. Therefore, the real and imaginary parts of the β_r can be obtained in terms of the initial conditions by a procedure analogous to that used in Section 13.10. Using $\beta_r = \mu_r + i\nu_r$, we have from Eq. (14.5),

$$q(x, 0) = \sum_r \mu_r \sin \frac{r\pi x}{L} \tag{14.6a}$$

$$\dot{q}(x, 0) = -\sum_r \omega_r \nu_r \sin \frac{r\pi x}{L} \tag{14.6b}$$

Next, we multiply each of these equations by $\sin(s\pi x/L)$ and integrate from $x = 0$ to $x = L$. We then make use of the trigonometric relation,

$$\int_0^L \sin \frac{r\pi x}{L} \sin \frac{s\pi x}{L}\, dx = \frac{L}{2}\, \delta_{rs} \tag{14.7}$$

from which we obtain

$$\mu_r = \frac{2}{L} \int_0^L q(x, 0) \sin \frac{r\pi x}{L}\, dx \tag{14.8a}$$

$$\nu_r = -\frac{2}{\omega_r L} \int_0^L \dot{q}(x, 0) \sin \frac{r\pi x}{L}\, dx \tag{14.8b}$$

The characteristic frequency ω_r may also be obtained as the limiting value of the result for the loaded string. From Eq. (13.101) we have

$$\omega_r = 2\sqrt{\frac{\tau}{md}} \sin\left[\frac{r\pi}{2(n+1)}\right] \tag{14.9}$$

which can be written as

$$\omega_r = \frac{2}{d}\sqrt{\frac{\tau}{\rho}} \sin \frac{r\pi d}{2L} \tag{14.10}$$

When $d \to 0$, we can approximate the sine term by its argument with the result

$$\omega_r = \frac{r\pi}{L}\sqrt{\frac{\tau}{\rho}} \tag{14.11}$$

▶ Example 14.2 Vibration of a Plucked String

A "plucked string" is one for which a point of the string is displaced (so that the string assumes a triangular shape) and then released from rest. Consider the case shown in Fig. 14-1 in which the center of the string is displaced a distance h. The initial conditions are

$$\left.\begin{aligned} q(x, 0) &= \begin{cases} \dfrac{2h}{L}x, & 0 \le x \le L/2 \\ \dfrac{2h}{L}(L-x), & L/2 \le x \le L \end{cases} \\ \dot{q}(x, 0) &= 0 \end{aligned}\right\} \tag{1}$$

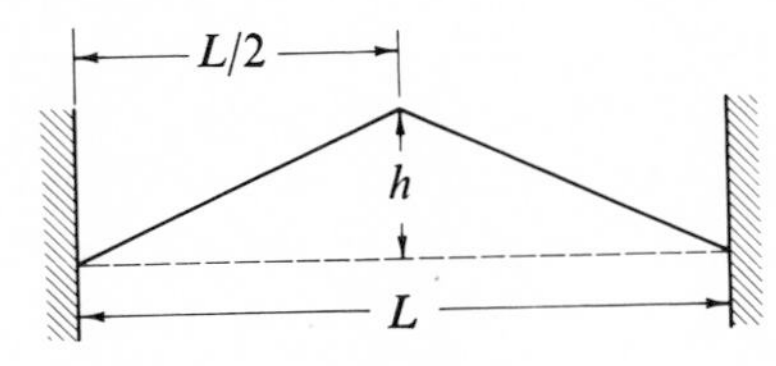

FIG. **14-1**

Since the string is released from rest, all the ν_r vanish. The μ_r are given by

$$\mu = \frac{4h}{L^2}\int_0^{L/2} x \sin\frac{r\pi x}{L}\,dx + \frac{4h}{L^2}\int_{L/2}^{L} (L-x)\sin\frac{r\pi x}{L}\,dx \tag{2}$$

Integrating,

$$\mu_r = \frac{8h}{r^2\pi^2}\sin\frac{r\pi}{2} \tag{3}$$

so that

$$\mu_r = \begin{cases} 0, & r \text{ even} \\ \dfrac{8h}{r^2\pi^2}(-1)^{\frac{1}{2}(r-1)} & r \text{ odd} \end{cases} \tag{4}$$

Therefore,

$$q(x, t) = \frac{8h}{\pi^2}\left[\sin\frac{\pi x}{L}\cos\omega_1 t - \frac{1}{9}\sin\frac{3\pi x}{L}\cos\omega_3 t + \cdots\right] \tag{5}$$

where the ω_r are proportional to r and are given by Eq. (14.11).

From Eq. (5) we see that the fundamental mode (with frequency ω_1) and all of the *odd* harmonics (with frequencies ω_3, ω_5, etc.) are excited but that none of the *even* harmonics are involved in the motion. Since the initial displacement was symmetric, the subsequent motion must also be symmetric so that none of the even modes (for which the center position of the string is a node) are excited. In general, if the string is plucked at some arbitrary point, none of the harmonics with nodes at that point will be excited.

As we shall prove in the next section, the energy in each of the excited modes is proportional to the square of the coefficient of the corresponding term in Eq. (5). Thus, the energy ratios for the fundamental, third harmonic, fifth harmonic, etc., are $1 : \frac{1}{81} : \frac{1}{625} : \cdots$. Therefore, the energy in the system (or the intensity of the emitted sound) is dominated by the fundamental. The third harmonic is 19 db* down from the fundamental and the fifth harmonic is down by 28 db.

14.3 Energy of a Vibrating String

Since we have made the assumption that frictional forces are not present, it is clear that the total energy of a vibrating string must remain constant. This we will now show explicitly; moreover, we will show that the energy of the string is expressed simply as the sum of contributions from each of the normal modes. According to Eq. (14.4), the displacement of the string is given by

$$q(x, t) = \sum_r \eta_r(t) \sin \frac{r\pi x}{L} \tag{14.12}$$

where the normal coordinates are

$$\eta_r(t) = \beta_r e^{i\omega_r t} \tag{14.13}$$

As always, the β_r are complex quantities and the physically meaningful normal coordinates are obtained by taking the *real part* of Eq. (14.13).

The kinetic energy of the string is obtained by calculating the kinetic energy for an element of the string, $\frac{1}{2}(\rho\, dx)\dot{q}^2$, and then integrating over the length. Thus,

$$T = \frac{1}{2}\rho \int_0^L \left(\frac{\partial q}{\partial t}\right)^2 dx \tag{14.14}$$

* The *decibel* (db) is a unit of relative sound intensity (or acoustic power). The intensity ratio of a sound with intensity I to a sound with intensity I_0 is given by $10 \log(I/I_0)$ db. Thus, for the fundamental (I_0) and third harmonic (I), we have $10 \log(1/81) = -19.1$ db or "19 db down" in intensity. A ratio of 3 db corresponds approximately to a factor of 2 in relative intensity.

or, using Eq. (14.12),

$$T = \frac{1}{2}\rho \int_0^L \left[\sum_r \dot{\eta}_r \sin\frac{r\pi x}{L}\right]^2 dx \tag{14.15}$$

The square of the series can be expressed as a double sum, this technique insuring that all cross terms are properly included:

$$T = \frac{1}{2}\rho \sum_{r,s} \dot{\eta}_r \dot{\eta}_s \int_0^L \sin\frac{r\pi x}{L} \sin\frac{s\pi x}{L}\, dx \tag{14.16}$$

The integral is now the same as that in Eq. (14.7), so

$$\begin{aligned} T &= \frac{\rho L}{4} \sum_{r,s} \dot{\eta}_r \dot{\eta}_s\, \delta_{rs} \\ &= \frac{\rho L}{4} \sum_r \dot{\eta}_r^2 \end{aligned} \tag{14.17}$$

In the evaluation of the kinetic energy we must be careful to take the product of *real* quantities. Therefore, we must compute the square of the *real* part of $\dot{\eta}_r$:

$$\begin{aligned} (\mathrm{Re}\, \dot{\eta}_r)^2 &= \left(\mathrm{Re} \frac{d}{dt}[(\mu_r + i\nu_r)(\cos\omega_r t + i \sin\omega_r t)]\right)^2 \\ &= (-\omega_r \mu_r \sin\omega_r t - \omega_r \nu_r \cos\omega_r t)^2 \end{aligned} \tag{14.18}$$

Therefore, the kinetic energy of the string is

$$T = \frac{\rho L}{4} \sum_r \omega_r^2 (\mu_r \sin\omega_r t + \nu_r \cos\omega_r t)^2 \tag{14.19}$$

The potential energy of the string can be calculated easily by writing down the expression for the loaded string and then passing to the limit of a continuous string. (Recall that we consider the potential energy to be only the elastic energy in the string.) For the loaded string,

$$U = \frac{1}{2}\frac{\tau}{d} \sum_j (q_{j-1} - q_j)^2 \tag{14.20}$$

Multiplying and dividing by d,

$$U = \frac{1}{2}\tau \sum_j \left(\frac{q_{j-1} - q_j}{d}\right)^2 d$$

In passing to the limit, $d \to 0$, the term in parentheses becomes just the partial derivative of $q(x, t)$ with respect to x and the sum (including the factor d) becomes an integral:

$$U = \frac{1}{2} \tau \int_0^L \left(\frac{\partial q}{\partial x} \right)^2 dx \tag{14.21}$$

Using Eq. (14.12), we have

$$\frac{\partial q}{\partial x} = \sum_r \frac{r\pi}{L} \eta_r \cos \frac{r\pi x}{L} \tag{14.22}$$

so that

$$U = \frac{1}{2} \tau \int_0^L \left[\sum_r \frac{r\pi}{L} \eta_r \cos \frac{r\pi x}{L} \right]^2 dx \tag{14.23}$$

Again, the squared term can be written as a double sum, and since the trigonometric relation, Eq. (14.7), applies for cosines as well as sines, we have

$$\begin{aligned} U &= \frac{\tau}{2} \sum_{r,s} \frac{r\pi}{L} \frac{s\pi}{L} \eta_r \eta_s \int_0^L \cos \frac{r\pi x}{L} \cos \frac{s\pi x}{L} \, dx \\ &= \frac{\tau}{2} \sum_{r,s} \frac{r\pi}{L} \frac{s\pi}{L} \eta_r \eta_s \cdot \frac{L}{2} \delta_{rs} \\ &= \frac{\tau}{2} \sum_r \frac{r^2\pi^2}{L^2} \cdot \frac{L}{2} \eta_r^2 \\ &= \frac{\rho L}{4} \sum_r \omega_r^2 \eta_r^2 \end{aligned} \tag{14.24}$$

where Eq. (14.11) has been used in the last line to express the result in terms of ω_r^2. Evaluating the square of the real part of η_r, we have, finally,

$$U = \frac{\rho L}{4} \sum_r \omega_r^2 (\mu_r \cos \omega_r t - \nu_r \sin \omega_r t)^2 \tag{14.25}$$

The total energy is now obtained by adding Eqs. (14.19) and (14.25), in which the cross terms cancel and the squared terms add to unity:

$$\begin{aligned} E &= T + U \\ &= \frac{\rho L}{4} \sum_r \omega_r^2 (\mu_r^2 + \nu_r^2) \end{aligned} \tag{14.26}$$

or,

$$E = \frac{\rho L}{4} \sum_r \omega_r^2 \, |\beta_r|^2 \tag{14.26a}$$

so that the total energy is constant in time and, furthermore, is given by a sum of contributions from each of the normal modes.

The kinetic and potential energies each vary with time, so it is sometimes useful to calculate the *time-averaged* kinetic and potential energies (i.e., averaged over one complete period of the fundamental vibration, $r = 1$). That is,

$$\langle T \rangle = \frac{\rho L}{4} \sum_r \omega_r^2 \langle (\mu_r \sin \omega_r t + \nu_r \cos \omega_r t)^2 \rangle \tag{14.27}$$

where the slanted brackets denote an average over the time interval $2\pi/\omega_1$. Now, the averages of $\sin^2 \omega_1 t$ or $\cos^2 \omega_1 t$ over this interval are each equal to $\frac{1}{2}$. Similarly, the averages of $\sin^2 \omega_r t$ and $\cos^2 \omega_r t$ for $r \geq 2$ are also $\frac{1}{2}$ since the period of the fundamental vibration is always some integer times the period of a higher harmonic vibration. The averages of the cross terms, $\cos \omega_r t \sin \omega_r t$, all vanish. Therefore,

$$\begin{aligned}\langle T \rangle &= \frac{\rho L}{8} \sum_r \omega_r^2 (\mu_r^2 + \nu_r^2)\\ &= \frac{\rho L}{8} \sum_r \omega_r^2 \, |\beta_r|^2\end{aligned} \tag{14.28}$$

For the time-averaged potential energy we have a similar result:

$$\begin{aligned}\langle U \rangle &= \frac{\rho L}{4} \sum_r \omega_r^2 \langle (\mu_r \cos \omega_r t - \nu_r \sin \omega_r t)^2 \rangle\\ &= \frac{\rho L}{8} \sum_r \omega_r^2 (\mu_r^2 + \nu_r^2)\\ &= \frac{\rho L}{8} \sum_r \omega_r^2 \, |\beta_r|^2\end{aligned} \tag{14.29}$$

We therefore have the important result that *the average kinetic energy of a vibrating string is equal to the average potential energy**:

$$\langle T \rangle = \langle U \rangle \tag{14.30}$$

We will find this conclusion of importance in the following section. Notice also, in Eqs. (14.28) and (14.29), another simplification that results from the use of normal coordinates, viz., both $\langle T \rangle$ and $\langle U \rangle$ are simple sums of contributions from each of the normal modes.

* This result also follows from the virial theorem.

14.4 Rayleigh's Principle

In a large class of vibratory systems the fundamental mode of oscillation is of dominant interest. This is not to imply that the higher harmonics are unimportant (most musical instruments, for example, depend upon harmonics to produce their characteristic sounds), but only that the greatest intensity is usually associated with the fundamental mode. Our previous method of determining the fundamental frequency of a vibrating system required solving the equation of motion. In some situations it is difficult (or even impossible) to obtain an exact solution to the problem. In order to cope with such cases, a method was devised by Lord Rayleigh in which it is necessary to know only a crude approximation to the solution in order to obtain a rather accurate value for the fundamental frequency.

Rayleigh's principle is stated as follows: In the fundamental mode of oscillation of a vibratory system, the kinetic and potential energies are distributed in such a way that the frequency (and, hence, the total energy) is a minimum.* We know from the results of the preceding section that the average kinetic energy is equal to the average potential energy when the exact solution $q(x, t)$ is used in the calculation. Therefore, if we *guess* (or establish by some systematic procedure) an approximate solution for the fundamental mode and then equate the calculated average kinetic and potential energies we will obtain a result for the fundamental frequency which will not be less than the true frequency. Indeed, successive approximation procedures may be used to approach as closely as desired to the true result.†

In order to make use of Rayleigh's method, we must first obtain convenient expressions for the average kinetic and potential energies. The exact solution for the fundamental mode of the vibrating string is

$$q(x, t) = \eta_1(t) \sin \frac{\pi x}{L} \tag{14.31}$$

We now replace the spatial function $\sin(\pi x/L)$ by an arbitrary function $g(x)$ which it will be our task to guess or choose conveniently in particular situations. Thus,

$$q(x, t) = \eta_1(t) g(x) \tag{14.32}$$

* We will not give a proof of this statement here [see, for example, Temple and Bickley (Te56, Chapter 3)].

† This procedure is closely related to the variational procedure used in quantum mechanics to establish the ground-state energy of a system by using trial wave functions.

According to Eq. (14.14) the kinetic energy is

$$T = \frac{1}{2}\rho\int_0^L \left(\frac{\partial q}{\partial t}\right)^2 dx$$

$$= \frac{1}{2}\rho\int_0^L \dot{\eta}_1^2 g^2\, dx \tag{14.33}$$

In order to obtain $\langle T\rangle$ we need average only $\dot{\eta}_1^2$:

$$\langle T\rangle = \langle\dot{\eta}_1^2\rangle\cdot\frac{1}{2}\rho\int_0^L g^2\, dx \tag{14.34}$$

Using the same method for averaging as in the preceding section, we find

$$\langle\dot{\eta}_1^2\rangle = \tfrac{1}{2}\omega_1^2\,|\beta_1|^2 \tag{14.35}$$

or, replacing the amplitude factor $|\beta_1|^2$ by the constant A^2,

$$\langle T\rangle = \frac{1}{4}\rho A^2\omega_1^2\int_0^L g^2\, dx \tag{14.36}$$

Similarly, for the potential energy, which again we consider to be only the elastic energy in the string,

$$U = \frac{1}{2}\tau\int_0^L \left(\frac{\partial q}{\partial x}\right)^2 dx$$

$$= \frac{1}{2}\tau\int_0^L \eta_1^2 g'^2\, dx \tag{14.37}$$

where $g' \equiv dg/dx$. Therefore, upon averaging we have

$$\langle U\rangle = \langle\eta_1^2\rangle\cdot\frac{1}{2}\tau\int_0^L g'^2\, dx$$

$$= \frac{1}{4}\tau A^2\int_0^L g'^2\, dx \tag{14.38}$$

Thus, equating $\langle T\rangle$ and $\langle U\rangle$, we find for the frequency of the fundamental mode,

$$\omega_1^2 = \frac{\tau\int_0^L g'^2\, dx}{\rho\int_0^L g^2\, dx} \tag{14.39}$$

If $g(x)$ is chosen so that it is different from the true solution, the frequency given by Eq. (14.39) will in fact be larger than the true fundamental frequency.

But if any function reasonably close to the true $g(x)$ is used, ω_1 will be only slightly different from the true frequency.

▶ Example 14.4(a) The Parabolic Approximation

If we did not already know the solution of the vibrating string, we might approach the problem in the following way. The ends of the string are fixed and when vibrating with the lowest possible frequency, the maximum displacement is observed to occur at the center. Therefore, a reasonable guess as to the displacement function would be a parabola:

$$g(x) = x(L - x)/L^2 \tag{1}$$

Squaring g and g' and substituting in Eq. (14.39) gives

$$\begin{aligned} \omega_1^2 &= \frac{\tau \int_0^L (L^2 - 4xL + 4x^2)\, dx}{\rho \int_0^L (x^2L^2 - 2x^3L + x^4)\, dx} \\ &= \frac{\tau \cdot \frac{1}{3}L^3}{\rho \cdot \frac{1}{30}L^5} \\ &= 10\frac{\tau}{\rho L^2} \end{aligned} \tag{2}$$

Thus,

$$\omega_1 = 3.1623 \sqrt{\frac{\tau}{\rho L^2}} \tag{3}$$

whereas the exact result is

$$\omega_1 = \pi \sqrt{\frac{\tau}{\rho L^2}} = 3.1416 \sqrt{\frac{\tau}{\rho L^2}} \tag{4}$$

Therefore, the approximate value is too large by 0.6%.

If we had assumed a triangular function for $g(x)$ with a maximum in the center (as in Fig. 14-1), we would have obtained a somewhat poorer approximation (understandably, since the triangular function is a poorer caricature than the parabola):

$$\omega_1 = \sqrt{12}\sqrt{\frac{\tau}{\rho L^2}} = 3.4641 \sqrt{\frac{\tau}{\rho L^2}} \tag{5}$$

or an error of 10%.

Because the accuracy of the frequency calculation can suffer if a poor choice is made for the approximate displacement function, Rayleigh devised a simple method for improving the result. An adjustable constant is included in the displacement function and this constant is varied to make ω_1 a minimum and therefore closer to the true fundamental frequency. In this example Rayleigh incorporated such a constant into the displacement function by writing*

$$g(x) = 1 - \left(\frac{2|x|}{L}\right)^n \qquad n > 1 \tag{6}$$

where n is the constant and where x is now to be measured from the *center* of the string. [When $n = 2$, this expression for $g(x)$ is the same as that in Eq. (1).] Calculating the frequency according to the prescription, we find

$$\omega_1^2 = \frac{2(n+1)(2n+1)}{2n-1} \cdot \frac{\tau}{\rho L^2} \tag{7}$$

The minimum value of ω_1 occurs when

$$n = \tfrac{1}{2}(\sqrt{6} + 1) = 1.7247 \tag{8}$$

so that the frequency is

$$\omega_1 = 3.1464 \sqrt{\frac{\tau}{\rho L^2}} \tag{9}$$

which is larger than the true result by 0.15%, an increase in accuracy of a factor of 4 over the value obtained in Eq. (3).

▶ Example 14.4(b) Vibrations of a String with a Mass Attached

Consider a string of mass m, length L, and density $\rho = m/L$, to which is attached a mass M at its center position. If M is small compared to m, the displacement function will be approximately that for $M = 0$, which we know is

$$g(x) = \sin \frac{\pi x}{L} \tag{1}$$

Using this expression, we can calculate from Eq. (14.36) the average kinetic energy of the string to which we must add the average kinetic energy of the mass:

$$\begin{aligned} \langle T_{\text{string}} \rangle &= \frac{1}{4} \rho A^2 \omega_1^2 \int_0^L \sin^2\left(\frac{\pi x}{L}\right) dx \\ &= \frac{1}{8} L\rho\, A^2\, \omega_1^2 \end{aligned} \tag{2}$$

Now, the mass M moves according to

$$q_M(t) = A \cos \omega_1 t \tag{3}$$

so that its average kinetic energy is

$$\begin{aligned} \langle T_{\text{mass}} \rangle &= \tfrac{1}{2} M \langle \dot{q}_M^2 \rangle \\ &= \tfrac{1}{4} M A^2 \omega_1^2 \end{aligned} \tag{4}$$

* See Lord Rayleigh (Ra94, Vol. 1, p. 113).

Therefore, the total average kinetic energy is

$$\begin{aligned}\langle T \rangle &= \langle T_{\text{string}} \rangle + \langle T_{\text{mass}} \rangle \\ &= \tfrac{1}{4}A^2\omega_1^2(\tfrac{1}{2}L\rho + M) \qquad (5)\end{aligned}$$

The potential energy, as usual, is considered to be only the elastic energy of the string, therefore, using Eq. (14.38),

$$\begin{aligned}\langle U \rangle &= \frac{1}{4}\tau A^2 \int_0^L \frac{\pi^2}{L^2}\cos^2\left(\frac{\pi x}{L}\right) dx \\ &= \frac{\pi^2}{8}\frac{\tau A^2}{L} \qquad (6)\end{aligned}$$

Equating $\langle T \rangle$ and $\langle U \rangle$, we obtain

$$\begin{aligned}\omega_1^2 &= \frac{\pi^2}{2}\frac{\tau}{L(\frac{1}{2}L\rho + M)} \\ &= \pi^2 \frac{\tau}{\rho L^2} \cdot \frac{1}{(1 + 2M/m)} \qquad (7)\end{aligned}$$

This formula obviously gives the correct result in the limit $M = 0$ [since we chose $g(x)$ to be the exact solution for that case]. For various values of M/m the frequencies are listed in Table 14.1 where they are compared with the exact results.

Table 14.1

M/m	$\omega_1/\sqrt{\tau\rho L^2}$	Exact	Percent error
0	3.1416	3.1416	0
0.1	2.8679	2.8577	0.36
0.2	2.6552	2.6277	1.05
0.5	2.2215	2.1542	3.12
1	1.8138	1.7198	5.47
5	0.9375	0.8657	8.29

The approximate result is accurate only for small values of M/m since as M increases, the string becomes seriously distorted from the sine function that it follows for $M = 0$.

14.5 The Wave Equation

Our procedure thus far has been to describe the motion of a continuous string as the limiting case of the loaded string for which we have a complete solution; we have not yet written down the fundamental equation of motion for the continuous case. We may accomplish this by returning to the loaded string and again using the limit technique but now on the equation of motion rather than on the solution. Equation (13.81) can be expressed as

$$\frac{m}{d}\ddot{q}_j = \frac{\tau}{d}\left(\frac{q_{j-1} - q_j}{d}\right) - \frac{\tau}{d}\left(\frac{q_j - q_{j+1}}{d}\right) \tag{14.40}$$

As d approaches zero, we have

$$\frac{q_j - q_{j+1}}{d} \to \frac{q(x) - q(x+d)}{d} \to -\left.\frac{\partial q}{\partial x}\right|_{x+d/2}$$

which is the derivative at $x + d/2$. For the other term in Eq. (14.40), we have

$$\frac{q_{j-1} - q_j}{d} \to \frac{q(x-d) - q(x)}{d} \to -\left.\frac{\partial q}{\partial x}\right|_{x-d/2}$$

which is the derivative at $x - d/2$. Therefore, the limiting value of the right-hand side of Eq. (14.40) is

$$\lim_{d\to 0} \tau\left(\frac{\left.\frac{\partial q}{\partial x}\right|_{x+d/2} - \left.\frac{\partial q}{\partial x}\right|_{x-d/2}}{d}\right) = \tau \left.\frac{\partial^2 q}{\partial x^2}\right|_x = \tau\frac{\partial^2 q}{\partial x^2}$$

Also in the limit, m/d becomes ρ, so that the equation of motion is

$$\rho\ddot{q} = \tau\frac{\partial^2 q}{\partial x^2} \tag{14.41}$$

or,

$$\boxed{\frac{\partial^2 q}{\partial x^2} = \frac{\rho}{\tau}\frac{\partial^2 q}{\partial t^2}} \tag{14.42}$$

This is the *wave equation* in one dimension. In the following chapter we will discuss the solutions to this equation.

14.6 The Nonuniform String—Orthogonal Functions and Perturbation Theory

In the derivation of the wave equation in the preceding section, it was implicitly assumed that the tension τ was a constant. If the string is non-uniform, however, so that $\rho = \rho(x)$, then in general the tension will also be a function of x: $\tau = \tau(x)$. Therefore, we must write, in passing to the limit on the right-hand side of Eq. (14.40),

$$\lim_{d \to 0} \frac{\left[\tau\left(x + \frac{d}{2}\right) \frac{\partial q}{\partial x}\bigg|_{x+d/2} - \tau\left(x - \frac{d}{2}\right) \frac{\partial q}{\partial x}\bigg|_{x-d/2} \right]}{d} = \frac{\partial}{\partial x}\left(\tau \frac{\partial q}{\partial x}\right)$$

Thus, the wave equation for this case takes the more complicated form*

$$\frac{\partial}{\partial x}\left(\tau \frac{\partial q}{\partial x}\right) - \rho \frac{\partial^2 q}{\partial t^2} = 0 \tag{14.43}$$

where both τ and ρ may be functions of x but not of the time. For arbitrary functions τ, ρ there is no general method of solution for equations of this type. Only for certain particular forms of the functions τ, ρ can solutions be obtained in terms of simple functions.

The problem of the nonuniform string is one of rather limited practical importance. The equation which describes the situation [Eq. (14.43)] is, however, of considerable interest, especially in many quantum mechanical problems. Consequently, there is some purpose in pursuing the solution in detail, even though the actual application of the results to vibrating strings is rarely made. Because of the fact that simple, general solutions to Eq. (14.43) do not exist, it is necessary to employ an approximation procedure, and we shall again turn to the powerful perturbation technique. Solutions will be developed in terms of *orthogonal functions*, and the method of establishing orthogonality follows closely that previously used on several occasions.

Suppose that we have a particular solution of Eq. (14.43) which has the form

$$q_r(x, t) = u_r(x)e^{i\omega_r t} \tag{14.44}$$

where the time dependence has been placed, as usual, in exponential form and where ω_r denotes the rth eigenfrequency. If the string were uniform, $u_r(x)$ would simply be $\sin(r\pi x/L)$, as in Eq. (14.5). The function $u_r(x)$ is called the rth *wave function* (or *eigenfunction* or *normal mode*) for the system. Equation (14.43) then becomes

$$\frac{d}{dx}\left(\tau \frac{du_r}{dx}\right) + \rho\omega_r^2 u_r = 0 \tag{14.45a}$$

* This type of equation occurs frequently in various areas of mathematical physics; in general, the right-hand side is not zero. In the form of Eq. (14.103), it is known as the *homogeneous Sturm–Liouville* equation. Investigations of the solution to equations of this type were made by Joseph Liouville (1809–1882) and Jacques Charles Francois Sturm (1803–1855) in the mid-1830's.

Also, for the sth eigenfrequency, we may write

$$\frac{d}{dx}\left(\tau \frac{du_s}{dx}\right) + \rho\omega_s^2 u_s = 0 \tag{14.45b}$$

In the same manner in which we have previously proved orthogonality [see, for example, Eqs. (12.57), Eqs. (13.58) or Eqs. (13.107) and following], we multiply Eq. (14.45a) by u_s and integrate from $x = 0$ to $x = L$, and we multiply Eq. (14.45b) by u_r and integrate. Upon subtracting these results, we have

$$\int_0^L \left[u_s \frac{d}{dx}\left(\tau \frac{du_r}{dx}\right) - u_r \frac{d}{dx}\left(\tau \frac{du_s}{dx}\right)\right] dx + (\omega_r^2 - \omega_s^2)\int_0^L \rho u_r u_s \, dx = 0 \tag{14.46}$$

The first term may be integrated by parts, with the result

$$\left[u_s \tau \frac{du_r}{dx} - u_r \tau \frac{du_s}{dx}\right]_{x=0}^{x=L} - \int_0^L \tau\left(\frac{du_r}{dx}\frac{du_s}{dx} - \frac{du_s}{dx}\frac{du_r}{dx}\right) dx$$
$$- (\omega_r^2 - \omega_s^2)\int_0^L \rho u_r u_s \, dx = 0$$

The term in brackets vanishes in view of the boundary condition on the wave functions, $u_p(0) = 0 = u_p(L)$. And since the second term obviously vanishes, we have

$$(\omega_r^2 - \omega_s^2)\int_0^L \rho u_r u_s \, dx = 0 \tag{14.47}$$

Thus, if the eigenfrequencies are distinct, then for $r \neq s$, the integral in Eq. (14.47) must vanish. And since the integral is undetermined for $r = s$, we may require that it equal unity. Hence, we have the orthonormality condition,

$$\int_0^L \rho u_r u_s \, dx = \delta_{rs} \tag{14.48}$$

This condition corresponds to that previously found for the sine functions [Eq. (13.108)] and to the requirement placed on the a_{jr} [Eq. (13.63)].

We could now write the general solution as

$$q(x, t) = \sum_r \beta_r u_r(x) e^{i\omega_r t} \tag{14.49}$$

and proceed, as before, to evaluate the real and imaginary parts of the β_r in terms of the initial conditions. This procedure is not particularly useful, however, since the functions $u_r(x)$ are not *numbers*, as were the a_{jr}, and their ratios are, therefore, not simple. We can obtain a general solution by applying the method of perturbations. We shall simplify the problem by assuming that

ρ is a slowly varying function of the coordinate x and that the tension τ is constant.* We therefore express ρ as a series expansion in powers of a perturbation parameter λ,

$$\rho(x) = \rho^{(0)} + \lambda\rho^{(1)}(x) + \lambda^2\rho^{(2)}(x) + \cdots \tag{14.50}$$

where $\rho^{(0)}$ is a constant and where λ is a small quantity. All of the quantities $\rho^{(p)}(x)$ are independent of λ. Now, if the string had a density $\rho = \rho^{(0)}$, then there would exist a set of wave functions $u_r^{(0)}(x)$ and eigenfrequencies $\omega_r^{(0)}$ that satisfy the equation

$$\tau \frac{d^2u_r^{(0)}}{dx^2} + (\omega_r^{(0)})^2\rho^{(0)}u_r^{(0)} = 0 \tag{14.51}$$

If the density of the string does not differ appreciably from $\rho^{(0)}$ (that is, if λ is indeed a small quantity), then the functions $u_r^{(0)}(x)$ and the frequencies $\omega_r^{(0)}$ will be close† to the eigenfunctions $u_r(x)$ and the true eigenfrequencies ω_r. We may then expand $u_r(x)$ and ω_r in powers of λ:

$$\begin{aligned} u_r(x) &= u_r^{(0)}(x) + \lambda u_r^{(1)}(x) + \lambda^2 u_r^{(2)}(x) + \cdots \\ \omega_r &= \omega_r^{(0)} + \lambda\omega_r^{(1)} + \lambda^2\omega_r^{(2)} + \cdots \end{aligned} \tag{14.52}$$

where the functions $u_r^{(p)}(x)$ and the quantities $\omega_r^{(p)}$ are independent of λ and are to be determined. The term $\omega_r^{(1)}$ is called the *first-order frequency correction*; $u_r^{(2)}$ is the *second-order correction to the wave function*, etc.

It is clear that we may establish an orthogonality relation for the zero-order wave functions in the same manner that the condition on the true wave functions was derived. We find

$$\int_0^L \rho^{(0)}u_r^{(0)}u_s^{(0)}\,dx = \delta_{rs} \tag{14.53}$$

Since the $u_r^{(0)}$ include the functions $\sin(r\pi x/L)$, in order that the normalization be proper, we must have [cf. Eq. (14.7)]:

$$u_r^{(0)}(x) = \sqrt{\frac{2}{\rho^{(0)}L}}\,\sin\frac{r\pi x}{L} \tag{14.54}$$

The equation satisfied by the $u_r(x)$ is

$$\tau\frac{d^2u_r}{dx^2} + \omega_r^2\rho u_r = 0 \tag{14.55}$$

Substituting the expansions for $u_r(x)$, $\rho(x)$, and ω_r into this equation, we obtain

* The additional complication of a varying tension can be handled in a manner equivalent to the development which follows (see Problem 14-17).

† However, recall the warning given in Section 5.5 regarding such statements.

$$\tau \frac{d^2u_r^{(0)}}{dx^2} + \tau\lambda \frac{d^2u_r^{(1)}}{dx^2} + \tau\lambda^2 \frac{d^2u_r^{(2)}}{dx^2} + \cdots + [(\omega_r^{(0)} + \lambda\omega_r^{(1)} + \lambda^2\omega_r^{(2)} + \cdots)^2 \cdot (\rho^{(0)} + \lambda\rho^{(1)} + \lambda^2\rho^{(2)} + \cdots) \cdot (u_r^{(0)} + \lambda u_r^{(1)} + \lambda^2 u_r^{(2)} + \cdots)] = 0 \tag{14.56}$$

In order for Eq. (14.56) to be satisfied for arbitrary λ, it is necessary that the coefficient of each power of λ vanish. The zero-order terms (i.e., those without a coefficient λ) satisfy Eq. (14.51) so that they sum to zero. The first-order terms are

$$\tau \frac{d^2u_r^{(1)}}{dx^2} + 2\omega_r^{(0)}\omega_r^{(1)}\rho^{(0)}u_r^{(0)} + (\omega_r^{(0)})^2\rho^{(1)}u_r^{(0)} + (\omega_r^{(0)})^2\rho^{(0)}u_r^{(1)} = 0 \tag{14.57}$$

Any arbitrary function can be expanded in terms of a complete set of orthogonal functions.* Since the $u_r^{(0)}$ are solutions of the wave equation for $\rho = \text{const.}$, they constitute such a set, and we may use them to expand the functions $u_r^{(1)}$, $u_r^{(2)}$, etc. Thus,

$$u_r^{(1)} = \sum_{s=1}^{\infty} C_{rs}\, u_s^{(0)} \tag{14.58}$$

where the C_{rs} are coefficients (also independent of λ) that are to be determined. Therefore, we have

$$\tau \frac{d^2u_r^{(1)}}{dx^2} = \sum_s C_{rs}\, \tau \frac{d^2u_s^{(0)}}{dx^2} \tag{14.59}$$

and using Eq. (14.51), this becomes

$$\tau \frac{d^2u_r^{(1)}}{dx^2} = -\sum_s C_{rs}(\omega_s^{(0)})^2\rho^{(0)}u_s^{(0)} \tag{14.60}$$

If we substitute the expansions for $u_r^{(1)}$ and its second derivative [Eqs. (14.58) and (14.60)] into Eq. (14.57), we obtain

$$\sum_s C_{rs}[(\omega_r^{(0)})^2 - (\omega_s^{(0)})^2]\rho^{(0)}u_s^{(0)} + \omega_r^{(0)}[2\omega_r^{(1)}\rho^{(0)} + \omega_r^{(0)}\rho^{(1)}]u_r^{(0)} = 0 \tag{14.61}$$

Now we multiply this equation by $u_t^{(0)}$ and integrate over the length of the string; we then have

$$\sum_s C_{rs}[(\omega_r^{(0)})^2 - (\omega_s^{(0)})^2] \int_0^L \rho^{(0)}u_s^{(0)}u_t^{(0)}\, dx + \omega_r^{(0)}\left[2\omega_r^{(1)} \int_0^L \rho^{(0)}u_r^{(0)}u_t^{(0)}\, dx + \omega_r^{(0)} \int_0^L \rho^{(1)}u_r^{(0)}u_t^{(0)}\, dx\right] = 0 \tag{14.62}$$

* See the discussion in Section 14.7.

We define

$$M_{rt}^{(1)} \equiv \int_0^L \rho^{(1)} u_r^{(0)} u_t^{(0)}\, dx \tag{14.63}$$

and also make use of the orthogonality condition [Eq. (14.53]) to write Eq. (14.62) as

$$\sum_s C_{rs}[(\omega_r^{(0)})^2 - (\omega_r^{(0)})^2]\delta_{st} + \omega_r^{(0)}[2\omega_r^{(1)}\delta_{rt} + \omega_r^{(0)} M_{rt}^{(1)}] = 0 \tag{14.64}$$

Carrying out the summation in the first term, we obtain

$$C_{rt}[(\omega_r^{(0)})^2 - (\omega_t^{(0)})^2] + \omega_r^{(0)}[2\omega_r^{(1)}\delta_{rt} + \omega_r^{(0)} M_{rt}^{(1)}] = 0 \tag{14.65}$$

Clearly, we now have two cases to investigate, $r = t$ and $r \neq t$. First, for $r = t$, the leading term vanishes, and

$$\omega_r^{(1)} = -\tfrac{1}{2}\omega_r^{(0)} M_{rr}^{(1)} \tag{14.66}$$

which yields the first-order frequency correction. If we write

$$\omega_r^2 = [\omega_r^{(0)} + \lambda\omega_r^{(1)} + \cdots]^2$$

and retain only terms up to first order in λ, we have

$$\omega_r^2 \cong (\omega_r^{(0)})^2 + 2\lambda\omega_r^{(0)}\omega_r^{(1)} \tag{14.67}$$

which, in view of Eq. (14.66), becomes

$$\begin{aligned} \omega_r^2 &\cong (\omega_r^{(0)})^2[1 - \lambda M_{rr}^{(1)}] \\ &= (\omega_r^{(0)})^2\left[1 - \lambda \int_0^L \rho^{(1)}(u_r^{(0)})^2\, dx\right] \end{aligned} \tag{14.68}$$

Now, from Eq. (14.54) we have

$$u_r^{(0)} \propto \sin\frac{r\pi x}{L}$$

so that the first-order frequency correction, Eq. (14.66), varies as

$$\begin{aligned} \omega_r^{(1)} &\propto \int_0^L \rho^{(1)}(u_r^{(0)})^2\, dx \\ &\propto \int_0^L \rho^{(1)} \sin^2\frac{r\pi x}{L}\, dx \end{aligned}$$

Therefore, we see that if the density perturbation is concentrated at the *nodes* for the rth mode (i.e., those points for which $\sin(r\pi x/L) = 0$, then there is zero correction for this particular eigenfrequency. On the other hand, the correction is a maximum if $\rho^{(1)}$ is concentrated at the antinodes.

Returning to Eq. (14.65) and considering the case $r \neq t$, we find that the first term in the second bracket vanishes, so that

$$C_{rt} = -\frac{(\omega_r^{(0)})^2 M_{rt}^{(1)}}{(\omega_r^{(0)})^2 - (\omega_t^{(0)})^2}, \qquad r \neq t \tag{14.69}$$

Therefore, the first-order correction to the wave function is

$$\begin{aligned} u_r^{(1)}(x) &= \sum_s C_{rs} u_s^{(0)} \\ &= -\sum_s{}' \frac{(\omega_r^{(0)})^2 M_{rs}^{(1)}}{(\omega_r^{(0)})^2 - (\omega_s^{(0)})^2} u_s^{(0)} \end{aligned} \tag{14.70}$$

where the prime on the summation indicates that the term with $r = s$ is excluded.

In Eq. (14.70), the summation extends over the range $s = 1, 2, \ldots, \infty$, and so it is not possible to evaluate the sum completely. But because of the form of the denominator, only those terms for which $\omega_r^{(0)} \approx \omega_s^{(0)}$ will give appreciable contributions. Therefore, the value of the infinite sum may be closely approximated by considering only a few terms. It is just this fact that makes the perturbation expansion a useful approximation procedure.

▶ Example 14.6 String with a Linear Density Variation

Consider a nonuniform string whose density is given by

$$\rho(x) = \rho_0 + \lambda x \tag{1}$$

where ρ_0 is a constant and where λ is a small quantity, $\lambda \ll \rho_0/L$. In order to find the first-order frequency correction, we must evaluate

$$\begin{aligned} \omega_r^{(1)} &= -\tfrac{1}{2}\omega_r^{(0)} M_{rr}^{(1)} \\ &= -\tfrac{1}{2}\omega_r^{(0)} \int_0^L \rho^{(1)} u_r^{(0)} u_r^{(0)}\, dx \end{aligned} \tag{2}$$

where we have

$$\rho^{(1)} = x \tag{3}$$

$$u_r^{(0)} = \sqrt{\frac{2}{\rho_0 L}} \sin\frac{r\pi x}{L} \tag{4}$$

and

$$\omega_r^{(0)} = \frac{r\pi}{L}\sqrt{\frac{\tau}{\rho_0}} \tag{5}$$

Thus,

$$\begin{aligned} \omega_r^{(1)} &= -\frac{1}{2}\frac{r\pi}{L}\sqrt{\frac{\tau}{\rho_0}} \cdot \frac{2}{\rho_0 L}\int_0^L x \sin^2\!\left(\frac{r\pi x}{L}\right) dx \\ &= -\frac{r\pi}{4\rho_0}\sqrt{\frac{\tau}{\rho_0}} \end{aligned} \tag{6}$$

The frequency, correct to first-order, is given by

$$\omega_r = \omega_r^{(0)} + \lambda\omega_r^{(1)}$$
$$= \frac{r\pi}{L}\sqrt{\frac{\tau}{\rho_0}}\left[1 - \lambda\frac{L}{4\rho_0}\right] \tag{7}$$

If we return to Eq. (14.56) and collect all of the terms which have a coefficient λ^2, we find

$$\tau\frac{d^2u_r^{(2)}}{dx^2} + (\omega_r^{(0)})^2[\rho^{(2)}u_r^{(0)} + \rho^{(1)}u_r^{(1)} + \rho^{(0)}u_r^{(2)}]$$
$$+ 2\omega_r^{(0)}\omega_r^{(1)}[\rho^{(1)}u_r^{(0)} + \rho^{(0)}u_r^{(1)}]$$
$$+ \rho^{(0)}u_r^{(0)}[2\omega_r^{(0)}\omega_r^{(2)} + (\omega_r^{(1)})^2] = 0 \tag{14.71}$$

As in Eq. (14.58), we expand the second-order wave function in terms of the zero-order functions:

$$u_r^{(2)} = \sum_{s=1}^{\infty} D_{rs}u_s^{(0)} \tag{14.72}$$

Following the previous development, we have

$$\tau\frac{d^2u_r^{(0)}}{dx^2} = \sum_s D_{rs}\tau\frac{d^2u_s^{(0)}}{dx^2}$$
$$= -\sum_s D_{rs}(\omega_s^{(0)})^2\rho^{(0)}u_s^{(0)} \tag{14.73}$$

Substituting Eq. (14.73) into Eq. (14.71), there results

$$\sum_s D_{rs}[(\omega_r^{(0)})^2 - (\omega_s^{(0)})^2]\rho^{(0)}u_s^{(0)} + (\omega_r^{(0)})^2[\rho^{(2)}u_r^{(0)} + \rho^{(1)}u_r^{(1)}]$$
$$+ 2\omega_r^{(0)}\omega_r^{(1)}[\rho^{(1)}u_r^{(0)} + \rho^{(0)}u_r^{(1)}] + \rho^{(0)}u_r^{(0)}[2\omega_r^{(0)}\omega_r^{(2)} + (\omega_r^{(1)})^2] = 0$$
$$\tag{14.74}$$

We can now use Eqs. (14.66) and (14.70) to substitute for $\omega_r^{(1)}$ and $u_r^{(1)}$ in Eq. (14.74). Thus,

$$\sum_s D_{rs}[(\omega_r^{(0)})^2 - (\omega_s^{(0)})^2]\rho^{(0)}u_s^{(0)}$$
$$+ \omega_r^{(0)}\Bigg[\omega_r^{(0)}\rho^{(2)}u_r^{(0)} - \omega_r^{(0)}\rho^{(1)}\sum_s{}'\frac{(\omega_r^{(0)})^2M_{rs}^{(1)}}{(\omega_r^{(0)})^2 - (\omega_s^{(0)})^2}u_s^{(0)}$$
$$- \omega_r^{(0)}\rho^{(1)}u_r^{(0)}M_{rr}^{(1)} + \omega_r^{(0)}\rho^{(0)}M_{rr}^{(1)}\sum_s{}'\frac{(\omega_r^{(0)})^2M_{rs}^{(1)}}{(\omega_r^{(0)})^2 - (\omega_s^{(0)})^2}u_s^{(0)}$$
$$+ 2\omega_r^{(2)}\rho^{(0)}u_r^{(0)} + \tfrac{1}{4}\omega_r^{(0)}\rho^{(0)}u_r^{(0)}(M_{rr}^{(1)})^2\Bigg] = 0 \tag{14.75}$$

Next we multiply this equation by $u_r^{(0)}\,dx$ and integrate over x in the range $0 \le x \le L$:

$$\sum_s D_{rs}[(\omega_r^{(0)})^2 - (\omega_s^{(0)})^2] \int_0^L \rho^{(0)} u_r^{(0)} u_s^{(0)}\,dx + \omega_r^{(0)} \Bigg[\omega_r^{(0)} \int_0^L \rho^{(2)} u_r^{(0)} u_r^{(0)}\,dx$$
$$- \omega_r^{(0)} \sum_s{}' \frac{(\omega_r^{(0)})^2 M_{rs}^{(1)}}{(\omega_r^{(0)})^2 - (\omega_s^{(0)})^2} \int_0^L \rho^{(1)} u_r^{(0)} u_s^{(0)}\,dx - \omega_r^{(0)} M_{rr}^{(1)} \int_0^L \rho^{(1)} u_r^{(0)} u_r^{(0)}\,dx$$
$$+ \omega_r^{(0)} M_{rr}^{(1)} \sum_s{}' \frac{(\omega_r^{(0)})^2 M_{rs}^{(1)}}{(\omega_r^{(0)})^2 - (\omega_s^{(0)})^2} \int_0^L \rho^{(0)} u_r^{(0)} u_s^{(0)}\,dx + 2\omega_r^{(2)} \int_0^L \rho^{(0)} u_r^{(0)} u_r^{(0)}\,dx$$
$$+ \tfrac{1}{4}\omega_r^{(0)} (M_{rr}^{(1)})^2 \int_0^L \rho^{(0)} u_r^{(0)} u_r^{(0)}\,dx \Bigg] = 0 \qquad (14.76)$$

In view of the orthogonality condition [Eq. (14.53)], the first and fifth terms of Eq. (14.76) vanish, and the integrals of the last two terms become unity. Moreover, the integral in the third term is just $M_{rs}^{(1)}$ and that in the fourth term is $M_{rr}^{(1)}$. Thus, upon dividing by $\omega_r^{(0)} \neq 0$, we have

$$\omega_r^{(0)} \int_0^L \rho^{(2)} u_r^{(0)} u_r^{(0)}\,dx - \omega_r^{(0)} \sum_s{}' \frac{(\omega_r^{(0)})^2 (M_{rs}^{(1)})^2}{(\omega_r^{(0)})^2 - (\omega_s^{(0)})^2}$$
$$- \omega_r^{(0)} (M_{rr}^{(1)})^2 + 2\omega_r^{(2)} + \tfrac{1}{4}\omega_r^{(0)} (M_{rr}^{(1)})^2 = 0 \qquad (14.77)$$

If we define

$$M_{rr}^{(2)} \equiv \int_0^L \rho^{(2)} u_r^{(0)} u_r^{(0)}\,dx \qquad (14.78)$$

then, solving for $\omega_r^{(2)}$, we find

$$\omega_r^{(2)} = \omega_r^{(0)} \left[\tfrac{3}{8} (M_{rr}^{(1)})^2 - \tfrac{1}{2} M_{rr}^{(2)} + \tfrac{1}{2} \sum_s{}' \frac{(\omega_r^{(0)})^2 (M_{rs}^{(1)})^2}{(\omega_r^{(0)})^2 - (\omega_s^{(0)})^2} \right] \qquad (14.79)$$

It is clear that the frequency corrections may be calculated to any order by this procedure, each succeeding correction being more complicated than the last.

14.7 Generalized Fourier Series

Trigonometric Fourier series were discussed briefly in Section 4.5 in connection with driven harmonic oscillators. In the preceding section we have again made use of the generalized technique of Fourier analysis although trigonometric functions were not employed. Indeed, a Fourier analysis of any particular problem can always be made in terms of any complete set of *orthogonal functions*.

We first need to establish some details of the properties of orthogonal functions. We know from our discussion of vectors (see Section 1.12) that it is possible to express any three-dimensional vector in the form

$$\mathbf{A} = A_1\mathbf{e}_1 + A_2\mathbf{e}_2 + A_3\mathbf{e}_3 = \sum_j A_j\mathbf{e}_j \tag{14.80}$$

where the A_j are the projections along the three orthogonal axes, the unit vectors along which are $\mathbf{e}_1$, $\mathbf{e}_2$, $\mathbf{e}_3$. Now, it is clearly impossible to represent a vector which has a component in the x_3 direction solely in terms of $\mathbf{e}_1$ and $\mathbf{e}_2$; the *basis set* of unit vectors must include $\mathbf{e}_3$ as well. Therefore, in order to represent an arbitrary vector, the basis set is required to be a *complete set*; that is, the set must include unit vectors for all possible directions.

The general method for determining the projection of $\mathbf{A}$ onto the x_j-axis is to compute the scalar product of $\mathbf{A}$ and the unit vector $\mathbf{e}_j$: $A_j = \mathbf{A} \cdot \mathbf{e}_j$. The unit vectors obey the orthogonality relation

$$\mathbf{e}_j \cdot \mathbf{e}_k = \delta_{jk} \tag{14.81}$$

We now wish to generalize the results for vectors in three-dimensional space to the case of *functions* in n-dimensional space (i.e., *function space*), and, in fact, where n is infinite.* That is, we have a set of orthogonal functions

$$\varphi_1(x), \quad \varphi_2(x), \quad \varphi_3(x), \ldots$$

which are defined within a certain fundamental interval, $-L < x < L$. If there is no function that is orthogonal to all of the $\varphi_j(x)$, then the functions $\varphi_j(x)$ constitute a *complete orthogonal set*, and any arbitrary function can be expanded in terms of this set. In analogy with Eq. (14.80), we have†

$$\left.\begin{aligned} f(x) &= a_1\varphi_1(x) + a_2\varphi_2(x) + \cdots \\ &= \sum_j a_j\varphi_j(x) \end{aligned}\right\} \quad -L < x < L \tag{14.82}$$

Such a series is called a *generalized Fourier series*.

The statement that the $\varphi_j(x)$ are orthogonal implies that none of these functions can be expressed in terms of the others; and, hence, the series in Eq. (14.82) is a linearly independent combination of the $\varphi_j(x)$. Thus, if $f(x)$ vanishes, each of the coefficients a_j must also vanish. The orthogonality

* The degree of the function space (i.e., n) must be *denumerably* infinite in order to properly specify the orthogonality condition, Eq. (14.83), since in that equation the indices j and k must be definable numbers.

† In making such an expansion we must always inquire as to the convergence of the series. We shall assume that the series does converge in all cases of interest here.

condition can be expressed in analogy with that for the unit vectors $\mathbf{e}_j$:

$$\int_{-L}^{+L} \varphi_j(x)\varphi_k(x)\,dx = \delta_{jk} \tag{14.83}$$

where an integration is now necessary since our "unit vectors" are functions of x, and where we have arbitrarily required the normalization

$$\int_{-L}^{+L} [\varphi_j(x)]^2\,dx = 1$$

Now, we can determine the coefficients a_j in the expansion for $f(x)$ [Eq. (14.82)] by multiplying this equation by $\varphi_k(x)$ and integrating:

$$\int_{-L}^{+L} f(x)\varphi_k(x)\,dx = \int_{-L}^{+L} \sum_j a_j\,\varphi_j(x)\varphi_k(x)\,dx \tag{14.84}$$

Interchanging the order of integration and summation, and applying the orthogonality condition, we obtain

$$\begin{aligned}\int_{-L}^{+L} f(x)\varphi_k(x)\,dx &= \sum_j a_j \int_{-L}^{+L} \varphi_j(x)\varphi_k(x)\,dx \\ &= \sum_j a_j\,\delta_{jk} \\ &= a_k\end{aligned} \tag{14.85}$$

Thus, we have

$$f(x) = \sum_j \varphi_j(x) \int_{-L}^{+L} f(x')\varphi_j(x')\,dx' \tag{14.86}$$

The functions $\varphi_j(x)$ are, of course, *any* complete set of orthogonal functions. Such functions include Legendre polynomials, Bessel functions, Hermite polynomials, etc., as well as the trigonometric functions.

Let us now consider the representation of a function $f(x)$ in the interval $-\pi < x < \pi$ by a trigonometric series composed of the orthogonal functions $\cos rx$ and $\sin rx$:

$$\boxed{f(x) = \frac{a_0}{2} + \sum_{r=1}^{\infty} (a_r \cos rx + b_r \sin rx)} \tag{14.87}$$

This is the most familiar example of a Fourier series* and is the expression

* Trigonometric series were first used by Daniel Bernoulli, and the integral expressions for the coefficients were first given by Euler. But these were treatments of special cases; it was Baron Jean Baptiste Joseph Fourier (1768–1830) who pointed out in 1807 that an *arbitrary* function, and, in fact, even a *discontinuous* function, could be represented by a trigonometric series. Fourier's 1807 paper, however, was not rigorous (it was, in fact, rejected by the Paris Academy); Dirichlet first gave a rigorous proof in 1829.

we obtained in Eq. (4.61). We note that $f(x)$ need not be a continuous function of x; $f(x)$ may have a finite number of finite discontinuities within the range $-\pi < x < \pi$. Discontinuous functions can be handled by dividing the interval into sub-intervals such that $f(x)$ is continuous within each sub-interval. At a point of discontinuity $x = x_0$ it can be shown that the Fourier series converges to the mean value; i.e.,

$$f(x_0) = \tfrac{1}{2} \lim_{\delta \to 0} [f(x_0 + \delta) + f(x_0 - \delta)] \tag{14.88}$$

In order to evaluate the coefficients a_r and b_r, we make use of some well-known trigonometric results. First, we have

$$\int_{-\pi}^{+\pi} \sin rx \cos sx \, dx = 0 \tag{14.89a}$$

for any integer or zero value of r and s. We also have the identities [cf. Eq. (14.7)]:

$$\int_{-\pi}^{+\pi} \sin rx \sin sx \, dx = \pi \delta_{rs}; \qquad r, s = 1, 2, \ldots \tag{14.89b}$$

$$\int_{-\pi}^{+\pi} \cos rx \cos sx \, dx = \pi \delta_{rs}; \qquad r, s = 1, 2, \ldots \tag{14.89c}$$

The functions $\sin rx$ and $\cos rx$, $r = 1, 2, 3, \ldots$, together with a constant term, represented by $a_0/2$ in Eq. (14.87), therefore constitute a complete orthogonal set which may be used to expand an arbitrary function of x (where, of course, we assume convergence).

We can therefore evaluate the coefficients a_s by multiplying Eq. (14.87) by $\cos sx$ and integrating from $-\pi$ to $+\pi$:

$$\boxed{a_s = \frac{1}{\pi} \int_{-\pi}^{+\pi} f(x) \cos sx \, dx} \tag{14.90a}$$

and similarly for the b_s:

$$\boxed{b_s = \frac{1}{\pi} \int_{-\pi}^{+\pi} f(x) \sin sx \, dx} \tag{14.90b}$$

If we compute the average value of $f(x)$ in the interval $-\pi \le x \le \pi$, we have

$$\overline{f(x)} = \frac{1}{2\pi} \int_{-\pi}^{+\pi} f(x) \, dx \tag{14.91}$$

Inserting the series expansion for $f(x)$ [Eq. (14.87)], we note that the integral of each cosine and sine term vanishes since the average value of $\cos rx$ or of $\sin rx$ is zero between $-\pi$ and $+\pi$ if r is an integer. Only the term involving a_0 is nonvanishing:

$$\frac{1}{2\pi}\int_{-\pi}^{+\pi}\frac{a_0}{2}\,dx = \frac{a_0}{2}$$

Thus,

$$\overline{f(x)} = \frac{a_0}{2} \tag{14.92}$$

and the leading term in the series represents the average value of the function within the interval under consideration.

The result for the complete series is:

$$\boxed{\begin{aligned} f(x) = \frac{1}{2\pi}\int_{-\pi}^{+\pi} f(x')\,dx' + \frac{1}{\pi}\sum_r \Bigg[\cos rx \int_{-\pi}^{+\pi} f(x')\cos rx'\,dx \\ + \sin rx \int_{-\pi}^{+\pi} f(x')\sin rx'\,dx'\Bigg] \end{aligned}} \tag{14.93}$$

We note that $\cos rx$ is an *even* function, whereas $\sin rx$ is an *odd* function; that is,

$$\left.\begin{aligned} \cos rx &= \cos r(-x) \\ \sin rx &= -\sin r(-x) \end{aligned}\right\} \tag{14.94}$$

Thus, if $f(x)$ is an even function, the integral

$$\int_{-\pi}^{+\pi} f(x')\sin rx'\,dx'$$

will vanish, and $f(x)$ will be represented by a cosine series (plus the leading, or constant, term). Similarly, if $f(x)$ is an odd function, only the sine portion of the series will remain (and the leading term vanishes).

We have considered thus far only the representation of the function $f(x)$ in the interval $-\pi < x < \pi$. Because the functions $\sin rx$ and $\cos rx$ have a periodicity of 2π, the series expansion for $f(x)$ [Eq. (14.93)] will also repeat in every interval of 2π. We can, however, alter the range by a simple change of variable. Thus, for the region $-L < x < L$, we make the substitution $x' \to \pi x'/L$ and find

$$\begin{aligned} f(x) = \frac{1}{2L}\int_{-L}^{+L} f(x')\,dx' + \frac{1}{L}\sum_r \Bigg[\cos\frac{r\pi x}{L}\int_{-L}^{+L} f(x')\cos\frac{r\pi x'}{L}\,dx' \\ + \sin\frac{r\pi x}{L}\int_{-L}^{+L} f(x')\sin\frac{r\pi x'}{L}\,dx'\Bigg] \end{aligned} \tag{14.95}$$

In addition, if $f(x)$ is an *odd* function for $-L < x < L$, we have a sine series*:

$$f(x) = \frac{2}{L}\sum_r \sin\frac{r\pi x}{L}\int_0^L f(x')\sin\frac{r\pi x'}{L}\,dx' \tag{14.96}$$

Similarly, if $f(x)$ is an *even* function for $-L < x < L$, the cosine series is*

$$f(x) = \frac{1}{L}\int_0^L f(x')\,dx' + \frac{2}{L}\sum_r \cos\frac{r\pi x}{L}\int_0^L f(x')\cos\frac{r\pi x'}{L}\,dx' \tag{14.97}$$

If $f(x)$ is defined only in the interval $0 < x < L$, and if it is immaterial whether the Fourier series that represents this function is even or odd in the interval $-L < x < L$, then the expansion can be made in terms of either a sine or a cosine series. Depending on the circumstances, one or the other of these choices may be more appropriate.

Suggested References

The vibrations of strings is discussed in most of the references listed in Chapter 13. See particularly Becker (Be54, Chapter 15), Morse (Mo48), and Slater and Frank (Sl47, Chapter 8).

Lord Rayleigh's discussion of vibrating strings (Ra94, Chapter 6) is still a standard. Rayleigh's principle is treated in detail by Temple and Bickley (Te56).

The vibrations of strings with variable density and tension is treated by Slater and Frank (Sl47, Chapter 10) and by Wangness (Wa63, Chapter 16). Perturbation theory is discussed, for example, by Symon (Sy60, Chapter 12).

The theory of Fourier series and general orthogonal functions is discussed in a variety of mathematical texts; see, for example, Arfken (Ar66, Chapter 14), Churchill (Ch41), Kaplan (Ka52, Chapter 7), and Pipes (Pi46, Chapter 3). A particularly useful modern treatment is that by Davis (Da63); orthogonal functions (Chapter 2), Fourier series (Chapter 3), and applications to vibrating strings (Chapter 6) are discussed in detail.

Problems

14-1. Discuss the motion of a continuous string when the initial conditions are $\dot{q}(x, 0) = 0$, $q(x, 0) = A\sin(3\pi x/L)$. Resolve the solution into normal modes.

14-2. Rework the problem in Example 14.2 in the event that the plucked point is a distance $L/3$ from one end. Comment on the nature of the allowed modes.

* Since the integral for $-L < x < 0$ is the same as that for $0 < x < L$, we need to consider only one part and double the result.

14-3. Refer to Example 14.2. Show by a numerical calculation that the initial displacement of the string is well represented by the first three terms of the series in Eq. (5). Sketch the shape of the string at intervals of time of $\frac{1}{8}$ of a period.

14-4. Discuss the motion of a string when the initial conditions are

$$q(x, 0) = 4x(L - x)/L^2; \qquad \dot{q}(x, 0) = 0$$

Find the characteristic frequencies and calculate the amplitude of the nth mode.

14-5. A string with no initial displacement is set into motion by being struck over a length $2s$ about its center. This center section is given an initial velocity v_0. Describe the subsequent motion.

14-6. A string is set into motion by being struck at a point $L/4$ from one end by a triangular hammer. The initial velocity is greatest at $x = L/4$ and decreases linearly to zero at $x = 0$ and $x = L/2$. The region $L/2 \leq x \leq L$ is initially undisturbed. Determine the subsequent motion of the string. Why are the fourth, eighth, etc., harmonics absent? How many decibels down from the fundamental are the second and third harmonics?

14-7. A string is pulled aside a distance h at a point $3L/7$ from one end. At a point $3L/7$ from the other end the string is pulled aside a distance h in the opposite direction. Discuss the vibrations in terms of normal modes.

14-8. Compare, by plotting a graph, the characteristic frequencies ω_r as a function of the mode number r for a loaded string consisting of 3, 5, and 10 particles and for a continuous string with the same values of τ and $m/d = \rho$. Comment on the results.

14-9. Perform the calculations necessary to obtain Eqs. (5), (7), and (8) in Example 14.4(a).

14-10. Work the problem in Example 14.4(b) by taking the displacement function to be a triangle. Show that this approximation yields accurate results for large values of M/m and becomes better than the results obtained in Example 14.4(b) for $M/m > 0.4$. Explain why this is reasonable. Without making a calculation, argue regarding the merit of choosing a parabolic displacement function compared to sinusoidal or triangular functions.

14-11. Three particles, each of mass M, are attached at equal intervals to a (massless) string of length L in which the tension is τ (as in Fig. 13-11). Use

Rayleigh's method to obtain an approximate result for the fundamental frequency. Choose the displacement function to be a parabola and proceed by calculating $\langle T \rangle$ for each of the particles and $\langle U \rangle$ for the string. Compare the approximate result with the true value.

14-12. Consider a uniform string with density $\rho(x) = \rho_0 = \text{const.}$ which is loaded with 3 small weights, equally spaced and each of mass m. Calculate the first-order frequency corrections for the first 5 modes of vibration. Comment on the physical significance of these results.

14-13. Refer to Example 14.6 for the string with a density $\rho(x) = \rho_0 + \lambda x$. Obtain an expression for the frequency ω_r of a string of the same mass, but with uniform density. Compare the result with the first-order result found in the example.

14-14. Obtain the second-order frequency correction for a string with density $\rho(x) = \rho_0 + \lambda x$.

14-15. Consider a string whose density is $\rho(x) = \rho_0 + \lambda \sin(\pi x/L)$. Calculate the elements of the tensor $\{\mathsf{m}\}$ and show that there is coupling between all of the *odd* modes and between all of the *even* modes. (Thus, all of the tensor elements m_{jk} vanish unless j and k are both odd or both even.)

14-16. Perform a perturbation calculation for the string of the preceding problem and obtain the first- and second-order frequency corrections.

14-17. Consider the case of the continuous string of constant density in which the tension is a slowly varying function of the distance along the string. Expand τ in a power series and calculate the first-order frequency correction for this case.

14-18. Refer to Example 14.6 for the string with density $\rho(x) = \rho_0 + \lambda x$. Calculate the first-order correction to the wave functions [see Eq. (14.70)] for the first 3 modes of vibration (i.e., $r = 1, 2, 3$). Examine the magnitudes of successive terms in each summation and retain only the first few large terms.

14-19. Find the Fourier sine series and the Fourier cosine series which represent the function

$$f(x) = e^x, \qquad 0 < x < \pi$$

14-20. Find the Fourier series which represents the function

$$f(x) = x^2, \qquad -L < x < L$$

14-21. Find the Fourier series which represents the function

$$f(x) = \begin{cases} 1, & 0 < x < 1 \\ 2, & 1 < x < 3 \end{cases}$$

$$f(x+3) = f(x), \quad \text{all } x$$

14-22. Expand the function $f(x) = 1$ $(0 < x < \pi)$ in a sine series and substitute $x = \pi/2$ into the result to show that

$$1 - \frac{1}{3} + \frac{1}{5} - \frac{1}{7} + \cdots = \frac{\pi}{4}$$

14-23. Consider a function $f(x)$ which is defined within the interval $-\pi \leq x \leq \pi$. Such a function may be represented by the Fourier series given in Eq. (14.87). The derivative of $f(x)$ may also be expanded as

$$f'(x) = \frac{A_0}{2} + \sum_{r=1}^{\infty} (A_r \cos rx + B_r \sin rx)$$

Show that

$$A_0 = \frac{1}{\pi} [f(\pi) - f(-\pi)]$$

$$A_r = rb_r + \frac{\cos r\pi}{\pi} [f(\pi) - f(-\pi)]$$

$$B_r = -ra_r$$

where the a_r and b_r are given by Eqs. (14.90). Hence, show that $f'(x)$ may be obtained by differentiating the series for $f(x)$ term by term only if

$$f(\pi) = f(-\pi).$$

Demonstrate this result by considering the functions $f_1(x) = x$ and $f_2(x) = x^2$.

CHAPTER 15

The Wave Equation in One Dimension

15.1 Introduction

In the preceding chapter we introduced the *wave equation*, one of the truly fundamental equations of mathematical physics. The solutions of this equation are in general subject to various limitations imposed by certain physical restrictions that are peculiar to a given problem. These limitations frequently take the form of conditions on the solution that must be met at the extremes of the intervals of space and time that are of physical interest. We must therefore deal with a *boundary-value problem* involving a partial differential equation. Indeed, such a description characterizes essentially the whole of what we call *mathematical physics*. We have already seen some indications of the solutions to this type of equation, and in this chapter we shall investigate in detail the solutions of the wave equation. We shall, however, confine ourselves to a discussion of waves in one dimension.* One-dimensional waves describe, for example, the motion of a vibrating string. On the other hand, the compression (or sound) waves which may be transmitted through an elastic medium, such as a gas, can also be approximated

* Three-dimensional waves are treated, for example, in Marion (Ma65b, Chapter 10).

as one-dimensional waves (if the medium is sufficiently large so that the edge effects are unimportant). In such a case, the condition of the medium is approximately the same at every point on a plane, and the properties of the wave motion are then functions only of the distance along a line normal to the plane. Such a wave in an extended medium is called a *plane wave*, and is mathematically identical to the one-dimensional waves which are treated here. We will consider in this chapter only mechanical waves; an extensive discussion of electromagnetic waves is given in Marion (Ma65b).

15.2 General Solutions of the Wave Equation

The one-dimensional wave equation which was found for the vibrating string is* [cf. Eq. (14.42]:

$$\frac{\partial^2 \Psi}{\partial x^2} - \frac{\rho}{\tau}\frac{\partial^2 \Psi}{\partial t^2} = 0 \tag{15.1}$$

where ρ is the linear mass density of the string and where τ is the tension. Now, the dimensions of ρ are $[ML^{-1}]$ and the dimensions of τ are those of a force, viz., $[MLT^{-2}]$. Therefore, the dimensions of ρ/τ are $[T^2L^{-2}]$, i.e., the dimensions of the reciprocal of a squared velocity. If we write $\sqrt{\tau/\rho} = v$, the wave equation becomes

$$\boxed{\frac{\partial^2 \Psi}{\partial x^2} - \frac{1}{v^2}\frac{\partial^2 \Psi}{\partial t^2} = 0} \tag{15.2}$$

It will be one of our tasks to give a physical interpretation of the velocity v; it is not sufficient to say that v is the "velocity of propagation" of the wave.

To show that Eq. (15.2) does indeed represent a general wave motion, we introduce two new variables,

$$\left.\begin{aligned} \xi &\equiv x + vt \\ \eta &\equiv x - vt \end{aligned}\right\} \tag{15.3}$$

Evaluating the derivatives of $\Psi = \Psi(x, t)$ which appear in Eq. (15.2), we have

$$\begin{aligned} \frac{\partial \Psi}{\partial x} &= \frac{\partial \Psi}{\partial \xi}\frac{\partial \xi}{\partial x} + \frac{\partial \Psi}{\partial \eta}\frac{\partial \eta}{\partial x} \\ &= \frac{\partial \Psi}{\partial \xi} + \frac{\partial \Psi}{\partial \eta} \end{aligned} \tag{15.4}$$

* We use the notation $\Psi = \Psi(x,t)$ to denote a *time-dependent* wave function and $\psi = \psi(x)$ to denote a *time-independent* wave function.

Then,

$$\frac{\partial^2\Psi}{\partial x^2} = \frac{\partial}{\partial x}\frac{\partial\Psi}{\partial x} = \frac{\partial}{\partial x}\left(\frac{\partial\Psi}{\partial\xi} + \frac{\partial\Psi}{\partial\eta}\right)$$

$$= \frac{\partial}{\partial\xi}\left(\frac{\partial\Psi}{\partial\xi} + \frac{\partial\Psi}{\partial\eta}\right)\frac{\partial\xi}{\partial x} + \frac{\partial}{\partial\eta}\left(\frac{\partial\Psi}{\partial\xi} + \frac{\partial\Psi}{\partial\eta}\right)\frac{\partial\eta}{\partial x}$$

$$= \frac{\partial^2\Psi}{\partial\xi^2} + 2\frac{\partial^2\Psi}{\partial\xi\,\partial\eta} + \frac{\partial^2\Psi}{\partial\eta^2} \tag{15.5}$$

Similarly, we find

$$\frac{1}{v}\frac{\partial\Psi}{\partial t} = \frac{\partial\Psi}{\partial\xi} - \frac{\partial\Psi}{\partial\eta} \tag{15.6}$$

and,

$$\frac{1}{v^2}\frac{\partial^2\Psi}{\partial t^2} = \frac{1}{v}\frac{\partial}{\partial t}\left(\frac{1}{v}\frac{\partial\Psi}{\partial t}\right) = \frac{1}{v}\frac{\partial}{\partial t}\left(\frac{\partial\Psi}{\partial\xi} - \frac{\partial\Psi}{\partial\eta}\right)$$

$$= \frac{\partial^2\Psi}{\partial\xi^2} - 2\frac{\partial^2\Psi}{\partial\xi\,\partial\eta} + \frac{\partial^2\Psi}{\partial\eta^2} \tag{15.7}$$

But, according to Eq. (15.2), the right-hand sides of Eqs. (15.5) and (15.7) must be equal. This can be true only if

$$\frac{\partial^2\Psi}{\partial\xi\,\partial\eta} \equiv 0 \tag{15.8}$$

The most general expression for ψ that can satisfy this equation is a sum of two terms, one of which depends only on ξ and the other only on η; no more complicated function of ξ and η will permit Eq. (15.8) to be valid. Thus,

$$\Psi = f(\xi) + g(\eta) \tag{15.9}$$

or, substituting for ξ and η,

$$\Psi = f(x + vt) + g(x - vt) \tag{15.9a}$$

where f and g are *arbitrary* functions of the variables $x + vt$ and $x - vt$, respectively, which are not necessarily of a periodic nature, although, of course, they may be.

As time increases, the value of x must also increase in order to maintain a constant value for $x - vt$. Therefore, the function g will retain its original form as time increases if we shift our viewpoint along the x-direction (in a positive sense) with a velocity v. Thus, the function g must represent a disturbance which moves to the right (i.e., to larger values of x) with a velocity v, whereas f represents the propagation of a disturbance to the left. We therefore conclude that Eq. (15.2) does indeed describe wave motion and, in general, a *traveling* (or *propagating*) wave.

Let us now attempt to interpret Eq. (15.9a) in terms of the motion of a stretched string. At the time $t = 0$, the displacement of the string will be described by

$$q(x, 0) = f(x) + g(x)$$

If we take identical triangular forms for $f(x)$ and $g(x)$, the shape of the string at $t = 0$ is as shown at the top of Fig. 15-1. As time increases the disturbance represented by $f(x + vt)$ propagates to the *left* whereas the disturbance

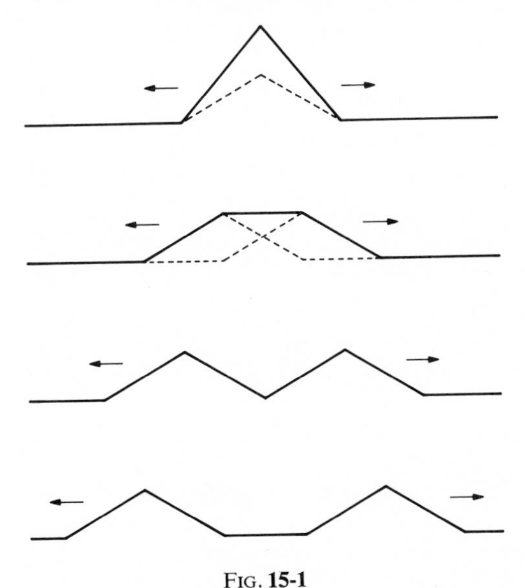

FIG. **15-1**

represented by $g(x - vt)$ propagates to the *right*. This propagation of the individual disturbances to the left and right is illustrated in Fig. 15-1.

Consider next the left-going disturbance alone. If we terminate the string (at $x = 0$) by attaching it to a rigid support, we will find the phenomenon of *reflection*. Because the support is *rigid*, we must have $f(vt) \equiv 0$ for all values of the time. Clearly, this condition cannot be met by the function f alone (unless it trivially vanishes). We can satisfy the condition at $x = 0$ if we consider, in addition to $f(x + vt)$, an imaginary disturbance, $-f(-x + vt)$, which approaches the boundary point from the left, as in Fig. 15-2. The disturbance $f(x + vt)$ continues to propagate to the left, even into the imaginary section of the string ($x < 0$), while the disturbance $-f(-x + vt)$

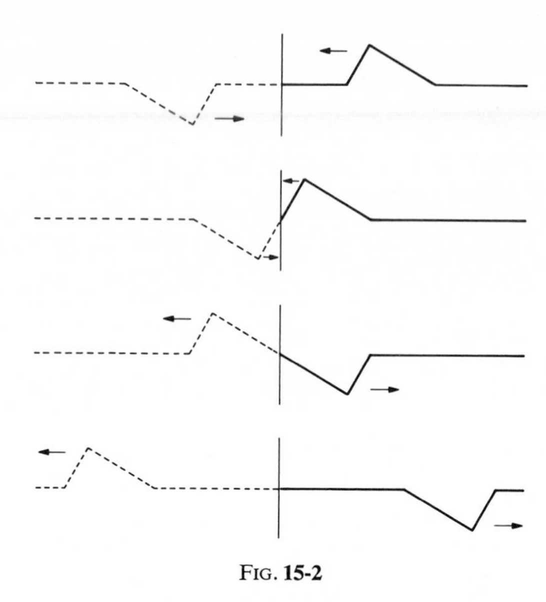

FIG. 15-2

propagates across the boundary and along the real string. The net effect is that the original disturbance is *reflected* at the support and thereafter propagates to the *right*.

If the string is terminated by rigid supports at $x = 0$ and also at $x = L$, the disturbance will propagate periodically back and forth with a period $2L/v$.

15.3 Separation of the Wave Equation

If we require a general solution of the wave equation which is harmonic (as for the vibrating string or, for that matter, for a large number of problems of physical interest), we can write

$$\Psi(x, t) = \psi(x)e^{i\omega t} \tag{15.10}$$

so that the one-dimensional wave equation [Eq. (15.2)] becomes

$$\frac{\partial^2 \psi}{\partial x^2} + \frac{\omega^2}{v^2}\psi = 0 \tag{15.11}$$

where ψ is now a function of x only.

The general wave motion of a system is not, of course, restricted to a single frequency ω. For a system with n degrees of freedom there are n possible characteristic frequencies, and for a continuous string there is an infinite set* of frequencies. If we designate the rth frequency by ω_r, the wave function corresponding to this frequency is

$$\Psi_r(x, t) = \psi_r(x)e^{i\omega_r t} \tag{15.12}$$

The complete wave function is a superposition (recall that we are dealing with a *linear* system) of all the particular wave functions (or *modes*). Thus,

$$\Psi(x, t) = \sum_r \Psi_r(x, t) = \sum_r \psi_r(x)e^{i\omega_r t} \tag{15.13}$$

In Eq. (15.10) we *assumed* that the wave function was periodic in time. But now we see that this assumption entails no real restriction at all (apart from the usual assumptions regarding the continuity of the functions and the convergence of the series), since the summation in Eq. (15.13) actually gives a Fourier representation of the wave function and is therefore the most general expression for the true wave function.†

We now wish to show that Eq. (15.12) results naturally from a powerful method that can often be used to obtain solutions to partial differential equations—the method of *separation of variables*. First, we express the solution as

$$\Psi(x, t) \equiv \psi(x) \cdot \chi(t) \tag{15.14}$$

That is, we assume that the variables are *separable* and therefore that the complete wave function can be expressed as the product of two functions, one of which is a spatial function only, and one of which is a temporal function only. It is not guaranteed that we will always be able to find such functions, but many of the partial differential equations encountered in physical problems are, fortunately, separable in at least one coordinate system; some (such as those involving the Laplacian operator) are separable in many coordinate systems. In short, the justification of the method of separation of variables, as is the case with many assumptions in physics, is in

* An infinite set of frequencies would exist for a truly *continuous* string, but since a real string is composed fundamentally of atoms, there does exist an upper limit for ω (see Section 15.4).

† Euler proved in 1748 that the wave equation for a continuous string was satisfied by an arbitrary function of $x \pm vt$, and Daniel Bernoulli showed in 1753 that the motion of a string was a superposition of its characteristic frequencies. These two results, taken together, indicated that an arbitrary function could be described by a superposition of trigonometric functions. This Euler could not believe, and so he (as well as Lagrange) rejected Bernoulli's superposition principle. The French mathematician Alexis Claude Clairaut (1713–1765) gave a proof in an obscure paper in 1754 that the results of Euler and Bernoulli were actually consistent, but it was not until Fourier gave his famous proof in 1807 that the question was settled.

its success in producing acceptable solutions to a problem. ("Acceptable" ≡ "experimentally verifiable.")

Substituting $\Psi = \psi\chi$ into Eq. (15.2), we have

$$\chi \frac{d^2\psi}{dx^2} - \frac{\psi}{v^2}\frac{d^2\chi}{dt^2} = 0$$

or,

$$\frac{v^2}{\psi}\frac{d^2\psi}{dx^2} = \frac{1}{\chi}\frac{d^2\chi}{dt^2} \tag{15.15}$$

But, in view of the definitions of $\psi(x)$ and $\chi(t)$, the left-hand side of Eq. (15.15) is a function of x alone, while the right-hand side is a function of t alone. This is possible only if each part of the equation is equal to the same constant. To be consistent with our previous notation, we choose this constant to be $-\omega^2$. Thus, we have

$$\frac{d^2\psi}{dx^2} + \frac{\omega^2}{v^2}\psi = 0 \tag{15.16a}$$

and,

$$\frac{d^2\chi}{dt^2} + \omega^2\chi = 0 \tag{15.16b}$$

These equations are of a familiar form, and we know that the solutions are

$$\psi(x) = Ae^{i(\omega/v)x} + Be^{-i(\omega/v)x} \tag{15.17a}$$

$$\chi(t) = Ce^{i\omega t} + De^{-i\omega t} \tag{15.17b}$$

where the constants A, B, C, D are, of course, determined by the boundary conditions. We may write the solution $\Psi(x, t)$ in a shorthand manner as

$$\begin{aligned}\Psi(x, t) = \psi(x)\chi(t) &\sim \exp[\pm i(\omega/v)x]\exp[\pm i\omega t] \\ &\sim \exp[\pm i(\omega/v)(x \pm vt)]\end{aligned} \tag{15.18}$$

This notation means that the wave function Ψ varies as a *linear combination* of the terms

$$\exp[i(\omega/v)(x + vt)]$$

$$\exp[i(\omega/v)(x - vt)]$$

$$\exp[-i(\omega/v)(x + vt)]$$

$$\exp[-i(\omega/v)(x - vt)]$$

The *separation constant* for Eq. (15.15) was chosen to be $-\omega^2$. There is nothing in the mathematics of the problem which indicates that there is a unique value of ω; hence, there must exist a set* of equally acceptable frequencies, ω_r. To each such frequency there corresponds a wave function:

$$\Psi_r(x, t) \sim \exp[\pm i(\omega_r/v)(x \pm vt)]$$

The general solution is therefore not only a linear combination of the harmonic terms, but also a sum over all possible frequencies:

$$\begin{aligned}\Psi(x, t) &\sim \sum_r a_r \Psi_r \\ &\sim \sum_r a_r \exp[\pm i(\omega_r/v)(x \pm vt)] \qquad (15.19)\end{aligned}$$

The general solution of the wave equation therefore leads to a very complicated wave function. There are, in fact, an infinite number of arbitrary constants a_r. This is a general result for partial differential equations; but this infinity of constants must satisfy the physical requirements of the problem (the *boundary conditions*), and therefore they can be evaluated in the same manner that the coefficients of an infinite Fourier expansion can be evaluated.

For much of our discussion it will be sufficient to consider only one of the four possible combinations expressed by Eq. (15.18); that is, we select a wave propagating in a particular direction and with a particular phase. Then we can write, for example,

$$\Psi_r(x, t) \sim \exp[-i(\omega_r/v)(x - vt)]$$

This is, of course, the rth *Fourier component* of the wave function, and the general solution will be a summation over all such components. However, the functional form of each component is the same, and they can be discussed separately. Thus, we shall usually write, for simplicity,

$$\Psi(x, t) \sim \exp[-i(\omega/v)(x - vt)] \qquad (15.20)$$

It will be understood that the general solution must be obtained by a summation over all frequencies that are allowed by the particular physical situation.

It is customary to write the differential equation for $\psi(x)$ as

$$\boxed{\frac{d^2\psi}{dx^2} + k^2\psi = 0} \qquad (15.21)$$

* At this stage of the development, the set is, in fact, infinite since no frequencies have yet been eliminated by imposing the boundary conditions.

which is the time-independent form of the one-dimensional wave equation, also called the *Helmholtz equation*,* and where

$$k^2 \equiv \frac{\omega^2}{v^2} \tag{15.22}$$

The quantity k is called the *propagation constant* or the *wave number* (i.e., the number of wavelengths per unit length) and has dimensions $[L^{-1}]$. The wavelength λ is the distance required for one complete vibration of the wave:

$$\lambda = \frac{v}{\nu} = \frac{2\pi v}{\omega}$$

and hence the relation between k and λ is

$$k = \frac{2\pi}{\lambda}$$

Therefore, we can write in general

$$\Psi_r(x, t) \sim e^{\pm ik_r(x \pm vt)}$$

or, for the simplified wave function,

$$\Psi(x, t) \sim e^{-ik(x-vt)} = e^{i(\omega t - kx)} \tag{15.23}$$

If we superimpose two traveling waves of the type given by Eq. (15.23), and if these waves are of equal magnitude (or *amplitude*) but moving in opposite directions, then

$$\Psi = \Psi_+ + \Psi_- = Ae^{-ik(x+vt)} + Ae^{-ik(x-vt)} \tag{15.24}$$

or,

$$\begin{aligned}\Psi &= Ae^{-ikx}(e^{i\omega t} + e^{-i\omega t}) \\ &= 2Ae^{-ikx} \cos \omega t\end{aligned}$$

The real part of which is

$$\Psi = 2A \cos kx \cos \omega t \tag{15.25}$$

Such a wave no longer has the property that it propagates; the wave form does not move forward with time. There are, in fact, certain positions which undergo no motion whatsoever. These positions are the *nodes* and result from the complete cancellation of one wave by the other. The nodes of the wave function given by Eq. (15.25) occur at $x = (2n + 1)\pi/2k$, where n

* Hermann von Helmholtz (1821–1894) used this form of the wave equation in his treatment of acoustic waves, 1859.

is an integer. Because there are fixed positions in waves of this type, they are called *standing waves*. Solutions to the problem of the vibrating string are of this form (but with a phase factor attached to the term kx so that the cosine is transformed into a sine function satisfying the boundary conditions).

15.4 Phase Velocity, Dispersion, and Attenuation

We have seen, in Eqs. (15.18), that the general solution to the wave equation produces, even in the one-dimensional case, a complicated system of exponential factors. For the purposes of further discussion, we shall restrict our attention to the particular combination

$$\boxed{(\Psi x, t) = Ae^{i(\omega t - kx)}} \tag{15.26}$$

This equation describes the propagation to the right (larger x) of a wave which possesses a well-defined frequency ω. Certain physical situations can be quite adequately approximated by a wave function of this type—e.g., the propagation of a monochromatic light wave in space, or the propagation of a sinusoidal wave on a long (strictly, *infinitely* long) string.

If the argument of the exponential in Eq. (15.26) remains constant, then the wave function $\Psi(x, t)$ will, of course, also remain constant. The argument of the exponential is called the *phase* φ of the wave,

$$\varphi \equiv \omega t - kx \tag{15.27}$$

and, if we move our viewpoint along the x-axis at a velocity such that the phase at every point is the same, we will always see a stationary wave of the same shape. The velocity V with which we must move is called the *phase velocity* of the wave and corresponds to the velocity with which the *waveform* propagates. To insure $\varphi = \text{const.}$, we set

$$d\varphi = 0 \tag{15.28}$$

or,

$$\omega\, dt = k\, dx$$

from which

$$V = \frac{dx}{dt} = \frac{\omega}{k} = v \tag{15.29}$$

so that the *phase velocity* in this case is just the quantity originally introduced as the *velocity*. It is important to note that it is possible to speak of a "phase velocity" only when the wave function has the same form throughout its length. This condition is necessary in order that we be able to measure the wavelength by taking the distance between *any* two successive wave crests (or between *any* two successive corresponding points on the wave). If the wave form were to change as a function of time or of distance along the wave, these measurements would not always yield the same results. As we shall see in Section 15.8, the only way in which we can insure that the wavelength will not be a function of time or space (i.e., that ω will be *pure*) is that the wave train be of infinite length. If the wave train is of finite length, there must be a spectrum of frequencies present in the wave, and the phase velocity has no meaning in the strict sense. However, we shall occasionally speak of the "phase velocity" for a wave which is not actually infinite in extent. This terminology will be convenient, even though it is not strictly correct.

Let us next return to the example of the loaded string and examine the properties of the phase velocity in that case. We have previously found [Eq. (13.101)] that the frequency for the rth mode of the loaded string when terminated at both ends is given by

$$\omega_r = 2\sqrt{\frac{\tau}{md}} \sin\left(\frac{r\pi}{2(n+1)}\right) \tag{15.30}$$

where the notation is the same as in Chapter 13. We recall also that we take only positive values for the frequencies. Now, when $r = 1$, there is a node at both ends, and none between; hence, the length of the string is one-half of a wavelength. Similarly, when $r = 2$, $L = \lambda$ and, in general, $\lambda_r = 2L/r$. Therefore,

$$\frac{r\pi}{2(n+1)} = \frac{r\pi d}{2d(n+1)} = \frac{r\pi d}{2L} = \frac{\pi d}{\lambda_r} = \frac{k_r d}{2} \tag{15.31}$$

and then

$$\omega_r = 2\sqrt{\frac{\tau}{md}} \sin\left(\frac{k_r d}{2}\right) \tag{15.32}$$

Since this expression no longer contains n or L, it applies equally well to a terminated or an infinite loaded string.

In order to study the propagation of a wave in the loaded string, we initiate a disturbance by selecting one of the particles, say the *zero*th one, and forcing it to move according to

$$q_0(t) = Ae^{i\omega t} \tag{15.33}$$

If the string contains many particles,* then any frequency less than $2\sqrt{\tau/md}$,

* Strictly, we need an infinite number of particles for this type of analysis, but we may approach as closely as desired to the ideal conditions by increasing the finite number of particles.

in particular, the frequency ω in Eq. (15.33), is an allowed frequency, i.e., an eigenfrequency. After the transient effects have subsided and the steady-state conditions attained, the phase velocity of the wave is given by*†

$$V = \frac{\omega}{k} = \sqrt{\frac{\tau d}{m}} \frac{|\sin(kd/2)|}{kd/2} = V(k) \tag{15.34}$$

Therefore, we have the result that the phase velocity is a function of the wave number; i.e., V is frequency-dependent. When $V = V(k)$ for a given medium, that medium is said to be *dispersive*, and the wave exhibits *dispersion*. The best known example of this phenomenon is the simple optical prism. The index of refraction of the prism is dependent on the wavelength of the incident light (i.e., the prism is a dispersive medium for optical light); and, upon passing through the prism, the light is separated into a spectrum of wavelengths (i.e., the light wave is *dispersed*).

From Eq. (15.34) we see that as the wavelength becomes very long ($\lambda \to \infty$ or $k \to 0$) the phase velocity approaches the constant value

$$V(\lambda \to \infty) = \sqrt{\frac{\tau d}{m}} \tag{15.35}$$

But, otherwise, $V = V(k)$, and the wave is dispersive. We note that the phase velocity for the continuous string is [see Eq. (15.2)]:

$$V_{\text{cont.}} = v = \sqrt{\frac{\tau}{\rho}} \tag{15.36}$$

and since m/d for the loaded string corresponds to ρ for the continuous string, the phase velocities for the two cases are equal in the long-wavelength limit (but *only* in this limit). This is a reasonable result since as λ becomes large compared with d, the properties of the wave are less sensitive to the spacing between particles, and, in the limit, d may vanish without affecting the phase velocity.

In Eq. (15.32), the restriction on r is $1 \leq r \leq n$. Then since $k_r = r\pi/L$, we see that the maximum value of k is

$$k_{\text{max}} = \pi/d \tag{15.37}$$

* In Eq. (15.30) the values of r are required to be $\leq n$ [see Eq. (13.94)], so that we automatically have $\omega_r \geq 0$ since $\sin[r\pi/2(n+1)] \geq 0$ for $0 \leq r \leq n$. We no longer have such a restriction on kd so that $\sin(kd/2)$ can become negative. We continue to consider only positive frequencies by always taking only the magnitude of $\sin(kd/2)$.

† This result was obtained by Baden-Powell in 1841, but it was William Thomson (Lord Kelvin) (1824–1907) who realized the full significance, 1881.

The corresponding frequency, from Eq. (15.34), is $2\sqrt{\tau/md}$. What will be the result if we force the string to vibrate at a frequency *greater* than $2\sqrt{\tau/md}$? For this purpose we allow k to become complex and inquire as to the consequences:

$$k \equiv \kappa - i\beta, \qquad \kappa, \beta > 0 \tag{15.38}$$

The expression for ω [Eq. (15.32)] then becomes

$$\begin{aligned}
\omega &= 2\sqrt{\frac{\tau}{md}} \sin \frac{d}{2}(\kappa - i\beta) \\
&= 2\sqrt{\frac{\tau}{md}} \left[\sin \frac{d\kappa}{2} \cos \frac{i\beta d}{2} - \cos \frac{\kappa d}{2} \sin \frac{i\beta d}{2} \right] \\
&= 2\sqrt{\frac{\tau}{md}} \left[\sin \frac{\kappa d}{2} \cosh \frac{\beta d}{2} - i \cos \frac{\kappa d}{2} \sinh \frac{\beta d}{2} \right]
\end{aligned} \tag{15.39}$$

If the frequency is to be a real quantity, the imaginary part of this expression must vanish. Thus, we may have either $\cos(\kappa d/2) = 0$ or $\sinh(\beta d/2) = 0$. But the latter choice would require $\beta = 0$, contrary to the necessity for k to be complex. Therefore, we have

$$\cos \frac{\kappa d}{2} = 0 \tag{15.40}$$

For this case we must also have

$$\sin \frac{\kappa d}{2} = 1 \tag{15.41}$$

Then the expression for the frequency becomes

$$\omega = 2\sqrt{\frac{\tau}{md}} \cosh \frac{\beta d}{2} \tag{15.42}$$

Thus, we have the result that for $\omega \leq 2\sqrt{\tau/md}$, the wave number k is real, and the relation between ω and k is given by Eq. (15.32); whereas for $\omega > 2\sqrt{\tau/md}$, k is complex with the real part κ fixed by Eq. (15.40) at the value $\kappa = \pi/d$ and with the imaginary part β given by Eq. (15.42). The situation is shown in Fig. 15-3.

What is the physical significance of a complex wave number? Our original wave function was of the form

$$\Psi = Ae^{i(\omega t - kx)}$$

but if $k = \kappa - i\beta$, then Ψ can be written as

$$\Psi = Ae^{-\beta x} e^{i(\omega t - \kappa x)} \tag{15.43}$$

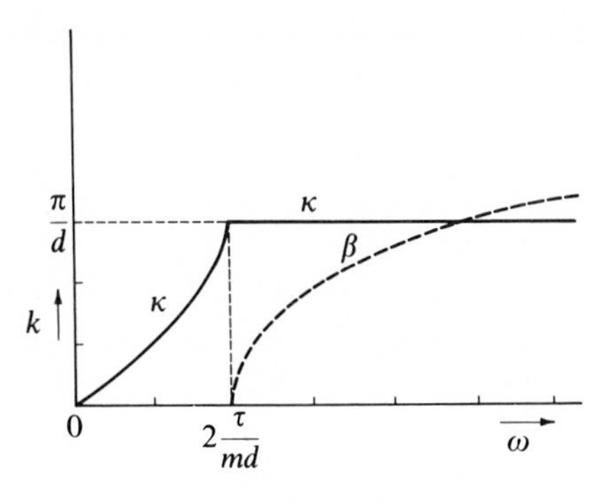

FIG. 15-3

and the factor $\exp(-\beta x)$ represents a damping or an *attenuation* of the wave with increasing distance x. We conclude, therefore, that the wave is propagated without attenuation for $\omega \leq 2\sqrt{\tau/md}$ (this region is called the *passing band* of frequencies), and that attenuation sets in at $\omega_c = 2\sqrt{\tau/md}$ (called the *critical* or *cutoff* frequency*) and increases with increasing frequency.

The physical significance of the real and imaginary parts of k is now apparent: β is the attenuation coefficient† (and exists only if $\omega > \omega_c$), whereas κ is the wave number in the sense that the phase velocity V' is given by

$$V' = \frac{\omega}{\kappa} = \frac{\omega}{\text{Re } k} \tag{15.44}$$

rather than by $V = \omega/k$. Of course, if k is real, these expressions for V and V' are identical.

This example emphasizes the fact that the fundamental definition of the phase velocity is based upon the requirement of the constancy of the phase and *not* upon the ratio ω/k. Thus, in general, the phase velocity V and the so-called wave velocity v are distinct quantities. We note also that if ω is real and if the wave number k is complex, then the *wave velocity* v must also be complex so that the product kv will yield a real quantity for the frequency through the relation $\omega = kv$. On the other hand, the *phase velocity*, which arises from the requirement that $\varphi = const.$, is necessarily always a real quantity.

* The occurrence of a cutoff frequency was discovered by Lord Kelvin, 1881.

† The reason for writing $k = \kappa - i\beta$ rather than $k = \kappa + i\beta$ in Eq. (15.38) is now clear; if $\beta > 0$ for the latter choice, then the amplitude of the wave would increase without limit rather than decrease toward zero.

In the preceding discussion we considered the system to be *conservative* and we argued that this required ω to be a real quantity.* We found that if ω exceeds the critical frequency ω_c, attenuation results and the wave number becomes complex. If we relax the condition that the system is conservative, the frequency may then be *complex* and the wave number *real.* In such a case the wave is damped in *time*, rather than in *space*. (See Problem 15-1.) Spatial attentuation (ω real, k complex) is of particular significance for traveling waves, whereas temporal attentuation (ω complex, k real) is important for standing waves.

Although attenuation occurs in the loaded string if $\omega > \omega_c$, the system is still conservative and no energy is lost. This seemingly anomalous situation results because the force applied to the particle in the attempt to initiate a traveling wave is (after the steady-state condition of an attenuated wave is set up) exactly 90° out of phase with the velocity of the particle so that the power transferred, $P = \mathbf{F} \cdot \mathbf{v}$, is zero.

In this treatment of the loaded string, we have tacitly assumed an ideal situation. That is, the system was assumed to be lossless. As a result, we found that there was attenuation for $\omega > \omega_c$, but none for $\omega < \omega_c$. Of course, every real system is subject to losses, so that, in fact, there is some attenuation even for $\omega < \omega_c$. We shall return to this point in the next section and make some qualitative remarks concerning *lossy* systems.

15.5 Electrical Analogies—Filtering Networks

The occurrence of filtering action and a cutoff frequency in the propagation of waves along a loaded string has an important analogy in the theory of electrical networks. Since the impedance presented to an alternating electric current by an inductance L is proportional to ωL and since the impedance presented by a capacitance C is proportional to $1/\omega C$, then by a proper arrangement of inductors and capacitors it is possible to selectively shunt either the high or low frequencies and pass the others.† For example, a high-pass filter network can be constructed as in Fig. 15-4, in which the low frequencies are shunted by the inductors while the high frequencies are transmitted through the capacitors.

Similarly, a low-pass filter can be constructed by interchanging the positions of the inductors and capacitors. Let us analyze such a case. Figure 15-5 shows a section of a transmission line with lumped inductive and capacitive

* See the discussion in Section 13.6 in the paragraph following Eq. (13.52).

† Oliver Heaviside (1850–1925) realized this fact in 1887, but it was not until 1900 that a successful low-pass filter was constructed by Pupin. A high-pass filter was built by Campbell in 1906.

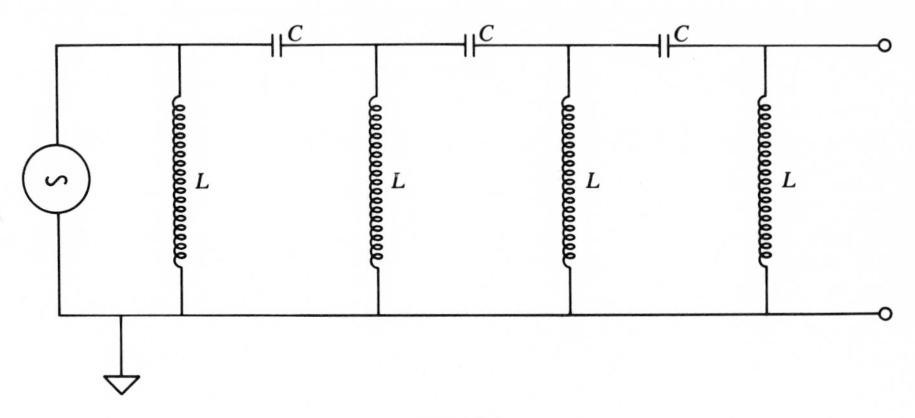

FIG. 15-4

elements. (We neglect, for the moment, the possibility of resistance in the line.) We let the current in the nth inductor be I_n, and the potential to ground at the point between the nth elements be V_n. We have, then,

$$V_n = \frac{Q_n}{C} \tag{15.45}$$

and

$$\begin{aligned} L \frac{dI_n}{dt} &= V_{n-1} - V_n \\ &= \frac{Q_{n-1}}{C} - \frac{Q_n}{C} \end{aligned} \tag{15.46}$$

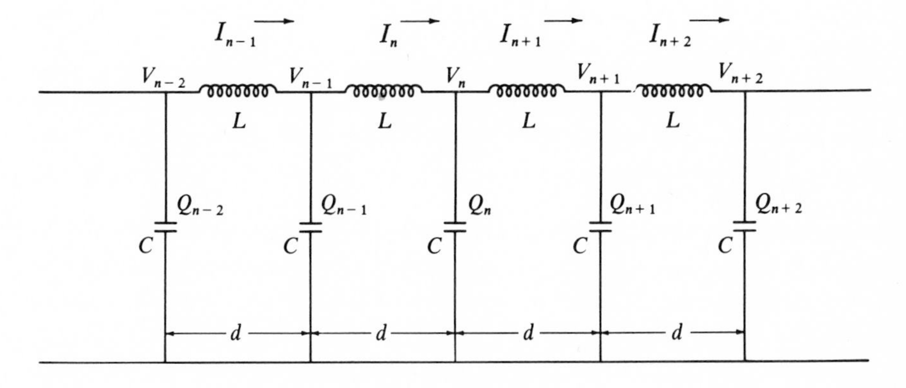

FIG. 15-5

We can also write

$$\frac{dQ_n}{dt} = I_n - I_{n+1} \tag{15.47}$$

Differentiating Eq. (15.46) with respect to time and then substituting for dQ_n/dt from Eq. (15.47), we obtain

$$\frac{d^2 I_n}{dt^2} = \frac{1}{LC}(I_{n-1} - 2I_n + I_{n+1}) \tag{15.48}$$

This equation is of exactly the same form as that we have previously obtained for the loaded string [Eq. (13.81)]:

$$\frac{d^2 q_n}{dt^2} = \frac{\tau}{md}(q_{n-1} - 2q_n + q_{n+1}) \tag{15.48a}$$

Therefore, the solution of the equation for the lumped-element electrical transmission line must be equivalent to that for the vibrating mechanical system. Thus, the phase velocity is [cf. Eq. (15.34)]:

$$V(k) = \frac{d}{\sqrt{LC}} \frac{|\sin(kd/2)|}{kd/2} \tag{15.49}$$

where d is the spacing between elements in the line. The velocity of propagation for very long wavelengths is therefore $d/\sqrt{LC}$.

According to the discussion in the last section, the cutoff frequency is equal to twice the square root of the coefficient of the right-hand side of Eq. (15.48); thus

$$\omega_c = \frac{2}{\sqrt{LC}} \tag{15.50}$$

so that all frequencies below ω_c are transmitted and those above ω_c are highly attenuated.

The k *versus* ω curve for the idealized lumped L-C transmission line is the same as Fig. 15-3. If, however, we consider a more realistic case in which the line has resistance (and, therefore, *losses*) in addition to inductance and capacitance, then the cutoff frequency is no longer sharp, and the k-ω curve has the general shape shown in Fig. 15.6.

It should be emphasized that the occurrence of the cutoff frequency is a result of the *discrete* nature of the inductive and capacitive elements. If these elements were distributed uniformly along the line, then a situation analogous to the continuous string would result. Since the velocity of propagation for long wavelengths along the lumped-element line is $d/\sqrt{LC}$, then for the uniformly distributed line we expect a propagation velocity of $1/\sqrt{L'C'}$,

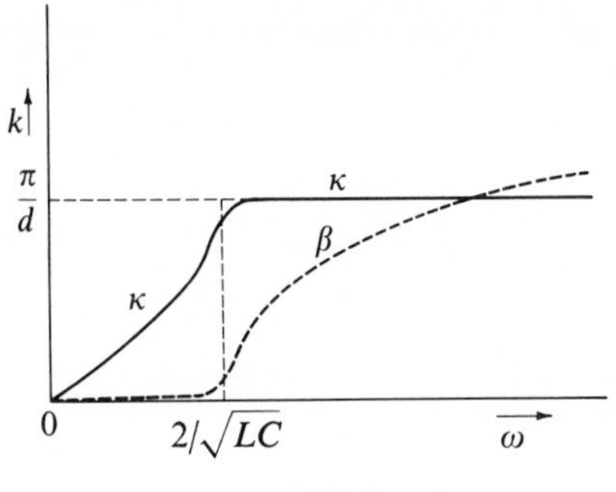

FIG. **15-6**

where L' and C' are the inductance and capitance per unit length, respectively. (See Problem 15-2.)

15.6 Group Velocity and Wave Packets

It was demonstrated in Section 6.4 that the superposition of various solutions of a linear differential equation is still a solution to the equation. Indeed, we formed the general solution to the problem of small oscillations [see Eq. (13.55)] by summing all of the particular solutions. Let us assume, therefore, that we have two almost equal solutions to the wave equation represented by the wave functions Ψ_1 and Ψ_2, each of which has the same amplitude,

$$\left.\begin{aligned}\Psi_1(x, t) &= Ae^{i(\omega t - kx)}\\ \Psi_2(x, t) &= Ae^{i(\Omega t - Kx)}\end{aligned}\right\} \tag{15.51}$$

but whose frequencies and wave numbers differ by only small amounts:

$$\left.\begin{aligned}\Omega &= \omega + \Delta\omega\\ K &= k + \Delta k\end{aligned}\right\} \tag{15.52}$$

Forming the solution which consists of the sum of Ψ_1 and Ψ_2, we have

$$\begin{aligned}\Psi(x, t) = \Psi_1 + \Psi_2 &= A[\exp\{i\omega t\} \exp\{-ikx\}\\ &\quad + \exp\{i(\omega + \Delta\omega)t\} \exp\{-i(k + \Delta k)x\}]\\ &= A\left[\exp\left\{i\left(\omega + \frac{\Delta\omega}{2}\right)t\right\} \exp\left\{-i\left(k + \frac{\Delta k}{2}\right)x\right\}\right]\\ &\quad \cdot \left[\exp\left\{-i\left(\frac{(\Delta\omega)t - (\Delta k)x}{2}\right)\right\} + \exp\left\{i\left(\frac{(\Delta\omega)t - (\Delta k)x}{2}\right)\right\}\right]\end{aligned}$$

The second bracket is just twice the cosine of the argument of the exponential, and the real part of the first bracket is also a cosine. Thus, the real part of the wave function is

$$\Psi(x, t) = 2A \cos\left[\frac{(\Delta\omega)t - (\Delta k)x}{2}\right] \cos\left[\left(\omega + \frac{\Delta\omega}{2}\right)t - \left(k + \frac{\Delta k}{2}\right)x\right] \quad (15.53)$$

This expression is similar to that obtained for the problem of the weakly-coupled oscillators (see Section 13.3), in which we found a slowly varying amplitude, corresponding to the term

$$2A \cos\left[\frac{(\Delta\omega)t - (\Delta k)x}{2}\right]$$

which modulates the wave function. The primary oscillation takes place at a frequency $\omega + (\Delta\omega/2)$ which, according to our assumption that $\Delta\omega$ is small, differs negligibly from ω. The varying amplitude gives rise to *beats*, as shown in Fig. 15-7.

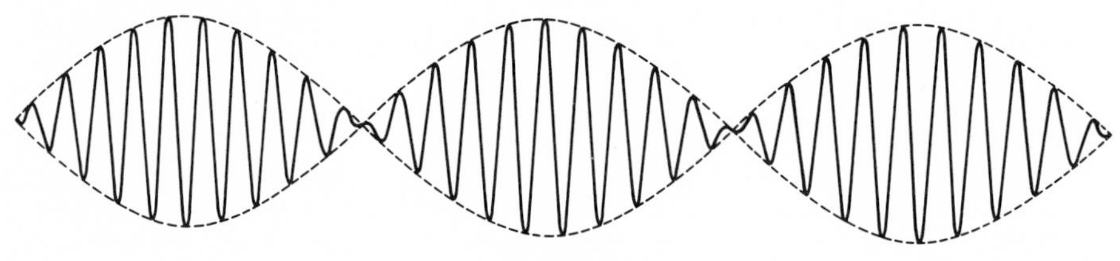

FIG. **15-7**

The velocity U (called the *group velocity**) with which the modulations (or groups of waves) propagate is given by the requirement that the phase of the amplitude term be constant. Thus,

$$U = \frac{dx}{dt} = \frac{\Delta\omega}{\Delta k} \quad (15.54)$$

In a nondispersive medium $\Delta\omega/\Delta k = V$, so that the group and phase velocities are identical.† If dispersion is present, however, U and V are distinct.

Thus far we have considered only the superposition of two waves. If we wish to superpose a system of n waves, we must write

$$\Psi(x, t) = \sum_{r=1}^{n} A_r \exp[i(\omega_r t - k_r x)] \quad (15.55)$$

* The concept of group velocity is due to Hamilton, 1839; the distinction between phase and group velocity was made clear by Lord Rayleigh (*Theory of Sound*, 1st edition, 1877; see Ra94).

† This is shown explicitly in Eq. (15.61).

where the A_r represent the amplitudes of the individual waves. In the event that n becomes very large (strictly, infinite), then the frequencies are continuously distributed, and we may replace the summation by an integration, obtaining*

$$\Psi(x, t) = \int_{-\infty}^{+\infty} A(k) e^{i(\omega t - kx)} \, dk \tag{15.55a}$$

where the factor $A(k)$ represents the distribution of amplitudes of the component waves with different frequencies, i.e., it is the *spectral distribution* of the waves. The most interesting cases occur when $A(k)$ has a significant value only in the neighborhood of a particular wave number, say k_0, and becomes vanishingly small for k outside a small range, denoted by $k_0 \pm \Delta k$. Then, the wave function can be written as

$$\Psi(x, t) = \int_{k_0 - \Delta k}^{k_0 + \Delta k} A(k) e^{i(\omega t - kx)} \, dk \tag{15.56}$$

A function of this type is called a *wave packet*.† The concept of a group velocity applies only to those cases which can be represented by a wave packet, i.e., to wave functions containing a small range (or *band*) of frequencies.

For the case of the wave packet represented by Eq. (15.56), the contributing frequencies will be restricted to those lying near $\omega(k_0)$. We can therefore expand $\omega(k)$ about $k = k_0$:

$$\omega(k) = \omega(k_0) + \left(\frac{d\omega}{dk}\right)_{k=k_0} \cdot (k - k_0) + \cdots \tag{15.57}$$

which we can abbreviate as

$$\omega = \omega_0 + \omega_0'(k - k_0) + \cdots \tag{15.57a}$$

The argument of the exponential in the wave packet integral becomes, approximately,

$$\omega t - kx = (\omega_0 t - k_0 x) + \omega_0'(k - k_0)t - (k - k_0)x$$

where we have added and subtracted the term $k_0 x$. Thus,

$$\omega t - kx = (\omega_0 t - k_0 x) + (k - k_0)(\omega_0' t - x) \tag{15.58}$$

* We have previously made the tacit assumption that $k \geq 0$. However, k is defined by $k^2 = \omega^2/v^2$ [see Eq. (15.22)], so that there is no mathematical reason why we may not also have $k < 0$. Therefore, we may extend the region of integration to include $-\infty < k < 0$ without mathematical difficulty. This procedure allows the identification of the integral representation of $\Psi(x, t)$ as a Fourier integral (see Section 15.7).

† The term "wave packet" is due to Erwin Schrödinger.

and Eq. (15.56) becomes

$$\Psi(x, t) = \int_{k_0 - \Delta k}^{k_0 + \Delta k} A(k) \exp[i(k - k_0)(\omega_0' t - x)] \exp[i(\omega_0 t - k_0 x)]\, dk \qquad (15.59)$$

The wave packet, expressed in this fashion, may be interpreted as follows. The quantity

$$A(k) \exp[i(k - k_0)(\omega_0' t - x)]$$

constitutes an effective amplitude which, because of the small quantity $(k - k_0)$ in the exponential, varies slowly with time and describes the motion of the wave packet (or envelope of a group of waves) in the same manner that the term

$$2A \cos\left[\frac{(\Delta\omega)t - (\Delta k)x}{2}\right]$$

described the propagation of the packet formed from two superposed waves. The requirement of constant phase for the amplitude term now leads to

$$U = \omega_0' = \left(\frac{d\omega}{dk}\right)_{k=k_0} \qquad (15.60)$$

for the group velocity. As stated earlier, only if the medium is dispersive does U differ from the phase velocity V. To show this explicitly, we write Eq. (15.60) as

$$\frac{1}{U} = \left(\frac{dk}{d\omega}\right)_0$$

where the subscript zero means "evaluated at $k = k_0$ or, equivalently, at $\omega = \omega_0$." Since $k = \omega/v$,

$$\frac{1}{U} = \left[\frac{d}{d\omega}\left(\frac{\omega}{v}\right)\right]_0 = \frac{v_0 - (\omega\, dv/d\omega)_0}{v_0^2}$$

Thus,

$$U = \frac{v_0}{1 - \dfrac{\omega_0}{v_0}\left(\dfrac{dv}{d\omega}\right)_0} \qquad (15.61)$$

If the medium is nondispersive, $v = V = \text{const.}$, so that $dv/d\omega = 0$ [see Eq. (15.29)]; hence, $U = v_0 = V$.

The remaining quantity in Eq. (15.59), $\exp[i(\omega_0 t - k_0 x)]$, varies rapidly with time; and, if this were the only factor in Ψ, it would describe an infinite wave train oscillating at a frequency ω_0 and traveling with a phase velocity $V = \omega_0/k_0$.

We should note that an infinite train of waves of a given frequency cannot transmit a signal or carry information from one point to another. This can be accomplished only if it is possible to start and stop a wave train and thereby impress a signal on the wave. This constitutes forming a wave packet. As a consequence of this fact, it is the group velocity, not the phase velocity, that corresponds to the velocity at which a signal may be transmitted.*† We shall make this point clear when, in Section 15.8, we consider the propagation of energy along a lattice or loaded string.

15.7 Fourier Integral Representation of Wave Packets

It was implicit in the discussion of the preceding section that we were considering the description of the wave function for a wave packet for which the amplitude distribution $A(k)$ was given. We may wish, on the other hand, to obtain the distribution function $A(k)$ which describes a given wave function. This problem is similar to that of calculating the Fourier series which represents a given function. For that case, we found [see Eq. (14.93)] for $-\pi < x < \pi$,

$$f(x) = \frac{a_0}{2} + \sum_{r=1}^{\infty} (a_r \cos rx + b_r \sin rx) \tag{15.62}$$

where [see Eqs. (14.96)]:

$$a_r = \frac{1}{\pi} \int_{-\pi}^{+\pi} f(x) \cos rx \, dx, \qquad r = 0, 1, 2, \ldots \tag{15.63a}$$

$$b_r = \frac{1}{\pi} \int_{-\pi}^{+\pi} f(x) \sin rx \, dx, \qquad r = 1, 2, 3, \ldots \tag{15.63b}$$

* The group velocity corresponds to the signal velocity only in media that are nondispersive (in which case the phase, group, and signal velocities are all equal) and in media of normal dispersion (in which case the phase velocity exceeds the group and signal velocities). In media with *anomalous* dispersion, the group velocity may exceed the signal velocity (and, in fact, may even become negative or infinite!). We need only note here that a medium in which the wave number k is complex exhibits attenuation, and the dispersion is said to be *anomalous*. If k is real, there is no attenuation, and the dispersion is *normal*. Due to an historical misconception, we now know that what is called anomalous dispersion is, in fact, normal (i.e., frequent), whereas normal dispersion is anomalous (i.e., rare). Dispersive effects are quite important in optical and electromagnetic phenomena.

† Detailed analyses of the interrelationship among phase, group, and signal velocities, were made by Arnold Sommerfeld and by Léon Brillouin in 1914. Translations of these papers are given in the book by Brillouin (Br60).

For our present purposes it will prove more convenient to express $f(x)$ in a complex exponential series,

$$f(x) = \sum_{r=-\infty}^{\infty} c_r e^{-irx}, \qquad -\pi < x < \pi \tag{15.64}$$

If we multiply this equation by $\exp(isx)$ and integrate over the range $-\pi \leq x \leq \pi$, we find (see Problem 15-5):

$$c_s = \frac{1}{2\pi} \int_{-\pi}^{+\pi} f(x) e^{isx}\, dx, \qquad s = 0, \pm 1, \pm 2, \ldots \tag{15.65}$$

We now effect an enlargement of the fundamental interval from $\pm\pi$ to $\pm L$ by the same method used in Eq. (14.102). In Eq. (15.65) we change the dummy variable from x to $\pi u/L$, and in Eq. (15.64), we make the substitution $x \to \pi x/L$. Thus,

$$c_r = \frac{1}{2L} \int_{-L}^{+L} f(u) \exp\left(i\frac{r\pi u}{L}\right) du \tag{15.66}$$

and

$$f(x) = \frac{1}{2L} \sum_{r=-\infty}^{\infty} \int_{-L}^{+L} f(u) \exp\left[-i\frac{r\pi}{L}(x-u)\right] du \tag{15.67}$$

It now seems plausible that by suitably passing to the limit we may further enlarge our interval from $\pm L$ to $\pm\infty$. For this to be possible, we must require, as before, that $f(x)$ be single-valued and have only a finite number of finite discontinuities in the range $-\infty < x < \infty$, and we must require in addition that $f(x)$ be absolutely convergent (i.e., the integral of $|f(x)|$ between $-\infty$ and $+\infty$ must exist). Under these conditions, if we write $\Delta k \equiv \pi/L$, $f(x)$ becomes

$$f(x) = \lim_{\substack{L\to\infty \\ \Delta k \to 0}} \frac{1}{2\pi} \sum_{r=-\infty}^{\infty} e^{-ir(\Delta k)x}\, \Delta k \int_{-L}^{+L} f(u) e^{ir(\Delta k)u}\, du \tag{15.68}$$

In the limit, $\Delta k \to dk$ and $r\Delta k \to k$, so that, according to the definition of an integral in terms of the limit of a sum, viz.,

$$\int_{-\infty}^{+\infty} g(k)\, dk \equiv \lim_{\Delta k \to 0} \sum_{r=-\infty}^{\infty} g(r\,\Delta k)\Delta k,$$

the sum over r in Eq. (15.68) becomes an integral:

$$f(x) = \frac{1}{2\pi} \int_{-\infty}^{+\infty} e^{-ikx}\, dk \int_{-\infty}^{+\infty} f(u) e^{iku}\, du \tag{15.69}$$

The function $f(x)$ is therefore expressed as a *Fourier integral*, which is a double, infinite integral over the dummy variables k and u. It should be noted that we have not *proved* that the integral representation in Eq. (15.69) actually describes the function $f(x)$; we have only shown that such a representation is plausible.*

By comparing Eq. (15.55a) (with $t = 0$) and Eq. (15.69), we see that the integral

$$\boxed{A(k) = \frac{1}{\sqrt{2\pi}} \int_{-\infty}^{+\infty} f(u) e^{iku}\, du} \tag{15.70}$$

represents the spectral distribution of the function $f(x)$ [if we supply the constant factor $(2\pi)^{-\frac{1}{2}}$]. $A(k)$ is called the *Fourier transform* of $f(u)$ and is sometimes written $F(k)$. A reciprocal relationship exists between $f(u)$ and $A(k)$ since

$$\boxed{f(x) = \frac{1}{\sqrt{2\pi}} \int_{-\infty}^{+\infty} A(k) e^{-ikx}\, dk} \tag{15.71}$$

We previously found [see Eqs. (14.102) and (14.103)] that if $f(x)$ is an even function of x, then the Fourier series consists of only a cosine series, and that if $f(x)$ is odd, only the sine series remains. We have a similar situation in the integral representation; and, furthermore, if $f(x)$ is real, we find, upon taking the real parts of the integrals in Eq. (15.69),

$$f(x) = \frac{2}{\pi} \int_0^\infty \cos kx\, dk \int_0^\infty f(u) \cos ku\, du, \qquad f(x) \text{ even} \tag{15.72a}$$

$$f(x) = \frac{2}{\pi} \int_0^\infty \sin kx\, dk \int_0^\infty f(u) \sin ku\, du, \qquad f(x) \text{ odd} \tag{15.72b}$$

It should be noted that we have, in all of these discussions, arbitrarily chosen to consider the wave number k as the fundamental quantity. This is particularly useful for discussions of the *spatial* variation of the wave function. An equally valid choice, especially suitable for discussing the time variation of the wave function, is to express the various quantities in terms of the frequency ω. Thus, the distribution function $A(\omega)$ is called a *spectral* distribution. We shall still use the term "spectral" even when we choose to write A as a function of the wave number, $A(k)$.

* For a proof of the Fourier integral theorem see, for example, Churchill (Ch41, p. 89), Davis (Da63, p. 320), or Morse and Feshbach (Mo53, p. 458).

If we wish to treat the *time* variation of the function $f(x, t)$ at a fixed position x, then the equations corresponding to Eqs. (15.70) and (15.71) are

$$A(\omega) = \frac{1}{\sqrt{2\pi}} \int_{-\infty}^{+\infty} f(t) e^{i\omega t}\, dt \tag{15.73a}$$

$$f(t) = \frac{1}{\sqrt{2\pi}} \int_{-\infty}^{+\infty} A(\omega) e^{-i\omega t}\, d\omega \tag{15.73b}$$

Therefore, if $A(\omega)$ is known, $f(t)$ can always be obtained by a *Fourier inversion*; similarly, $A(\omega)$ can always be calculated from a knowledge of $f(t)$.

▶ Example 15.7 Spectral Representation of a Wave Packet

Let us consider the calculation of the spectral distribution which is necessary to represent a finite portion of a wave train that oscillates with a single frequency $\omega_0 = k_0 v$. That is, we initiate the oscillations at some time and then at a later time terminate the oscillations. Thus, the wave function is

$$\Psi(x, t) = \begin{cases} \cos(k_0 x - \omega_0 t) = \cos k_0(x - vt), & |x - vt| < L \\ 0, & |x - vt| > L \end{cases} \tag{1}$$

where the total time of oscillation of the device which initiates the disturbance is $2L/v$. Such an oscillatory pulse is shown in Fig. 15-8.

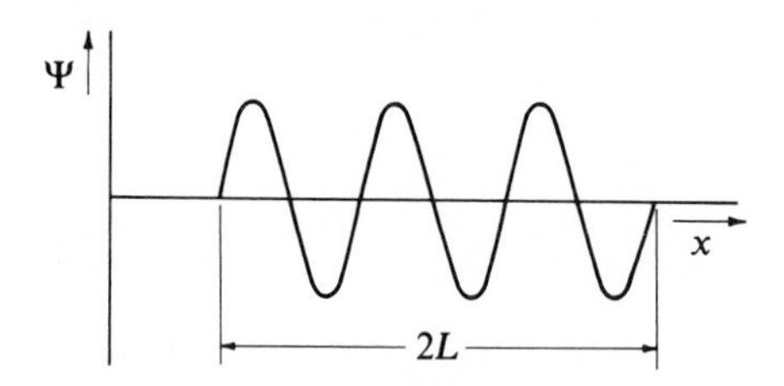

FIG. **15-8**

Since the wave function is real, the distribution function must also be real, and we can write Eq. (15.70) as

$$A(k) = \frac{1}{\sqrt{2\pi}} \int_{-\infty}^{+\infty} f(u) \cos ku\, du \tag{2}$$

Now,

$$\Psi(x, t) = \cos k_0(x - vt) = f(x - vt) = f(u) \tag{3}$$

or,

$$f(u) = \cos k_0 u, \qquad u \equiv x - vt \tag{4}$$

and since $f(u)$ is an even function, we can write

$$A(k) = \sqrt{\frac{2}{\pi}} \int_0^L \cos k_0 u \cos ku \, du \tag{5}$$

Then, if we use the identity

$$\cos k_0 u \cos ku = \tfrac{1}{2}[\cos(k_0 + k)u + \cos(k_0 - k)u] \tag{6}$$

we obtain

$$A(k) = \frac{1}{\sqrt{2\pi}} \left[\frac{\sin(k_0 + k)L}{k_0 + k} + \frac{\sin(k_0 - k)L}{k_0 - k} \right] \tag{7}$$

For wave numbers k in the vicinity of k_0, the second term will dominate, so that we have, approximately,

$$A(k) \propto \frac{\sin(k_0 - k)L}{k_0 - k}, \qquad k \cong k_0 \tag{8}$$

This function is shown in Fig. 15-9, and it is seen that there is a large maximum at $k = k_0$ with subsidiary maxima of decreasing magnitude as $|k_0 - k|$ increases.

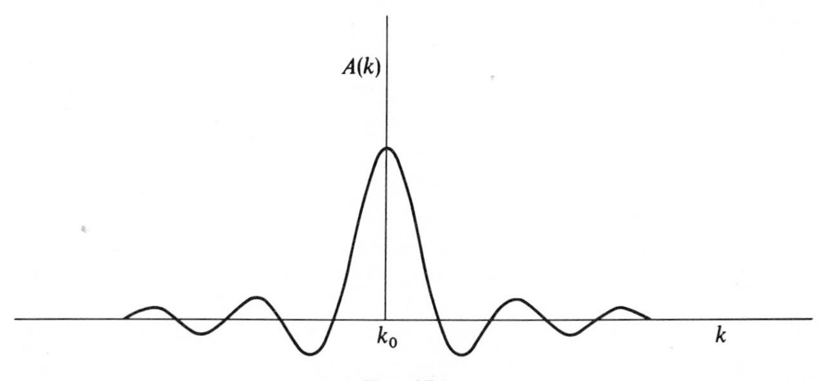

FIG. **15-9**

Equation (8) and Fig. 15-9 give the *amplitudes* of the waves with wave numbers k that must be superposed in order to generate the finite pulse shown in Fig. 15-8. Note that negative amplitudes merely mean that these waves are *out of phase*. The *intensity* $I(k)$ of these waves is given by the square of $A(k)$:

$$I(k) = |A(k)|^2 \propto \left| \frac{\sin(k_0 - k)L}{k_0 - k} \right|^2, \qquad k \cong k_0 \tag{9}$$

This function is shown in Fig. 15-10, where the secondary maxima (dashed curve) have been multiplied by 10 in order to exhibit these portions of the functions more clearly.

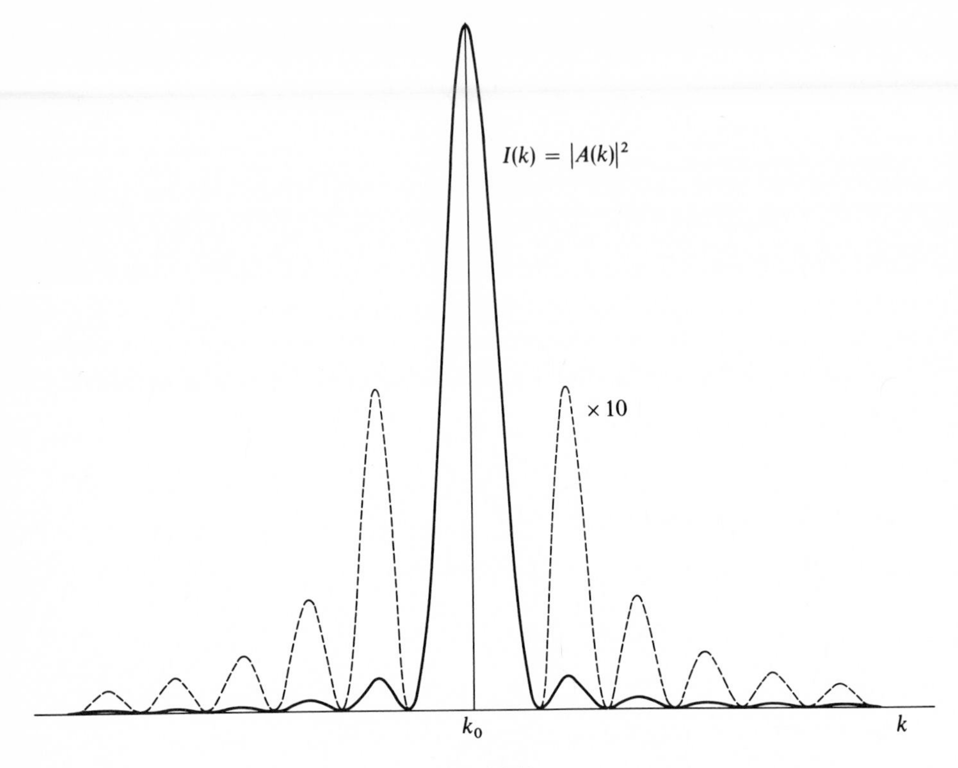

FIG. **15-10**

It is evident from Eq. (8) that $A(k)$ vanishes for values of k such that

$$k_0 - k = \frac{r\pi}{L}, \qquad r = \pm 1, \pm 2, \ldots \tag{10}$$

Thus, the main contribution to $A(k)$ arises from the region near $k = k_0$, that is, for

$$k = k_0 \pm \Delta k \tag{11}$$

where

$$\Delta k \cong \frac{\pi}{L} \tag{12}$$

Now, the length of the pulse is $2L$; and, if we denote this length by Δx, we have

$$\Delta k \, \Delta x \cong 2\pi \tag{13}$$

This is an interesting and important result since it implies that as we shorten the pulse length (i.e., Δx decreases), then we must include more and more frequencies (i.e., Δk increases) in order to give a Fourier representation of the wave function. On the other hand, if we allow the pulse length to become very long, then the spectral distribution becomes quite pure (i.e., only a narrow range of frequencies are necessary to represent the pulse). We may state this result in other terms by saying that if the frequency of the pulse is well defined (Δk small), then it is not possible to localize the pulse in space (i.e., Δx large). If we require localization, then the frequency is not well defined. A wave which consists of a *single* frequency (i.e., a *monochromatic* wave) must therefore be of infinite extent. Consequently, a truly monochromatic wave is an idealization and can never actually be physically realized.*

In modern physics the relation between Δk and Δx is expressed by the *Heisenberg uncertainty principle*, which states that for two conjugate dynamical variables† P and Q,

$$\Delta P \, \Delta Q \cong h \tag{14}$$

where h is Planck's constant. Now, position (x) and momentum (p) are conjugate variables. Thus,

$$\Delta p \, \Delta x \cong h \tag{15}$$

But, according to the hypothesis of de Broglie, the momentum associated with the wave number k is $hk/2\pi$, so that

$$\Delta(hk/2\pi) \, \Delta x \cong h \tag{16}$$

or,

$$\Delta k \, \Delta x \cong 2\pi \tag{17}$$

15.8 Energy Propagation in the Loaded String

We now wish to return again to the case of the infinite loaded string in order to investigate the connection between the group velocity and the rate of energy propagation along the string. We can express the kinetic and potential energies associated with the jth particle on the string as‡§ [cf. Eqs. (13.84) and (13.82)]:

* For further discussion see Marion (Ma65b, Section 11.3 ff).

† Two quantities are *conjugate* if their product has the dimensions of *action*, i.e. (energy) × (time). (See Section 7.12.)

‡ In this chapter we use the symbol W to denote the potential energy in order to avoid confusion with the group velocity U.

§ We adopt here a slightly different approach from that used in Chapter 13 in that we now define the potential energy for a *single* particle, whereas we had previously been interested in the total potential energy. We can no longer use the latter quantity since, of course, it is infinite for the infinite loaded string. Thus, we associate with the jth particle the potential energy which resides in the string between the jth and the $(j+1)$th particles. This is an arbitrary choice since we could equally well use the value for the element of string between the $(j-1)$th and the jth particles, or some combination of the two. We require only that our definition of W_j be consistent and that we have for the total potential energy $W = \sum_j W_j$. These requirements are met by Eq. (15.47b).

$$T_j = \frac{1}{2} m\dot{q}_j^2 \tag{15.74a}$$

$$W_j = \frac{1}{2}\frac{\tau}{d}(q_j - q_{j+1})^2 \tag{15.74b}$$

But now we wish to consider the time averages of the kinetic and potential energy densities, $\mathscr{T}$ and $\mathscr{W}$, where $\mathscr{T}$ is the kinetic energy per unit length of the string and similarly for $\mathscr{W}$. Since the fundamental unit of length for the loaded string is the spacing between particles d, we have

$$\langle \mathscr{T} \rangle = \frac{1}{2}\frac{m}{d}\langle \dot{q}_j^2 \rangle \tag{15.75a}$$

$$\langle \mathscr{W} \rangle = \frac{1}{2}\frac{\tau}{d^2}\langle (q_j - q_{j+1})^2 \rangle \tag{15.75b}$$

If we were considering an infinitely long *continuous* string, the wave function could be written in standard form as

$$\Psi(x, t) = Ae^{i(\omega t - kx)}$$

For the *loaded string*, the variable x is just jd, where the index j identifies the particle and d is the spacing. Thus, for an oscillation at the particular frequency ω, we have for the displacement of the jth particle,

$$q_j(t) = Ae^{i(\omega t - jkd)} \tag{15.76}$$

where A is the amplitude of the vibration. (Compare this with the previous result for the loaded string, Section 13.10.)

Differentiating, we have

$$\begin{aligned} \dot{q}_j &= i\omega Ae^{i(\omega t - jkd)} \\ &= -\omega A \sin(\omega t - jkd) \end{aligned} \tag{15.77}$$

where the last line is the real part of the previous expression. Therefore,

$$\langle \mathscr{T} \rangle = \frac{m\omega^2 A^2}{2d}\langle \sin^2(\omega t - jkd) \rangle \tag{15.78}$$

Now, since we are considering only a single frequency ω, we may compute the time average in this equation by averaging over one complete period of the oscillation. (If a *range* of frequencies were involved, as in a wave packet, then the average would have to be computed over a very long period of time—strictly, an infinite time.) The average over one period of the square of a sine function is $\frac{1}{2}$; therefore,

$$\langle \mathscr{T} \rangle = \frac{m\omega^2 A^2}{4d} \tag{15.79}$$

We also have [see Eq. (15.32)]:

$$\omega^2 = 4\frac{\tau}{md}\sin^2\frac{kd}{2} \tag{15.80}$$

so that the average kinetic energy density can be expressed as

$$\langle \mathscr{T} \rangle = \frac{\tau A^2}{d^2}\sin^2\frac{kd}{2} \tag{15.81}$$

In order to compute the average potential energy density we must consider the quantity

$$\begin{aligned} q_j - q_{j+1} &= Ae^{i\omega t}[e^{-ijkd} - e^{-i(j+1)kd}] \\ &= Ae^{i\omega t}e^{-ijkd}[1 - e^{-ikd}] \\ &= Ae^{i(\omega t - jkd)}e^{-i(kd/2)}\left(2i\sin\frac{kd}{2}\right) \\ &= -2A\sin\frac{kd}{2}\sin\left(\omega t - jkd - \frac{kd}{2}\right) \end{aligned} \tag{15.82}$$

where we have again taken only the real part of the expression. Averaging the square of this quantity, we obtain

$$\langle (q_j - q_{j+1})^2 \rangle = 2A^2\sin^2\frac{kd}{2} \tag{15.83}$$

and then,

$$\langle \mathscr{W} \rangle = \frac{\tau A^2}{d^2}\sin^2\frac{kd}{2} = \langle \mathscr{T} \rangle \tag{15.84}$$

so that the average potential and kinetic energy densities are equal, just as we found in Section 14.3 for the case of the continuous string.

Therefore, the average total energy density $\langle \mathscr{E} \rangle$ is

$$\langle \mathscr{E} \rangle = \langle \mathscr{T} \rangle + \langle \mathscr{W} \rangle = \frac{2\tau A^2}{d^2}\sin^2\frac{kd}{2} \tag{15.85}$$

Let us now consider the string to be initially at rest in its equilibrium position. Then, if we select the rth particle and force it to oscillate with a frequency ω and an amplitude A, the displacement of this particle is

$$q_r(t) = Ae^{i\omega t} \tag{15.86}$$

The motion of the rth particle will exert a force on the $(r+1)$th particle and cause it to oscillate at the frequency ω. Thus, the motion is propagated from particle to particle down the string. We wish to calculate the average input power which is applied to the rth particle and which causes the propagation of energy to the particles indexed by $r+1$, $r+2, \ldots$. (An equal amount of power is required for the propagation to $r-1$, $r-2, \ldots$, but we shall consider only the energy flowing to the right.) The input power to the rth particle may be expressed as the product of the velocity of the rth particle and the reaction force of the $(r+1)$th particle:

$$\langle P \rangle = \langle F_{r+1}\, \dot{q}_r \rangle \tag{15.87}$$

In analogy with the first term in Eq. (13.80), we write

$$F_{r+1} = -\frac{\tau}{d}(q_{r+1} - q_r) \tag{15.88}$$

Using Eq. (15.82), this becomes

$$\begin{aligned} F_{r+1} &= -\frac{2\tau A}{d} \sin\frac{kd}{2} \sin\left(\omega t - rkd - \frac{kd}{2}\right) \\ &= -\frac{2\tau A}{d} \sin\frac{kd}{2}\left[\sin(\omega t - rkd)\cos\frac{kd}{2} - \cos(\omega t - rkd)\sin\frac{kd}{2}\right] \end{aligned} \tag{15.89}$$

Equation (15.77) gives the expression for $\dot{q}_r$, and then the average input power is

$$\begin{aligned} \langle P \rangle = \frac{2\tau\omega A^2}{d} \sin\frac{kd}{2}\bigg[&\langle \sin^2(\omega t - rkd)\rangle \cos\frac{kd}{2} \\ &- \langle \sin(\omega t - rkd)\cos(\omega t - rkd)\rangle \sin\frac{kd}{2}\bigg] \end{aligned} \tag{15.90}$$

Now, the average value of $\sin\theta\cos\theta$ over one period vanishes, so there is no contribution from the second term in the brackets of Eq. (15.90). Computing the time average of the first term, we find

$$\langle P \rangle = \frac{\tau\omega A^2}{d} \sin\frac{kd}{2}\cos\frac{kd}{2} \tag{15.91}$$

Substituting ω from Eq. (15.80), we have

$$\langle P \rangle = 2\sqrt{\frac{\tau}{md}} \cdot \frac{\tau A^2}{d} \sin^2\frac{kd}{2}\cos\frac{kd}{2} \tag{15.92}$$

Now, the ratio $\langle P\rangle/\langle\mathscr{E}\rangle$ has the dimensions of *velocity* and can be interpreted as representing the average velocity of the flow of energy along the string. Therefore, using Eqs. (15.85) and (15.92), we obtain

$$\frac{\langle P\rangle}{\langle\mathscr{E}\rangle} = \sqrt{\frac{\tau d}{m}}\cos\frac{kd}{2} \tag{15.93}$$

From Eq. (15.80) we can compute the derivative of ω with respect to k:

$$\frac{d\omega}{dk} = \sqrt{\frac{\tau d}{m}}\cos\frac{kd}{2} \tag{15.94}$$

But $d\omega/dk$ is just the group velocity U [see Eq. (15.60)], so we conclude that

$$U = \frac{\langle P\rangle}{\langle\mathscr{E}\rangle} \tag{15.95}$$

which indicates that energy is propagated along the loaded string with the group velocity U.

We must recall that if the frequency impressed upon the string exceeds the critical frequency ω_c, then attenuation sets in, and the wave number can no longer be identified with the velocity of energy propagation. There is in fact no propagation of energy in this particular case. (See the comments at the end of Section 15.4.)

15.9 Reflected and Transmitted Waves

An important problem in the study of wave phenomena is the behavior of waves when they strike a boundary between two media. The effects that occur are particularly important in the case of electromagnetic waves [see the discussion in Marion (Ma65b, Chapter 6)]. The example of the vibrating string affords a convenient means of discussing the reflection and transmission of waves as well as allowing the introduction of the calculational methods in a simple form.

We shall consider the "boundary" between our two "media" as a point of discontinuity in the linear mass density of the string. That is, for $x < 0$, we have a density ρ_1, and for $x > 0$, we have a density ρ_2. If we allow a continuous wave train to be incident from the left (i.e., from negative values of x), then at the point of discontinuity we expect a portion of the wave to be transmitted and a portion to be reflected. Thus, in region 1 ($x < 0$) we will have the superposition of the incident and reflected waves, and in region 2 ($x > 0$) we will have only the transmitted wave:

$$\left.\begin{aligned} \Psi_1(x, t) &= \Psi_{\text{inc}} + \Psi_{\text{refl}} = Ae^{i(\omega t - k_1 x)} + Be^{i(\omega t + k_1 x)} \\ \Psi_2(x, t) &= \Psi_{\text{trans}} = Ce^{i(\omega t - k_2 x)} \end{aligned}\right\} \tag{15.96}$$

In Eqs. (15.96) we have explicitly taken into account the fact that the waves in both regions will have the same frequency. But since the wave velocity on a string is given by

$$v = \sqrt{\frac{\tau}{\rho}} \tag{15.97}$$

we will have $v_1 \neq v_2$, and therefore $k_1 \neq k_2$. We have, of course,

$$k = \frac{\omega}{v} = \omega\sqrt{\frac{\rho}{\tau}} \tag{15.98}$$

so that, in terms of the wave number of the incident wave,

$$k_2 = k_1\sqrt{\frac{\rho_2}{\rho_1}} \tag{15.99}$$

The amplitude A of the incident wave [see Eqs. (15.96)] is considered given and will be a real quantity. We must then obtain the amplitudes B and C of the reflected and transmitted waves to complete the solution of the problem. There are as yet no restrictions on B and C, and they may be complex quantities.

The physical requirements on the problem may be stated in terms of the *boundary conditions*. These are, simply, that the total wave function $\Psi = \Psi_1 + \Psi_2$ and its derivative must be continuous across the boundary. The continuity of Ψ results, of course, from the fact that the string is continuous. The condition on the derivative prevents the occurrence of a "kink" in the string, for if $\partial\Psi/\partial x_{0+} \neq \partial\Psi/\partial x_{0-}$, then $\partial^2\Psi/\partial x^2$ would be infinite at $x = 0$; but the wave equation relates $\partial^2\Psi/\partial x^2$ and $\partial^2\Psi/\partial t^2$; and, if the former is infinite, this implies an infinite acceleration which is clearly not allowed by the physical situation. We have, therefore, for all values of the time t,

$$\Psi_1|_{x=0} = \Psi_2|_{x=0} \tag{15.100a}$$

$$\left.\frac{\partial\Psi_1}{\partial x}\right|_{x=0} = \left.\frac{\partial\Psi_2}{\partial x}\right|_{x=0} \tag{15.100b}$$

From Eqs. (15.96) and (15.100a) we have

$$A + B = C \tag{15.101a}$$

and from Eqs. (15.96) and (15.100b) we obtain

$$-k_1 A + k_1 B = -k_2 C \tag{15.101b}$$

The solution of this pair of equations yields

$$B = \frac{k_1 - k_2}{k_1 + k_2} A \tag{15.102a}$$

and

$$C = \frac{2k_1}{k_1 + k_2} A \tag{15.102b}$$

Now, the wave numbers, k_1 and k_2, are both real, so that the amplitudes B and C are likewise real. Furthermore, k_1, k_2, and A are all positive, so C is always positive. Thus, the transmitted wave is always *in phase* with the incident wave. Similarly, if $k_1 > k_2$, then the incident and reflected waves are *in phase*, but they are *out of phase* for $k_2 > k_1$, i.e., for $\rho_2 > \rho_1$.

The *reflection coefficient* R is defined to be the ratio of the squared magnitudes of the amplitudes of the reflected and incident waves,

$$R \equiv \frac{|B|^2}{|A|^2} = \left(\frac{k_1 - k_2}{k_1 + k_2}\right)^2 \tag{15.103}$$

Since the energy content of a wave is proportional to the square of the amplitude of the wave function, R represents the ratio of the reflected energy to the incident energy. The quantity $|B|^2$ represents the *intensity* of the reflected wave.

Since no energy can be stored in the junction of the two strings, the incident energy must be equal to the sum of the reflected and transmitted energies. That is, $R + T = 1$. Thus,

$$T = 1 - R = \frac{4k_1 k_2}{(k_1 + k_2)^2} \tag{15.104}$$

or,

$$T = \frac{k_2}{k_1} \frac{|C|^2}{|A|^2} \tag{15.104a}$$

In the study of the reflection and transmission of electromagnetic waves we find quite similar expressions for R and T.

■ 15.10 Damped Plane Waves

In all of the preceding discussions we have assumed the absence of any frictional or dissipative forces. As was pointed out in Chapters 2 and 3, the forces most amenable to calculation are those which are proportional to the instantaneous velocity. Such forces are therefore widely used to approximate the dissipative effects in real physical situations. If we introduce such a force into the problem of the propagation of a plane wave, the wave equation can be written as

$$\frac{\partial^2 \Psi}{\partial x^2} - \beta \frac{\partial \Psi}{\partial t} - \frac{1}{v^2}\frac{\partial^2 \Psi}{\partial t^2} = 0 \tag{15.105}$$

We can effect a solution by the method of separation of variables by writing, as before

$$\Psi(x, t) = \psi(x)\chi(t) \tag{15.106}$$

The wave equation then becomes

$$\chi \frac{d^2\psi}{dx^2} - \beta\psi \frac{d\chi}{dt} - \frac{\psi}{v^2}\frac{d^2\chi}{dt^2} = 0 \tag{15.107}$$

or,

$$\frac{1}{\psi}\frac{d^2\psi}{dx^2} = \frac{\beta}{\chi}\frac{d\chi}{dt} + \frac{1}{\chi v^2}\frac{d^2\chi}{dt^2} = -k^2 \tag{15.108}$$

where the separation constant has been set equal to $-k^2$. The separated equations are therefore

$$\frac{d^2\psi}{dx^2} + k^2\psi = 0 \tag{15.109}$$

and

$$\frac{d^2\chi}{dt^2} + v^2\beta \frac{d\chi}{dt} + k^2 v^2 \chi = 0 \tag{15.110}$$

The solution for $\psi(x)$ is clearly

$$\psi(x) \sim e^{\pm ikx} \tag{15.111}$$

We are more interested, however, in the solution for $\chi(t)$. If we make the following definitions,

$$2\gamma \equiv v^2\beta; \qquad \omega_0^2 \equiv k^2 v^2 \tag{15.112}$$

then, Eq. (15.110) becomes

$$\ddot{\chi} + 2\gamma\dot{\chi} + \omega_0^2 \chi = 0 \tag{15.113}$$

which is exactly the equation for the damped harmonic oscillator found in Chapter 3 [see Eq. (3.38)]. The solution is [see Eq. (3.40)]:

$$\chi(t) = \exp(-\gamma t)[B \exp(\sqrt{\gamma^2 - \omega_0^2}\, t) + C \exp(-\sqrt{\gamma^2 - \omega_0^2}\, t)] \tag{15.114}$$

If we consider the case of *underdamping*, then $\gamma^2 < \omega_0^2$, so that the exponentials are imaginary and represent oscillatory motion. Define

$$\sqrt{\gamma^2 - \omega_0^2} \equiv i\Omega \tag{15.115}$$

where Ω is real. Then, we have

$$\chi(t) = e^{-\gamma t}[Be^{i\Omega t} + Ce^{-i\Omega t}] \tag{15.116}$$

The real part of the terms in the brackets can be written as either a sine or a cosine if the proper phase factor is included. Let us consider the sine representation and suppress the phase for simplicity. Then, we can write

$$\chi(t) = De^{-\gamma t}\sin\Omega t \tag{15.117}$$

Thus, the motion is oscillatory with a frequency $\Omega = \sqrt{\psi^2 - \gamma^2}$ and is damped in time. Let us consider a case in which a disturbance is initiated at a time $t = 0$. Then, the time-dependent part of the wave function can be written as

$$\chi(t) = \begin{cases} 0, & t < 0 \\ De^{-\gamma t}\sin\Omega t, & t \geq 0 \end{cases} \tag{15.118}$$

If we ask for the spectral representation of such a wave, we must use Eq. (15.73a) in order to calculate the Fourier transform of $\chi(t)$:

$$\begin{aligned} A(\omega) &= \frac{1}{\sqrt{2\pi}}\int_{-\infty}^{+\infty}\chi(t)e^{i\omega t}\,dt \\ &= \frac{D}{\sqrt{2\pi}}\int_{-\infty}^{+\infty}e^{-\gamma t}e^{i\omega t}\sin\Omega t\,dt \end{aligned} \tag{15.119}$$

Writing $\sin\Omega t$ in terms of exponentials, we have

$$A(\omega) = \frac{D}{2i\sqrt{2\pi}}\int_0^{\infty}[e^{i(\omega+\Omega+i\gamma)t} - e^{i(\omega-\Omega+i\gamma)t}]\,dt$$

This integrates immediately to

$$\begin{aligned} A(\omega) &= \frac{D}{2\sqrt{2\pi}}\left[\frac{1}{\omega+\Omega+i\gamma} - \frac{1}{\omega-\Omega+i\gamma}\right] \\ &= \frac{D}{2\sqrt{2\pi}}\frac{-2\Omega}{(\omega+\Omega+i\gamma)(\omega-\Omega+i\gamma)} \\ &= -\frac{D}{\sqrt{2\pi}}\frac{\Omega}{(\omega^2-\Omega^2-\gamma^2+2i\gamma\omega)} \end{aligned} \tag{15.120}$$

But, $\Omega^2 = \omega_0^2 - \gamma^2$, so that we have

$$A(\omega) = -\frac{D}{\sqrt{2\pi}} \frac{\Omega}{(\omega^2 - \omega_0^2) + 2i\gamma\omega}$$

and, multiplying numerator and denominator by the complex conjugate of the denominator, we obtain

$$A(\omega) = -\frac{D\Omega}{\sqrt{2\pi}} \frac{(\omega^2 - \omega_0^2) - 2i\gamma\omega}{(\omega^2 - \omega_0^2)^2 + 4\gamma^2\omega^2} \tag{15.121}$$

This equation gives the spectrum of frequencies necessary to produce a time-decaying oscillation initiated at $t = 0$. The imaginary portion of this expression results entirely from the fact that attenuation is present. (Recall that an imaginary frequency component always produces temporal damping.) If the system producing the wave were an atom in which an electron is oscillating, radiation in the form of light would be emitted and the system would lose energy and would therefore be damped in time. Such radiation cannot be monochromatic and, indeed, the investigation of the shapes of atomic spectral lines (with high dispersion spectrographs) reveals a peak at the frequency ω_0 and a rapidly decreasing intensity for higher and lower frequencies. The shapes of spectral lines are well described by Eq. (15.121), called the *Lorentzian line shape*.*

Suggested References

General treatments of wave motion in mechanical systems and with applications to vibrating strings will be found in Becker (Be54, Chapter 15), Bradbury (Br68, Chapter 12), Coulson (Co49, Chapters 1 and 2), Hauser (Ha65, Chapter 12), Konopinski (Ko69, Chapter 12), Menzel (Me53, Part 3), Slater and Frank (Sl47, Chapter 9), Symon (Sy60, Chapter 8), and Wangsness (Wa63, Chapter 15).

Most texts on electromagnetism discuss wave propagation, and some introduce the topic with mechanical concepts; see, for example, Corson and Lorrain (Co62, Appendix G).

Lindsay (Li60) discusses many aspects of mechanical wave motion; see, particularly Sections 3.7 and 3.8 (wave packets and group velocity), Chapter 4 (waves on strings), and Section 6.6 (wave propagation in a one-dimensional lattice).

* After Hendrik Antoon Lorentz (1853–1928), the Dutch physicist who developed much of the early theory of atomic radiation.

The topic of wave propagation on the loaded string is given an excellent treatment by Brillouin (Br46); several electrical analogies are also discussed.
Fourier integrals are discussed, for example, by Churchill (Ch41, Chapter 5), Davis (Da63, Chapter 6), Pipes (Pi46, Chapter 3), and Stratton (St41, Section 5.7).

Problems

15-1. Consider the simplified wave function

$$\Psi(x, t) = Ae^{i(\omega t - kx)}$$

and assume that ω and v are complex quantities and that k is real:

$$\omega = \alpha + i\beta$$

$$v = u + iw$$

Show that the wave is damped in time. Use the fact that $k^2 = \omega^2/v^2$ to obtain expressions for α and β in terms of u and w. Find the phase velocity for this case.

15-2. Consider an electrical transmission line that has a uniform inductance per unit length L and a uniform capacitance per unit length C. Show that an alternating current I in such a line obeys the wave equation*

$$\frac{\partial^2 I}{\partial x^2} - LC\frac{\partial^2 I}{\partial t^2} = 0$$

so that the wave velocity is $v = 1/\sqrt{LC}$.

15-3. Consider a transmission line constructed of inductive and capacitive elements and arranged as in Fig. 15-4. Analyze this case in detail and show that all frequencies *above* a cutoff frequency are transmitted, while all those *below* are attenuated.

15-4. Consider the superposition of two infinitely long wave trains which have almost the same frequency, but which have different amplitudes. Show that the phenomenon of beats occurs, but that the waves never beat to zero amplitude.

* This equation is called the *telegraphy equation* and was first derived in 1876 by Oliver Heaviside. Lord Kelvin had made an investigation of inductanceless transmission lines in 1854.

15-5. Derive Eq. (15.65), and then relate the c_r to the a_r and b_r of Eqs. (15.63).

15-6. Calculate and plot the Fourier transform of the function

$$\Psi(x, t) = \begin{cases} 1 + \cos(k_0 x - \omega_0 t), & \left| x - \dfrac{\omega_0}{k_0} t \right| < L \\ 0, & \left| x - \dfrac{\omega_0}{k_0} t \right| > L \end{cases}$$

Compare the results with Fig. 15-9.

15-7. Consider the periodic square wave shown in the figure. Each pulse is of unit magnitude and of unit duration; successive pulses are separated by a time interval T. Write a complex Fourier series as

$$f(t) = \sum_{r=-\infty}^{\infty} c_r e^{-i\omega_r t}$$

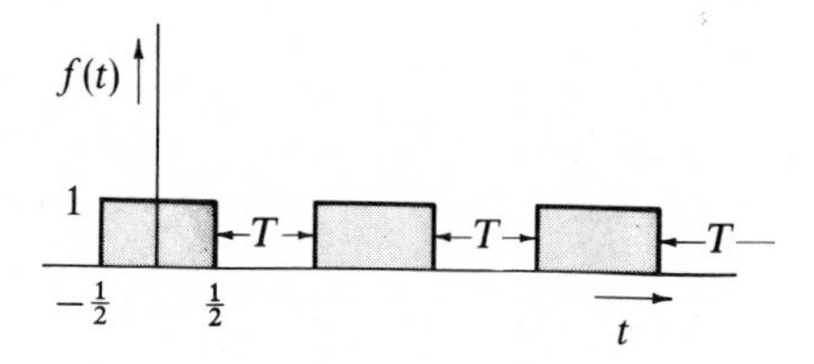

Take $T = 1$ and obtain the Fourier coefficients. Represent the results graphically by plotting the values of the c_r as vertical lines versus the frequency. Such a plot is therefore the frequency spectrum of the square wave. Next, allow $T \to \infty$ and again obtain a frequency spectrum. (It will be necessary to pass to the limit of an integral expression for the Fourier coefficients.) Compare the results with each other and with Fig. 15-9.

15-8. Consider a wave packet in which the amplitude distribution is given by

$$A(k) = \begin{cases} 1, & |k - k_0| < \Delta k \\ 0, & \text{otherwise} \end{cases}$$

Show that the wave function is

$$\Psi(x, t) = \frac{2 \sin[(\omega_0' t - x)\, \Delta k]}{\omega_0' t - x}\, e^{i(\omega_0 t - k_0 x)}$$

Sketch the shape of the wave packet (choose $t = 0$ for simplicity).

15-9. Consider a wave packet with a Gaussian amplitude distribution

$$A(k) = B\exp[-\sigma(k - k_0)^2]$$

where $2/\sqrt{\sigma}$ is equal to the $1/e$ width* of the packet. Using this function for $A(k)$, show that

$$\Psi(x, 0) = B\int_{-\infty}^{+\infty} \exp[-\sigma(k - k_0)^2]\exp[-ikx]\,dk$$

$$= B\sqrt{\frac{\pi}{\sigma}}\exp[-x^2/4\sigma]\exp[-ik_0 x]$$

Sketch the shape of this wave packet. Next, expand $\omega(k)$ in a Taylor series, retain the first two terms, and integrate the wave packet equation to obtain the general result

$$\Psi(x, t) = B\sqrt{\frac{\pi}{\sigma}}\exp[-(\omega_0' t - x)^2/4\sigma]\exp[i(\omega_0 t - k_0 x)]$$

Finally, take one additional term in the Taylor series expansion of $\omega(k)$ and show that σ now is replaced by a complex quantity. Find the expression for the $1/e$ width of the packet as a function of time for this case and show that the packet moves with the same group velocity as before but now spreads in width as it moves. Illustrate this result with a sketch.

15-10. Find the spectral distribution function which is necessary to represent the Gaussian wave function

$$\Psi(x, 0) = B\exp[-a^2x^2/2]$$

15-11. Treat the problem of wave propagation along a string loaded with particles of two different masses, m' and m'', which alternate in placement; i.e.,

$$m_j = \begin{cases} m', & \text{for } j \text{ even} \\ m'', & \text{for } j \text{ odd} \end{cases}$$

Show that the ω-k curve has two branches in this case, and show that there is attenuation for frequencies between the branches as well as for frequencies above the upper branch.

15-12. Sketch the phase velocity $V(k)$ and the group velocity $U(k)$, for the propagation of waves along a loaded string, in the range of wave numbers

* That is, at the points $k = k_0 \pm 1/\sqrt{\sigma}$, the amplitude distribution is $1/e$ of its maximum value $A(k_0)$. Thus $2/\sqrt{\sigma}$ is the width of the curve at the $1/e$ height.

$0 \leq k \leq \pi/d$. Show that $U(\pi/d) = 0$ whereas $V(\pi/d)$ does not vanish. What is the interpretation of this result in terms of the behavior of the waves?

15-13. Consider the function $A(k)$ defined by

$$A(k) = \begin{cases} 0, & k < -a \\ 1, & -a < k < a \\ 0, & k > a \end{cases}$$

Show that the Fourier transform of this function is

$$f(x) = \sqrt{\frac{2}{\pi}} \frac{\sin ax}{x}$$

Use this result to show

$$\int_0^\infty \frac{\sin ax}{x} dx = \begin{cases} \pi/2, & a > 0 \\ 0, & a = 0 \\ -\pi/2, & a < 0 \end{cases}$$

[Hint: set $k = 0$.]

15-14. Compare graphically the spectral distribution functions $A(\omega)$ for the following two wave functions:

(a) $$\Psi_a(x, t) = \begin{cases} 1, & |t| < t_1 \\ 0, & |t| > t_1 \end{cases}$$

(b) $$\Psi_b(x, t) = \begin{cases} 2\left(1 - \dfrac{|t|}{t_1}\right), & |t| < t_1 \\ 0, & |t| > t_1 \end{cases}$$

(The two wave functions are normalized to have the same time integral.)

15-15. Consider an infinitely long continuous string which, for $x < 0$ and for $x > L$, has a linear mass density ρ_1, but has a density $\rho_2 > \rho_1$ for $0 < x < L$. If a wave train, oscillating with a frequency ω, is incident from the left on the high-density section of the string, find the reflected and transmitted intensities for the various portions of the string. Find a value of L which will allow a maximum transmission through the high-density section. Discuss briefly the relationship of this problem to the application of nonreflective coatings to optical lenses.

15-16. Consider an infinitely long continuous string in which the tension is τ. A mass M is attached to the string at $x = 0$. If a wave train with velocity ω/k is incident from the left, show that reflection and transmission occur at $x = 0$ and that the coefficients R and T are given by

$$R = \sin^2 \theta; \qquad T = \cos^2 \theta$$

where

$$\tan \theta = \frac{M\omega^2}{4\pi k\tau}$$

Consider carefully the boundary condition on the derivatives of the wave functions at $x = 0$. What are the phase changes for the reflected and transmitted waves?

"*Better is the end of a thing than the beginning thereof.*"—Ecclesiastes

APPENDIX A

Taylor's Theorem

A theorem of considerable importance in mathematical physics is *Taylor's theorem** which relates to the expansion of an arbitrary function in a power series. We will find many instances in which it is necessary to employ this theorem in order to simplify a problem to a tractable form.

Consider a function $f(x)$ which has continuous derivatives of all orders within a certain interval of the independent variable x. Then, if this interval includes $x_0 \leq x \leq x_0 + h$, we may write

$$I \equiv \int_{x_0}^{x_0+h} f'(x)\, dx = f(x_0 + h) - f(x_0) \tag{A.1}$$

where $f'(x)$ is the derivative of $f(x)$ with respect to x. If we make the change of variable

$$x = x_0 + h - t \tag{A.2}$$

we have

$$I = \int_0^h f'(x_0 + h - t)\, dt \tag{A.3}$$

* First published in 1715 by the English mathematician Brook Taylor (1685–1731).

Integrating by parts

$$I = t f'(x_0 + h - t)\Big|_0^h + \int_0^h t f''(x_0 + h - t)\, dt$$

$$= h f'(x_0) + \int_0^h t f''(x_0 + h - t)\, dt \tag{A.4}$$

Now integrating the second term by parts, we find

$$I = h f'(x_0) + \frac{h^2}{2!} f''(x_0) + \int_0^h \frac{t^2}{2!} f'''(x_0 + h - t)\, dt \tag{A.5}$$

Continuing this process we will generate an infinite series for I. From the definition of I we then have

$$\boxed{f(x_0 + h) = f(x_0) + h f'(x_0) + \frac{h^2}{2!} f''(x_0) + \cdots} \tag{A.6}$$

This is the Taylor series expansion* of the function $f(x_0 + h)$. A more common form of the series results if we set $x_0 = 0$ and $h = x$ (i.e., the function $f(x)$ is expanded about the origin):

$$\boxed{\begin{aligned} f(x) = f(0) + x f'(0) + \frac{x^2}{2!} f''(0) + \frac{x^3}{3!} f'''(0) + \cdots \\ + \frac{x^n}{n!} f^{(n)}(0) + \cdots \end{aligned}} \tag{A.7}$$

where

$$f^{(n)}(0) \equiv \frac{d^n}{dx^n} f(x)\Big|_{x=0} \tag{A.8}$$

Equation (A.7) is usually called *Maclaurin's series*† for the function $f(x)$.

▶ **Example A.1.** Since the derivative of exp x of any order is just exp x, the exponential series is

$$\boxed{e^x = 1 + x + \frac{x^2}{2!} + \frac{x^3}{3!} + \cdots} \tag{A.9}$$

This result is of considerable importance and it will be used often.

* The *remainder* term which remains if the series is terminated after a finite number of terms is discussed, for example, by Kaplan (Ka52, pp. 357–360).

† Discovered by James Stirling in 1717 and published by Colin Maclaurin in 1742.

▶ **Example A.2.** In order to expand $f(x) = \sin x$, we need

$$\begin{aligned} f(x) &= \sin x; & f(0) &= 0 \\ f'(x) &= \cos x; & f'(0) &= 1 \\ f''(x) &= -\sin x; & f''(0) &= 0 \\ f'''(x) &= -\cos x; & f'''(0) &= -1 \end{aligned}$$

Therefore,

$$\boxed{\sin x = x - \frac{x^3}{3!} + \frac{x^5}{5!} - \cdots} \tag{A.10}$$

Similarly,

$$\boxed{\cos x = 1 - \frac{x^2}{2!} + \frac{x^4}{4!} - \cdots} \tag{A.11}$$

▶ **Example A.3.** A series expansion can often be profitably used in the evaluation of a definite integral. (This is particularly true for those cases in which the indefinite integral cannot be found in closed form.)

$$\int_0^x \frac{dt}{1+t} = \int_0^x (1 - t^2 - t^3 + \cdots)\, dt, \qquad |t| < 1 \tag{1}$$

Integrating term by term, we find

$$\int_0^x \frac{dt}{1+t} = x - \frac{x^2}{2} + \frac{x^3}{3} - \cdots \tag{2}$$

Since

$$\frac{d}{dx} \ln(1+x) = \frac{1}{1+x} \tag{3}$$

we also have the result

$$\ln(1+x) = x - \frac{x^2}{2} + \frac{x^3}{3} - \cdots \tag{4}$$

Exercises

A-1. Show by division and by direct expansion in a Taylor series that

$$\frac{1}{1-x} = 1 + x + x^2 + x^3 + \cdots + x^n + \cdots$$

For what range of x is the series valid?

A-2. Expand $\cos x$ about the point $x = \pi/4$.

A-3. Use a series expansion to show that

$$\int_0^1 \frac{e^x - e^{-x}}{x}\, dx = 2.1145 \ldots$$

A-4. Use a Taylor series to expand $\sin^{-1} x$. Verify the result by expanding the integral in the relation

$$\sin^{-1} x = \int_0^x \frac{dt}{\sqrt{1 - t^2}}$$

A-5. Evaluate to three decimal places:

$$\int_0^1 \exp(-x^2/2)\, dx$$

Compare the result with that determined from tables of the probability integral.

A-6. Show that if $f(x) = (1 + x)^n$ (with $|x| < 1$) is expanded in a Taylor series, the result is the same as a binomial expansion.

APPENDIX B

Complex Numbers

B.1 Complex Numbers

A complex number† consists of an ordered pair of real numbers, x, y, and is customarily written in the form $x + iy$, where $i \equiv \sqrt{-1}$. The *modulus* or *absolute value* of a complex number $z = x + iy$ is defined to be

$$|z| \equiv +\sqrt{x^2 + y^2} \tag{B.1}$$

In the complex number z, we call x the *real part* and y the *imaginary part*:

$$\text{Re } z = x; \qquad \text{Im } z = y \tag{B.2}$$

The *complex conjugate* of z is denoted by z^* and is formed from z by changing the sign of the imaginary part:

$$\begin{aligned} z &= x + iy \\ z^* &= x - iy \end{aligned} \tag{B.3}$$

† The terminology "complex number" is due to Gauss, 1831.

We may easily verify that

$$zz^* = |z|^2 \tag{B.4a}$$

$$z + z^* = 2 \operatorname{Re} z = 2x \tag{B.4b}$$

$$z - z^* = 2i \operatorname{Im} z = 2iy \tag{B.4c}$$

B.2 Geometrical Representation of Complex Numbers

Since a complex number is an ordered pair of real numbers, we may represent such a combination by a point on a two-dimensional diagram. We, may, in fact, elect to connect such a point (x, y) with the origin and represent the complex number by the line segment drawn in this manner. Figure B-1 shows a geometrical representation of the complex number $z_1 = x_1 + iy_1$. Such a representation is called an *Argand diagram*.†

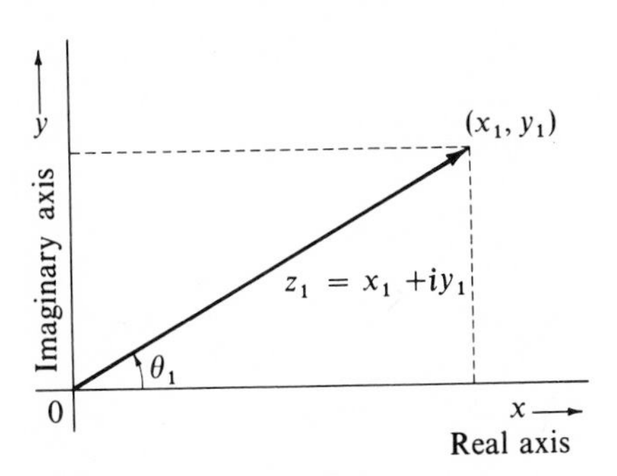

FIG. **B-1**

The addition of two complex numbers, z_1 and z_2, may be obtained geometrically by the parallelogram construction shown in Fig. B-2. The sum is given by

$$z = z_1 + z_2 = (x_1 + x_2) + i(y_1 + y_2) \tag{B.5}$$

Subtraction may be represented in an equivalent manner.

If we choose plane polar coordinates for the representation of a complex number, then the point (x, y) may be designated by (r, θ), where

$$|z| = r \tag{B.6a}$$

$$z = x + iy = r(\cos\theta + i \sin\theta) \tag{B.6b}$$

† After J. R. Argand (1806), but the construction had previously been given by C. Wessel and by Gauss (1797) and possibly had been discovered as early as 1673 by J. Wallis.

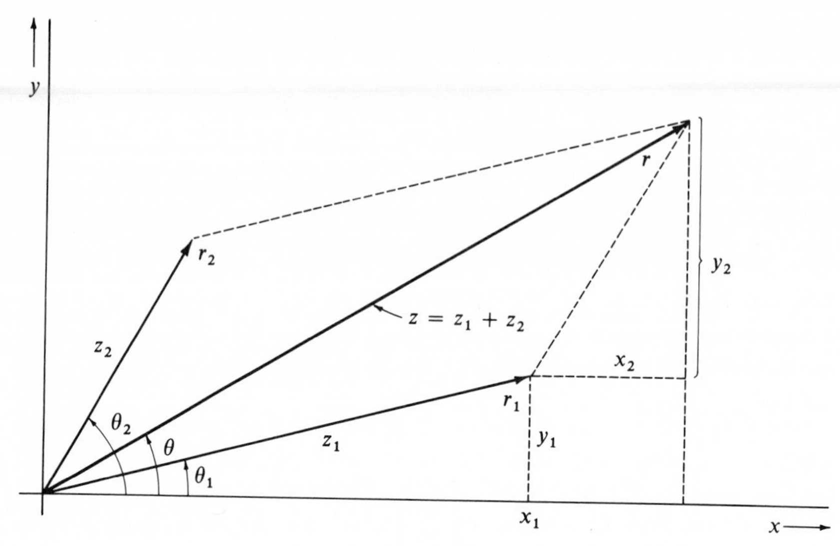

FIG. **B-2**

The following relations are easily verified:

$$z_1 z_2 = r_1 r_2[\cos(\theta_1 + \theta_2) + i \sin(\theta_1 + \theta_2)] \tag{B.7}$$

$$\frac{z_1}{z_2} = \frac{r_1}{r_2}[\cos(\theta_1 - \theta_2) + i \sin(\theta_1 - \theta_2)] \tag{B.8}$$

$$z^n = r^n (\cos n\theta + i \sin n\theta) \qquad \text{(De Moivre's formula)} \tag{B.9}$$

The angle θ is called the *argument* of z (and is sometimes written $\theta = \arg z$). We have

$$\tan \theta = \frac{\text{Im } z}{\text{Re } z} = \frac{y}{x} \tag{B.10}$$

Since we may increase θ by 2π and obtain the same point in the x-y or r-θ diagram, we must establish a convention in order to uniquely specify the argument of a complex quantity. For our purposes here we will always choose $0 \leq \theta < 2\pi$.

B.3 Trigonometric Functions of Complex Variables

In the theory of *real* variables, the exponential function of x, denoted by e^x or $\exp x$, may be represented by an infinite series [see Eq. (A.9)]:

$$e^x = 1 + x + \frac{x^2}{2!} + \frac{x^3}{3!} + \cdots \tag{B.11}$$

In a similar fashion, we may write

$$\begin{aligned} e^{iy} &= 1 + (iy) + \frac{(iy)^2}{2!} + \frac{(iy)^3}{3!} + \cdots \\ &= 1 + iy - \frac{y^2}{2!} - i\frac{y^3}{3!} + \cdots \\ &= \left[1 - \frac{y^2}{2!} + \frac{y^4}{4!} - \cdots\right] + i\left[y - \frac{y^3}{3!} + \frac{y^5}{5!} - \cdots\right] \end{aligned} \tag{B.12}$$

But the series in the first square brackets is just the expression for $\cos y$ [Eq. (A.11)] and that in the second is just $\sin y$ [Eq. (A.10)]; hence,

$$\boxed{e^{iy} = \cos y + i \sin y} \tag{B.13a}$$

Similarly,

$$e^{-iy} = \cos y - i \sin y \tag{B.13b}$$

Upon adding (or subtracting) Eqs. (B.13a) and (B.13b) and solving for $\cos y$ (or $\sin y$), we find

$$\cos y = \frac{e^{iy} + e^{-iy}}{2} \tag{B.14a}$$

$$\sin y = \frac{e^{iy} - e^{-iy}}{2i} \tag{B.14b}$$

These are called the *Euler relations.**

If we combine Eqs. (B.6b) and (B.13a), we see that we may always write any complex number as

$$z = x + iy = Ae^{i\varphi} \tag{B.15}$$

where $A = |z|$ is called the *amplitude*, and where $\varphi = \tan^{-1}(y/x)$ is the *phase*.

* After the eminent Swiss mathematician, Leonard Euler (1707–1783).

B.4 Hyperbolic Functions

If we calculate the cosine of $z = x + iy$, we find, using the Euler relations,

$$\cos z = \tfrac{1}{2}[e^{i(x+iy)} + e^{-i(x+iy)}]$$
$$= \tfrac{1}{2}e^{-y}(\cos x + i \sin x) + \tfrac{1}{2}e^{y}(\cos x - i \sin x)$$
$$= \frac{e^{y} + e^{-y}}{2} \cos x - i \frac{e^{y} - e^{-y}}{2} \sin x \tag{B.16}$$

The coefficients of $\cos x$ and $i \sin x$ are defined as the *hyperbolic* functions:

$$\boxed{\begin{aligned} \cosh y &\equiv \frac{e^{y} + e^{-y}}{2} \\ \sinh y &\equiv \frac{e^{y} - e^{-y}}{2} \end{aligned}} \tag{B.17}$$

Equation (B.16) then becomes

$$\cos z = \cosh y \cos x - i \sinh y \sin x \tag{B.18}$$

We may also expand $\sin z$ with the result

$$\sin z = \cosh y \sin x + i \sinh y \cos x \tag{B.19}$$

If we let $x = 0$ in Eqs. (B.18) and (B.19), we obtain expressions for the sine and cosine of a pure imaginary argument:

$$\left.\begin{aligned} \sin(iy) &= i \sinh y \\ \cos(iy) &= \cosh y \end{aligned}\right\} \tag{B.20}$$

Similarly, we may verify the following relations:

$$\left.\begin{aligned} \sinh(iy) &= i \sin y \\ \cosh(iy) &= \cos y \end{aligned}\right\} \tag{B.21}$$

The derivatives of the hyperbolic functions are:

$$\left.\begin{aligned} \frac{d}{dy} \sinh y &= \cosh y \\ \frac{d}{dy} \cosh y &= \sinh y \end{aligned}\right\} \tag{B.22}$$

Exercises

B-1. Is it possible to ascribe a meaning to the inequality $z_1 < z_2$? Explain. Does the inequality $|z_1| < |z_2|$ have a different meaning?

B-2. Solve the following equations:

(a) $z^2 + 2z + 2 = 0$ (b) $2z^2 + z + 2 = 0$

B-3. Express the following in polar form:

(a) $z_1 = i$ (b) $z_2 = -1$
(c) $z_3 = 1 + i\sqrt{3}$ (d) $z_4 = 1 + 2i$
(e) Find the product $z_1 z_2$ (f) Find the product $z_1 z_3$
(g) Find the product $z_3 z_4$

B-4. Express $(z^2 - 1)^{-\frac{1}{2}}$ in polar form.

B-5. If the function $w = \sin^{-1} z$ is defined as the inverse of $z = \sin w$, then use the Euler relation for $\sin w$ to find an equation for $\exp(iw)$. Solve this equation and obtain the result

$$w = \sin^{-1} z = -i \ln\left(iz + \sqrt{1 - z^2}\right)$$

B-6. Show that

$$y = Ae^{ix} + Be^{-ix}$$

can be written as

$$y = C \cos(x - \delta)$$

where A and B are *complex* but where C and δ are *real.*

B-7. Show that

(a) $\sinh(x_1 + x_2) = \sinh x_1 \cosh x_2 + \cosh x_1 \sinh x_2$
(b) $\cosh(x_1 + x_2) = \cosh x_1 \cosh x_2 + \sinh x_1 \sinh x_2$

B-8. Prepare graphs of the functions $\sinh x$ and $\cosh x$.

APPENDIX C

*Ordinary Differential Equations of Second Order**

C.1 Linear Homogeneous Equations

By far, the most important type of ordinary differential equation encountered in problems in mathematical physics is the second-order linear equation with constant coefficients. Equations of this type have the form

$$\frac{d^2y}{dx^2} + a\frac{dy}{dx} + by = f(x) \tag{C.1}$$

or, denoting derivatives by primes.

$$y'' + ay' + by = f(x) \tag{C.1a}$$

A particularly important class of such equations are those for which $f(x) = 0$. These equations (called *homogeneous equations*) are important not only in themselves but also as *reduced* equations in the solution of the more general type of equation (C.1).

* A standard treatise on differential equations is that of Ince (In27). A listing of many types of equations and their solutions is given by Murphy (Mu60). A modern viewpoint is contained in the book by Hochstadt (Ho64).

The linear homogeneous second-order equation with constant coefficients will be considered first*:

$$y'' + ay' + by = 0 \tag{C.2}$$

These equations have the following important properties:

(a) If $y_1(x)$ is a solution of (C.2) then $c_1 y_1(x)$ is also a solution.

(b) If $y_1(x)$ and $y_2(x)$ are solutions, then $y_1(x) + y_2(x)$ is also a solution (principle of *superposition*).

(c) If $y_1(x)$ and $y_2(x)$ are *linearly* independent solutions, then the *general* solution to the equation is given by $c_1 y_1(y) + c_2 y_2(x)$. (The general solution always contains two arbitrary constants.)

[The functions $y_1(x)$ and $y_2(x)$ are *linearly independent* if and only if the equation

$$\lambda y_1(x) + \mu y_2(x) \equiv 0 \tag{C.3}$$

is satisfied only by $\lambda = \mu = 0$. If Eq. (C.3) can be satisfied with λ and μ different from zero, then $y_1(x)$ and $y_2(x)$ are said to be *linearly dependent.*

The general condition (i.e., the necessary and sufficient condition) that a set of functions $y_1, y_2, y_3, \ldots$ be linearly dependent is that the *Wronskian determinant* of these functions vanish identically:

$$W = \begin{vmatrix} y_1 & y_2 & y_3 & \cdots & y_n \\ y_1' & y_2' & y_3' & \cdots & y_n' \\ y_1'' & y_2'' & y_3'' & \cdots & y_n'' \\ \vdots & & & & \\ y_1^{(n-1)} & y_2^{(n-1)} & y_3^{(n-1)} & \cdots & y_n^{(n-1)} \end{vmatrix} = 0 \tag{C.4}$$

where $y^{(n)}$ is the nth derivative of y with respect to x.]

The properties (a) and (b) can be verified by direct substitution, but (c) will only be asserted here to yield the general solution. It should be noted that the properties listed above apply *only* to the homogeneous equation (C.2) and *not* to the general equation (C.1).

Equations of the type (C.2) are reducible through the substitution

$$y = e^{rx} \tag{C.5}$$

Now,

$$y' = re^{rx}; \qquad y'' = r^2 e^{rx} \tag{C.6}$$

* The first published solution of an equation of this type was by Euler, 1743; but the solution appears to have been known to Daniel and Johann Bernoulli, 1739.

Using these expressions for y' and y'' in Eq. (C.2), there results an algebraic equation called the *auxiliary equation*:

$$r^2 + ar + b = 0 \tag{C.7}$$

The solution of this quadratic in r is

$$r = -\frac{a}{2} \pm \tfrac{1}{2}\sqrt{a^2 - 4b} \tag{C.8}$$

We first assume that the two roots, denoted by r_1 and r_2, are not identical, and write the solution as

$$y = e^{r_1 x} + e^{r_2 x} \tag{C.9}$$

Since the Wronskian determinant of $\exp(r_1 x)$ and $\exp(r_2 x)$ does not vanish, these functions are linearly independent. Thus, the general solution is

$$\boxed{y = c_1 e^{r_1 x} + c_2 e^{r_2 x}, \qquad r_1 \neq r_2} \tag{C.10}$$

If it happens that $r_1 = r_2 = r$, then it can be verified by direct substitution that $x \exp(rx)$ is also a solution, and since $(\exp)rx$ and $x \exp(rx)$ are linearly independent, the general solution for identical roots is given by

$$\boxed{y = c_1 e^{rx} + c_2 x e^{rx}, \qquad r_1 = r_2 \equiv r} \tag{C.11}$$

▶ **Example C.1.** Solve the equation:

$$y'' - 2y' - 3y = 0 \tag{1}$$

The auxiliary equation is

$$r^2 - 2r - 3 = (r - 3)(r + 1) = 0 \tag{2}$$

The roots are

$$r_1 = 3; \qquad r_2 = -1 \tag{3}$$

Therefore, the general solution is

$$y = c_1 e^{3x} + c_2 e^{-x} \tag{4}$$

▶ **Example C.2.** Solve the equation:

$$y'' + 4y' + 4y = 0 \tag{1}$$

The auxiliary equation is

$$r^2 + 4r + 4 = (r + 2)^2 = 0 \tag{2}$$

The roots are equal, and $r = -2$. Therefore, the general solution is

$$y = c_1 e^{-2x} + c_2 x e^{-2x} \tag{3}$$

If the roots, r_1 and r_2, of the auxiliary equation are imaginary, the solutions given by $c_1 \exp(r_1 x)$ and $c_2 \exp(r_2 x)$ are still correct.

In order to give the solutions entirely in terms of real quantities, we use the Euler relations to express the exponentials. Then,

$$\left.\begin{aligned} e^{r_1 x} &= e^{\alpha x} e^{i\beta x} = e^{\alpha x}(\cos \beta x + i \sin \beta x) \\ e^{r_2 x} &= e^{\alpha x} e^{-i\beta x} = e^{\alpha x}(\cos \beta x - i \sin \beta x) \end{aligned}\right\} \tag{C.12}$$

and the general solution is

$$\begin{aligned} y &= c_1 e^{r_1 x} + c_2 e^{r_2 x} \\ &= e^{\alpha x}[(c_1 + c_2) \cos \beta x + i(c_1 - c_2) \sin \beta x] \end{aligned} \tag{C.13}$$

Now, c_1 and c_2 are arbitrary, but these constants may be complex. However, not all the four elements can be independent (since there would be *four* arbitrary constants rather than *two*). The number of independent elements can be reduced to the required *two* by making c_1 and c_2 complex conjugates. Then, the combinations $A \equiv c_1 + c_2$ and $B \equiv i(c_1 - c_2)$ become a pair of arbitrary, real constants. Using these quantities in the solution we have

$$y = e^{\alpha x}(A \cos \beta x + B \sin \beta x) \tag{C.14}$$

Equation (C.14) may be put into a form that is sometimes more convenient by multiplying and dividing by $\mu = \sqrt{A^2 + B^2}$:

$$y = \mu e^{\alpha x}[(A/\mu) \cos \beta x + (B/\mu) \sin \beta x] \tag{C.15}$$

Next, we define an angle δ (see Fig. C-1) such that

$$\sin \delta = A/\mu; \qquad \cos \delta = B/\mu; \qquad \tan \delta = A/B \tag{C.16}$$

Then the solution becomes

$$\begin{aligned} y &= \mu e^{\alpha x}(\sin \delta \cos \beta x + \cos \delta \sin \beta x) \\ &= \mu e^{\alpha x} \sin(\beta x + \delta) \end{aligned}$$

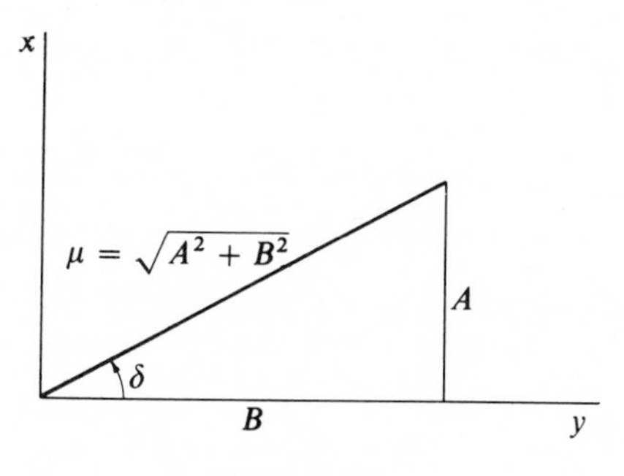

FIG. C-1

Depending on the exact definition of the phase δ, we may write the solution alternatively as

$$y = \mu e^{\alpha x} \sin(\beta x + \delta) \tag{C.17a}$$

$$y = \mu e^{\alpha x} \cos(\beta x + \delta) \tag{C.17b}$$

▶ **Example C.3.** Solve the equation:

$$y'' + 2y' + 4y = 0 \tag{1}$$

The auxiliary equation is

$$r^2 + 2r + 4 = 0 \tag{2}$$

with

$$r = \frac{-2 \pm \sqrt{4 - 16}}{2} = -1 \pm i\sqrt{3} \tag{3}$$

Hence,

$$\alpha = -1; \qquad \beta = \sqrt{3} \tag{4}$$

and the general solution is

$$y = e^{-x}(c_1 \cos \sqrt{3}x + c_2 \sin \sqrt{3}x) \tag{5}$$

or,

$$y = \mu e^{-x} \sin[(\sqrt{3}x + \delta)] \tag{6}$$

Summarizing, then, there are three possible types of general solutions to homogeneous second-order linear differential equations, as indicated in Table C.1.

Table C.1

If the roots of the auxiliary equations are	then the general solution is $y =$
Real, unequal ($r_1 \neq r_2$)	$c_1 e^{r_1 x} + c_2 e^{r_2 x}$
Real, equal ($r_1 = r_2 \equiv r$)	$c_1 e^{rx} + c_2 x e^{rx}$
Imaginary ($\alpha \pm i\beta$)	$e^{\alpha x}(c_1 \cos \beta x + c_2 \sin \beta x)$ or, $\mu e^{\alpha x} \sin(\beta x + \delta)$

C.2 Linear Inhomogeneous Equations

In order to solve the general (i.e., inhomogeneous) second-order linear differential equation, consider the following. Let $y = u$ be the *general solution* of

$$y'' + ay' + by = 0 \tag{C.18}$$

and let $y = v$ be *any* solution of

$$y'' + ay' + by = f(x) \tag{C.19}$$

then $y = u + v$ is a solution of (C.19), since

$$\begin{aligned} y'' + ay' + by &= (u'' + au' + bu) + (v'' + av' + bv) \\ &= 0 + f(x) \end{aligned}$$

Now, u contains the two arbitrary constants c_1 and c_2, and therefore the combination $u + v$ satisfies all of the requirements of the general solution to Eq. (C.19). The function u is the *complementary function* and v is the *particular integral* of the equation. Since a general method of finding u has been given above, it only remains to find, by inspection or by trial, some function v that will satisfy

$$v'' + av' + bv = f(x) \tag{C.20}$$

▶ **Example C.4.** Solve the equation:

$$y'' + 5y' + 6y = x^2 + 2x \tag{1}$$

The auxiliary equation is

$$r^2 + 5r + 6 = (r + 3)(r + 2) = 0 \tag{2}$$

$$r_1 = -3; \qquad r_2 = -2 \tag{3}$$

So that the complementary function is

$$u = c_1 e^{-3x} + c_2 e^{-2x} \tag{4}$$

Since the right-hand side of the original equation is a second-degree polynomial, we guess a particular integral of the form

$$v = Ax^2 + Bx + C \tag{5}$$

Then,

$$v' = 2Ax + B \tag{6}$$

$$v'' = 2A \tag{7}$$

Substituting into the differential equation, we have

$$2A + 5(2Ax + B) + 6(Ax^2 + Bx + C) = x^2 + 2x \tag{8}$$

or,

$$(6A)x^2 + (10A + 6B)x + (2A + 5B + 6C) = x^2 + 2x \tag{9}$$

Equating coefficients of like powers of x:

$$\begin{aligned} 6A &= 1 \\ 10A + 6B &= 2 \\ 2A + 5B + 6C &= 0 \end{aligned} \tag{10}$$

Solving,

$$A = \tfrac{1}{6}; \qquad B = \tfrac{1}{18}; \qquad C = -\tfrac{11}{108} \tag{11}$$

Hence,

$$v = \frac{1}{6}x^2 + \frac{1}{18}x - \frac{11}{108}$$

$$= \frac{18x^2 + 6x - 11}{108} \tag{12}$$

Thus, the general solution is

$$y = u + v = c_1 e^{-3x} + c_2 e^{-2x} + \frac{18x^2 + 6x - 11}{108} \tag{13}$$

The type of solution illustrated in the above example is called the *method of undetermined coefficients.*

▶ **Example C.5.** Solve the equation:

$$y'' + 4y = 3x \cos x \tag{1}$$

The auxiliary equation is

$$r^2 + 4 = (r + 2i)(r - 2i) = 0 \tag{2}$$

with roots

$$\left.\begin{aligned} r_1 &= \alpha + i\beta = 0 + 2i \\ r_2 &= \alpha - i\beta = 0 - 2i \end{aligned}\right\} \tag{3}$$

so that

$$\alpha = 0; \qquad \beta = 2 \tag{4}$$

and the complementary function is

$$\begin{aligned} u &= e^{\alpha x}(c_1 \cos \beta x + c_2 \sin \beta x) \\ &= c_1 \cos 2x + c_2 \sin 2x \end{aligned} \tag{5}$$

To find a particular integral, we note that from $x \cos x$ and its derivatives, it is possible to generate only terms involving the following functions:

$$x \cos x; \qquad x \sin x; \qquad \cos x; \qquad \sin x$$

Therefore, since all of these functions are linearly independent, the trial particular integral is

$$v = Ax \cos x + Bx \sin x + C \cos x + D \sin x \tag{6}$$

$$\begin{aligned} v' = A\,(\cos x - x \sin x) + B\,(\sin x + x \cos x) \\ - C \sin x + D \cos x \end{aligned} \tag{7}$$

$$\begin{aligned} v'' = -A\,(2 \sin x + x \cos x) + B\,(2 \cos x - x \sin x) \\ - C \cos x - D \sin x \end{aligned} \tag{8}$$

Substituting into the original differential equation,

$$(3D - 2A) \sin x + (2B + 3C) \cos x + 3(A - 1)x \cos x + (3B)x \sin x = 0 \quad (9)$$

The coefficient of each term must vanish (because of the linear independence of the terms):

$$3D = 2A; \qquad 2B = -3C; \qquad A = 1; \qquad 3B = 0 \quad (10)$$

from which,

$$A = 1; \qquad B = 0; \qquad C = 0; \qquad D = \tfrac{2}{3} \quad (11)$$

Hence, the general solution is

$$y = c_1 \sin 2x + c_2 \cos 2x + x \cos x + \tfrac{2}{3} \sin x \quad (12)$$

If the right-hand side, $f(x)$, of the general equation [Eq. (C.1) or (C.19)] is such that $f(x)$ and its first two derivatives (only second-order equations are being considered) contain only linearly independent functions, then a linear combination of these functions will constitute the trial particular integral. In the event that the trial function contains a term which already appears in the complementary function, use the term multiplied by x; if this combination also appears in the complementary function, use the term multiplied by x^2. No higher powers will be needed since only second-order equations are being considered and only $\exp(rx)$ or $x \exp(rx)$ occur as solutions to the reduced equation; $(x^2) \exp(rx)$ never occurs.

Exercises

C-1. Solve the following homogeneous second-order equations:

(a) $y'' + 2y' - 3y = 0$ (b) $y'' + y = 0$
(c) $y'' - 2y' + 2y = 0$ (d) $y'' - 2y' + 5y = 0$

C-2. Solve the following inhomogeneous equations by the method of undetermined coefficients:

(a) $y'' + 2y' - 8y = 16x$ (b) $y'' - 2y' + y = 2e^{2x}$
(c) $y'' + y = \sin x$ (d) $y'' - 2y' + y = 3xe^x$
(e) $y'' - 4y' + 5y = e^{2x} + 4 \sin x$

C-3. Use a Taylor series expansion to obtain the solution of

$$y'' + y^2 = x^2$$

that obeys the conditions $y(0) = 1$ and $y'(0) = 0$. (Differentiate the equation successively to obtain the derivatives that occur in the Taylor series.)

APPENDIX D

*Useful Formulas**

D.1 Binomial Expansion

$$(1 + x)^n = 1 + nx + \frac{n(n-1)}{2!}x^2 + \frac{n(n-1)(n-2)}{3!}x^3$$
$$+ \cdots + \binom{n}{r}x^r + \cdots, \qquad |x| < 1 \qquad \text{(D.1)}$$

$$(1 - x)^n = 1 - nx + \frac{n(n-1)}{2!}x^2 - \frac{n(n-1)(n-2)}{3!}x^3$$
$$+ \cdots + (-1)^r\binom{n}{r}x^r + \cdots, \qquad -|x| < 1 \qquad \text{(D.2)}$$

where the *binomial coefficient* is

$$\binom{n}{r} \equiv \frac{n!}{(n-r)!\,r!} \qquad \text{(D.3)}$$

* An extensive list may be found, for example, in Dwight (Dw61).

Some particularly useful cases of the above are:

$$(1 \pm x)^{\frac{1}{2}} = 1 \pm \tfrac{1}{2}x - \tfrac{1}{8}x^2 \pm \tfrac{1}{16}x^3 - \cdots \tag{D.4}$$

$$(1 \pm x)^{\frac{1}{3}} = 1 \pm \tfrac{1}{3}x - \tfrac{1}{9}x^2 \pm \tfrac{5}{81}x^3 - \cdots \tag{D.5}$$

$$(1 \pm x)^{-\frac{1}{2}} = 1 \mp \tfrac{1}{2}x + \tfrac{3}{8}x^2 \mp \tfrac{5}{16}x^3 + \cdots \tag{D.6}$$

$$(1 \pm x)^{-\frac{1}{3}} = 1 \mp \tfrac{1}{3}x + \tfrac{2}{9}x^2 \mp \tfrac{14}{81}x^3 + \cdots \tag{D.7}$$

$$(1 \pm x)^{-1} = 1 \mp x + x^2 \mp x^3 + \cdots \tag{D.8}$$

$$(1 \pm x)^{-2} = 1 \mp 2x + 3x^2 \mp 4x^3 + \cdots \tag{D.9}$$

$$(1 \pm x)^{-3} = 1 \mp 3x + 6x^2 \mp 10x^3 + \cdots \tag{D.10}$$

For convergence of *all* of the above series, we must have $|x| < 1$.

D.2 Trigonometric Relations

$$\sin(A \pm B) = \sin A \cos B \pm \cos A \sin B \tag{D.11}$$

$$\cos(A \pm B) = \cos A \cos B \mp \sin A \sin B \tag{D.12}$$

$$\sin 2A = 2 \sin A \cos A = \frac{2 \tan A}{1 + \tan^2 A} \tag{D.13}$$

$$\cos 2A = 2 \cos^2 A - 1 \tag{D.14}$$

$$\sin^2 \frac{A}{2} = \tfrac{1}{2}(1 - \cos A) \tag{D.15}$$

$$\cos^2 \frac{A}{2} = \tfrac{1}{2}(1 + \cos A) \tag{D.16}$$

$$\sin^2 A = \tfrac{1}{2}(1 - \cos 2A) \tag{D.17}$$

$$\sin^3 A = \tfrac{1}{4}(3 \sin A - \sin 3A) \tag{D.18}$$

$$\sin^4 A = \tfrac{1}{8}(3 - 4 \cos 2A + \cos 4A) \tag{D.19}$$

$$\cos^2 A = \tfrac{1}{2}(1 + \cos 2A) \tag{D.20}$$

$$\cos^3 A = \tfrac{1}{4}(3 \cos A + \cos 3A) \tag{D.21}$$

$$\cos^4 A = \tfrac{1}{8}(3 + 4 \cos 2A + \cos 4A) \tag{D.22}$$

$$\tan(A + B) = \frac{\tan A + \tan B}{1 - \tan A \tan B} \tag{D.23}$$

$$\tan^2 \frac{A}{2} = \frac{1 - \cos A}{1 + \cos A} \tag{D.24}$$

$$\sin x = \frac{e^{ix} - e^{-ix}}{2i} \tag{D.25}$$

$$\cos x = \frac{e^{ix} + e^{-ix}}{2} \tag{D.26}$$

$$e^{ix} = \cos x + i \sin x \tag{D.27}$$

D.3 Trigonometric Series

$$\sin x = x - \frac{x^3}{3!} + \frac{x^5}{5!} - \frac{x^7}{7!} + \cdots \tag{D.28}$$

$$\cos x = 1 - \frac{x^2}{2!} + \frac{x^4}{4!} - \frac{x^6}{6!} + \cdots \tag{D.29}$$

$$\tan x = x + \frac{x^3}{3} + \frac{2}{15} x^5 + \cdots, \qquad |x| < \pi/2 \tag{D.30}$$

$$\sin^{-1} x = x + \frac{x^3}{6} + \frac{3}{40} x^5 + \cdots, \qquad \begin{cases} |x| < 1 \\ |\sin^{-1} x| < \pi/2 \end{cases} \tag{D.31}$$

$$\cos^{-1} x = \frac{\pi}{2} - x - \frac{x^3}{6} - \frac{3}{40} x^5 - \cdots, \qquad \begin{cases} |x| < 1 \\ 0 < \cos^{-1} x < \pi \end{cases} \tag{D.32}$$

$$\tan^{-1} x = x - \frac{x^3}{3} + \frac{x^5}{5} - \frac{x^7}{7} + \cdots, \qquad |x| < 1 \tag{D.33}$$

D.4 Exponential and Logarithmic Series

$$e^x = 1 + x + \frac{x^2}{2!} + \frac{x^3}{3!} + \cdots = \sum_{n=0}^{\infty} \frac{x^n}{n!} \tag{D.34}$$

$$\ln(1 + x) = x - \frac{x^2}{2} + \frac{x^3}{3} - \frac{x^4}{4} + \cdots, \qquad |x| < 1; \quad x = 1 \tag{D.35}$$

$$\ln[\sqrt{(x^2/a^2) + 1} + (x/a)] = \sinh^{-1} x/a \tag{D.36}$$

$$= -\ln[\sqrt{(x^2/a^2) + 1} - (x/a)] \tag{D.37}$$

D.5 Hyperbolic Functions

$$\sinh x = \frac{e^x - e^{-x}}{2} \tag{D.38}$$

$$\cosh x = \frac{e^x + e^{-x}}{2} \tag{D.39}$$

$$\tanh x = \frac{e^{2x} - 1}{e^{2x} + 1} \tag{D.40}$$

$$\sin ix = i \sinh x \tag{D.41}$$

$$\cos ix = \cosh x \tag{D.42}$$

$$\sinh ix = i \sin x \tag{D.43}$$

$$\cosh ix = \cos x \tag{D.44}$$

$$\sinh^{-1} x = \tanh^{-1}\left[\frac{x}{\sqrt{x^2+1}}\right] \tag{D.45}$$

$$= \ln[x + \sqrt{x^2+1}] \tag{D.46}$$

$$= \cosh^{-1}[\sqrt{x^2+1}], \qquad \begin{cases} >0, & x>0 \\ <0, & x<0 \end{cases} \tag{D.47}$$

$$\cosh^{-1} x = \pm \tanh^{-1}\left[\frac{\sqrt{x^2-1}}{x}\right], \qquad x > 1 \tag{D.48}$$

$$= \pm \ln x[+ \sqrt{x^2-1}], \qquad x > 1 \tag{D.49}$$

$$= \pm \sinh^{-1}[\sqrt{x^2-1}], \qquad x > 1 \tag{D.50}$$

APPENDIX E

*Useful Integrals**†

E.1 Algebraic Functions

$$\int \frac{dx}{a^2 + x^2} = \frac{1}{a} \tan^{-1}\left(\frac{x}{a}\right), \qquad \left|\tan^{-1}\left(\frac{x}{a}\right)\right| < \frac{\pi}{2} \tag{E.1}$$

$$\int \frac{x\,dx}{a^2 + x^2} = \tfrac{1}{2} \ln(a^2 + x^2) \tag{E.2}$$

$$\int \frac{dx}{x(a^2 + x^2)} = \frac{1}{2a^2} \ln\left(\frac{x^2}{a^2 + x^2}\right) \tag{E.3}$$

$$\int \frac{dx}{a^2x^2 - b^2} = \frac{1}{2ab} \ln\left[\frac{ax - b}{ax + b}\right] \tag{E.4a}$$

$$= -\frac{1}{ab} \coth^{-1}\left(\frac{ax}{b}\right) \tag{E.4b}$$

$$= -\frac{1}{ab} \tanh^{-1}\left(\frac{ax}{b}\right), \qquad a^2x^2 < b^2 \tag{E.4c}$$

* Extremely useful compilations are, for example, Pierce and Foster (Pi57) and Dwight (Dw61).

† This list is confined to those (nontrivial) integrals which arise in the text and in the problems.

$$\int \frac{dx}{\sqrt{a+bx}} = \frac{2}{b}\sqrt{a+bx} \tag{E.5}$$

$$\int \frac{dx}{\sqrt{x^2+a^2}} = \ln[x+\sqrt{x^2+a^2}] \tag{E.6}$$

$$\int \frac{x^2\,dx}{\sqrt{(a^2-x^2)}} = -\frac{x}{2}\sqrt{(a^2-x^2)} + \frac{a^2}{2}\sin^{-1}\frac{x}{a} \tag{E.7}$$

$$\int \frac{dx}{\sqrt{ax^2+bx+c}} = \frac{1}{\sqrt{a}}\ln[2\sqrt{a}\sqrt{ax^2+bx+c}+2ax+b], \quad a>0 \tag{E.8a}$$

$$= \frac{1}{\sqrt{a}}\sinh^{-1}\left[\frac{2ax+b}{\sqrt{4ac-b^2}}\right], \quad \begin{cases} a>0 \\ 4ac>b^2 \end{cases} \tag{E.8b}$$

$$= -\frac{1}{\sqrt{-a}}\sin^{-1}\left[\frac{2ax+b}{\sqrt{b^2-4ac}}\right], \quad \begin{cases} a<0 \\ b>4ac \\ |2ax+b|<\sqrt{b^2-4ac} \end{cases} \tag{E.8c}$$

$$\int \frac{x\,dx}{\sqrt{ax^2+bx+c}} = \frac{1}{a}\sqrt{ax^2+bx+c} - \frac{b}{2a}\int\frac{dx}{\sqrt{ax^2+bx+c}} \tag{E.9}$$

$$\int \frac{dx}{x\sqrt{ax^2+bx+c}} = -\frac{1}{\sqrt{c}}\sinh^{-1}\left[\frac{bx+2c}{|x|\sqrt{4ac-b^2}}\right], \quad \begin{cases} c>0 \\ 4ac>b^2 \end{cases} \tag{E.10a}$$

$$= \frac{1}{\sqrt{-c}}\sin^{-1}\left[\frac{bx+2c}{|x|\sqrt{b^2-4ac}}\right], \quad \begin{cases} c<0 \\ b^2>4ac \end{cases} \tag{E.10b}$$

$$= -\frac{1}{\sqrt{c}}\ln\left[\frac{2\sqrt{c}}{x}\sqrt{ax^2+bx+c}+\frac{2c}{x}+b\right], \quad c>0 \tag{E.10c}$$

$$\int \sqrt{ax^2+bx+c}\,dx = \frac{2ax+b}{4a}\sqrt{ax^2+bx+c} + \frac{4ac-b^2}{8a}\int\frac{dx}{\sqrt{ax^2+bx+c}} \tag{E.11}$$

E.2 Trigonometric Functions

$$\int \sin^2 x\,dx = \frac{x}{2} - \tfrac{1}{4}\sin 2x \tag{E.12}$$

$$\int \cos^2 x\,dx = \frac{x}{2} + \tfrac{1}{2}\sin 2x \tag{E.13}$$

$$\int \frac{dx}{a + b\sin x} = \frac{2}{\sqrt{a^2 - b^2}} \tan^{-1}\left[\frac{a\tan(x/2) + b}{\sqrt{a^2 - b^2}}\right], \qquad a^2 > b^2 \tag{E.14}$$

$$\int \frac{dx}{a + b\cos x} = \frac{2}{\sqrt{a^2 - b^2}} \tan^{-1}\left[\frac{(a - b)\tan(x/2)}{\sqrt{a^2 - b^2}}\right], \qquad a^2 > b^2 \tag{E.15}$$

$$\int \frac{dx}{(a + b\cos x)^2} = \frac{b\sin x}{(b^2 - a^2)(a + b\cos x)} - \frac{a}{b^2 - a^2}\int \frac{dx}{a + b\cos x} \tag{E.16}$$

$$\int \tan x\,dx = -\ln|\cos x| \tag{E.17}$$

$$\int \tanh x\,dx = \ln\cosh x \tag{E.17a}$$

$$\int e^{ax}\sin x\,dx = \frac{e^{ax}}{a^2 + 1}(a\sin x - \cos x) \tag{E.18}$$

$$\int e^{ax}\sin^2 x\,dx = \frac{e^{ax}}{a^2 + 4}\left(a\sin^2 x - 2\sin x\cos x + \frac{2}{a}\right) \tag{E.18a}$$

$$\int_{-\infty}^{\infty} e^{-ax^2}\,dx = \sqrt{\pi/a} \tag{E.18b}$$

E.3 Gamma Functions

$$\Gamma(n) = \int_0^{\infty} x^{n-1}e^{-x}\,dx \tag{E.19a}$$

$$= \int_0^{1} [\ln(1/x)]^{n-1}\,dx \tag{E.19b}$$

$$\Gamma(n) = (n - 1)!, \qquad \text{for} \quad n = \text{positive integer} \tag{E.19c}$$

$$n\Gamma(n) = \Gamma(n + 1) \tag{E.20}$$

$$\Gamma(\tfrac{1}{2}) = \sqrt{\pi} \tag{E.21}$$

$$\Gamma(1) = 1 \tag{E.22}$$

$$\Gamma(1\tfrac{1}{4}) = 0.906 \tag{E.23}$$

$$\Gamma(1\tfrac{3}{4}) = 0.919 \tag{E.24}$$

$$\Gamma(2) = 1 \tag{E.25}$$

$$\int_0^1 \frac{dx}{\sqrt{1-x^n}} = \frac{\sqrt{\pi}}{n}\,\frac{\Gamma\left(\frac{1}{n}\right)}{\Gamma\left(\frac{1}{n}+\frac{1}{2}\right)} \tag{E.26}$$

$$\int_0^1 x^m(1-x^2)^n\,dx = \frac{\Gamma(n+1)\Gamma\left(\frac{m+1}{2}\right)}{2\Gamma\left(n+\frac{m+3}{2}\right)} \tag{E.27}$$

$$\int_0^{\pi/2} \cos^n x\,dx = \frac{\sqrt{\pi}}{2}\,\frac{\Gamma\left(\frac{n+1}{2}\right)}{\Gamma\left(\frac{n}{2}+1\right)}, \qquad n > -1 \tag{E.27a}$$

E.4 Elliptic Integrals*

Elliptic integral of the *first kind*:

$$F(\varphi, k) = \int_0^{\varphi} \frac{d\varphi}{\sqrt{1-k^2\sin^2\varphi}}, \qquad |k| < 1 \tag{E.28a}$$

$$= \int_0^{x} \frac{dx}{\sqrt{(1-x^2)(1-k^2x^2)}}, \qquad x = \sin\varphi \tag{E.28b}$$

$$F\left(\frac{\pi}{2}, k\right) = \frac{\pi}{2}\left[1 + \frac{1}{4}k^2 + \frac{9}{64}k^4 + \cdots\right] \tag{E.29}$$

Elliptic integral of the *second kind*:

$$E(\varphi, k) = \int_0^{\varphi} \sqrt{1-k^2\sin^2\varphi}\,d\varphi \tag{E.30a}$$

$$= \int_0^{x} \frac{\sqrt{1-k^2x^2}}{\sqrt{1-x^2}}\,dx, \qquad x = \sin\varphi \tag{E.30b}$$

* Extensive numerical tables are given, for example, by Adams and Hippisley (Ad22); short tables may be found in Dwight (Dw61).

APPENDIX F

Differential Relations in Different Coordinate Systems

F.1 Rectangular Coordinates

$$\mathbf{grad}\, U = \nabla U = \sum_i \mathbf{e}_i \frac{\partial U}{\partial x_i} \tag{F.1}$$

$$\operatorname{div} \mathbf{A} = \nabla \cdot \mathbf{A} = \sum_i \frac{\partial A_i}{\partial x_i} \tag{F.2}$$

$$\mathbf{curl}\, \mathbf{A} = \nabla \times \mathbf{A} = \sum_{i,j,k} \varepsilon_{ijk} \frac{\partial A_k}{\partial x_j} \mathbf{e}_i \tag{F.3}$$

$$\nabla^2 U = \nabla \cdot \nabla U = \sum_i \frac{\partial^2 U}{\partial x_i^2} \tag{F.4}$$

F.2 Cylindrical Coordinates

Refer to Figs. F-1 and F-2.

$$x_1 = r \cos \theta; \qquad x_2 = r \sin \theta; \qquad x_3 = z \tag{F.5}$$

$$r = \sqrt{x_1^2 + x_2^2}; \qquad \theta = \tan^{-1} \frac{x_2}{x_1}; \qquad z = x_3 \tag{F.6}$$

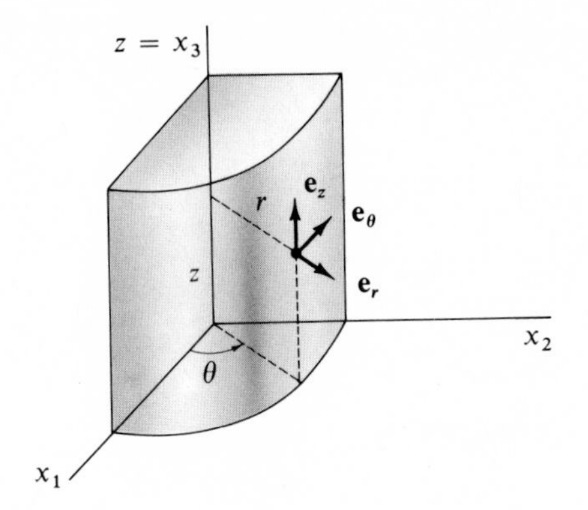

FIG. **F-1**

$$ds^2 = dr^2 + r^2\, d\theta^2 + dz^2 \tag{F.7}$$

$$dv = r\, dr\, d\theta\, dz \tag{F.8}$$

$$\mathbf{grad}\ \psi = \mathbf{e}_r \frac{\partial \psi}{\partial r} + \mathbf{e}_\theta \frac{1}{r} \frac{\partial \psi}{\partial \theta} + \mathbf{e}_z \frac{\partial \psi}{\partial z} \tag{F.9}$$

$$\operatorname{div} A = \frac{1}{r} \frac{\partial}{\partial r} (r A_r) + \frac{1}{r} \frac{\partial A_\theta}{\partial \theta} + \frac{\partial A_z}{\partial z} \tag{F.10}$$

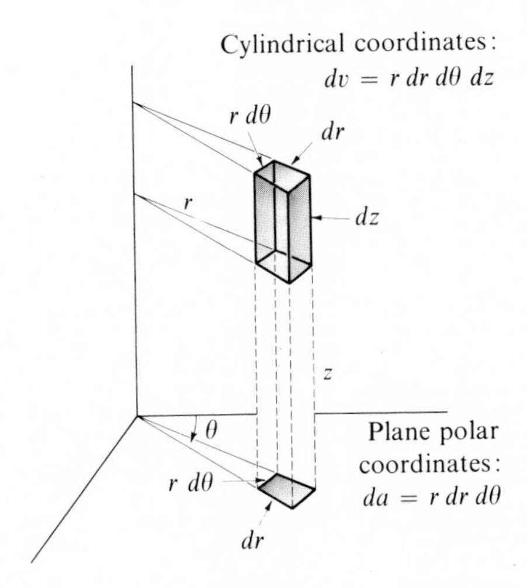

FIG. **F-2**

$$\mathbf{curl\,A} = \mathbf{e}_r\left(\frac{1}{r}\frac{\partial A_z}{\partial \theta} - \frac{\partial A_\theta}{\partial z}\right) + \mathbf{e}_\theta\left(\frac{\partial A_r}{\partial \theta} - \frac{\partial A_z}{\partial r}\right)$$

$$+ \mathbf{e}_z\left(\frac{1}{r}\frac{\partial}{\partial r}(rA_\theta) - \frac{1}{r}\frac{\partial A_r}{\partial \theta}\right) \tag{F.11}$$

$$\nabla^2\psi = \frac{1}{r}\frac{\partial}{\partial r}\left(r\frac{\partial \psi}{\partial r}\right) + \frac{1}{r^2}\frac{\partial^2\psi}{\partial \theta^2} + \frac{\partial^2\psi}{\partial z^2} \tag{F.12}$$

F.3 Spherical Coordinates

Refer to Figs. F-3 and F-4.

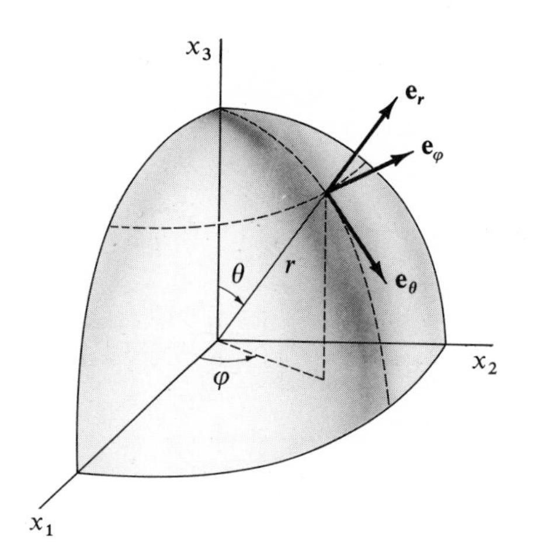

FIG. **F-3**

$$x_1 = r\sin\theta\cos\varphi; \qquad x_2 = r\sin\theta\sin\varphi; \qquad x_3 = r\cos\theta \tag{F.13}$$

$$r = \sqrt{x_1^2 + x_2^2 + x_3^2}; \qquad \theta = \cos^{-1}\frac{x_3}{r}; \qquad \varphi = \tan^{-1}\frac{x_2}{x_1} \tag{F.14}$$

$$ds^2 = dr^2 + r^2\,d\theta^2 + r^2\sin^2\theta\,d\varphi^2 \tag{F.15}$$

$$dv = r^2\sin\theta\,dr\,d\theta\,d\varphi \tag{F.16}$$

$$\mathbf{grad}\,\psi = \mathbf{e}_r\frac{\partial\psi}{\partial r} + \mathbf{e}_\theta\frac{1}{r}\frac{\partial\psi}{\partial\theta} + \mathbf{e}_\varphi\frac{1}{r\sin\theta}\frac{\partial\psi}{\partial\varphi} \tag{F.17}$$

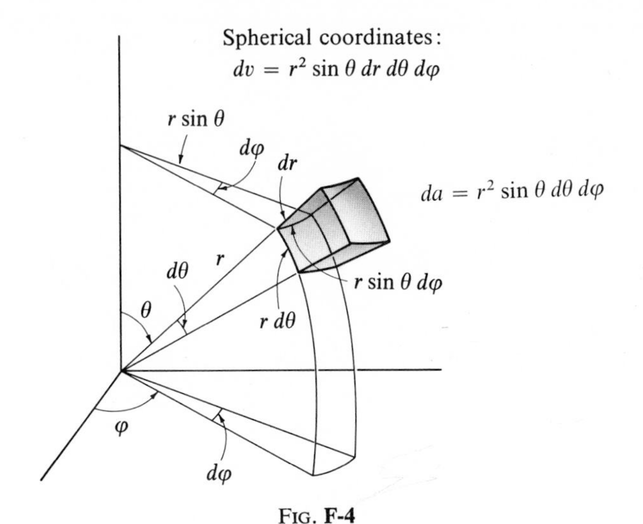

FIG. **F-4**

$$\operatorname{div} A = \frac{1}{r^2}\frac{\partial}{\partial r}(r^2 A_r) + \frac{1}{r\sin\theta}\frac{\partial}{\partial\theta}(A_\theta \sin\theta) + \frac{1}{r\sin\theta}\frac{\partial A_\varphi}{\partial_\varphi} \tag{F.18}$$

$$\begin{aligned}\mathbf{curl\ A} = \mathbf{e}_r \frac{1}{r\sin\theta}\left(\frac{\partial}{\partial\theta}(A_\varphi \sin\theta) - \frac{\partial A_\theta}{\partial\varphi}\right) \\ + \mathbf{e}_\theta \frac{1}{r\sin\theta}\left(\frac{\partial A_r}{\partial\varphi} - \sin\theta \frac{\partial}{\partial r}(rA_\varphi)\right) \\ + \mathbf{e}_\varphi \frac{1}{r}\left(\frac{\partial}{\partial r}(rA_\theta) - \frac{\partial A_r}{\partial\theta}\right)\end{aligned} \tag{F.19}$$

$$\nabla^2\psi = \frac{1}{r^2}\frac{\partial}{\partial r}\left(r^2\frac{\partial\psi}{\partial r}\right) + \frac{1}{r^2\sin\theta}\frac{\partial}{\partial\theta}\left(\sin\theta\frac{\partial\psi}{\partial\theta}\right) + \frac{1}{r^2\sin^2\theta}\frac{\partial^2\psi}{\partial\varphi^2} \tag{F.20}$$

APPENDIX G

A "Proof" of the Relation $\sum_\mu x_\mu^2 = \sum_\mu x_\mu'^2$

Consider the two inertial systems K and K' which are moving relative to one another with a velocity v. At the instant when the two origins coincide ($t = 0$, $t' = 0$), let a light pulse be emitted from the common origin. The equations that describe the propagation of the wave fronts are required, by the second Einstein postulate, to be of the same form in the two systems:

$$\sum_j x_j^2 - c^2t^2 = \sum_\mu x_\mu^2 \equiv s^2 = 0 \qquad \text{in } K \tag{G.1a}$$

$$\sum_j x_j'^2 - c^2t'^2 = \sum_\mu x_\mu'^2 \equiv s'^2 = 0 \qquad \text{in } K' \tag{G.1b}$$

These equations state that the vanishing of the four-dimensional interval between two events in one inertial reference frame implies the vanishing of the interval between the same two events in any other inertial reference frame. But we need more than this; we must show, in fact that $s^2 = s'^2$, in general.

Now, if we require that the motion of a particle, observed to be *linear* in the system K, also be linear in the system K', then the equations of transformation that connect the x_μ and the x_μ' must themselves be linear. In such a case, the quadratic forms s^2 and s'^2 can be connected by, at most, a proportionality factor:

$$s'^2 = \kappa s^2 \tag{G.2}$$

The factor κ could conceivably depend on the coordinates, the time, and the relative velocity of the two systems. As pointed out in Section 2.3, the space and time associated with an inertial reference frame are *homogeneous*, so that the relation between s^2 and s'^2 cannot be different at different points in space nor at different instants of time. Therefore, the factor κ cannot depend on either the coordinates or the time. A dependence on v is still allowed, however, but the *isotropy* of space forbids a dependence on the *direction* of v. We have therefore reduced the possible dependence of s'^2 on s^2 to a factor which involves at most the magnitude of the velocity v. That is, we have

$$s'^2 = \kappa(v)s^2 \tag{G.2a}$$

If we make the transformation from K' back to K, we have the result

$$s^2 = \kappa(-v)s'^2$$

where $-v$ occurs since the velocity of K relative to K' is the negative of the velocity of K' relative to K. But we have already argued that the factor κ can depend only on the *magnitude* of v. Therefore, we have the two equations

$$s'^2 = \kappa(v)s^2; \qquad s^2 = \kappa(v)s'^2 \tag{G.3}$$

combining these equations, we conclude that $\kappa^2 = 1$, or $\kappa(v) = \pm 1$. Now, the value of $\kappa(v)$ must not be a discontinuous function of v. That is, if we change v at some rate, κ cannot suddenly jump from $+1$ to -1. Clearly, in the limit of zero velocity, the systems K and K' become identical, so that $\kappa(v = 0) = +1$. Hence,

$$\kappa = +1 \tag{G.4}$$

for all values of the velocity, and we have, finally,

$$s^2 = s'^2 \tag{G.5}$$

This important result states that the four-dimensional interval between two events is the same in all inertial reference frames.

SELECTED REFERENCES

The following texts are particularly recommended as general sources of collateral reading material.

A. General Theoretical Physics

Blass (Bl62), *Theoretical Physics.*
Lindsay and Margenau (Li36), *Foundations of Physics.*
Wangsness (Wa63), *Introduction to Theoretical Physics.*

B. Intermediate Mechanics

Becker (Be54), *Introduction to Theoretical Mechanics.*
Lindsay (Li61), *Physical Mechanics.*
Slater and Frank (Sl47), *Mechanics.*
Symon (Sy60), *Mechanics.*

C. Advanced Mechanics

Goldstein, (Go50), *Classical Mechanics.*
Landau and Lifshitz (La60), *Mechanics.*
McCuskey (Mc59), *An Introduction to Advanced Dynamics.*

D. Mathematical Methods

Arfken (Ar66), *Mathematical Methods for Physicists.*
Churchill (Ch41), *Fourier Series and Boundary Value Problems.*
Davis (Da61), *Introduction to Vector Analysis.*
Dwight (Dw61), *Tables of Integrals and Other Mathematical Data.*
Kaplan (Ka52), *Advanced Calculus.*
Pipes (Pi46), *Applied Mathematics for Engineers and Physicists.*

BIBLIOGRAPHY

Ab50 M. Abraham and R. Becker, *The Classical Theory of Electricity and Magnetism*; transl. of the 14th German ed. Hafner, New York, 1950.

Ad22 E. P. Adams and R. L. Hippisley, *Smithsonian Mathematical Formulae and Tables of Elliptic Functions*. Smithsonian Institution, Washington, D.C., 1922.

Am63 American Association of Physics Teachers, *Special Relativity Theory, Selected Reprints*. American Institute of Physics, New York, 1963.

An49 A. A. Andronow and C. E. Chaikin, *Theory of Oscillations*, transl. of the 1937 Russian ed. Princeton Univ. Press, Princeton, New Jersey, 1949.

Ar66 G. Arfken, *Mathematical Methods for Physicists*. Academic Press, New York, 1966.

Ba59 R. H. Baker, *Astronomy*, 7th ed. Van Nostrand, Princeton, New Jersey, 1959.

Ba60 R. M. L. Baker, Jr., and M. W. Makemson, *An Introduction to Astrodynamics*. Academic Press, New York, 1960.

Ba61 A. Baldin, V. I. Goldanskii, and I. L. Rozental, *Kinematics of Nuclear Reactions*. Oxford Univ. Press, London and New York, 1961.

Ba63 R. G. Barnett and J. N. Fujii, *Vectors*. Wiley, New York, 1963.

Be46 P. G. Bergmann, *Introduction to the Theory of Relativity*. Prentice-Hall, Englewood Cliffs, New Jersey, 1946.

Be54 R. A. Becker, *Introduction to Theoretical Mechanics*. McGraw-Hill, New York, 1954.

Be64 R. Bellman, *Perturbation Techniques in Mathematics, Physics, and Engineering*. Holt, New York, 1964.

Bl46 G. A. Bliss, *Lectures on the Calculus of Variations*. Univ. of Chicago Press, Chicago, Illinois, 1946.

Bl61 V. M. Blanco and S. W. McCuskey, *Basic Physics of the Solar System*. Addison-Wesley, Reading, Massachusetts, 1961.

Bl62 G. A. Blass, *Theoretical Physics*. Appleton-Century-Crofts, New York, 1962.

Bo64 H. Bondi, *Relativity and Common Sense*. Doubleday, New York, 1964.

Bo65 D. Bohm, *The Special Theory of Relativity*. Benjamin, New York, 1965.

Br46 L. Brillouin, *Wave Propagation in Periodic Structures*. McGraw-Hill, New York, 1946; reprinted by Dover, New York, 1953.

Br60 L. Brillouin, *Wave Propagation and Group Velocity*. Academic Press, New York, 1960.

Br61 D. Brauwer and G. M. Clemence, *Methods of Celestial Mechanics.* Academic Press, New York, 1961.

Br68 T. C. Bradbury, *Theoretical Mechanics.* Wiley, New York, 1968.

By13 W. E. Byerly, *An Introduction to the Use of Generalized Coordinates in Mechanics and Physics.* Ginn, Boston, Massachusetts, 1913.

Ca41 H. S. Carslaw and J. C. Jaeger, *Operational Methods in Applied Mathematics.* Oxford Univ. Press, London and New York, 1941.

Ch41 R. V. Churchill, *Fourier Series and Boundary Value Problems.* McGraw-Hill, New York, 1941.

Ch44 R. V. Churchill, *Operational Methods in Engineering.* McGraw-Hill, New York, 1944.

Ch58 R. V. Churchill, *Operational Mathematics*, 2nd ed. McGraw-Hill, New York, 1958.

Co49 C. A. Coulson, *Waves.* Wiley (Interscience), New York, 1949.

Co53 R. Courant and D. Hilbert, *Methods of Mathematical Physics*, Vol. 1. Wiley (Interscience), New York, 1953.

Co54 F. W. Constant, *Theoretical Physics: Mechanics of Particles, Rigid and Elastic Bodies, Fluids, and Heat Flow.* Addison-Wesley, Reading, Massachusetts, 1954.

Co60 H. C. Corben and P. Stehle, *Classical Mechanics*, 2nd ed. Wiley, New York, 1960.

Co62 D. R. Corson and P. Lorrain, *Introduction to Electromagnetic Fields and Waves.* Freeman, San Francisco, California, 1962.

Cr63 F. H. Crawford, *Heat, Thermodynamics, and Statistical Physics.* Harcourt, Brace, and World, New York, 1963.

Da61 H. F. Davis, *Introduction to Vector Analysis.* Allyn & Bacon, Boston, Massachusetts, 1961.

Da62 J. M. A. Danby, *Fundamentals of Celestial Mechanics.* Macmillan, New York, 1962.

Da63 H. F. Davis, *Fourier Series and Orthogonal Functions.* Allyn & Bacon, Boston, Massachusetts, 1963.

De62 J. W. Dettman, *Mathematical Methods in Physics and Engineering.* McGraw-Hill, New York, 1962.

Dw61 H. B. Dwight, *Tables of Integrals and Other Mathematical Data*, 4th ed. Macmillan, New York, 1961.

Ed30 Sir A. S. Eddington, *The Nature of the Physical World.* Macmillan, New York, 1930.

Ei63 R. L. Eisenman, *Matrix Vector Analysis.* McGraw-Hill, New York, 1963.

Er54 A. Erdélyi, *Tables of Integral Transforms*, Vol. 1. McGraw-Hill, New York, 1954.

Fe59 N. Feather, *The Physics of Mass, Length, and Time.* Edinburgh Univ. Press, Edinburgh, 1959.

Fe61 N. Feather, *An Introduction to the Physics of Vibrations and Waves.* Edinburgh Univ. Press, Edinburgh, 1961.

Fo62 G. R. Fowles, *Analytical Mechanics.* Holt, New York, 1962.

Go50 H. Goldstein, *Classical Mechanics*. Addison-Wesley, Reading, Massachusetts, 1950.

Go68 R. H. Good, *Basic Concepts of Relativity*. Reinhold, New York, 1968.

Ha62 J. Haag, *Oscillatory Motions*. Wadsworth, Belmont, California, 1962.

Ha62a R. L. Halfman, *Dynamics*, 2 vols. Addison-Wesley, Reading, Massachusetts, 1962.

Ha64 C. Hayashi, *Nonlinear Oscillations in Physical Systems*. McGraw-Hill, New York, 1964.

Ha65 W. Hauser, *Introduction to the Principles of Mechanics*. Addison-Wesley, Reading, Massachussetts, 1965.

Ho48 W. V. Houston, *Principles of Mathematical Physics*, 2nd ed. McGraw-Hill, New York, 1948.

Ho64 H. Hochstadt, *Differential Equations—A Modern Approach*. Holt, New York, 1964.

In27 E. L. Ince, *Ordinary Differential Equations*. Longmans, Green, New York, 1927; reprinted by Dover, New York, 1944.

Ir59 J. Irving and N. Mullineux, *Mathematics in Physics and Engineering*. Academic Press, New York, 1959.

Ja60 B. Jaffe, *Michelson and the Speed of Light*. Doubleday, New York, 1960.

Ja62 J. D. Jackson, *Classical Electrodynamics*. Wiley, New York, 1962.

Je46 H. Jeffreys and B. S. Jeffreys, *Methods of Mathematical Physics*. Cambridge Univ. Press, London and New York, 1946.

Jo50 G. Joos and I. M. Freeman, *Theoretical Physics*, 2nd ed. Hafner, New York, 1950.

Ka52 W. Kaplan, *Advanced Calculus*. Addison-Wesley, Reading, Massachusetts, 1952.

Ka63 P. van de Kamp, *Elements of Astrodynamics*. Freeman, San Francisco, California, 1963.

Ka64 R. Katz, *An Introduction to the Special Theory of Relativity*. Van Nostrand, Princeton, New Jersey, 1964.

Ka67 C. Kacser, *Introduction to the Special Theory of Relativity*. Prentice-Hall, Englewood Cliffs, New Jersey, 1967.

Ke29 O. D. Kellog, *Foundations of Potential Theory*. Springer, Berlin, 1929; reprinted by Dover, New York, 1953.

Ko69 E. J. Konopinski, *Classical Descriptions of Motion*. Freeman, San Francisco, 1969.

Ku58 Y. H. Ku, *Analysis and Control of Nonlinear Systems*. Ronald Press, New York, 1958.

La49 C. Lanczos, *The Variational Principles of Mechanics*. Univ. of Toronto Press, Toronto, 1949.

La51 L. D. Landau and E. M. Lifshitz, *The Classical Theory of Fields*. Addison-Wesley, Reading, Massachusetts, 1951.

La60 L. D. Landau and E. M. Lifshitz, *Mechanics*. Addison-Wesley, Reading, Massachusetts, 1960.

Le59 R. B. Leighton, *Principles of Modern Physics*. McGraw-Hill, New York, 1959.

Li36 R. B. Lindsay and H. Margenau, *Foundations of Physics*. Wiley, New York, 1936; reprinted by Dover, New York, 1957.

Li51 R. B. Lindsay, *Concepts and Methods of Theoretical Physics*. Van Nostrand, Princeton, New Jersey, 1951.

Li60 R. B. Lindsay, *Mechanical Radiation*. McGraw-Hill, New York, 1960.

Li61 R. B. Lindsay, *Physical Mechanics*, 3rd ed. Van Nostrand, Princeton, New Jersey, 1961.

Li64 B. W. Lindgren, *Vector Calculus*. Macmillan, New York, 1964.

Lo23 H. A. Lorentz, A. Einstein, H. Minowski, and H. Weyl, *The Principle of Relativity*, original papers. Translated in 1923; reprinted by Dover, New York, 1952.

Ma35 W. F. Magie, *A Source Book in Physics*. McGraw-Hill, New York, 1935.

Ma43 H. Margenau and G. M. Murphy, *The Mathematics of Physics and Chemistry*. Van Nostrand, Princeton, New Jersey, 1943.

Ma50 H. Margenau, *The Nature of Physical Reality*. McGraw-Hill, New York, 1950.

Ma58 K. B. Mather and P. Swan, *Nuclear Scattering*. Cambridge Univ. Press, London and New York, 1958.

Ma59 J. B. Marion, T. I. Arnette, and H. C. Owens, "Tables for the Transformation Between the Laboratory and Center-of-mass Coordinate Systems and for the Calculation of the Energies of Reaction Products." Oak Ridge National Lab. Rept. ORNL-2574, 1959.

Ma60 E. Mach, *The Science of Mechanics*. 6th Am. ed. Open Court, LaSalle, Illinois, 1960. (Original German edition published 1883.)

Ma65a J. B. Marion, *Principles of Vector Analysis*. Academic Press, New York, 1965.

Ma65b J. B. Marion, *Classical Electromagnetic Radiation*. Academic Press, New York, 1965.

Ma65c K. Magnus, *Vibrations*. Blackie, Glasgow and London, 1965.

Mc51 N. W. McLachlan, *Theory of Vibrations*. Dover, New York, 1951.

Mc59 S. W. McCuskey, *An Introduction to Advanced Dynamics*. Addison-Wesley, Reading, Massachussetts, 1959.

Mc61 D. B. McLaughlin, *Introduction to Astronomy*. Houghton-Mifflin, Boston, Massachusetts, 1961.

Mc63 S. W. McCuskey, *Introduction to Celestial Mechanics*. Addison-Wesley, Reading, Massachusetts, 1963.

Me53 D. H. Menzel, *Mathematical Physics*. Prentice-Hall, Englewood Cliffs, New Jersey, 1953.

Mi47 N. Minorsky, *Introduction to Non-linear Mechanics*. Edwards, Ann Arbor, Michigan, 1947.

Mo48 P. M. Morse, *Vibration and Sound*, 2nd ed. McGraw-Hill, New York, 1948.

Mø52 C. Møller, *Theory of Relativity*. Oxford Univ. Press, London and New York, 1952.

Mo53 P. M. Morse and H. Feshbach, *Methods of Theoretical Physics*, 2 vols. McGraw-Hill, New York, 1953.

Mo53a P. Morrison, "A Survey of Nuclear Reactions," in *Experimental Nuclear Physics* (E. Segré, ed.), Vol. II. Wiley, New York, 1953.

Mo58 F. F. Moulton, *An Introduction to Celestial Mechanics*, 2nd ed. Macmillan, New York, 1958.

Mo60 J. Monahan, in *Fast Neutron Physics* (J. B. Marion and J. L. Fowler, eds.), Part I, Chap. I.B. Wiley (Interscience), New York, 1960.

Mo62 P. M. Morse, *Thermal Physics*. Benjamin, New York, 1962.

Mu60 G. M. Murphy, *Ordinary Differential Equations and Their Solutions*. Van Nostrand, Princeton, New Jersey, 1960.

Ne62 E. P. Ney, *Electromagnetism and Relativity*. Harper, New York, 1962.

Ow63 G. E. Owen, *Introduction to Electromagnetic Theory*. Allyn & Bacon, Boston, Massachusetts, 1963.

Pa52 L. Page, *Introduction to Theoretical Physics*, 3rd, ed. Van Nostrand, Princeton, New Jersey, 1952.

Pa58 W. Pauli, *Theory of Relativity*. Macmillan (Pergamon Press), New York, 1958 (translated from the original publication, 1921).

Pa62 W. K. H. Panofsky and M. Phillips, *Classical Electricity and Magnetism*. Addison-Wesley, Reading, Massachusetts, 1962.

Pe53 E. R. Peck, *Electricity and Magnetism*. McGraw-Hill, New York, 1953.

Ph33 H. B. Phillips, *Vector Analysis*. Wiley, New York, 1933.

Pi46 L. A. Pipes, *Applied Mathematics for Engineers and Physicists*. McGraw-Hill, New York, 1946.

Pi57 B. O. Pierce and R. M. Foster, *A Short Table of Integrals*, 4th ed. Ginn, Boston, Massachusetts, 1957.

Ra40 A. S. Ramsey, *An Introduction to the Theory of Newtonian Attraction*. Cambridge Univ. Press, London and New York, 1940.

Ra94 J. W. S. Rayleigh, *The Theory of Sound*, 2nd ed., 2 vols. Macmillan, London, 1894; reprinted by Dover, New York, 1945.

Re60 J. R. Reitz and F. J. Milford, *Foundations of Electromagnetic Theory*. Addison-Wesley, Reading, Massachusetts, 1960.

Re68 R. Resnick, *Introduction to Special Relativity*. Wiley, New York, 1968.

Ro64 W. G. V. Rosser, *An Introduction to the Theory of Relativity*. Butterworth, London and Washington, D.C., 1964.

Ri69 F. K. Richtmyer, E. H. Kennard, and J. N. Cooper, *Introduction to Modern Physics*, 6th ed. McGraw-Hill, New York, 1969.

Sc55 L. I. Schiff, *Quantum Mechanics*, 2nd ed. McGraw-Hill, New York, 1955.

Sc59 W. T. Scott, *The Physics of Electricity and Magnetism*. Wiley, New York, 1959.

Se58 F. W. Sears, *Mechanics, Wave Motion, and Heat*. Addison-Wesley, Reading. Massachusetts, 1958.

Sh61 C. W. Sherwin, *Basic Concepts of Physics*. Holt, New York, 1961.

Sh63 R. V. Sharman, *Vibrations and Waves*. Butterworth, London and Washington, D.C., 1963.

Sl47 J. C. Slater and N. H. Frank, *Mechanics*. McGraw-Hill, New York, 1947.

Sm53 W. M. Smart, *Celestial Mechanics*. Longmans, Green, New York, 1953.

So50 A. Sommerfeld, *Mechanics*. Academic Press, New York, 1950.

Sp59 M. R. Spiegel, *Vector Analysis and Introduction to Tensor Analysis*. Schaum, New York, 1959.

St41 J. A. Stratton, *Electromagnetic Theory*. McGraw-Hill, New York, 1941.

St50 J. J. Stoker, *Nonlinear Vibrations in Mechanical and Electrical Systems*. Wiley (Interscience), New York, 1950.

Sy60 K. R. Symon, *Mechanics*, 2nd ed. Addison-Wesley, Reading, Massachusetts, 1960.

Ta63 E. F. Taylor, *Introductory Mechanics*. Wiley, New York, 1963.

Ta66 E. F. Taylor and J. A. Wheeler, *Spacetime Physics*. Freeman, San Francisco, 1966.

Te56 G. Temple and W. G. Bickley, *Rayleigh's Principle and Its Application to Engineering*. Dover, New York, 1956.

To34 R. C. Tolman, *Relativity, Thermodynamics, and Cosmology*. Oxford Univ. Press, London and New York, 1934.

To38 R. C. Tolman, *The Principles of Statistical Mechanics*. Oxford Univ. Press, London and New York, 1938.

Tu04 H. H. Turner, *Astronomical Discovery*. Arnold, London, 1904; reprinted by the University of California Press, Berkeley, California, 1963.

va60 B. van der Pol, *Selected Scientific Papers* (H. Bremmer and C. J. Bouwkamp, eds.), 2 vols. North-Holland Publ., Amsterdam, 1960.

Wa63 R. K. Wangsness, *Introduction to Theoretical Physics*. Wiley, New York, 1963.

Wa63a R. K. Wangsness, *Introductory Topics in Theoretical Physics*. Wiley, New York, 1963.

We52 R. Weinstock, *Calculus of Variations*. McGraw-Hill, New York, 1952.

We61 J. Weber, *General Relativity and Gravitational Waves*. Wiley (Interscience), New York, 1961.

Wh37 E. T. Whittaker, *A Treatise on the Analytical Dynamics of Particles and Rigid Bodies*, 4th ed. Cambridge Univ. Press, London and New York, 1937; reprinted by Dover, New York, 1944.

Wh53 E. T. Whittaker, *A History of the Theories of Aether and Electricity*; Vol. II: *The Modern Theories*. Nelson, London, 1953; reprinted by Harper and Bros., New York, 1960.

Wi41 A. Wintner, *The Analytical Foundations of Celestial Mechanics*. Princeton Univ. Press, Princeton, New Jersey, 1941.

Ye64 N. V. Yefimov, *Quadratic Forms and Matrices*. Academic Press, New York, 1964.

Yo60 W. Yourgrau and S. Mandelstam, *Variational Principles in Dynamics and Quantum Theory*. Pitman, New York, 1960.

Index

J

K

L

M

S

W

Z